ADVANCES IN
X-RAY ANALYSIS
Volume 27

ADVANCES IN X-RAY ANALYSIS

Volume 27

Edited by

Jerome B. Cohen
Northwestern University
Evanston, Illinois

John C. Russ
North Carolina State University
Raleigh, North Carolina

Donald E. Leyden
Colorado State University
Fort Collins, Colorado

and

Charles S. Barrett
and Paul K. Predecki
University of Denver
Denver, Colorado

Sponsored by
University of Denver Research Institute
and
JCPDS — International Centre for Diffraction Data

PLENUM PRESS • NEW YORK AND LONDON

The Library of Congress cataloged the first volume of this title as follows:

Conference on Application of X-ray Analysis.
Proceedings 6th–   1957–   [Denver]

    v. illus. 24-28 cm. annual.
    No proceedings published for the first 5 conferences.
    Vols. for 1958–   called also: Advances in X-ray analysis, v. 2-
    Proceedings for 1957 issued by the conference under an earlier name: Conference on
Industrial Applications of X-ray Analysis. Other slight variations in name of con-
ference.
    Vol. for 1957 published by the University of Denver, Denver Research Institute,
Metallurgy Division.
    Vols. for 1958–   distributed by Plenum Press, New York.
    Conferences sponsored by University of Denver, Denver Research Institute.
    1. X-rays—Industrial applications—Congresses.   I. Denver University. Denver
Research Institute   II. Title: Advances in X-ray analysis.
TA406.5.C6   58-35928

Library of Congress Catalog Card Number 58-35928
ISBN-13: 978-1-4612-9713-0   e-ISBN-13: 978-1-4613-2775-2
DOI: 10.1007/978-1-4613-2775-2

Proceedings of the 1983 Denver Conference on the Applications of
X-Ray Analysis, held August 1-5, 1983, in Snowmass,
Colorado

© 1984 University of Denver
Plenum Press is a division of Plenum Publishing Corporation
233 Spring Street, New York, N.Y. 10013

Softcover reprint of the hardcover 1st edition 1984

PREFACE

This volume constitutes the proceedings of the 1983 Denver Conference on Applications of X-ray Analysis and is the 27th in the series. The conference was held jointly with the American Crystallographic Association at Snowmass Resort, Colorado, from August 1 to 5, 1983. The papers appearing in this volume are only from predominantly Denver Conference (DC) sessions and from joint DC/ACA sessions.

The early plans for holding a joint conference were initiated some three years ago by Q. C. Johnson of Lawrence Livermore Lab, J. B. Cohen of Northwestern University and P. K. Predecki of the University of Denver and were eventually brought to fruition by a joint organizing committee consisting of: O. P. Anderson, Colorado State University (ACA), D. E. Leyden, Colorado State University (DC), R. D. Witters, Colorado School of Mines (ACA) and P. K. Predecki (DC). We take this opportunity to thank the committee members and the early planners for their vision, ingenuity and hard work without which the conference would not have materialized.

There was no plenary session in 1983, instead a number of special sessions were organized and chaired by various individuals. These were as follows:

1. Hanawalt Award Session on Search/Match Methods, chaired by L. K. Frevel, Johns Hopkins University, and W. Parrish, IBM
2. John Gilfrich Session on Trends in XRF Instrumentation, chaired by J. Gilfrich, Naval Research Laboratory and D. E.Leyden, Colorado State University
3. X-ray Strain and Stress Determination, chaired by J. B. Cohen, Northwestern University and P. K. Predecki, University of Denver
4. Application of XRF to Archeological, Geochemical and Industrial Materials, chaired by B. Fabbi, Valencia, Calif., and R. Vane, Kevex Corp.
5. Position Sensitive Detectors, chaired by R. W. Hendricks, Technology for Energy Corp. and C. O. Ruud, The Pennsylvania State Univ. and C. R. Hubbard, NBS

6. Quantitative X-ray Phase Analysis, chaired by G. J. McCarthy,
    North Dakota State Univ.
7. Forensic and Regulatory Applications of Powder Diffraction,
    chaired by W. H. de Camp, Food and Drug Admin.

Tutorial workshops on various XRD and XRF topics were held at
various times during the conference. These are listed below with
the names of the workshop organizers and instructors.

1D. "Automated Data Collection Strategies." M. C. Nichols--Sandia
     Natl. Lab (chair), R. P. Goehner--General Electric Co. (chair),
     H. E. Goebel--Siemens, West Germany, R. L. Snyder--Alfred U.,
     J. Ringland--Sandia Natl. Lab, W. Parrish--IBM, C. R. Hubbard--
     NBS
2D. "JCPDS Workshop on Computer Assisted Phase Identification
     Using the Powder Diffraction File and Crystal Data Databases."
     G. J. McCarthy--N. Dakota State Univ. (chair), A. Mighell--NBS,
     G. G. Johnson, Jr.--The Pennsylvania State Univ.
3D. "Quantitative X-Ray Powder Diffraction." C. R. Hubbard--NBS
     (chair), R. L. Snyder--Alfred Univ. (chair), R. P. Goehner--
     General Electric Co., M. F. Garbauskas--General Electric Co.,
     R. Jenkins--Philips Electronic Instruments, G. J. McCarthy--
     N. Dakota State Univ.
4D. "Use and Implementation of the Rietveld Method in Powder Dif-
     fraction Studies." D. K. Smith--The Pennsylvania State Univ.
     (chair), M. E. Mueller--Argonne Natl. Lab. (chair), R. B.
     Von Dreele--Arizona State Univ., A. Santoro--NBS, J. D.
     Jorgensen--Argonne Natl. Lab., J. C. Taylor--CSIRO, Australia,
     J. Faber--Argonne Natl. Lab, E. Prince--NBS, D. E. Cox--
     Brookhaven Natl. Lab
1F. "XRF Sample Preparation." V. E. Buhrke--The Buhrke Co. (chair),
     J. Renault--New Mexico Bur. Mines and Min. Resources, R. Vane--
     Kevex Corp., F. Claisse--Laval Univ.
2F. Continuation of 1F.
3F. "Qualitative XRF Analysis--Manual and Computer Assisted."
     J. C. Russ--N. Carolina State Univ. (chair), D. Monk--Kevex
     Corp., J. Croke--Philips Electronic Instruments
4F. "Fundamental Parameters XRF." J. W. Criss--Criss Software,
     Inc. (chair), G. D. Bowling--Owens-Corning Fiberglas Corp.,
     E. M. Kenny--Halcon Research.

A highlight of the conference was the presentation of the first
J. D. Hanawalt Powder Diffraction Award to Ludo K. Frevel of Johns
Hopkins University and Ludo's subsequent award lecture. The Hanawalt
Award was established by the JCPDS to provide recognition for
excellence in powder diffraction work. The award was made by
J. D. Hanawalt himself.

The conference also marked the occasion of the retirement of
John V. Gilfrich from the Naval Research Lab. A plaque honoring

his many contributions to the field of X-ray fluorescence and his active participation in the Denver Conference was presented to him at a session in his honor.

The total number registered for the conference was 498, of whom 210 were ACA affiliated, 260 were Denver Conference affiliated and 28 were neither or both. A total of 31 students registered and 360 people attended the dinner. The dinner marked the first appearance of a barbershop quartet group known as "The Powdermen" (J. E. Edmonds, R. Jenkins, G. J. McCarthy and W. Schreiner, with S. Weissmann on the piano). Their repertoire included a song sung in honor the first Hanawalt Award winner. We reproduce herewith the words of this ditty for the benefit of those who could not hear it.

> *Ludo Frevel, we know him well,*
> *For 50 years, allayed our fears*
> *Of things unknown, our faith has grown*
> *In XRD, the place to be.*
> *He took a phase, not one but more,*
> *He mixed them up, our minds unsure,*
> *His search technique works at the peak,*
> *It sorts them out, now let's all shout.*
> *The method grew, and problems too,*
> *And now Ben Post, he is the most,*
> *His films to read, we are agreed,*
> *The way to go, in future years.*
>
> *(Sung to the tune of "Bill Grogan's Goat")*

Exhibits of diffraction and fluorescence equipment were provided by 22 companies and were open for the duration of the conference. Attendance at all eight of the tutorial workshops was outstanding with 100-200 participants at each.

On behalf of the organizing committee I would like to sincerely thank the workshop chairmen and instructors, the special session chairmen, the contributed session chairmen and the authors for their contributions. The efforts of these people made the Snowmass Conference a great success.

A special word of thanks to my staff: Joan Fitzpatrick, Penny Hudson, Cynthia King, Rich Miller, Stewart Pressnall, Dorothy Predecki and the conference secretary: Mildred Cain, all of whom worked long and unusual hours to make sure the conference ran smoothly.

Paul Predecki
For the conference committee

Unpublished papers

The following papers were presented at DC or joint DC/ACA sessions but are not published here for various reasons.

"Phase Decomposition of Cu-Ti Metallic Glass," B. Mozer, S. P. Singhal and R. D. Shull, NBS, Washington, DC 20234; and A. Maeland, Allied Corp., Morristown, NJ 07960

"Energy Dispersive X-ray Diffraction Measurements under High Pressure and Temperature Using Diamond-Anvil Cell and Synchrotron Radiation," M. H. Manghnani and L. C. Ming, Univ. of Hawaii, Honolulu, HI 96822; E. F. Skelton, Naval Research Lab, Washington, DC 20375; and S. B. Qadri, Stanford Synchrotron Radiation Lab, Stanford, CA 94305

"X-Ray Diffraction Studies on Shock-Modified Zirconia," E. J. Graeber, B. Morosin and R. A. Graham, Sandia Natl. Labs, Albuquerque, NM 87185

"Low Temperature Powder X-Ray Diffraction (PXRD) Studies of Uranium Hydride and Deuteride," J. S. Cantrell, Miami Univ., Oxford, OH 45056; and D. B. Sullenger, Monsanto Research Corp., Miamisburg, OH 45342

"3-D Shaded Perspective Display of Pole-Figure Data," W. J. Dallas, Philips GmbH, Hamburg 54, West Germany

"Applications of X-Ray Powder Diffraction in Forensic Drug Analysis," R. A. Henderson, Drug Enforcement Admin., New York, NY 10019

"X-Ray Powder Diffraction for Pharmaceutical Quality Control," R. S. Chao, K. C. Vail and D. A. Hatzenbuhler, The Upjohn Co., Kalamazoo, MI 49001

"Determination of Polymorphic Purity with X-Ray Powder Diffraction," P. Sutton and S. R. Byrn, Purdue Univ., West Lafayette, IN 49707

"Trace Identification of Drugs by X-Ray Diffraction," D. V. Canfield, Univ. of Southern Mississippi, Hattiesburg, MS 39406

"Findings of a Collaborative Test of the NIOSH XRD method for the Determination of Airborne Silica," M. T. Abell, NIOSH, Cincinnati, OH 45226; and C. C. Anderson, SRI International, Menlo Park, CA

"X-Ray Diffraction Phase Analysis of Process and Pollution Control Device Samples," F. Briden, USEPA, Research Triangle Park, NC 27711

"Analysis of Vanadium Oxides by X-Ray Diffraction," T. P. Carsey, NIOSH, Cincinnati, OH 45226

"The Psi-Differential and Integral Methods for Residual Stress Measurements by X-Ray Diffraction," C. N. J. Wagner and M. S. Boldrick, Univ. of California, Los Angeles, CA 90024

"New Developments in Instrument Design for EDXRF Analysis," W. E. Drummond, Tracor Xray, Inc., Mountain View, CA 94043

"The Use of Polarized X-rays in Energy Dispersive X-Ray Spectrometry," R. W. Ryon, Lawrence Livermore Natl. Lab., Livermore, CA 94550; and J. D. Zahrt, Northern Arizona Univ., Flagstaff, AZ 86001

"The New Concept in WDX Instrumentation: Combined Sequential/ Multichannel Spectrometer," Y. M. Gurvich, B. J. Price, W. Vogel and B. Fabbi, Bausch & Lomb Inc., 9545 Wentworth St., Sunland, CA 91040

"An X-Ray Method for Measuring Plating Thicknesses on Small Parts," T. H. Briggs, Western Electric Co., Allentown, PA 18103

"Quality Control Analysis of Nickel Base Alloys Using Energy-Dispersive X-Ray Fluorescence and the Fundamental Parameters Method," J. P. Nelson, Hoskins Mfg. Co., Hamburg, MI 48139

"XRF for Analysis of Multilayer Coatings," H. E. Marr and W. M. Silverman, UPA Technology, Inc., Syosset, NY 11791

"An On-Line Production Method for the X-Ray Fluorescence Analysis of Permalloy Films for Composition and Thickness," E. R. Babcock, Storage Technology, Inc., Louisville, CO 80028

"Production Environment Quantitative Analysis of Plating Thickness Using X-Ray Fluorescence for Small Parts Manufacturing," E. S. Woster, Seiko Instruments, Torrance, CA 90501

"Recognition of Patterns in Powder Diffraction," G. S. Smith, Lawrence Livermore National Lab, Livermore, CA 94550; and M. C. Nichols, Sandia National Lab, Livermore, CA 94550

"A Survey of X-Ray Powder Diffraction File Users," M. C. Nichols, Sandia Natl Lab, Livermore, CA 94550

"X-Ray Fluorescence of Archaeological Pottery Sherds," D. E. Stelz and J. N. Gundersen, Wichita State Univ., Wichita, KS 67208

"Choosing the Composition of Crystals Used to Scatter X-Rays," B. Greenberg, Philips Laboratories, Briarcliff Manor, NY 10510

"Improvements in Sequential XRF Analysis," B. J. Price, Bausch & Lomb, Ecublens, Switzerland, and J. A. Anzelmo, Bausch & Lomb, Dearborn, MI 48120

"Quantitative XRD Methods for Glass Content in Highly Crystalline SiO_2 Powders," C. R. Hubbard, NBS, Washington, DC 20234

"Standard Reference Materials for X-Ray Analysis," S. D. Rasberry, NBS, Washington, DC 20234

"Modelling XRF by Monte Carlo Simulation," D. Segal, Israel AEC; and A. Notea, Technion, Israel

"Some Novel Applications of X-Ray Analysis," K. Das Gupta, Texas Tech Univ., Lubbock, TX 79409

"New Capabilities Using Conventional X-Ray Equipment," P. J. Harget, Celanese Research Co., Summit, NJ 07901

"Observation of a Fine Structure in L_3-Absorption Spectra of Lead," K. Das Gupta and T. Henry, Texas Tech Univ., Lubbock, TX 79409

"Linear Position-Sensitive Proportional Counters (PSPC) in Focussing X-Ray Systems," H. E. Goebel, Siemens AG, W. Germany

"The Accuracy and Detection Limits of Position-Sensitive Detectors in Scanning Powder Diffractometry," R. L. Snyder, Alfred University, Alfred, NY 14802; and H. E. Goebel, Siemens AG, W. Germany

"The Use of a Linear Curved Sensitive Proportional Detector (PSD) in Powder Diffractometry," E. R. Wölfel, Stoe Applications Lab, Darmstadt, W. Germany

"Reticon Position Sensitive Detector in Energy Dispersive X-Ray Absorption Spectroscopy," Z. U. Rek, R. P. Phizackerley, G. B. Stephenson, S. D. Conradson and K. O. Hodgson, Stanford Univ., Stanford, CA 94305; and T. Matsushita and H. Oyanagi, Photon Factory, Japan

"Synchrotron Radiation Plus Photodiode Array: EXAFS in Dispersive Mode for Fast Microanalysis," E. Dartyge, A. M. Flank, A. Fontaine, A. Jucha and D. Raoux, L.U.R.E., Orsay, France

"Quantitative Amorphous/Crystalline Content in High Temperature Ceramic Powders," M. J. Downey and G. P. Hamill, GTE Laboratories, Inc., Waltham, MA 02254

ACKNOWLEDGEMENTS

    We acknowledge with gratitude sponsorship of the Conference by the University of Denver through its Research Institute, by the JCPDS International Centre for Diffraction Data and by Colorado State University Department of Chemistry through the participation of D. E. Leyden

CONTENTS

III. POSITION SENSITIVE DETECTORS AND X-RAY INSTRUMENTATION

IV.   QUANTITATIVE PHASE ANALYSIS BY XRD

V.   OTHER XRD APPLICATIONS

IX.   OTHER XRF APPLICATIONS

PRESENTATION OF THE FIRST J. D. HANAWALT

POWDER DIFFRACTION AWARD TO LUDO K. FREVEL

From left to right:

J. D. Hanawalt, former Director and Chairman of JCPDS

L. K. Frevel, recipient of the First Hanawalt Award

G. J. McCarthy. present Chairman of JCPDS

ESTABLISHMENT OF THE J. D. HANAWALT AWARD FOR EXCELLENCE IN

POWDER DIFFRACTION AND ITS FIRST PRESENTATION TO LUDO K. FREVEL

It is with a profound sense of the historical context of this morning that I share this podium with three people who are very much responsible for our ability to rapidly identify and characterize solids. In order to do such solids analysis, we need an instrument that provides the necessary diffraction data and a set of reference patterns for comparison to those we measure, along with a means of rapidly accessing those patterns to make our identifications. I do not need to tell this audience that I am alluding to the diffractometer developed by our Session Chairman, Bill Parrish, in the 1940's and to the milestone publications by Hanawalt, Rinn and Frevel in the 1930's that gave us the first 1000 patterns of the database that was to become the Powder Diffraction File, and the search method later called the Hanawalt Method.

In 1980, the JCPDS-International Centre for Diffraction Data established an endowed award named after its long time Director, Search Methods Committee Chairman and past Board Chairman, J. Donald Hanawalt. Don was born in 1902 near Philadelphia, Pennsylvania. After earning a BA in Physics at Oberlin College, he was awarded a Ph.D. in Physics from the University of Wisconsin in 1929. Following post-doctoral fellowships at the Rockefeller Foundation, University of Michigan and the University of Groningen in the Netherlands, Don went to work for Dow Chemical Company in 1931. He worked his way from a position as a research physicist to Vice-President of Dow Metal Products Company through pioneering work with magnesium metallurgy, work with which few of us in the crystallographic and analytical chemistry communities are familiar. It was in his position as Director of the Dow's Spectroscopy and X-ray Laboratory from 1934-1940 that Don, along with Ludo Frevel and the late Sid Rinn, made those first crucial contributions to analytical X-ray diffraction with which we are so familiar. It is certainly an honor and pleasure to have Don here to present the first Award to his long time friend and colleague, Ludo K. Frevel.

1

Ludo Karl Frevel was born in Frankfurt am Main, Germany in 1910. In 1934, Johns Hopkins University awarded him a Ph.D. in Physical Chemistry. After two years of post-doctoral research in crystallography with Linus Pauling at the California Institute of Technology. Ludo joined Dow as a Research Chemist. Ludo made many important contributions in catalysis for the petrochemical industry and holds 50 U.S. patents in this area. He retired as Director of the Laboratory for Chemical Physics in 1974. But he certainly did not retire from scientific research, as evidenced by a steady stream of publications in the Journal of Applied Crystallography and Analytical Chemistry right up to the present. Ludo has returned each year to his alma mater, Johns Hopkins, as a Fellow in the Chemistry Department. He has also held a Distinguished Visiting Professorship at Penn State and is a consultant for Dow Corning Corporation.

From among Ludo's many contributions in analytical X-ray analysis, starting with his vital contribution to the previously mentioned 1938 Hanawalt, Rinn and Frevel paper, the one that has probably had the greatest impact on the way we do things today is his pioneering work in the mid-1960's through the early 1970's on the use of computers in both the searching and matching steps of crystalline phase identification. Ludo was the first scientist to put the power of the digital computer to work in phase identification with his ZRD computer program first reported in 1965. The well-known search/match systems of Jerry Johnson and Monte Nichols which followed soon after benefited from Ludo's trailblazing work. One need only visit the instrument exhibit and note the titles of so many papers at this Conference to begin to appreciate the magnitude of this contribution.

I give you the first Hanawalt Award recipient, Dr. Ludo K. Frevel.

G. J. McCarthy, Chairman
JCPDS - International Centre for
   Diffraction Data
August 4, 1983

J. D. HANAWALT POWDER DIFFRACTION AWARD LECTURE

L. K. Frevel

Department of Chemistry
The Johns Hopkins University
Baltimore, Maryland 21218

ABSTRACT

A condensed chronology of computer SEARCH/MATCH programs for the analysis of multiphase crystalline powders is presented covering the years 1965 to 1981. For this period the various algorithms for searching a data base of standard powder diffraction patterns, $\{d_s,(I/I_1)_s\}$, are predicated on empirical "fingerprint" matching of the experimental powder data, $\{d_v,I_v\}$, with one or more of the standard patterns. Within the past year an interactive computer SEARCH/MATCH program has been developed based on a <u>structure-sensitive</u> SEARCH procedure.

The evolution of computer SEARCH/MATCH procedures undoubtedly will continue and will be influenced by marked improvements in the quality of digitized, machine-readable, powder diffraction data. Three conjectures are offered on future developments in x-ray diffractometry.

EVOLUTION OF COMPUTER SEARCH/MATCH PROCEDURES

The mid-sixties saw the beginnings of computer SEARCH/MATCH methods for the phase identification of crystalline powders[1,2,3,4]. In 1965, the ZRD method[1] was described which required as minimum input the digitized powder data $\{d_v,I_v\}$ of a sample and the set of its chief elements, $\{Z\}$, (Z standing for the atomic number of an element). The initial data base consisted of 1359 abbreviated standard patterns comprising only the 10 most intense diffraction lines for each standard. The search files stored on magnetic tape were expanded from 25 elements to 84 elements and encompassed some 14000 standard patterns. The full ZRD-files proved very effective

3

in unravelling complex mixtures of four or more phases. Unfortunately, in 1968 some misguided zealots from the Beaver 55 Group broke into Dow's Computation Laboratory in Midland, Michigan, and demagnetized most of the magnetic tapes in the tape library-the ZRD SEARCH/MATCH program was one of the casualties. Since no back-up tapes were stored in a separate location, further progress on computer SEARCH/MATCH programs was interrupted at The Dow Chemical Company until 1974[5,6,7]. Fortunately, Monte Nichols[2,8,9] as well as Gerry Johnson[3,4,10] advanced their respective programs first reported on in 1966 at the twenty-fourth Pittsburgh Diffraction Conference. Gerry Johnson has updated at least once a year the Johnson-Vand program so that there have been some 20 versions to date.

Tapping by phone into large mainframe computers was an obvious solution to the problem faced by those analytical diffractionists without adequate computer facilities. Diffraction Data Tele-Search (2dTS) was designed by JCPDS for that purpose[10,11]. Chemical Information System's Powder Diffraction SEARCH/MATCH (CIS/PDSM) became the second example of making available to a customer rapid identification of multiphase mixtures via a remote mainframe computer facility[12,13]. However, with any time-sharing computer-operation, accessibility to the central computer and I/O times become frustratingly long as the number of customers approaches the limit of the system. For that reason, dedicated computers with efficient programs have a strong appeal to the users of automated powder diffractometers. Within the past two years, considerable effort has been expended by the various manufacturers of powder diffractometers and by computer companies in devising efficient algorithms for SEARCH/MATCH procedures. As a most recent example, T. C. Huang, W. Parrish and B. Post[14,15] have devised a simplified effective computer SEARCH/MATCH method in which no more than the 12 largest d-values (regardless of their intensities) are used for the Powder Diffraction File Standards. They tested this procedure with the IBM Series SEARCH/MATCH method which employs all the necessary factors to determine the goodness of match as a Figure-of-Merit.

All the above-mentioned procedures for powder diffraction analysis are predicated on empirical "fingerprint" matching. If one uses the comprehensive Powder Diffraction File (PDF) of the Joint Committee on Powder Diffraction Standards[16], one finds that current SEARCH/MATCH programs often are frustrating and, in the absence of supplemental data such as elemental data, lead to a sizable subfile (10-100 standards) for matching the pattern of a multiphase genuine unknown. Some of the difficulty can be attributed to the marginal accuracy of early data obtained with 57.3 mm Debye-Scherrer cameras. More troublesome are the ubiquitous cases of dilute solid solutions encountered with minerals, corrosion products, scales, precipitates, alloys, catalysts, etc. When solid solution is extensive (>10 mol %), the current search procedures

usually fail either by not finding a matching standard or by finding
one or more pseudomatches with pure crystalline phases. A third
difficulty pertains to the case when the sought standard is not in
PDF. This situation occurs rather frequently in an industrial
analytical laboratory. The late H. W. Rinn of The Dow Chemical
Company kept a record of unknown patterns (UP) he and his co-workers
encountered in actual practice. At the time of his death (1971),
the UP file contained more than 2000 entries. This number exceeds
the 1045 standards required for the 22720 non-repetitive phase-
identifications carried out at Dow's Chemical Physics Research
Laboratory during the years 1965 through 1969[5]. To remedy this
persistent difficulty, a new SEARCH/MATCH approach has been attempted
by shifting from an empirical clerical basis to an x-ray crystallographic basis[17].

If $d(I_1)_S$ and $d(I_2)_S$, respectively, denote the interplanar
spacings for the most intense reflection and the second most intense
reflection of a standard S, and $d(I_1)_X$ and $d(I_2)_X$ denote the corresponding spacings (same indices $h_1k_1l_1$ and $h_2k_2l_2$) for the solid solution of standard S or for an isomorphous pure phase not in the PDF
data base, then one can form a structurally significant quasi-
invariant (Expression 1) based on the respective reciprocal cells

$$\frac{d*^2(I_1)_S - d*^2(I_2)_S}{d*^2(I_1)_S} \approx \frac{d*^2(I_1)_X - d*^2(I_2)_X}{d*^2(I_1)_X} \qquad (1)$$

The validity of this assertion follows from the proviso that
$a*_X:b*_X:c*_X \approx a*_S:b*_S:c*_S$ with the interaxial angles of the
respective unit cells remaining unchanged. For dilute solid
solutions (<10 mol %), one usually observes a fairly uniform dilation
or contraction of the unit cell of the host phase. Likewise for two
neighboring members of an isomorphous series the pair-index,
$1-d^2(I_1)d^{-2}(I_2)$, is essentially constant. For the cubic system the
pair-index is a true invariant and reduces to a ratio of intergers.
For the uniaxial systems the pairing of prism reflections results
in similar true invariants[17]. In all other cases the dimensionless
pair-index is only a quasi-invariant. The fact the sum of the
known cubic substances and the known uniaxial substances[18,19]
exceeds the number of known biaxial substances augers well for the
utility of the pair-index invariant. Table 1 illustrates how the
isomorphs of the rocksalt structure and the isomorphs of the barite
structure are distributed in the Pair-Index SEARCH file. The second
column of Table 1 lists the Pair-Index ranging from a theoretical
value of $-\infty$ to 1. Underlined values of the Pair-Index pertain to
cubic phases. The third column lists the calculated values of $d(I_1)$
based on published unit cell data[16]. Calculated d-spacings are also
shown for $d(I_2)$, $d(I_3)$ and $d(I_i)$, the innermost observed reflection
(or observable reflection consistent with space group criteria).
Subscripts to the four calculated d-spacings denote the corresponding

published experimental relative intensities; superscripts give the
respective hkl indices. Column 7 lists the cube edge in the case
of a cubic phase; the axial ratio $(c/a)_T$ for a tetragonal phase;
$(c/a)_H$ or $(c/a)_R$ for a hexagonal (rhombohedral) phase; a/b and c/b
for a biaxial phase, and the corresponding non-orthogonal interaxial
angles α, β, and γ. Column 8 lists the Strukturbericht symbols for
the simpler crystal structures. Column 10 gives the standard's
reference number of the Powder Diffraction File (PDF) of the Joint
Committee on Powder Diffraction Standards[16]. With the collaboration
of Dan Filsinger of the Analytical Department of the Dow Corning
Corporation, the semi-manual Structure SEARCH/MATCH program[17] has
been programmed for the 3354 Hewlett-Packard computer. The experimental data of Table II[17] was processed by this program and required
3 min I/O time to identify $KI_{1-x}Br_x$ and KBr_yI_{1-y} as the two phases
in the mixture. Details of this computer program will be published
at a later date. Suffice it to say that the automated Structure
SEARCH/MATCH program is more efficient than the latest version of
the ZRD SEARCH/MATCH program. Unquestionably the evolution of
computer SEARCH/MATCH programs will continue and will accelerate
with better machine-readable powder data.

FUTURE DEVELOPMENTS IN X-RAY DIFFRACTOMETRY

One way to circumvent predicting the future is to create the
future. To that end some incipient experimental results are presented that may lead to innovative improvements over current
practices in powder diffraction.

1.   Direct Determination of the Number of Phases in Powder Samples

A non-destructive micro-technique for determining the number
of phases in a powder sample would be a boon to any analytical
diffractionist. One such method would be to utilize an intense
micro beam of filtered $CuK\alpha$ radiation to obtain a two-dimensionally
spotty powder pattern from a stationary micro-thin cylindrical
sample of powder or from a micro-thin layer of powder. By visual
inspection of the degree of spottiness of the various discontinuous
"rings" under a binocular microscope, the size(s) and number of
spots for each diffraction ring could be ascertained and the various
rings segregated by phases. Figure 1 reveals directly the three
distinct sets of rings belonging to three phases in a stationary
0.03 mm thin sample: (a) the smooth arc pattern showing preferred
orientation of a 0.25 mil Mylar film; (b) the six very spotty
diffraction rings attributable to the crystallites of 325 mesh
hyperpure Si; and (c) the three much less spotty rings pertaining
to LiF. Figure 2 shows a threefold enlargement of the three
characteristic crystallite-size distributions. The 2Θ value corresponding to the interplanar spacing for any micro spot can be computed from the (x,z) coordinates measured from the central spot:

Table 1.　Abbreviated sections of the pair-index SEARCH file.
Pair-index $P = 1 - d^2(I_1)d^{-2}(I_2)$.

n	P	$d(I_1)$, Å	$d(I_2)$, Å	$d(I_3)$, Å	$d(I_4)$, Å	a/b c/b c/a　α, β, γ	S	Formula	PDF
297	$\underline{-1.00\dot{0}}$	3.5328^{200}	2.4980^{220}_{70}	4.0793^{111}_{42}	4.0793^{111}	7.0655Å	B1	KI	4-471
308	//	2.1065	1.4895^{220}_{52}	1.2162^{222}_{12}	2.4324^{111}_{10}	4.213Å		MgO	4-829
434	$\underline{-0.33\dot{3}}$	4.1923^{111}	3.6307^{200}_{95}	2.5673^{220}_{70}	4.1923^{111}_{100}	7.2613Å	B1	NH$_4$I	6-174
437	//	2.8232	$2.4450^{}_{75}$	$1.7289^{}_{60}$	$2.8252^{}_{100}$	4.890Å		NaH	2-809
502	$-0.23(1)$	3.4885^{021}	3.1438^{121}_{80}	2.8874^{211}_{65}	5.6120^{110}_{14}	0.8190 0.6394	H2	KCl O$_4$	7-211
503	-0.2316	3.4436	$3.1030^{}_{95}$	$2.1215^{311,140}_{80}$	5.5725^{110}_{2}	0.8058 0.6141		BaSO$_4$	24-1035
504	-0.2295	3.4153	$3.0801^{}_{100}$	2.8150^{002}_{80}	$4.3572^{101,020}_{20}$	0.8297 0.6553		PbSeO$_4$	14-431
739	0.1871	2.9700^{121}	3.2941^{021}_{98}	2.7310^{211}_{62}	4.2211^{101}_{11}	0.8214 0.6403	H2	SrSO$_4$	5-593
794	$\underline{0.25\dot{0}}$	3.2365^{200}	3.7372^{111}_{80}	2.2886^{220}_{65}	3.7372^{111}_{80}	6.473Å	B1	NaI	6-302
806	//	2.0131	$2.3245^{}_{95}$	$1.4235^{}_{48}$	$2.3245^{}_{95}$	4.0262Å		LiF	4-857

$$\cos 2\Theta = \frac{r \cos(x/r)}{(r^2+z^2)^{\frac{1}{2}}} \tag{2}$$

where r = camera radius (mm); x = abscissa (mm); z = ordinate (mm); and the angle, x/r, is expressed in radians. Controlled motion of the micro sample can be employed to increase the number of spots per ring; e.g., controllable oscillation[20], reciprocating translation along the cylinder axis of the camera, or a combination of these two motions. To realize the full potential of this micro examination of diffraction rings from micro-thin samples, one must exercise due care in the carefully controlled comminution or mechanical segregation of the as-received sample.

A sealed-off microfocus tube with a circular focal spot (10 μm diameter) would be a welcome new x-ray source for this proposed technique.

2. 4π Laue Patterns of Micro Crystals

Extending the above technique to its limit, one can pick out of a mixture an individual particle as small as 15 μm in diameter and mount such a particle (usually a single crystal) on a 1-mil W-wire or on a taut 10-denier Kevlar fiber. Utilizing a micro beam of unfiltered CuK or WL radiation, one can obtain an extensive Laue pattern in a "cup" camera having a cylindrical film and a flat circular film forming the base to the cylindrical film. Figure 3 shows such a Laue pattern for a 20 μm Si particle picked out of a mixture. Figure 4 shows the Laue pattern for a 0.15 mm α-Al_2O_3 sphere exposed in a Gandolfi camera with a flat film held on the cover plate. Since the vector components of the unit normals for all the reflecting planes can be determined directly from the coordinates of the registered micro Laue-spots, one can determine the Laue symmetry group (by visually observing a computer-programed rotation of the set of unit normals) and the interaxial angles of the particular crystal system. Oscillation about any dominant zone will produce layer lines for determining the axial lengths of the selected unit cell. This micro technique could be very useful for identifying very minor phases in a powder sample. It would also aid in determining the unit cell of a completely unknown phase.

3. Novel Diffractometers

Three principal disadvantages of conventional diffractometers are: (1) only a small fraction of the total diffraction cone for any particular reflection is measured; (2) the accuracy of the interplanar spacing measurements is markedly dependent on the Bragg angle; and (3) the intrinsic difficulty of achieving random orientation of the powder particles in a flat-faced briquet poses an ever-present problem. Additionally, maximum line intensity and optimum resolution of closely spaced reflections are not simul-

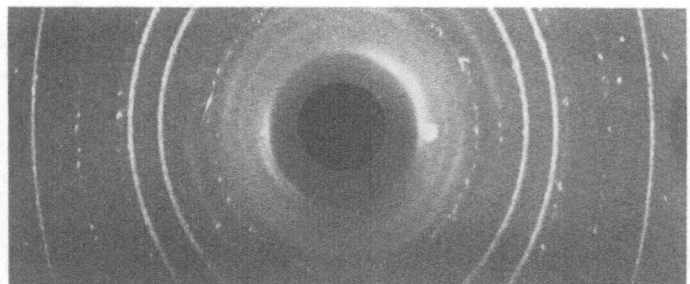

Fig. 1. Debye–Scherrer pattern of 0.03 mm layer of a powder
mixture of LiF and Si supported on 0.25 mil Mylar film.
Exposure: CuKα 40 Kv·20 ma·4h; 57.3 mm camera.

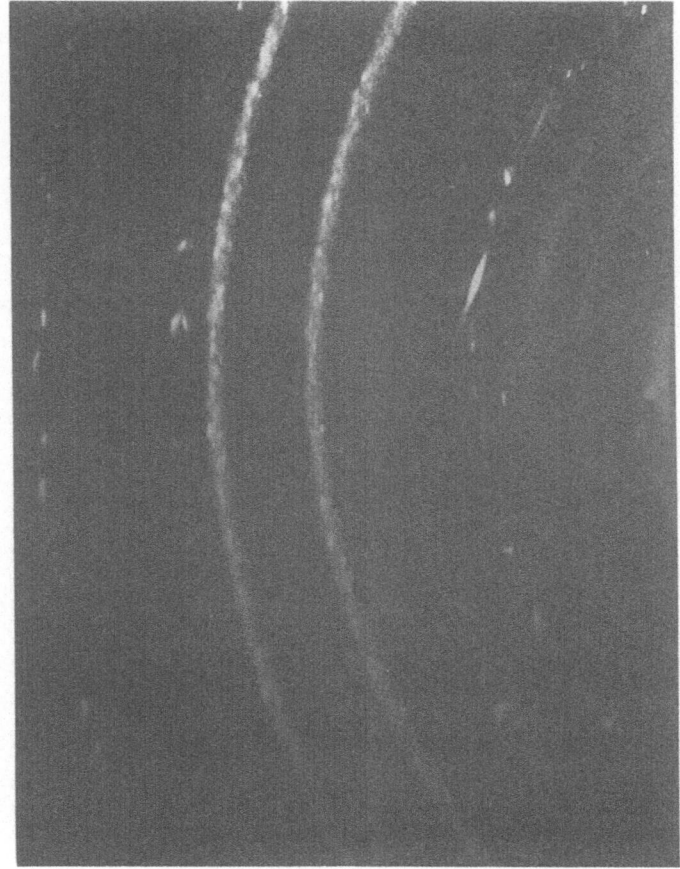

Fig. 2. Threefold enlargement of the three characteristic types
of diffraction "rings".

Fig. 3. Laue spots from a 20 μm Si-particle mounted on 1-mil
         W-wire. Exposure: CuK (40 Kv·20 ma·4h) in a modified
         Philip's camera[20]. The two round spots flanking the
         central hole are fiducial marks.

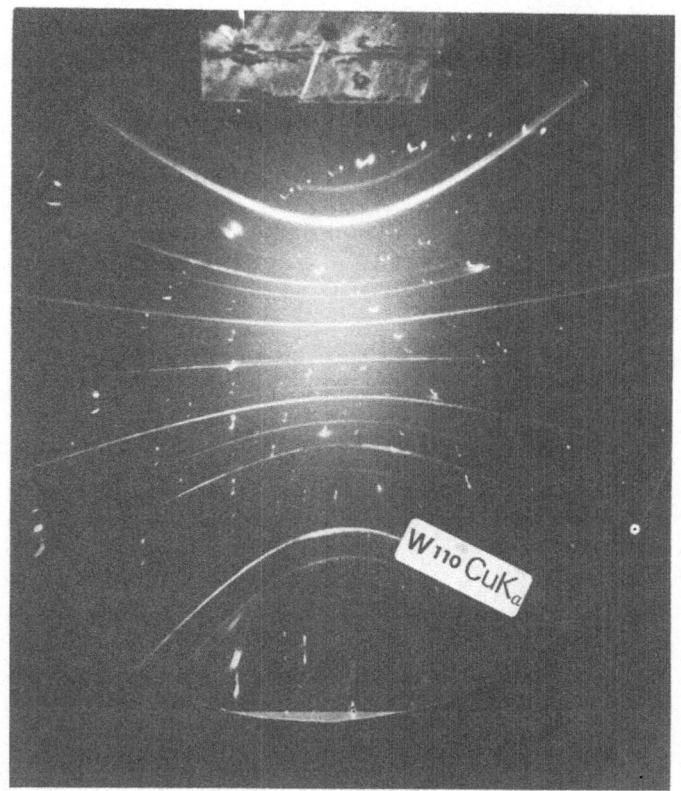

Fig. 4. Laue spots from 0.15 mm sphere of α-Al$_2$O$_3$ mounted on 1-mil
         W-wire. Exposure: CuK (40 Kv·20 ma·5h) in a Gandolfi
         camera with a flat circular film held on cover plate.

taneously achieved with known diffractometers. A proposed novel
extension of a cone-focus diffractometer[21] which avoids said
difficulties is shown in Figure 5. The cone of WL radiation emanating from the 10 μm focal spot S_W is precisely defined by the 20 μm-
wide ring slit so that the conically collimated rays of WL radiation
impinging on the 0.05 mm thin, correctly oriented, conically bent Ge
crystal are monochromatized (by the 111 plane) for $WL\alpha_1$ radiation
forming a cylinder of parallel rays of monochromatic x-rays grazing
the thin (0.05 mm) rim of powder sample held by the tapered sample
disk. The sample disk should be constructed of a rigid material of
high dimensional stability and having a low coefficient of thermal
expansion such as W or Ta. By moving the receiver-aperture (10 μm
diameter) with its concomitant x-ray sensor along the central axis,
one intercepts the focal apex of any diffracted cone of $WL\alpha_1$ radiation. The diameter of the sample disk, which determines the size
of the whole diffractometer, has a practical limit of 5-10 cm.
Figure 6 illustrates the random distribution of powder particles
adhering to the beveled knife edge of the rim of a prototype sample
disk (Figure 7). Since the colinearity of the focal spot, center
of the ring slit, cone axis of the Ge crystal, center of the sample
disk, and the center of the receiver-aperture has to be maintained
within stringent limits, it is imperative to maintain a constant
temperature (±0.2°C) within the housing of the diffractometer. Contrary to conventional diffractometers, the accuracy of relatively
large interplanar spacings (>8 Å) as determined by the proposed
parafocusing diffractometer is greater than for small interplanar
spacings (<2 Å) as seen from equations (3) and (4),

$$d = \frac{\lambda}{2 \sin[0.5 \tan^{-1}(R/s)]} = \frac{\lambda}{[2(1 - \frac{s}{\sqrt{R^2+s^2}})]^{\frac{1}{2}}} \tag{3}$$

$$\frac{\Delta d}{d} = \frac{R^2 \Delta s}{2(R^2+s^2)[(R^2+s^2)^{\frac{1}{2}} - s]} \tag{4}$$

where R is the radius of the sample disk and s is the variable
distance from the center of the disk to the receiver-aperture.
Moreover, the measured intensity of a particular reflection results
from the sum of all the diffracted radiation from many crystallites
of a very thin rim of powder sample. Thus better relative intensities are obtained than by measuring only a small portion of the
diffraction cone as is done by present-day instruments.

Instead of using a movable x-ray sensor right after the
receiver-aperture, one could position at 50 cm from the sample disk
a multiwire proportional counter[22] for photon counting and for
position-sensitive coverage of all diffracted $WL\alpha_1$ radiation passing
through the receiver-aperture. With such a device the spottiness
of the different rings could be quantitatively described by the
appropriate count-histograms.

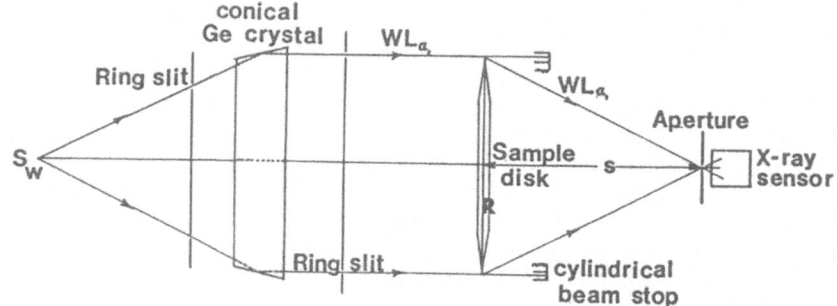

Fig. 5.   Schematic diagram of cone-focus powder diffractometer.

Fig. 6.   Cu$_2$O powder mounted on knife edge of sample disk.

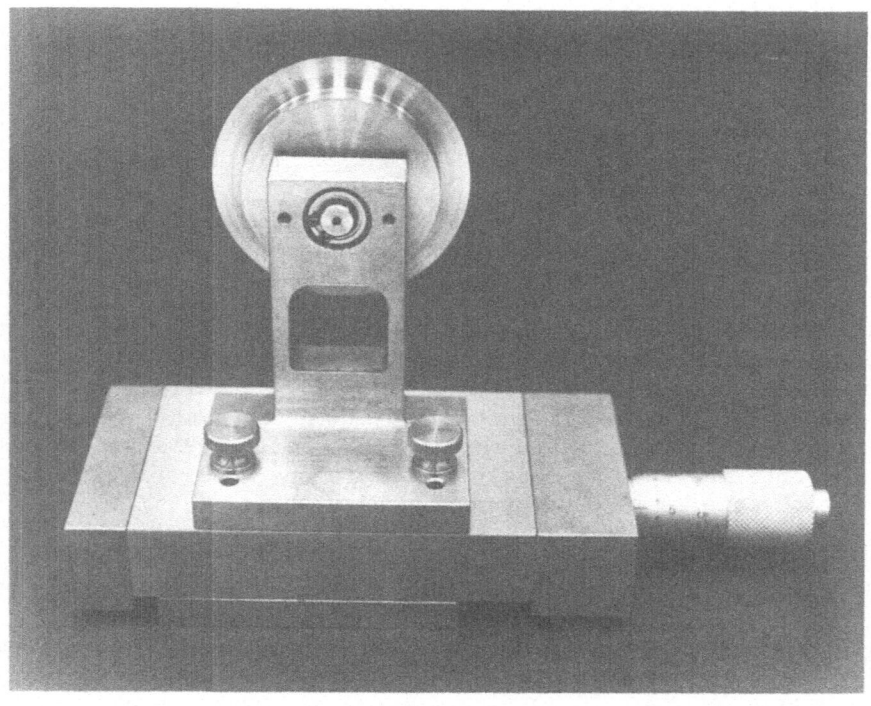

Fig. 7.   Sample disk mounted on alignment glide.

Lastly, the concept of a color powder pattern is worth exploring. By reversing the normal procedure of determining the various grating constants (d-spacings) of a 3-dimensional diffraction grating with monochromatic radiation (λ fixed), one varies λ monotonically until Bragg's law is satisfied for a particular d-spacing. The first attempt to use white x-rays for energy-dispersive powder diffraction (EDXRD) was published in 1968 by Giessen and Gordon[23]. In actual practice it was found that EDXRD was applicable only to very simple powder patterns[24,25] inasmuch as the detectable "color" peaks were more than an order of magnitude wider than in conventional diffractometry with filtered $CuK\alpha$ radiation. Moreover, certain geometrical and physical aberrations are encountered as pointed out by Wilson[26]. Nonetheless, Buras[27] has demonstrated that synchroton radiation with its continuous smooth spectrum is useful for studying fast phase-transitions and other time-related phenomena.

A rather novel approach to achieving well resolved color peaks would be to generate white x-rays having a monotonically decreasing minimum wavelength (λ min) corresponding to a precisely controlled ripple-free voltage V of the x-ray generator; i.e.,

$$\lambda_{min} = \frac{hc}{eV} \tag{5}$$

as first demonstrated by Duane and Hunt[28]. The second innovation would be to have this minimum wavelength satisfy Bragg's law at $\theta=90°$ as expressed in (6),

$$d = \frac{\lambda_{min}}{2} = \frac{6199}{V} \text{ Å} \tag{6}$$

where V is expressed in ordinary volts. Thus the descending interplanar spacings of a crystalline phase would be registered in descending wavelengths. Figure 8 depicts a conceptual configuration embodying these novel features. In a high vacuum ($<10^{-7}$ Torr) a beam of monoenergetic electrons is focused onto the 0.1 mm thin section of a concave tungsten target to form a line focus (2 mm length x 0.1 mm width). At a take-off angle of 1.5° the end-view of the 2 mm line focus appears as a point source of white x-rays (Point F in Figure 8). The beam of x-rays, passing through the narrow rectangular slit S_1 and scatter slit S_2, is collimated by two Soller slits, S_3 and S_4 (at right angles to each other), so as to define a rectangular divergent array of pencils of white x-rays. This collimated beam impinges at $\theta=90°$ on a thin, uniform, cylindrically curved layer of powder at P. Such a thin uniform layer (30 mm x 2 mm x 0.10 mm) held in a recessed Be holder can be prepared according to the procedure of Frevel and Roth[29]. The distance FP is 600 mm so as to restrict the over-all divergence of the x-ray beam to ∿2° (the divergence for any pencil of x-rays is only 0.06° for Soller plates 0.5 mm apart). All crystallites, with

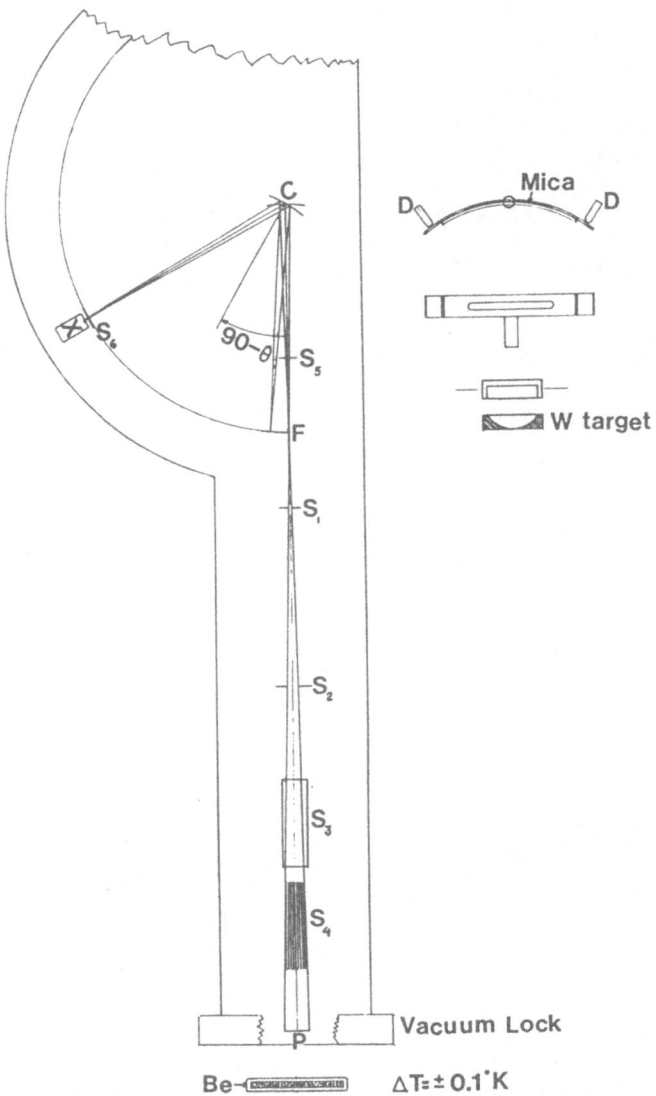

Fig. 8. Conceptual configuration of powder spectrodiffractometer.

any crystallographic planes normal to any of the pencils of white
x-rays, will diffract those x-rays for which $\lambda = 2d_\nu$, where d_ν is
the νth interplanar spacing of the powder specimen. The diffracted
x-rays of different "colors" pass through the Soller slits, S_4 and
S_3, to form a convergent beam of discrete wavelengths coming to a
focus at F with one half of the diffracted beam passing underneath
the 0.1 mm thin section of the W-target and the upper half entering
the x-ray generator. The unobstructed portion of the diffracted
x-rays passes through the scatter slit S_5 and then impinges on the
correctly bent mica crystal at C turned to the scanning angle of
$(90-\theta_\lambda)°$. For each discrete wavelength the x-ray sensor X will
detect a discontinuous jump in intensity at $(180-2\theta_\lambda)°$ according
to (7)

$$2\theta_\lambda = 2 \sin^{-1}\left[\frac{\lambda}{2d_{002}(1-4d_{002}^2\delta\cdot\lambda^{-2})}\right] \qquad (7)$$

where d_{002} (10.007Å) is the basal reflection for mica $2M_1$ and δ is
the refraction correction. If $\underline{R}$ (equal to CF) is the radius of the
goniometer circle, then the radius of the focusing circle of the
correctly bent mica crystal is $\underline{R}/2 \sin\theta_\lambda$. The precise continuous
bending of a thin (<0.1 mm) mica ribbon can be accomplished by
the accurate protrusion of the symmetrically positioned displacement pins D causing the continuous bending as θ_λ decreases from
87.5° (λ=19.995Å) to 29.97° ($\lambda/2$=9.997Å). To avoid exciting
$\lambda_{max}/2$, the constant ripple-free voltage of the x-ray generator
should be kept slightly below 1.24 Kv for the first scan. For the
second scan from 29.97° (λ=4.998Å) to 87.5° (2λ=9.997Å), one uses
the 004 reflection of mica and sets the excitation voltage slightly
below 2.48 Kv and employs pulse height discrimination to eliminate
the 2λ contribution. For the third scan the voltage should be kept
below 3.72 Kv. To compensate for the low excitation voltages, one
can increase the milliamperage of the x-ray generator to 200 ma
or higher. By programming the milliamperage as a function of θ_λ,
one can maintain a constant intensity-plateau for the back-scattered
white x-rays emanating from the empty Be holder. The only discontinuities in I_λ(Be) (see Figure 9) are the tungsten M-emission lines
and their Compton-radiation components as well as the d-spacings for
Be and the sharp faint Renninger reflection (2.4508Å) for the strong
101 reflection of Be (provided the diffracted x-rays are not completely polarized):

$$\lambda_R(Be) = 2d_{101} \sin 45° \qquad (8)$$

By contrast, when a powder specimen like U-3.6[1],[30] is run under
the same regimen, one will generally note five differences: (1)
a change in the flat background, (2) the appearance of characteristic profiles for the various interplanar spacings of the different
crystalline phases, (3) the sharp discontinuities pertaining to the

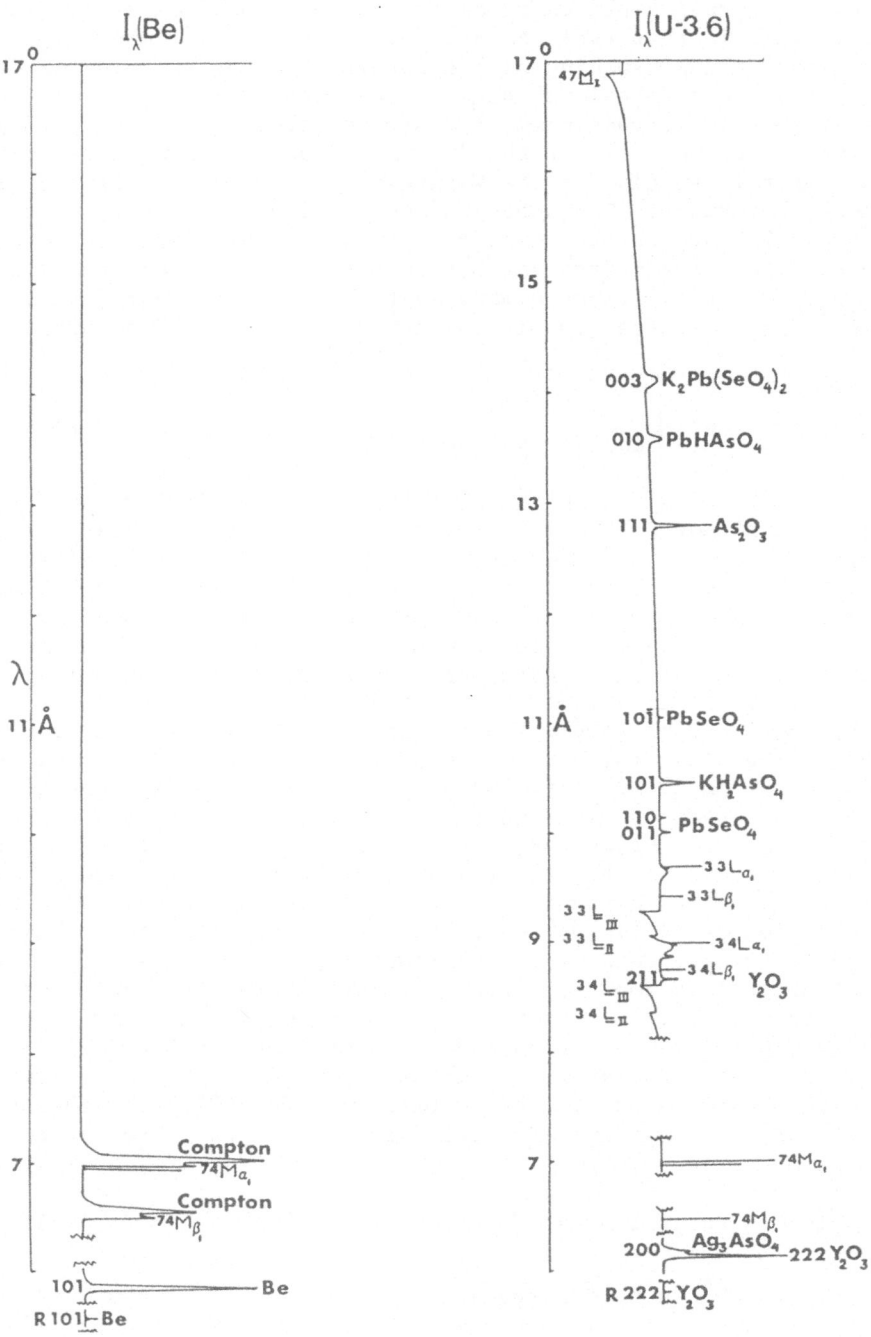

Fig. 9.   Simulated "color" diffractograms for empty Be sample-
holder and for 7-phase mixture U-3.6[1],[30].

secondary x-ray emission lines from the elements in the powder
specimens, (4) the appearance of specific absorption edges for the
elements present (Z > 8), and (5) a few weak sharp Renninger
reflections for the most intense reflections. Thus the proposed
x-ray spectro-diffractometer combines x-ray powder diffraction, x-ray
fluorescence, and x-ray absorption into one analytical procedure
that could lead to accurate quantitative assays of the crystalline
phases in a powder sample. The two unique advantages of this
spectrometric diffraction powder method are: first, that the combined Lorentz and polarization factor approaches ∞ as θ approaches
$90°$; and, second, that the line profiles for all interplanar
spacings are measured with minimum geometrical aberrations.

ACKNOWLEDGEMENTS

     The author is sincerely grateful to the International Centre
for Diffraction Data for establishing the J. D. Hanawalt Powder
Diffraction Award. He also wishes to acknowledge the cheerful
assistance he received from his capable coworkers at The Dow
Chemical Company: the late Harold W. Rinn, Lorenzo Sturkey,
June W. Turley, LaVerne R. Ruhberg, Philip P. North, Carole
Engbrecht Adams, Monte C. Nichols, James E. Edmonds, Timothy G.
Fawcett, Robert Newman and Pamela M. Kirchhoff. Dan Filsinger and
William C. Roth of the Analytical Department of the Dow Corning
Corporation have been most helpful with the author's more recent
research efforts in powder diffraction. To Prof. Thomas J.
Kistenmacher of the Johns Hopkins University he is especially
indebted for carefully reviewing the author's manuscripts written
during the past eight years.

REFERENCES

     1. L. K. Frevel, Anal. Chem., 37, 471-482 (1965).
     2. M. C. Nichols and R. A. Bideaux, Twenty-fourth Pittsburgh
Diffraction Conference, Pittsburgh, PA, Paper No. B-3 (1966).
     3. G. G. Johnson, Jr. and V. Vand, Twenty-fourth Pittsburgh
Diffraction Conference, Pittsburgh, PA, Paper No. B-1 (1966).
     4. G. G. Johnson, Jr. and V. Vand, Ind. Eng. Chem. 59,
19-31 (1967).
     5. L. K. Frevel, C. E. Adams and L. R. Ruhberg, J. Appl.
Cryst., 9, 199-204 (1976).
     6. J. W. Edmonds and W. W. Henslee, Adv. in X-ray Analysis,
.22, 143-150 (1979).
     7. J. W. Edmonds, J. Appl. Cryst., 13, 191-192 (1980).
     8. M. C. Nichols and R. C. Basinger, American Crystallographic
Association Meeting, Berkeley, CA, Paper L10, March 24-28 (1974).
     9. M. C. Nichols, American Crystallographic Association
Meeting, Charlottesville, Virginia, Paper F11, March 9-13 (1975).
     10. G. G. Johnson, Jr., "User Guide; Data Base and Search
Program", JCPDS, Swarthmore, PA (1974).

11.   J. D. Hanawalt, Adv. in X-ray Analysis, 20, 63-73 (1977).

12.   R. G. Marquardt, I. Katsnelson, G. W. A. Milne, S. R. Heller, G. G. Johnson, Jr., and R. Jenkins, J. Appl. Cryst., 12, 3-8 (1979).

13.   G. J. McCarthy, PDF Workbook for Computer SEARCH/MATCH Methods, JCPDS, Swarthmore, PA (1981).

14.   T. C. Huang, W. Parrish and B. Post, American Crystallographic Association Meeting, National Bureau of Standards, Gaithersburg, Maryland, Paper M4, March 29-April 2 (1982).

15.   T. C. Huang, W. Parrish and B. Post, Adv. in X-ray Analysis, 26,

16.   Powder Diffraction File of the Joint Committee on Powder Diffraction Standards, Sets 1-32, published by the International Centre for Diffraction Data, 1601 Park Lane, Swarthmore, PA 19081.

17.   L. K. Frevel, Anal. Chem., 54, 691-697 (1982).

18.   J. D. H. Donnay, G. Donnay, E. G. Cox, O. Kennard, and M. V. King, "Crystal Data Determination Tables," 2nd Ed., ACA Monograph No. 5, American Crystallographic Association (1963).

19.   J. D. H. Donnay, K. M. Ondik, A. D. Mighell, M. E. Mrose, C. R. Robbins, J. K. Stalik, D. A. Hansen, G. M. Wolten, and R. J. Boreni, "Crystal Data Determinative Tables," 3rd Ed. Vol. 4 Inorganic Compounds, published jointly by the U.S. Department of Commerce, National Bureau of Standards and the JCPDS-International Centre for Diffraction Data (1978).

20.   L. K. Frevel, D. C. DeLeeuw and W. R. Albe, Norelco Reporter Vol. 29, Number 2, 38-39 (1982).

21.   U.S. Patent 4,247,771 (1981).

22.   R. Hamlin, "Proceedings of the Symposium on New Crystallographic Detectors," Transactions of the American Crystallographic Association, Vol. 18, 95-123 (1982).

23.   B. C. Giessen and G. E. Gordon, Science, 159, 973-975 (1968).

24.   T. Fukamachi, S. Hosoya and O. Terasaki, J. Appl. Cryst., 6, 117-122 (1973).

25.   M. Mantler and W. Parrish, Adv. in X-ray Analysis, 20, 171-186 (1977).

26.   A. J. C. Wilson, J. Appl. Cryst., 6, 230-237 (1973).

27.   B. Buras, National Bureau of Standards, Special Publication 567, Proceedings of Symposium on Accuracy in Powder Diffraction held at NBS, Gaithersburg, MD, June 11-15, 1979, pp. 33-54 (1980).

28.   W. Duane and F. L. Hunt, Phys. Rev., 6, 166-171 (1915).

29.   L. K. Frevel and W. C. Roth, Anal. Chem., 54, 677-682, (1982).

30.   L. K. Frevel, Anal. Chem., 38, 1914-1920, (1966).

PRESENTATION OF THE J. D. HANAWALT AWARD

J. D. Hanawalt

The University of Michigan

Ann Arbor, Michigan 48109

First, let me say that I am deeply honored by the action of
my friends of the Joint Committee in establishing the Hanawalt
Award. I have greatly enjoyed my association over the years with
the members of the Joint Committee. So many brilliant and talented
people have contributed so much to the work of the Joint Committee.
By comparison, my contributions have been minor and I feel very
humble to be so honored. I want to thank you all for your part in
the proceedings today. It is a very high point in my life.

I would like to congratulate the Award Committee on their choice
of Ludo Frevel as the first recipient of this Award. It is a very
special pleasure to me to have the privilege of presenting this
Award to Ludo because, besides being a very worthy recipient, he
and his wife, Ruth, have been my very good friends for fifty years.
In fact, Ruth was a member of our Pot Luck Club in Midland for
about five years before Ludo came to town--and took over from the
local swains. I can take the credit for introducing Ruth and Ludo,
but Ludo did the rest. There is a side to Ludo you all may not
know about. You know him as a modest, quiet, intellectual person
and as a deep thinker--and that is true, but he is also a man of
action when the occasion calls for it. For example, our first
intimation that Ludo and Ruth were married was a wire from some
far-away place announcing that they had eloped.

You have, no doubt, noticed as I have that Ludo seems to be
getting younger year by year. He is reversing the clock. Early
in his career, he went to Cal Tech from Johns Hopkins and then to
Dow from Cal Tech, and now he is back at Johns Hopkins. I predict
a second distinguished career is just beginning. You may appreciate that prediction better when I tell you that Ludo's mother,

who lives in Baltimore, travels back to her native Germany <u>alone</u> each summer to celebrate her birthday--and this year it was her 100th birthday.

    Ludo himself is always in great shape. I don't think he ever missed a day, rain or shine, when he didn't briskly walk the two miles to and from work. Ludo is a perfectionist, whether it be writing a technical paper or playing golf. It is, indeed a great pleasure to present this Award to my friend, Ludo. Along with this certificate goes a check for $1,000. Ludo knows he doesn't really need another thousand dollars. On the other hand, I'm not so sure. I hear by the grapevine that our wives have been spending a lot of time shopping for these great bargains at Snowmass, so it may be Ludo will find out he does need it more than he thinks.

COMPUTER TECHNIQUES FOR FASTER X-RAY DIFFRACTION PHASE IDENTIFICATION

Raymond P. Goehner and Mary F. Garbauskas

General Electric Company
Corporate Research and Development
PO Box 8
Schenectady, New York   12301

ABSTRACT

This paper describes the procedures used to retrieve JCPDS powder diffraction data by certain characteristics. These characteristics may include chemistry, mineral name, highest intensity dspacing, largest dspacing, PDF number, etc. The storage scheme used for the powder data and the procedures used to enhance the retrieval speed are described.

INTRODUCTION

A set of programs for computer-aided x-ray powder diffraction phase analysis was previously described by the authors (1,2). This paper describes the procedures used to enhance the speed of these programs and thus make them more interactive. A computer program becomes an interactive tool when results can be obtained and displayed within a few seconds. If the diffractionist has to wait several minutes for the results, the major reason for automated searches, that of speed, will be negated. The need for an interactive tool was the primary design criteria for the "PDIDENT" computer software.

DISCUSSION

The powder diffraction data base is maintained and distributed by the JCPDS (3). It consists of about 40,000 phases, 27,500 of which are inorganic and 12,500 of which are organic. The JCPDS supplies the PDF numbers of various subfiles such as common phases, metals and alloys, minerals, and forensic phases. The data supplied on magnetic tape consist of PDF numbers, chemical formulae, data quality marks, subfile class indicators, I/I corundum values, chemical/mineral names, dspacings, and intensities. Figure 1 shows how this information is stored on disks. Each PDF set has a separate direct access binary file with 64 byte records. Thus, there are currently 32 files of various lengths comprising the primary data base. The dspacings are stored in 2 byte integers as $100*2\theta$ in degrees for $CuK\alpha_1$ radiation. The relative intensities are stored as 1 byte integers. There are some cards in the file with relative intensities greater than 127 (the maximum value that can be stored in one byte). These intensities are stored as 127 and this number then is a signal to the user that there is a peculiarity with the card. The quality mark, chemical formula, and chemical/mineral name are stored in ASCII and the I/Ic value as a 4 byte real number. The number of 2θ, I pairs and the number of chemical symbols are each stored in one byte. There are several advantages in packing the data base in this manner. First this packing scheme allows the data base to be built and accessed using only DEC FORTRAN IV routines. Thus there is no need for writing and maintaining any assembly language code. Second, the primary data takes only 8.5 Mbytes of disk space, as opposed to the approximately 15 Mbytes of ASC11 data supplied by the JCPDS. Third, since $2\theta*100$ are stored as two byte integers, a constant 2θ error can be assigned, and all comparisons of unknown and standard patterns can be performed using integer arithmetic. This procedure has a dramatic effect on speed, especially on small systems without a hardware floating point processor.

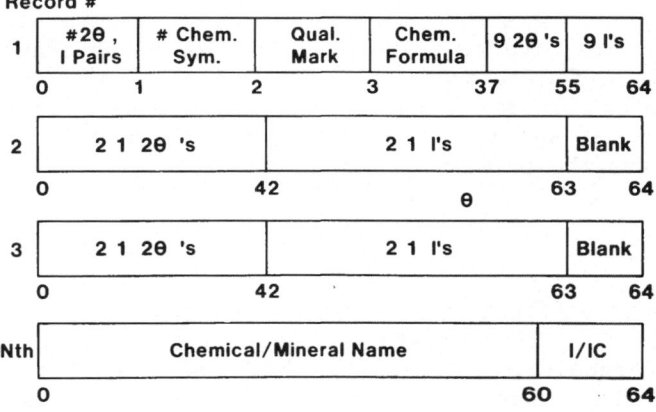

Figure 1.   JCPDS data base packing. 32 direct access files with 64 byte records.

Although this packing scheme is an efficient use of disk space and utilizes integer arithmetic, accessing information from these files would be extremely costly in terms of time. For this reason, cross-index files have been constructed to retrieve a card from the data base with the desired attribute. The concept of using cross-indices (extended data bases) is the same as that used by libraries. Instead of searching the entire book collection in a library for a book or books on x-ray powder diffraction, the exact location of the book or books can be determined using the card catalog.

The first type of index is designed to allow rapid access to the data base by its Powder Diffraction File (PDF) number. This PDF index is a binary direct access file where each entry contains a pointer to a card in the JCPDS file. There are presently 32 records in this index corresponding to the 32 PDF files. Each record contains 2500 two-byte integers corresponding to card numbers 1 to 2500 for that particular set. The value of the number stored in, for example, the 490th integer in record #5 contains the address in the binary PDF file #5 where the data for card 5-490 starts. Figure 2 is a graphical representation of this retrieval process. If the address is a negative number, the data have been classified by the JCPDS as organic and the actual address is the negative of this number. If the data have been classified as inorganic, the address is stored as a positive number. Any cards which have been deleted or are otherwise not present in the data base have an address of zero. This cross-index functions to provide the diffractionist instantaneous access to any card in the data base when its PDF number is known.

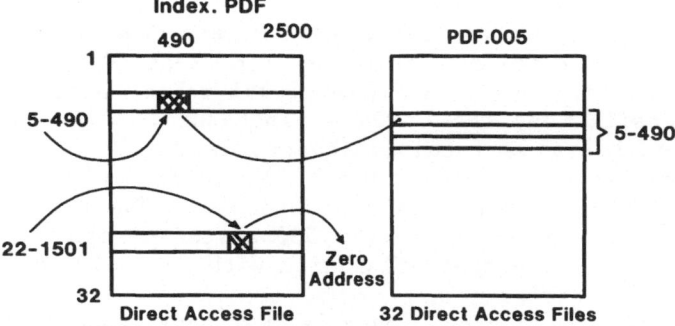

Figure 2.   Graphical representation of card retrieval by PDF number.

Most analysts, however, do not generally think in terms of PDF number but in terms of chemistry and in some cases chemical or mineral name as well. The second type of cross-index allows the same instantaneous access by chemical information that the first type of cross-index allows for PDF number. The chemical cross-index is actually a subset of indices to the inorganic JCPDS data base. Each subset corresponds to a particular atomic number, where that atomic number corresponds to the highest atomic number present in the chemical formula. Thus, Fe_3O_4 and $FeCr_2O_4$ would both be contained in chemical subset #26 since Fe is the highest Z in both of these chemical formulas. However, $NiFe_2O_4$ would be contained in chemical subset #28 since Ni is the highest atomic number in this formula. All of the inorganic cards in the PDF file have been assigned to a subset in this manner.

Each record of the subset contains up to 16 bytes of information about the particular phase. The first byte of information corresponds to the number of unique symbols in the chemical formula; e.g., $Na_2(Al_2Si_3)O_{10} \cdot H_2O$ contains 5 chemical symbols – Na, Al, Si, O and H. The next 9 bytes contain the atomic numbers, 1 per byte, of these chemical symbols in decreasing order. The highest atomic number is not stored since the file itself is determined by this atomic number. In the $Na_2(Al_2Si_3)O_{10} \cdot H_2O$ example, bytes 2 through 10 of the file SMO:CHEM.014 would contain 13, 11, 8 and 1 followed by 4 zeros. If any chemical formula contains more than 10 chemical symbols, the lowest atomic numbers are not stored. At present, there are currently no phases in the data base with more than ten unique elements. There are several phases where rare earth substitutions may occur. These phases contain the symbol Ln and this symbol has been assigned atomic number 104 in our chemical cross-index. A double precision real number is equivalenced with bytes 2 through 9 of each record and the entire subindex is ordered from low to high by this number. The number of phases in each subindex is stored in a master subindex. The remaining 6 bytes of the record contain the PDF file number, card number, and record number of the card address in the direct access PDF file.

The retrieval process reduces to binary searching the appropriate subindex for the real number that corresponds to the desired chemistry. Limitations to this retrieval scheme are: 1) the card must be classified by the JCPDS as inorganic, 2) no information about stoichiometry is retained (hence, Fe_2O_3, Fe_3O_4 and FeO all have the same retrieval number) and 3) unique retrieval of cards with more than 10 elements is not possible. The entire chemical cross-index occupies approximately 500 kbytes of storage with the largest subindex being that of uranium with over 1100 phases.

The third type of cross-index is that based upon a strong line of the pattern and is utilized primarily by the search programs to identify an unknown pattern. The cross-index is a binary direct-

access file with 8 byte records. The first two bytes contain the 2θ
position of the strongest line. The remaining 6 bytes contain the
PDF file number, card number, and address of the data in the binary
PDF file. The strong line cross-index contains the strongest line
of the pattern that occurs between 10 and 60 degrees 2θ with Cu
radiation for inorganic phases and between 5 and 60° 2θ for organic
phases. If more than one line occurs with the same intensity the
lowest 2θ is stored. If no line occurs (a few PDF cards have back
reflection data only) a zero is stored. The subfiles of minerals,
common phases, metals and alloys, common materials, common
minerals, and NBS patterns furnished by the JCPDS have been
repacked in an identical manner. All of these strong line cross-
indices occupy approximately 400 kbytes of storage and function to
increase speed in the computer search/match programs.

The fourth type of cross-index is based on the mineral name of
the mineral subfile. This index was built in order to enable the
user to access the data base by mineral name and it consists of a
direct access file with 16 byte records. The bytes 1-10 contain the
first 10 characters of the mineral name, while the file, card, and
record numbers occupy two bytes each. The file is ordered by the
composite number obtained by equivalencing the first eight bytes of
each record with a double precision real number. The retrieval
process thus reduces to a binary search of this cross-index by
composite number. For each entry with the correct composite number
a comparison is done on the last two bytes. There are several
limitations to the utility of the index. First, the mineral name
may not be unique in the first ten characters. This is only a
problem for a few minerals. Second, the current PDF data supplied
on magnetic tape has many of the mineral names truncated. This is a
serious problem and tends to negate the utility of this index.

CONCLUSION

There are several disadvantages to using index files. The
first being extended data bases use additional disk space. The
cross-indices described in this paper take 1.1 Mbytes of storage in
addition to the 8.5 Mbytes occupied by the data files. Since the
database is growing, the storage requirements will increase each
year. With the declining cost of disk storage, the relatively small
increased space requirements of the cross-indices should present no
real problem. Maintenance of the data base becomes a difficult
problem. If a card has to be changed because of an error or a user
card added to the data base, the appropriate cross-index also has be
changed and reordered. This represents a problem which the authors
to date have not fully resolved. The third disadvantage is that the
FORTRAN programs become more complex.

The major advantages of using an extended data base is that
access to the data becomes instantaneous with respect to a human

time frame. This means the analyst can actively interact with the
JCPDS file and thus truely use the computer as a retrieval tool.

REFERENCES

1.   R.P. Goehner and M.F. Garbauskas, "Computer-Aided Qualitative
         X-ray Powder Diffraction Phase Analysis," Adv. in X-ray
         Analysis, 26, pp. 81-86, 1983.
2.   R.P. Goehner and M.F. Garbauskas, "PDIDENT - A Set of Complete
         Programs For Powder Diffraction Phase Identification",
         General Electric Technical Information Series, 83CRD062,
         Schenectady, NY 1983.
3.   Joint Committee on Powder Diffraction Standards, International
         Centre For Diffraction Data, 1601 Park Lane, Swarthmore,
         PA 19081.

FUZZY SETS AND INVERTED SEARCH –

TWO CONCEPTS FOR COMPOUND.IDENTIFICATION IN SPECTRA

Thomas Blaffert

Philips GmbH Forschungslaboratorium Hamburg
Vogt-Kölln-Strasse 30
D-2000 Hamburg 54, F.R.G.

ABSTRACT

This paper describes a new approach to the interpretation of
spectra with 'fuzzy sets' and a fast search algorithm for spectrum
libraries, using a combination of this theory and the inverse
search method. The concepts are applied in a computer program
named CIF (Compound identification with Inverted search and Fuzzy
sets) which is used in the current application for the interpretation of X-ray powder diffraction signals. The methods are readily
applicable to other kinds of spectra such as IR, UV, chromatro-
graphy etc.

COMPUTER ASSISTED INTERPRETATION OF SPECTRA

A hierarchical classification is often used in computer
assisted interpretation of spectra. This classification reduces the
number of possible compounds in a 'search' step by looking at the
line positions and then investigates the remaining ones more carefully in a 'match' step. The 'identify' step compares combinations
of references with the sample spectrum [1,2].

Because line positions are most relevant, the introduction to
the fuzzy sets is described in terms of this feature, but, in
practice, it is applied to intensities and other features as well.

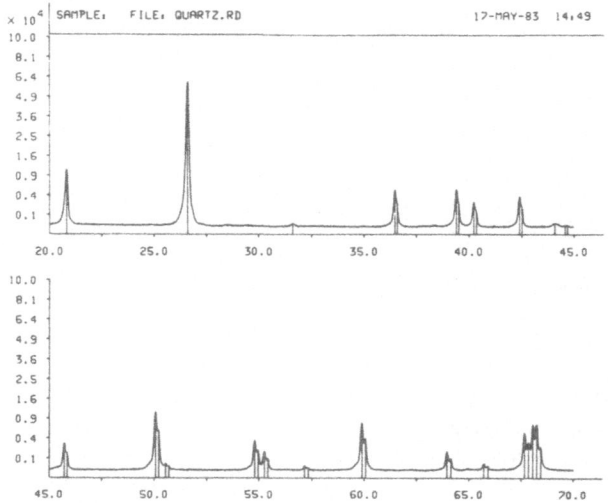

Fig. 1. Diffractogram of Quartz, the peaks are located by an automated peak search program and marked by lines in the diagram.

COMPARISON OF SPECTRA FORMULATED IN CONVENTIONAL AND FUZZY SET THEORY

As an example the diffractogram of quartz is shown in Fig. 1. All measured peaks have been located by a peak search program and are marked by vertical lines in the diagram. The spectrum is now characterized by a collection of features such as position (degr. two theta, wavenumber, time etc.), height and width. It is important to treat these features as a set of features rather than a feature vector. The reason is that in a vector peaks have to be assigned to vector components, and the comparison between sample and reference is done componentwise. Since this assignement requires extra rules like "assign reference peak to the nearest sample peak", the vector approach is disadvantageous, particularly in a multicomponent sample with more sample lines than lines in a single component reference. In Fig. 2a the elements of the set are illustrated by single lines. A reference spectrum is also a set of reference lines (Fig. 2b).

In conventional set theory the number of reference lines in the sample spectrum is given by counting the number of elements in an intersection of the two sets. This is the conventional set power of the intersection. The resulting intersection sets are usually small or may even be empty, as in Fig. 2c, because the measured data or the data in the comparison pattern are distorted with small (perhaps random) variations.

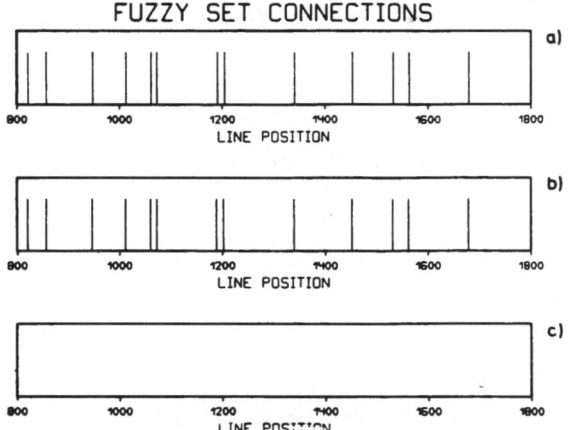

FUZZY SET CONNECTIONS

Fig. 2.
Comparison of measured
set of lines (a) and
reference set of lines
(b) using conventional
intersection. Since
there are always small
differences between
sample and reference
lines, no lines are
contained in the
intersection set (c).

This random variation is taken into account in the present
search/match method by introducing the theory of fuzzy sets. In
fuzzy set theory it is not only valid to say 'a spectrum line
either belongs to a set or it does not' but you can also say 'a
spectrum line belongs to a set to a certain degree'. So a reference
line which has a slightly different position from a measured line
does not abruptly disappear in the intersection, it is still
present, but with a lower degree of membership. A visual
illustration is given in Fig. 3a-c.

ELEMENTS OF FUZZY SET THEORY

A fuzzy set A, as introduces by Zadeh[3], is characterized by a
membership function

$$f_A : X \rightarrow [0,1] \quad . \tag{1}$$

The function $f_A(x)$ assignes a degree of membership to each
point x in the position range X. The membership function f_A can
be defined objectively by probability distributions, but it can
also be used to incorporate personal experience.

An intersection and unification of two fuzzy sets A and B is
defined as the membership functions of A ∩ B and A ∪ B given by

$$f_{A \cap B}(x) = \min[f_A(x), f_B(x)] \quad . \tag{2}$$

$$f_{A \cup B}(x) = \max[f_A(x), f_B(x)] \quad . \tag{3}$$

(∩ is the intersection symbol, ∪ is the unification symbol,
$f_{A \cap B}$ is the membership function of the intersection of A and B.)

The <u>power</u> of a fuzzy set A, for a finite universe X, is

$$N_A = \sum_{x \in X} f_A(x) \quad , \tag{4}$$

where $x \in X$ means that the sum is over all line positions in the considered position range.

The last definitions are consistent with conventional set theory if f is limited to the two valued sets {0;1}.

The power of a fuzzy set is a confidence measure which can be normalized for a match score.

The random errors in the sample spectrum are modelled by <u>fuzzyfying</u> the spectral lines, e.g. according to

$$f_{S}{}^{j}(x) = \exp\left[-(x^{j}-x)^2/2\sigma^2\right] \quad , \quad x^{j} \text{ is position of line } j \quad . \tag{5}$$

σ determines the broadness of the fuzzification.

The fuzzified spectrum is the unification of all fuzzified lines:

$$f_S(x) = \max_{j}\left[\exp\left[-(x^{j}-x)^2/2\sigma^2\right]\right] \quad . \tag{6}$$

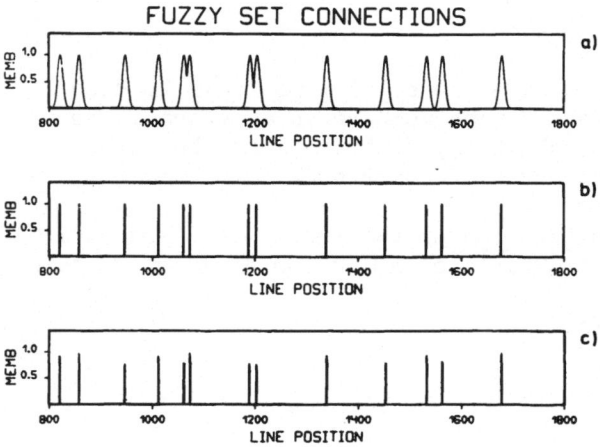

Fig. 3.
Comparison of a fuzzified measured set of lines (a) and a reference set of lines (b) using the fuzzy intersection. The intersect set (c) reflects all lines, but some of them have a degree of membership smaller than 1.

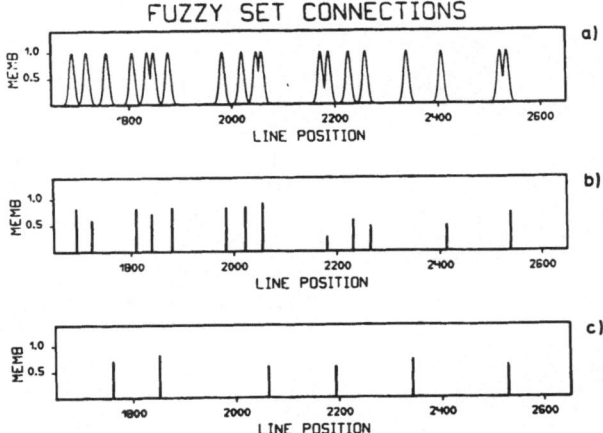

Fig. 4.
Comparison of reference with a two component unknown sample. (a) represents the measured spectrum, (b) and (c) are the two reference sets after a fuzzy intersection.

Fig. 3a shows a fuzzified spectrum. Notice that the Gaussians in the fuzzified spectrum must be distinguished from a line profile, they represent NOT an intensity- but a position-dependent error distribution. The fuzzifying of lines expresses the possibility of locating reference lines at slightly different positions than sample lines and is determined by instrumental misalignments, errors in the peak search etc. The broadness of lines in the diffractogram is a property of the sample and is not directly related to the fuzzification.

ADVANTAGES OF THE FUZZY SET APPROACH

Clearly the described method can be used if the spectrum is obtained from a mixture of different compounds (Fig. 4). The method implies naturally which sample and reference lines are related. The intersection finds all constituent compounds in the search.

The effect of missing lines resulting from background noise or preferred orientation of the sample can also be naturally included in the fuzzy set concept. Missing lines will decrease the set power by 1 and lower the position of a correct reference pattern in the confidence score list, but the pattern has not totally disappeared.

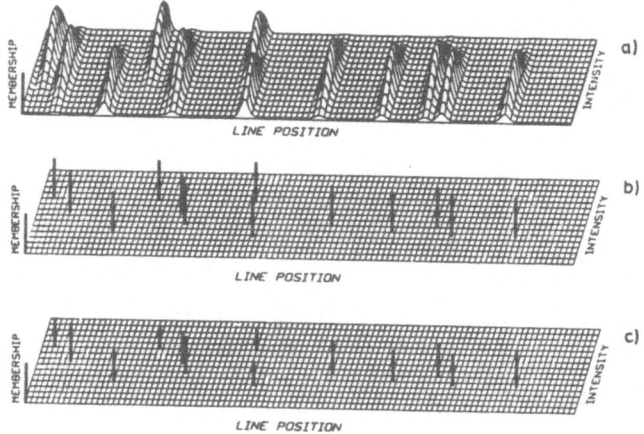

Fig. 5.
Line comparison with fuzzy sets including line intensities. A fuzzified set (a) intersected by a reference (b) yields the set (c).

EXTENSION OF THE FUZZY SET CONCEPT TO LINE INTENSITIES

The example in Fig. 5 illustrates the match between a spectrum and a reference on positions and line intensities. The 1-dimensional axis of peak positions is replaced by a 2-dimensional plane, where each point denotes a peak with a certain position and a certain intensity. The measured lines are, analogously to the 1-dimensional case, fuzzified in x- and i-direction (Fig. 5a).

The comparison between reference and sample spectrum in the 'match' step and the comparison of reference combinations with the sample spectrum in the 'identify' step can be modelled with 2-dimensional fuzzy sets.

INVERTED LIBRARY SEARCH WITH FUZZY SETS

The algorithm for the 'search' step in the CIF program contains in its nucleus the computation of the fuzzy intersection between a fuzzified spectrum and every reference pattern, and the computation of its set power. A speed improvement was obtained in CIF by using the <u>inverted search</u> method.

For comparison of reference lines it is not necessary to investigate all references k, since only those which contain a line near a sample line make a contribution to the set power. The numbers of those references can easily be found in a sorted array klist [l], which clusters reference numbers according to the line position (Fig. 6). The entry points of every cluster are stored in an entry point list entry [x].

The inverted search gets faster if only those parts of the pattern number list klist [l] are searched where the membership function $f_S(x)$ is significant. The structure klist [l] can be stored in a file, which is called the inverted reference file.

COMPARISONS WITH EXISTING SEARCH/MATCH PROGRAMS

The inverted search has been used by Johnson and Vand[4], Frevel et al.[5,6], Nichols and Basinger, and Schreiner et al.[2] All of these have used a singly inverted search file, in which the patterns are listed according to the position of the most intense reference line. In the SANDMAN search/match/identify program Schreiner et al. have extended the file by including two other lines, selected according to the Hanawalt and Fink method. SANDMAN also uses a variable pattern scoring rather than fixed window scoring[7]. A program which uses a score variable for every reference pattern in a fully inverted search, was described by Tian-Hui et al.[8]

In contrast to these programs CIF combines the fully inverted search with the fuzzy set approach. The searching and scoring of references is done in one step by using membership values between 0 and 1.

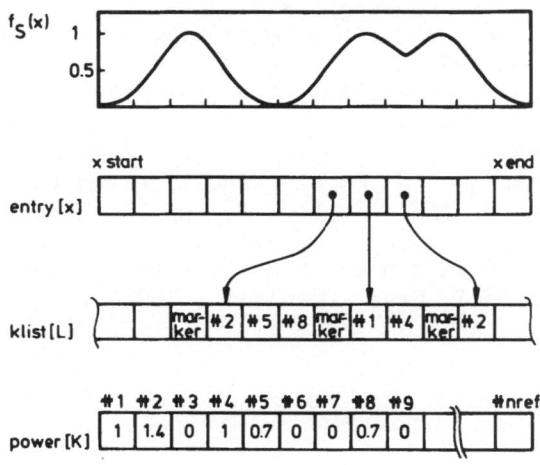

Fig. 6.
Access to the inverted reference file during the inverted search.

SUMMARY

With the fuzzy set theory a new approach to spectrum library
search/match/identify problems was developed. The inverted search
method was combined with this concept to obtain a fast and reliable
search step. Fuzzy sets with a two-dimensionsl support were used to
describe membership of spectrum lines including positions and
heights in the match and identify steps.

ACKNOWLEDGEMENTS

The author would like to thank W.J. Dallas for the helpful
discussions and P. Smit for his advice in crystallographic problems
and the preparation of samples for the program tests. The author
also appreciates the work and helpful comments of R. Jenkins and
W.N. Schreiber, which provided a basis for the CIF implementation
with the SANDMAN search/match/identify program. The project was
supported by the BMFT (German department of research and
technology).

REFERENCES

1. M.C. Nichols (1980), Minicomputer Search/Matching: The Pros
    and Cons, Norelco Reporter, vol. 27, No. 1, pp 23-26.
2. W.N. Schreiner, C. Surdukowski, R. Jenkins (1982), A New
    Search/Match/Identify Program for Qualitative Analysis with
    the Powder Diffractometer, J. Appl. Cryst., 15, 513-523.
3. L.A. Zadeh (1965), Fuzzy Sets, Inform. Control, vol. 8,
    338-353.
4. G.G. Johnson, V. Vand (1967), A Computerized Powder
    Diffraction System, Ind. Eng. Chem., vol. 59, pp 19-31.
5. L.K. Frevel, C.E. Adams, L.R. Ruhberg (1975), A Fast
    Search-Match Program for Powder Diffraction Analysis,
    J. Appl. Cryst. vol. 9, pp 199-204.
6. M.C. Nichols, R.C. Basinger (1974), American Crystallo-
    graphic Meeting, Berkeley, C.A.
7. W.N. Schreiner, C. Surdukowski, R. Jenkins (1982), A Proba-
    bility-Based Scoring Technique for Phase Identification in
    X-ray Powder Diffraction, J. Appl. Cryst., 15, pp 524-530.
8. L. Tian-Hui, Z. Sai-Zhu, C. Li-Jun, C. Xin-Xing (1983), An
    improved program for searching and matching of X-ray powder
    diffraction patterns, J. Appl. Cryst., vol. 16, 150-154.

PDF SEARCH AND MATCH ALGORITHMS FOR THE CYBER 205 VECTOR PROCESSOR

Dale Winder and Bruce Loftis

Physics Department
Colorado State University
Fort Collins, CO 80523

The CYBER 205 has a number of features applicable to the manipulation of the powder diffraction file (PDF). The FORTRAN language has been extended to include many vector procedures and calls to internal system procedures (e.g., the system clock is accessible and timing of routines is accurate to a microsecond). A VECTOR is specified by its address and its length. A pointer to a vector, a DESCRIPTOR, can be assigned and then used in arithmetic and relational expressions. A BIT vector is a binary switch such that a bit value (ON or OFF) can control the disposition of a corresponding element of a vector. The maximum size of a vector is 65535 elements.

These features are illustrated in Table 1, which presents the algorithm of search and match for the PDF. In the version of the PDF utilized for this project, sets 1 through 20, the first step was to split the PDF into 3 equal-length, ordered disk files: D, I, CN. The search begins by reading a maximum of 65535 entries each from the disk into vectors DSPACE, INTENSE, and CARDNBR which are assigned the descriptors, DSPACED, INTENSED, and CARDNBRD. The unknown d-value to be found, DSOLVE(D) is bracketed by WINDOW, and the main statement is exhibited. SEARCHBD is a descriptor for the bit-vector, SEARCHB, whose elements (bits) will be turned ON if the relational expression is satisfied and DSOLVE(D) is inside the bracket. Next lists are formed, first the sequential numbers of found DSPACE elements by Q8VCMPRS which puts into LLIST those numbers from NUMBER for which SEARCHB is ON. Then LLIST is used to gather (Q8VGATHR) corresponding d-spaces, intensities and card numbers into the 3 lists, DLIST, ILIST and CLIST. This idea is continued for the match section in which d-spaces belonging to a found card number are extracted from DSPACE and compared with the unknown set of d-values. If enough matches occur, the results are reported. The process is repeated for remaining unfound d-values. Intensity values can easily be included in the same pattern. Finally the name-formula file is searched by card number.

The time and cost estimates depend on the number of items in active core and the number of refills of core, the size of WINDOW and the efficiency of the algorithm. For typical runs with the search mode, the search time per entry is 34 nanoseconds and the cost is 32 nanodollars. The total time per entry including set up and output overhead is 12 microseconds and 43 microdollars for a WINDOW = .002 and 10 refills of the active core.

This effort was supported by a grant from the Institute for Computational Studies at Colorado State University.

Table 1. Search and Match Algorithm

Procedural Steps	CYBER 205 Statements
Loop over memory pages	DO 150 M = 1, NPAGES
	EPAGE = ENTRIES(M)
Assign descriptors	ASSIGN SEARCHBD, SEARCHB(1;EPAGE)
	ASSIGN DSPACED, DSPACE (1;EPAGE)
	ASSIGN INTENSED, INTENSE(1;EPAGE)
	ASSIGN CARDNBRD, CARDNBR(1;EPAGE)
Fill active vectors from disk	READ (2) (DSPACE(E), E= 1, EPAGE)
	READ (3) (INTENSE(E), E= 1, EPAGE)
	READ (4) (CARDNBR(E), E= 1, EPAGE)
Loop over unknown d-values	DO 140 D = 1, DCOUNT
Bracket d-value with window	DSUPPER = DSOLVE(D) + WINDOW
	DSLOWER = DSOLVE(D) - WINDOW
Form control vector if d fits	SEARCHBD = DSPACED .GT. DSLOWER
	.AND. DSPACED .LT. DSUPPER
Form corresponding lists	LLISTD = Q8VCMPRS(NUMBERD,SEARCHBD;
of found data	LLISTD)
	DLISTD = Q8VGATHR(DSPACED,LLISTD;
	DLISTD)
	ILISTD = Q8VGATHR(INTENSED,LLISTD;
	CLISTD)
Extract all d-spaces for	ASSIGN CSETD, CARDNBR(WSTART;WLENGTH)
each found cardnumber	SETBD = CSETD .EQ. CLIST(D)
	XSETD = Q8VCMPRS (DSETD,SETBD;XSETD)
Compare these d-spaces with	MATCHBD=VABS(XSETD - DSOLVD; WLENGTH)
unknown set	.LE. WINDOW
Report matched sets	NMATCH = Q8SCNT (MATCHBD)
	IF (NMATCH .GT. MINMATCH) PRINT ...

A METHOD OF BACKGROUND SUBTRACTION

FOR THE ANALYSIS OF BROADENED PROFILES

S. Enzo* and W. Parrish

IBM Research Laboratory

San Jose, CA 95193

ABSTRACT

This method for precisely determining the background level of broadened x-ray profiles assumes that the background under the peak can be approximated by a straight line, and the decay in the tails several profile widths from the peak follows a Cauchy-like law. It uses a plot of $I(s)s^2$ vs. s^2 where $I(s)$ is the intensity and the scattering vector $s = 2\cos\theta_0 \sin(\theta-\theta_0)/\lambda$ where θ_0 is the centroid, θ the Bragg angle and λ the x-ray wavelength. The true background is determined by the slope of the linear portion of the plot and its extension to $s^2=0$ gives the rate of decay of the tails. Results on synthesized and experimental profiles show that the method is useful for Warren-Averbach analysis, because it avoids spurious oscillations and the "hook" effect in the plot of the corrected Fourier coefficients, and makes it possible to correct for overlapping tails of adjacent reflections.

INTRODUCTION

The background determination and subtraction from a real spectrum is a common problem in many techniques. It is widely recognized that small errors in its determination can have a profound influence in the final parameters, which are used to derive physical data, especially when carrying out deconvolutions.

In x-ray line profile analysis the simplest method of background determination is the drawing of a continuous straight line connecting the minima on both sides of the peak. It is also common to make an

---

* Permanent address: Istituto di Chimica Fisica, Calle Larga S. Marta 2137, 30123 Venezia, Italy.

enlarged plot of the intensity vs. 2θ but it is difficult to exactly
determine the decay of the tails. However these methods are empirical and subjective. Langford and Wilson (1) have concluded that the
background determined in this way is usually considerably higher
than the true level. They based their choice of background on the
variance of the line profile, while Mitra (2) developed a parallel
method which is based on the integral breadth.

The background subtraction method described here was developed for
use with the Warren-Averbach method for crystallite size and strain
determination (3). It is well known that in applying the
Warren-Averbach method the first few corrected Fourier coefficients
of a broadened profile may be biased to some extent by a wrong evaluation of the background, and the plot displays the well-known "hook"
effect (4). This can cause an incorrect determination of the average
particle size, which is generally calculated as the negative reciprocal of the first derivative at zero in the plot of the corrected
Fourier coefficients. Moreover, the size distribution computed by
taking the second derivative of the Fourier coefficients may have negative values, which have no physical meaning (5).

The knowledge of the decay of the tails of the profile is very useful information which allows extrapolation of data points beyond the
experimental range. This has two main advantages: i) the contributions from tails of nearby reflections are avoided; ii) truncation
errors are almost totally eliminated in the Fourier analysis.

METHOD

The method contains two basic assumptions. The first is that a
background is a straight line with or without slope (6). This does
not take into account thermal diffuse scattering, which may have a
small peak under the Bragg peak. The second is that the decay of the
tails beginning several profile widths from the peak follows a Cauchy
law, i. e., K/s^2 where K is a constant and s the distance of each
point from the centroid of the profile; this has been accepted for a
long time (7-9).

- Description

The method requires a plot of $I(s)s^2$ vs. s^2, where $I(s)$ is the in-
tensity and the scattering vector $s = 2\cos\theta_o \sin(\theta - \theta_o)/\lambda$ in which λ is
the wavelength, $2\theta_o$ the centroid of the profile and 2θ the scattering
angle at each experimental point in the profile. In this plot the
term responsible for the decay which varies as K/s^2 in the intensity
profile becomes a constant K while the included background becomes
the slope of a linearly varying term.

To illustrate how the plot was used a computer synthesized profile
in the form of a pseudo-Voigt function (10) was analyzed, Figure 1.
This profile shape commonly occurs in the broadened peaks generally
analysed with this method. It differs only slightly from a true
Voigt function (11) and is written as:

$$I(2\theta)=I_p\left\{X\times\exp\left[-C^2\ln2(2\theta-2\theta_0)^2\right]+(1-X)/\left[1+C^2(2\theta-2\theta_0)^2\right]\right\}+B \quad (1)$$

where I_p is the peak intensity in counts, X is a value between 0 (for pure Cauchy) and 1 (for pure Gauss profiles) and expresses the relative Gauss and/or Cauchy content of the peak, C is related to the inverse of the full width at half maximum (C=2/W) and B is a background term. Only the Cauchy term is responsible for the inverse square decay, the Gauss term vanishing rapidly to zero in the tails.

Equation 1 was used to make a plot of $I(s)s^2$ vs. s^2. Intensities were computed at 2θ-steps of 0.01° from 38.9° to 41.3° using the following parameters: I =10,000, X=0.5, C=5, $2\theta_0$ =40.10°, B=1,000 and each intensity point was multiplied by the corresponding s^2. As an example we use the last experimental point at 41.30° (circled in Figure 1), in which s=2cos(20.05)sin(0.6)/1.54178 (CuKα)=1.2761×10^{-2} and obtain s^2=1.6284×10^{-4}. Using the observed intensity at this angle $I(s)s^2$=1110×s^2=0.1807.

The intensity of the true background B is determined by the slope of the linear portion of the plot, which, when extended to s^2=0 (the centroid of the profile) also gives the intercept K that determines the rate of decay of the tails. Thus B=(A-K)/s^2 where A and K are determined from the right side ordinate scale. Alternatively the slope can be obtained from any two points of the linear portion but this would not give the K-value.

The background correction method thus has three steps: i) After data collection a plot of $I(s)s^2$ vs. s^2 is prepared; ii) The linear range of the plot and the slopes and intercepts at s^2=0 are computed by least squares; iii) The tails are extended and the background subtracted from the experimental data.

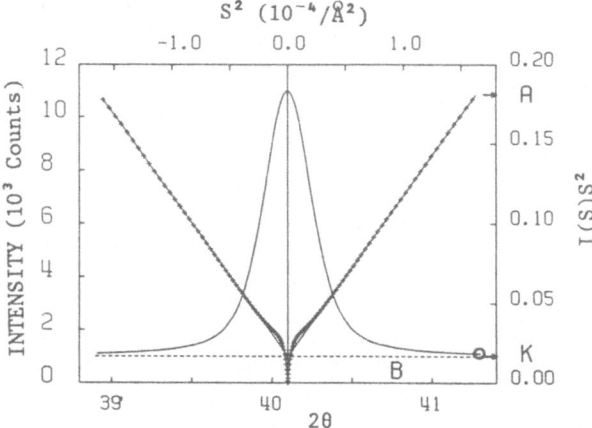

Figure 1. The intensity distribution of a pseudo-Voigt function from Equation 1 (full line) with a background B added (dashed line) and the second moment intensity distribution (crosses). Symbols explained in the text. Note that points on the profile have different positions on the variance plot. The negative values on the s^2 scale refer the low 2θ-side of the peak.

- Tests of the Method

The method was first tried using a series of profiles synthesized on the computer because all the parameters are exactly known. Although many kinds of synthetic profiles have been considered, only two different shapes will be described: a pure Cauchy peak (by letting X=0 in equation 1) and a mixed peak 50% Cauchy and 50% Gauss (X=0.5). Each profile was defined with a background B=10% of the peak intensity value I', full width at one-half of the maximum W=0.2° and the peak and centroid located at $2\theta_O$=40.10° for these symmetrical profiles. The range of definition of the peaks extended 2W on each side.

The results are summarized in Figure 2. In (a) the intensity distribution for the Cauchy profile (curve 1) is shown together with the same peak to which a background was added (curve 2). The corresponding $I(s)s^2$ vs. s^2 plots in (b) are very sensitive to the background addition as shown by the large difference in the slopes of the linear portions of the curves. Application of straight line least square analysis of curve 2 of Figure 2(b) in the range $7\times10^{-6} \leq s^2 \leq 1.7\times10^{-5}$ gave a value of B=1017 counts, which is only 1.7% greater than the original value of 1000. By extending the tails, the background determination becomes even closer to the true value. The same kind of analysis applied to the mixed function profiles (c) and their plots (d) show a behavior clearly distinguishable from the pure Cauchy function (b). The plots increase monotonically starting from s^2=0 for a Cauchy

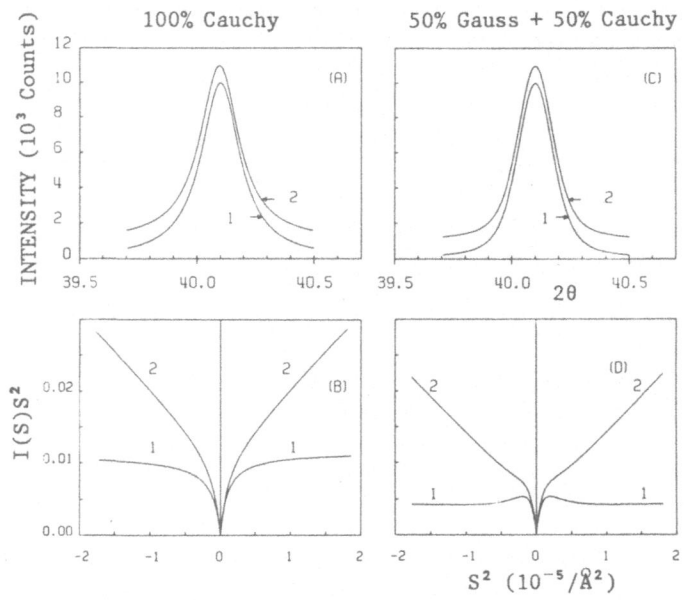

Figure 2(a). Computer synthesized Cauchy profiles without background (1) and with 10% of the peak intensity added as background (2). (b) Second moment plots of data in (a). (c) Synthesized mixed profiles without (1) and with background (2). (d) Second moment plot of (c).

function but a maximum occurs before the asymptotic value is reached. The least squares fitting procedure applied to curve 2 of Figure 2(d) in the same range as for (b) gave B=1008, which is less than 1% from the true value.

APPLICATION TO EXPERIMENTAL DATA

The background subtraction method has been used successfully in a number of analyses and a few typical cases will be described. Figure 3(a) shows the experimental (200) line profile of a pure palladium powder sample. The reflection was measured over a wide angular range making it possible to precisely determine the decay of the tails without the calculated extensions. Although the central portion of this profile is nearly symmetrical the tails have different background levels with the low 2θ side being higher because of the presence of the tail of the very strong (111) reflection at $40.12°$. This effect is shown in (b), in which the plots have different slopes (subscript l = left side of the profile, r = right side) and hence give different values of B on each side but have about the same values of K.

It is evident that the counting statistical precision has a major effect in determining the precision of the data derived from the plots. In the case of Figure 3, the standard deviation σ of $B_l=3\%$ and $B_r=6\%$. A better experimental strategy to improve the precision of the tails would have been to increase the step size and the counting time and the overall time would have been the same. It is only necessary to have a sufficient number of points to properly define the slope. Satisfactory results can be obtained in practice with the following minimal conditions: σ in the background region $\approx5\%$, step size about 0.1W and the angular range $\approx4W$ on each side of the peak.

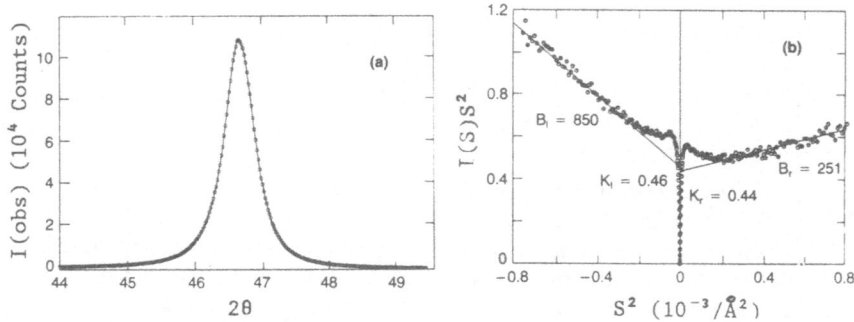

Figure 3(a). Experimental profile of palladium powder (200) reflection obtained with divergence slit = 1°, receiving slit = 0.11°, diffracted beam curved graphite monochromator, step size = 0.02°. (b) Second moment plot of (a) to determine B and intercept K. The slope and intercept are determined by least squares.

• Effect of Angular Range

   The angular range of the background used in the analysis may determine the presence and magnitude of the "hook" effect in the Fourier coefficient plot. This is illustrated by the broadened (111)
reflection (W=0.34°) of the same palladium powder sample used above.
Figure 4(a) shows this profile together with the unbroadened tungsten powder (110) profile that was used to determine the instrument
function. The lower portion of the palladium profile is enlarged in
Figure 4(b) to show the angular ranges used in the analysis which
varied from 1.5 to 8W on each side of the centroid position. Fourier
analyses using the Wagner program (12) were made for each of the
ranges with the background taken as straight lines drawn between the
various limits. The derived Fourier coefficients are shown as a function of crystallite length L (Å) in Figure 5(a). The "hook" effect is
larger for the small ranges and gradually disappears when the range
is increased to 8W on each side. The same data and ranges were then
used with our background correction method. For each truncated range
the background was determined and subtracted after the tails were extrapolated to 8W on each side using the $I(s)s^2$ vs. s^2 plots. The
Fourier coefficient plot, Figure 5(b) shows that all the ranges gave
the same values and confirms the reliability and reproducibility of
the method.

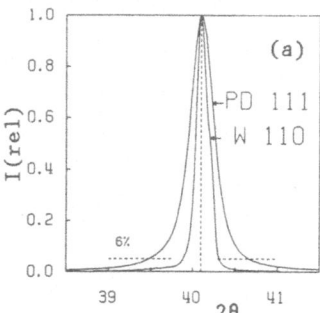 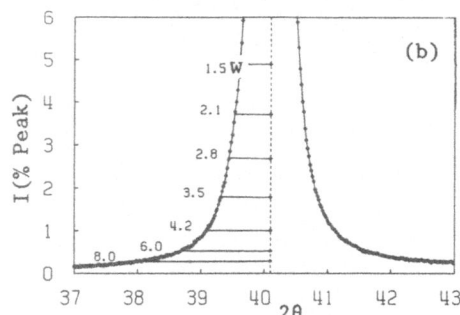

Figure 4(a). Experimental profile of palladium powder (111) reflection with extended angular range used to determine behavior of
tails. The unbroadened tungsten (110) profile was used to determine
the diffractometer instrument function. Both peaks are scaled to the
same height and 2θ position. (b) Enlarged section of portion of
palladium (111) profile showing various limits in terms of the width
at one-half peak height (W=0.34°).

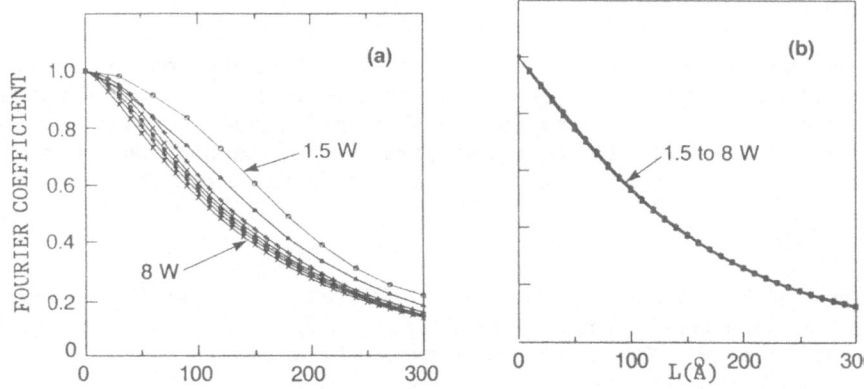

Figure 5. The Fourier coefficients of the (111) Pd powder profile (after instrumental correction) computed starting from peaks defined over the ranges shown in Figure 4(b). (b) Results using extrapolation and background correction as described in the text.

- Overlapped Tails

The separation of the particle size and strain component in the Warren-Averbach method requires the use of at least two orders of the same reflection. In practice the second order is often close to another reflection and the overlapping tails make it impossible to obtain a sufficiently long experimental data range, as shown in Figure 6. The use of the $I(s)s^2$ vs. s^2 plot made it possible to extrapolate the tails as shown by the solid lines to extend the range and to do the Fourier analysis without "hook" effect.

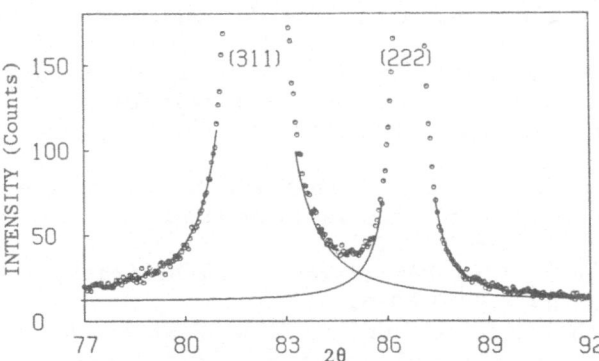

Figure 6. Experimental data (points) and extrapolated tails (solids lines) determined by method described.

CONCLUSIONS

The method of using a plot of $I(s)s^2$ vs. s^2 to determine the background and to extrapolate the tails in the form of K/s^2 provides a method of obtaining good data for the Warren-Averbach Fourier analysis determination of particle size and strain. The method has been used successfully in the analysis of powders, catalysts and thin films.

ACKNOWLEDGEMENTS

Thanks are due to the personnel of the Crystallography Department of the IBM San Jose Laboratory, particularly to T. C. Huang and G. Lim for helpful discussions. The aid of the staff of the User Service Computing Facility is also appreciated.

REFERENCES

(1) J. I. Langford and A. J. C. Wilson, in Crystallography and Crystal Perfection, edited by G. M. Ramchandran (Academic Press, London, 1963) pp. 207-222
(2) G. B. Mitra, "X-Ray Diffraction Profiles from Deformed Metals", Brit. J. Appl. Phys. 16 (1965) 77-84
(3) B. E. Warren and B. L. Averbach, "The Effect of Cold-work Distortion on X-ray Patterns", J. Appl. Phys. 23 (1950) 595-599
(4) B. E. Warren, X-ray Diffraction, Addison-Wesley, Reading, Massachusets 1969
(5) P. Pausescu, R. Manaila, M. Popescu and E. Jijovici, "Crystallite Size Distribution in Supported Catalysts", J. Appl. Cryst. 7 (1974) 281-286
(6) R. A. Young, R. J. Gerdes, A. J. C. Wilson, "Propagation of Some Systematic Errors in X-ray Line Profile Analysis", Acta Cryst. 22 (1967) 155-162
(7) G. Allegra, "Crystal Powder Statistics. IV. Calculation of the Line Profile Using the Sampling Line Method", Acta Cryst. A38 (1982) 863-867
(8) F. de Bergevin and P. Germi, "Corrections des Oscillations dans les Distributions de Tailles des Particules Obtenuees a Partir des Profils des Raies de Diffraction", J. Appl. Cryst. 5 (1972) 416-420
(9) I. S. Szanto and L. Varga, "An X-ray Method for the Determination of the Domain Size from the Tails of the Diffraction Profiles", J. Appl. Cryst. 2 (1969) 72-76
(10) M. Hecq, "A Fitting Method for X-ray Diffraction Profiles", J. Appl. Cryst. 14 (1981) 60-61
(11) G. K. Wertheim, M. A. Butler, K. W. West and D. N. E. Buchanan, "Determination of the Gaussian and Lorentzian Content of Experimental Line Shapes", Rev. Sci. Instrum. 45 (1974) 1369-1371
(12) C. N. J. Wagner, "Analysis of the Broadening and Changes in Position of the Peaks in an X-ray Powder Pattern", in Local Atomic Arrangement Studied by X-ray Diffraction, edited by J. B. Cohen and J. E. Hilliard, Gordon & Breach, New York, 1966

A COMBINED DERIVATIVE METHOD FOR PEAK SEARCH ANALYSIS

T. C. Huang and W. Parrish

IBM Research Laboratory

5600 Cottle Road, San Jose, CA. 95193

ABSTRACT

A comprehensive study of derivative methods for the peak search analysis of X-ray diffraction data was made to determine the relative merits of the methods. The peak positions were best determined by the cubic first derivative method which had an intrinsic error $\leq 0.001°$, and random error $\sim \pm 0.003°$ to $0.02°$ depending on the counting statistical noise. The quadratic/cubic second derivative method had the highest resolution with a separation limit $\geq \frac{1}{2}W$ (W = full width at half maximum). An effective algorithm combining the cubic first derivative and the quadratic/cubic second derivative methods was developed for high precision and resolution. The method uses a full screen menu for parameter selection, and the entire peak search analysis including peak identification and position determination, and graphic and numeric display of results at the color terminal is completed in a few seconds using a time sharing mode on an IBM 3083 central processing unit. The combined derivative method should be also applicable to other spectra such as gamma-rays, X-ray fluorescence, optical, infrared, ESCA, Mossbauer, etc.

INTRODUCTION

Peak identification and 2θ determination is one of the basic data reduction processes in X-ray diffraction analysis. Among many existing algorithms, peak search by derivatives (first, second, or higher order) is most commonly used[1,2]. To obtain the maximum advantage of using computers for instrument automation requires algorithms which can precisely and rapidly do the data reduction, and the derivative methods are well-suited for this use.

The purpose of this study was to evaluate the performances of peak search methods by derivatives in terms of their accuracy, precision and resolution in order to develop an optimum algorithm for high precision and high resolution peak search analysis of X-ray diffraction data. An interactive time sharing computer method was developed using a new peak search algorithm combining the first and second derivatives. The peak intensity was determined by fitting a Lorentzian curve to the highest points.

CHARACTERISTICS OF DERIVATIVE METHODS

The relative merits of the derivative methods can be best determined using computer simulated diffraction patterns. One of the advantages of using calculated patterns is that their peak positions $2\theta_0$ are exactly known, and therefore the performance of the methods can be precisely determined. To simulate reliable X-ray powder patterns, the profile fitting method was used to generate the calculated profiles whose shapes were identical to those obtained experimentally.[3] Counting statistical random noises were added when required. The Savitzky and Golay Procedure[4] was used to calculate the values of the derivatives.

• Accuracy

The various derivative methods were tested with three diffraction peaks at 20°, 50° and 80° ($2\theta_0$). For each peak, five profiles with no counting statistical noise were simulated by computer with $\Delta2\theta$ steps of 0.01°, 0.02° ... 0.05°. The peak position results ($2\theta_{PS}$s) from analyzing the profiles of the 20° peak by the derivative methods using various numbers of convolution points (N = 5, 7, 9... 25) are plotted vs. the convolution range (CR = $\Delta2\theta$ × N) in Figure 1 in which the quadratic first derivative results† are plotted as circles, the cubic first as triangles and the quadratic/cubic (Q/C) second derivative as squares. Each derivative method forms a curve which varies smoothly with CR, i.e., the value of the calculated $2\theta_{PS}$ depends only on the value of CR rather than $\Delta2\theta$ or N. Results obtained by the cubic first derivatives are closest to the true value of $2\theta_0$=20.000°. It should be noted that this 20° peak is markedly asymmetric.

Peak search accuracies for peaks at 20°, 50° and 80° are listed in Table 1 where CR is given as a multiple of W. The discrepancies (Δ's) between $2\theta_{PS}$ and $2\theta_0$ depend on the derivative method chosen and the values of CR. This is especially true for asymmetric peaks located at small 2θ (e.g. 20° and 50°). For nearly symmetric $K\alpha_1$ peaks at high 2θ (e.g. 80°), the Δ's gradually approach each other. Since the Δ's were obtained from ideal calculated profiles with zero counting statistical noise, these discrepancies were introduced purely by the derivative method used, and therefore they are the intrinsic errors associated with the method. (The term "systematic error" is not used here so that it can be distinguished from the well known systematic errors arising from instrument misalignment, geometrical aberrations, etc. in the X-ray patterns[5]). Comparing results among these three methods, those obtained by the cubic first derivatives have the smallest discrepancies with intrinsic errors ≤ 0.001° on the average.

---

† $2\theta_{PS}$s were obtained from the first derivatives of a second-order (i.e., quadratic) polynomial.[4]

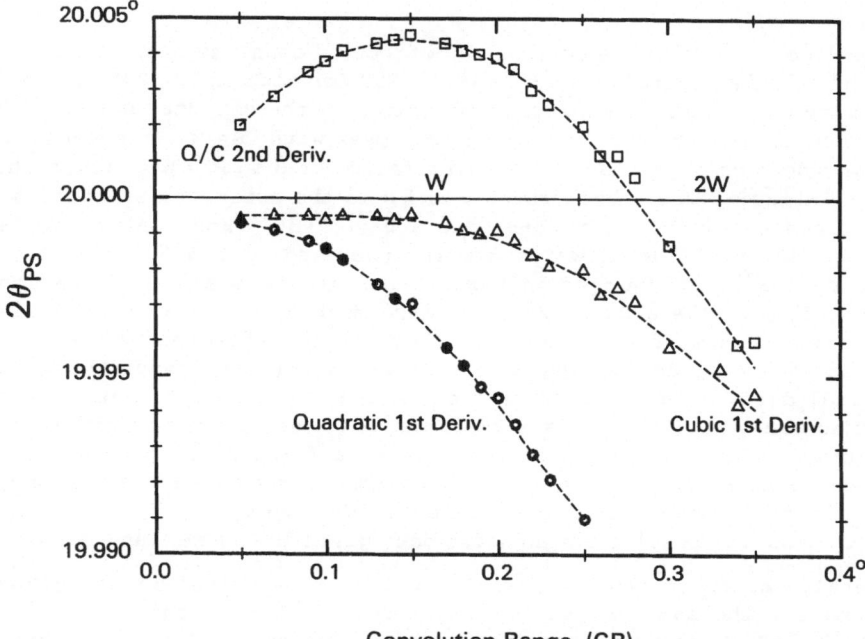

Figure 1.   Peak search results for derivative methods.

Table 1.   Peak Search Accuracy

$(\Delta = 2\theta_{PS} - 2\theta_o)$

$2\theta_o$	CR	Quadratic 1st Deriv.	Cubic 1st Deriv.	Q/C 2nd Deriv.
20°	0.5W	$\Delta=-0.001°$	-0.001	0.003
	1.0	-4	-1	4
	1.5	-9	-2	2
50	0.5	1	<0	-5
	1.0	4	>0	-6
	1.5	14	2	-7
80	0.5	-1	-1	-1
	1.0	<0	.-1	-1
	1.5	<0	-1	-1

• Average Deviation $(\Sigma|\Delta|/3)$:

	0.5	1	1	3
	1.0	3	>0	4
	1.5	8	1	4

• Precision

In order to obtain reliable data on the effect of counting statistics a large number of calculated profiles ($360 \times 48 = 17,280$ for each of the $20°$, $50°$ and $80°$ peaks) with counting noise randomly generated by the computer were analyzed. Table 2 summarizes the results for the $20°$ peak with CR=1W. It covers a range of three orders of magnitude in P/B and a factor of 8 in P/σ where P is the the net peak intensity above background B and σ is the estimated standard deviation for the total intensity P+B, i.e., $\sigma = \sqrt{P+B}$. Each peak value is the average of 360 profiles of which ⅓ were generated with $\Delta 2\theta$ steps of $0.01°$, $0.02°$ and $0.03°$. The number in parentheses is the standard deviation for these 360 $2\theta_{PS}$s. The average $2\theta_{PS}$s of each method remain approximately the same for all values of P/B and P/σ. They are not affected by the counting noise and therefore correspond to the intrinsic errors associated with the individual derivative methods. The cubic first derivative method has the highest accuracy with the average $2\theta_{PS}$s generally determined to within $0.001°$ of the true $20.000°$ value. Both the quadratic first derivative and the Q/C second derivative methods have about four times larger intrinsic errors, the former giving lower and the latter higher 2θ-values. These results are in agreement with the results for no noise data given in Figure 1 and Table 1.

In general, values of the standard deviation of each individual method are approximately the same for diffraction profiles having the same P/σ, even through their P/Bs differ by three orders of magnitude. Standard deviations increase as P/σs decrease. They represent the random errors associated with the derivative methods. As expected, the precision is poorer for profiles with greater noise. The Q/C second derivative method had difficulty in analyzing profiles having small values of P/σ and P/B especially those with large step size ($\Delta 2\theta = 0.03°$). The standard deviations for these cases are very large (lower right hand corner of Table 2.)

Table 2. Effect of Counting Statistics
(CR = 1W; $2\theta_o = 20°$)

• Quadratic First Derivative:

	P/σ=25		10		5		3	
P/B=100	19.996	(2)	19.996	(4)	19.996	(9)	19.997	(15)
10	6	(2)	7	(5)	6	(9)	95	(14)
1	6	(2)	7	(5)	5	(9)	97	(17)
0.1	6	(2)	6	(5)	96	(10)	95	(17)

• Cubic First Derivative:

	P/σ=25		10		5		3	
P/B=100	20.000	(3)	20.000	(8)	20.000	(15)	19.999	(20)
10	19.999	(3)	0	(7)	19.999	(14)	20.000	(20)
1	20.000	(3)	0	(8)	98	(14)	00	(21)
0.1	19.999	(3)	19.999	(8)	20.000	(14)	19.998	(21)

• Q/C Second Derivative:

	P/σ=25		10		5		3	
P/B=100	20.005	(5)	20.004	(13)	20.004	(23)	20.000	(31)
10	5	(5)	05	(12)	02	(23)	01	(31)
1	5	(5)	05	(14)	19.996	(63)	19.999	(98)
0.1	5	(6)	04	(23)	20.002	(76)	20.005	(159)

Further peak search analysis with higher order derivatives were also conducted. In general, the lower the order of derivative, the smaller the intrinsic and random errors. In the case of the polynomial the higher the degree, the smaller the intrinsic error, but the larger the random error. Taking into account both accuracy and precision, the cubic first derivative method gives the best results with intrinsic error of $\leq 0.001°$ and random error of $\pm 0.003°$ to $0.02°$ depending on P/σ.

• Resolution

Another important characteristic in peak search analysis is the resolution of overlaps in a cluster. Figure 2 shows a cluster (solid curve) of three overlapping peaks (dash curves) with separation $\Delta = W$, and relative intensity ratios of 1:1:0.5. The quadratic first, cubic first and second derivative curves obtained with CR=0.5W, 1W and 1.5W for this cluster are also plotted. The locations of peaks found by these methods are marked by circles. For the quadratic first derivatives with CR=0.5W and 1W, two peaks are resolved from this cluster, and the third peak is unresolved. As CR increases to 1.5W, only one peak can be found and it fails completely to resolve this cluster. For the cubic first derivatives, the results are slightly better but the third peak is still missing. The second derivative method is the only method that was successful in resolving all three overlaps.

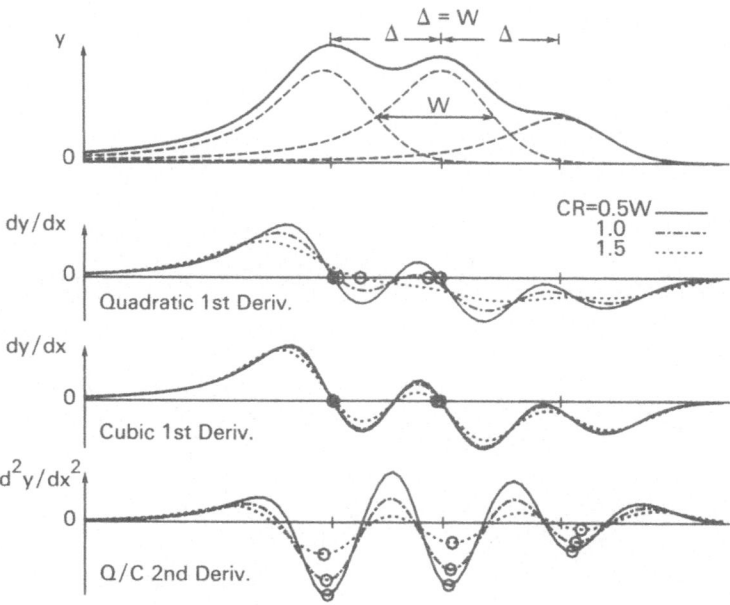

Figure 2.   Overlap resolution for derivative methods.

Table 3.   Resolution Characteristics of Derivative Methods.

Separation	Quadratic 1st Deriv.		Cubic 1st Deriv.		Q/C 2nd Deriv.	
	A	B	A	B	A	B
$\Delta = 1\frac{1}{4}W$	Y	Y	Y	Y	Y	Y
1	N	CR≤W	N	Y	Y	Y
¾	N	N	N	N	≤W	≤1.25W
½	N	N	N	N	N	≤0.5W

The resolution of these three methods is summarized in Table 3. Resolution depends not only on the separation (Δ) between overlaps, but also their relative intensities. Two types of relative intensity ratios: 2:1 (type A) and 1:1 (type B) are included in this table. All three methods are successful in resolving overlaps with $\Delta \geq 1\frac{1}{4}W$. As separation decreases, the quadratic first derivative is the first to fail. At $\Delta = 1W$, successful resolution is obtained by the quadratic first derivative method with CR $\leq$ W for overlaps of approximately equal intensities. The cubic first derivative method is the next to fail (also see Figure 2).

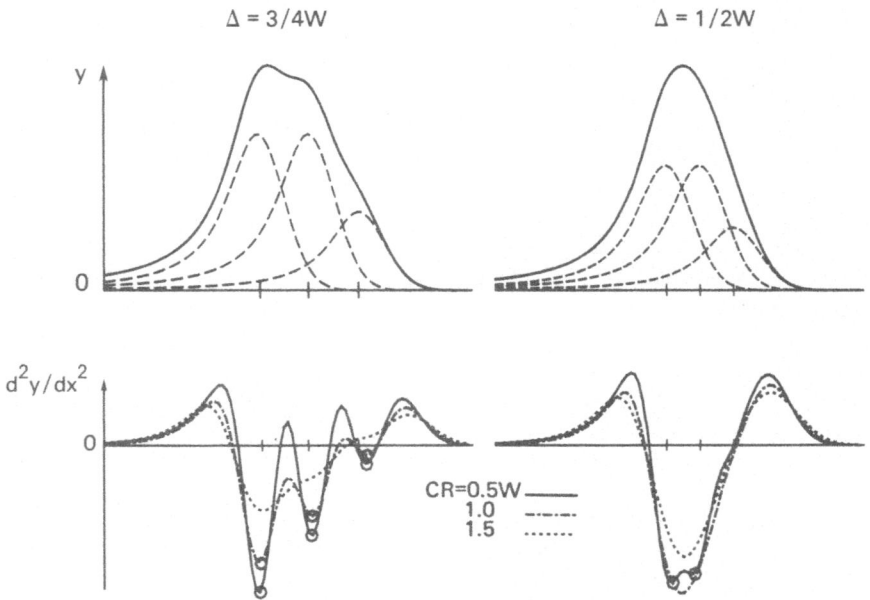

Figure 3.   Resolution of the Q/C second derivative method.

As Δ reduces to ≤ ¾W, only the Q/C second derivative method remains operational. At Δ = ¾W, resolution of type A overlap is obtained with CR ≤ W and type B overlap with CR ≤ 1.25W. The smallest separation for Q/C second derivative is with Δ = ½W for overlaps having approximately equal intensities. Details on the analysis of overlaps with Δ = ¾ and ½W by the Q/C second derivative method are plotted in Figure 3. The overlaps with Δ = ½W are so close that no visible evidence can be observed from the cluster (solid curve at the top right hand side of Figure 3). The study using higher orders of derivative (e.g., quartic, quintic etc.) to improve resolution further was also conducted. We found that these methods are too sensitive to counting noise and false peaks are commonly identified along with true overlaps. Therefore, the Q/C second derivative method is best for resolving overlaps in X-ray diffraction.

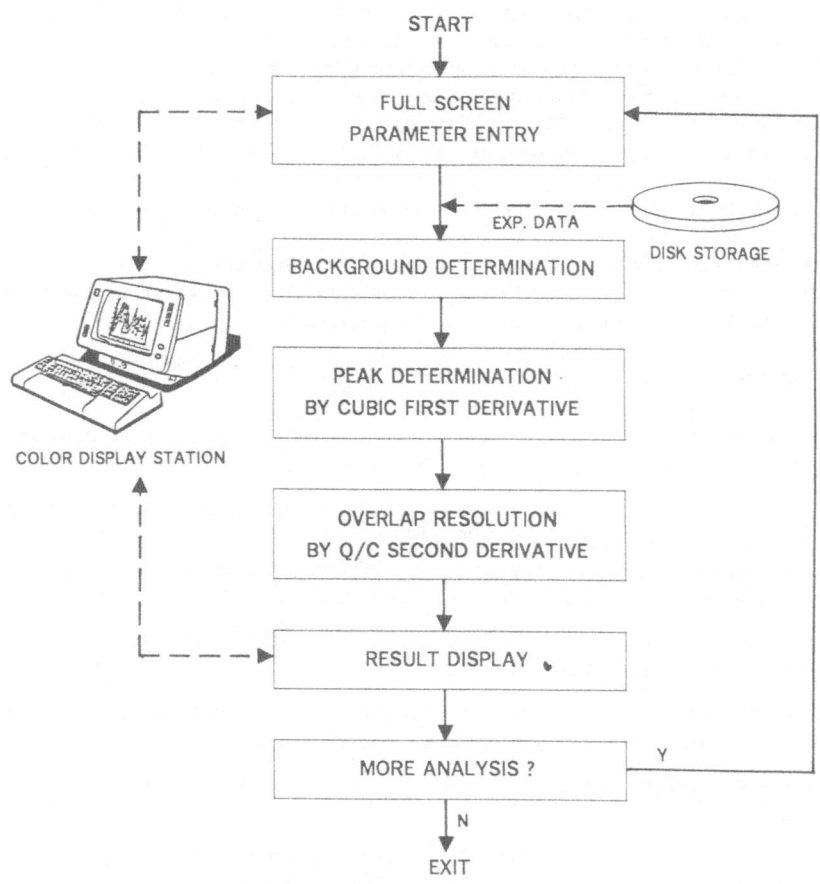

Figure 4.   Procedure for the IBM San Jose peak search system.

• Optimum Method for Peak Search

Based on the above evaluation, we believe that the optimum method for peak search analysis of X-ray diffraction data is a method which has the accuracy and precision of the cubic first derivatives and resolution power of the second derivatives. A new algorithm combining the cubic first derivatives and quadratic/cubic second derivatives has therefore been developed. In X-ray powder diffractometry almost all the peaks in the front reflection region have asymmetric profiles and the combined derivative method is essential for precision analysis.

NEW IBM SAN JOSE PEAK SEARCH SYSTEM

A new interactive computer system was developed and it is currently run on an IBM 3083 central processing unit under time sharing mode of operation. As shown in Figure 4, a full screen menu is used for parameter entry. Experimental data can be rapidly accessed from an on-line memory. Data reduction includes the determination of linear or other form of background, peak identification and 2θ determination by the cubic first derivative, and severe overlaps which were not resolved by the cubic first derivative are further examined and identified by the second derivative. Peak search results are then displayed graphically and numerically at the color terminal. The entire analysis is very rapid and it takes only a few seconds after parameter entry to have the final result display. Results can be conveniently examined using interactive graphics. This system is also arranged for the rapid return to the full screen menu for the entry of new or revision of the original parameters. A description of the parameter entry with the full screen menu and the application of the system for the analyses of various experimental data collected in our laboratory will be given in the following paper[6].

REFERENCES

1. E. J. Sonneveld and J. W. Visser, Automatic Collection of Powder Data from Photographs, J. Appl. Cryst. 8:1 (1975).

2. S. V. N. Naidu and C. R. Houska, Profile Separation in Complex Powder Patterns, J. Appl. Cryst. 15:190 (1982).
   (Due to limited space, only one of the earliest and one of the recent papers related to peak search by derivative method are listed here).

3. W. Parrish, G. L. Ayers and T. C. Huang, Computer Simulation of Powder Patterns, Adv. X-Ray Anal. 27 (1984).

4. A. Savitzky and J. E. Golay, Smoothing and Differentiation of Data by Simplified Least Squares Procedures, Anal. Chem. 36:1627 (1964).

5. W. Parrish, "X-Ray Analysis Papers", Centrex Publishing Co., Eindhoven (1965).

6. T. C. Huang, W. Parrish and G. Lim, Experimental Study of Precise Peak Determination in Powder Diffraction, Adv. X-Ray Anal. 27 (following paper) (1984).

EXPERIMENTAL STUDY OF PRECISE PEAK DETERMINATION IN X-RAY POWDER DIFFRACTION

T. C. Huang, W. Parrish and G. Lim

IBM Research Laboratory

5600 Cottle Road, San Jose, CA. 95193

ABSTRACT

The combined derivative method (accompanying paper) was tested with a large number of experimental patterns to illustrate its use in various difficult problems commonly arising in peak search analysis of X-ray diffraction data. Patterns obtained with various step sizes, resolution, counting statistical noise, and profile widths were used. The precision in 2θ determination and overlap resolution are in good agreement with those previously obtained from calculated profiles. False identification of noise as diffraction peaks was eliminated by using a convolution range proportional to the full width at half maximum. Peak search results (both 2θ and intensity) were also compared to those obtained by profile fitting to illustrate the different characteristics of these two methods.

INTRODUCTION

A comprehensive study of the results using various derivative methods for the peak search analysis of calculated X-ray powder diffractometer data showed that peak positions and overlap resolution are best determined by the cubic first and quadratic/cubic (Q/C) second derivative methods.[1] A new interactive computer system using an algorithm combining these two methods was developed for high precision and high resolution peak search analysis. The purpose of this paper was to study the peak search results obtained by this system on various types of experimental patterns.

A modified Norelco focussing diffractometer with θ-2θ scanning in the vertical plane and Cu Kα X-rays were used. The experimental parameters included a long fine focus X-ray tube with $12°$ take-off angle, diffracted beam curved graphite monochromator, vacuum path, and scintillation counter with pulse amplitude discrimination. The IBM Series/1 minicomputer X-ray analysis automation system was used for data collection.[2] The experimental data collected were automatically transferred to the main memory of the host IBM 3083 processor for peak search analysis.

PARAMETER ENTRY

Peak search parameters were entered into the computer using a full screen menu, Figure 1. Dataset name, X-ray unit and log numbers are used to locate the origin of the experimental data to be analyzed. Only one peak search parameter, the convolution range (CR), is required for this method. The angular range of CR used for peak search in this method is a multiple of the full width at half maximum (W); only the multiple is entered. W is calculated by computer from the strongest peak in the diffraction pattern. Results may be plotted graphically and numerically at the same terminal. A listing of the peaks with intensities above a selected threshold, in term of % of the strongest peak P(max), can also be displayed. Any particular form of background may be used by entering its dataset name to the final entry on the menu. If this entry is left blank, a linear background will be assumed. At the end of an analysis, the system may be returned to the parameter entry menu to use new parameters or revise existing parameters which are carried forward from the previous analysis. This menu is also used for parameter entry for profile fitting and standard profiles W*G determinations.

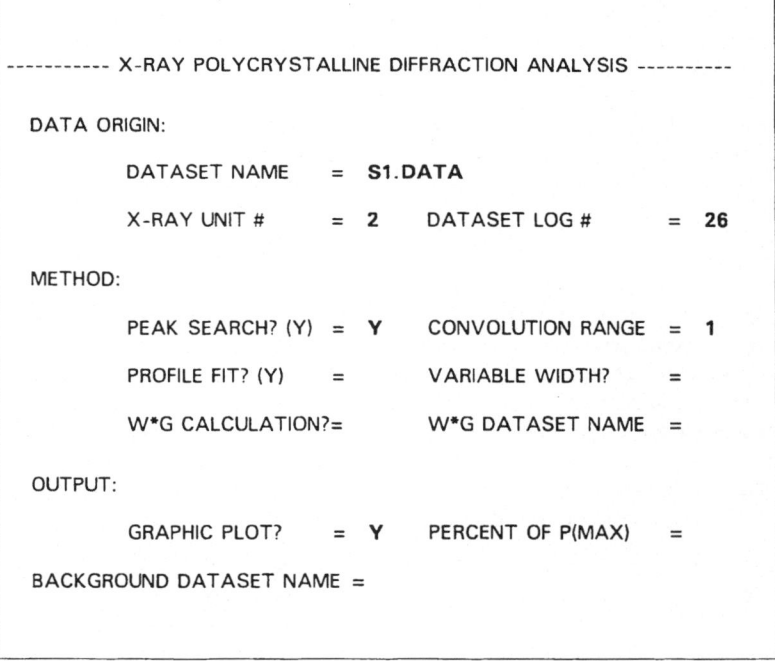

Figure 1. Full screen parameter menu for peak search analysis. Entries are shown in bold face.

PEAK IDENTIFICATION AND RESOLUTION

One of the major tests of the method was made with a specimen containing roughly equal proportions of the minerals quartz, orthoclase and albite which occur in many granites. The $8°(2\theta)$-sections shown in Figure 2 has a large number of peaks with various degrees of overlaps and intensities. The first analysis was done with good statistical data, $\Delta 2\theta=0.01°$ and count time 10 seconds per step. Linear background was first determined from areas away from peaks ($2\theta\sim20°$, $21.5°$, $25.1°$ and $28.2°$ in this case).

The effect of convolution range on peak identification is illustrated in Figure 2. The number of peaks found increases as the value of CR decreases. All reflections of the three compounds in this mixture that occur in this angular range were identified with CR=0.5W. In general, all peaks that are easily visible in the pattern were found by the cubic first derivatives. The remaining unresolved overlaps were then identified by the Q/C second derivatives. The resolution achieved depends on the value of CR, the smaller the CR, the better the resolution. For example, the second peak in Figure 2 (marked by x at $20.889°$) is separated by $0.06°$ ($\sim\frac{1}{2}$W) from the first peak and was resolved by the Q/C second derivatives with CR=0.5W. The third peak (at $20.999°$) has a larger separation and was identified by the cubic first derivative with CR=1W or 0.5W; however when the CR was increased to 2W only the Q/C second derivative could resolve it. The same resolution limits were obtained previously from calculated profiles.[1]

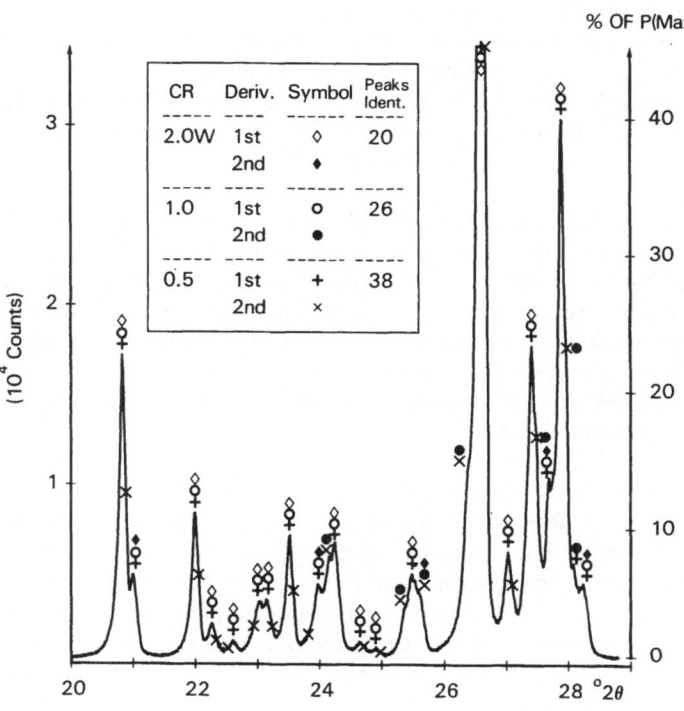

Figure 2.
Pattern of mixture of quartz, orthoclase and albite recorded with $\Delta 2\theta=0.01°$ and 10 seconds count time per step. The effects of convolution range CR and derivative method are shown by the symbols in the insert.

CR	Deriv.	Symbol	Peaks Ident.
2.0W	1st	◊	20
	2nd	◆	
1.0	1st	○	26
	2nd	●	
0.5	1st	+	38
	2nd	×	

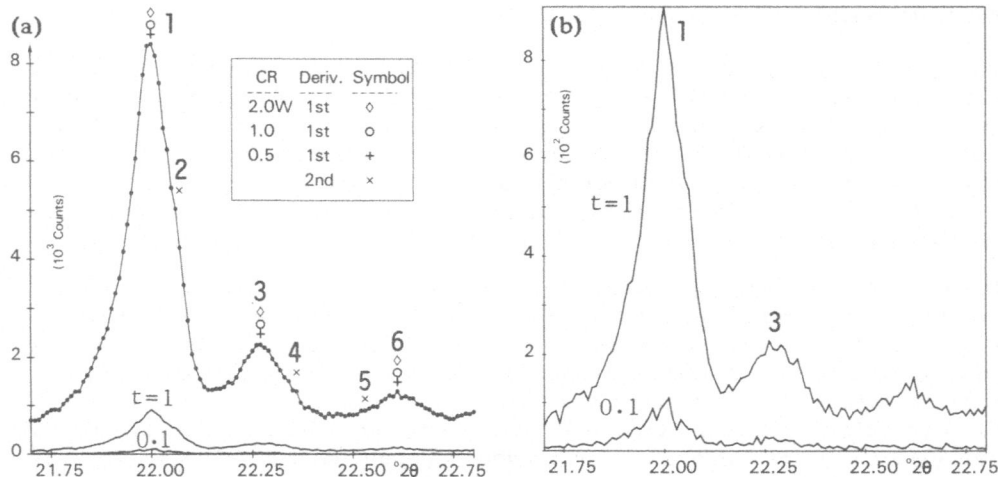

Figure 3   (a). Enlargement of a section of Figure 2.
            (b). Same section with intensity axis expanded by 10 times.

Figure 3a is an enlarged portion of Figure 2 and shows better the capability of the method as well as a possible drawback of using a small CR. Experimental data are plotted as dots and connected by a solid line (the lower curves will be described later). Three major peaks (#1, 3 and 6) with separations of ~1½W were identified by the cubic first derivatives with CR=2W, 1W, and 0.5W. The other three peaks (#2, 4 and 5) were so close to their main peaks (Δ~½W) that only the Q/C second derivatives with CR=0.5W could resolve them. Peaks #2 and 4 are the $K\alpha_2$ components of #1 and 3, respectively. The small CR identified a peak at #5. This is a noise peak which can be eliminated with a larger CR as described later.

The step size used in collecting data affects the resolution of the experimental pattern. As the step size increases details of the overlapping profiles become washed out and thus fewer peaks become visible and a smaller number of peaks are identified. It was also shown above that the smallest CR gave the maximum resolution in the data reduction. However, the minimum value of CR is limited by the requirement that at least five convolution points are needed in the Savitzky and Golay procedure[3] and thus the smallest CR allowed increases with increasing $\Delta 2\theta$.

To determine the effect of step size on resolution, the pattern of Figure 2 was analyzed with steps of $\Delta 2\theta$ = 0.01° to 0.04° using the smallest CR allowed for each step size and the results are shown in Figure 4. The number of peaks identified increases with decreasing $\Delta 2\theta$ and CR. If it is necessary to identify all peaks in a severely overlapped pattern it is essential to use the maximum diffractometer resolution, small $\Delta 2\theta$ steps, good counting statistics and a small convolution range. The effect of step size on precision of 2θ is described below.

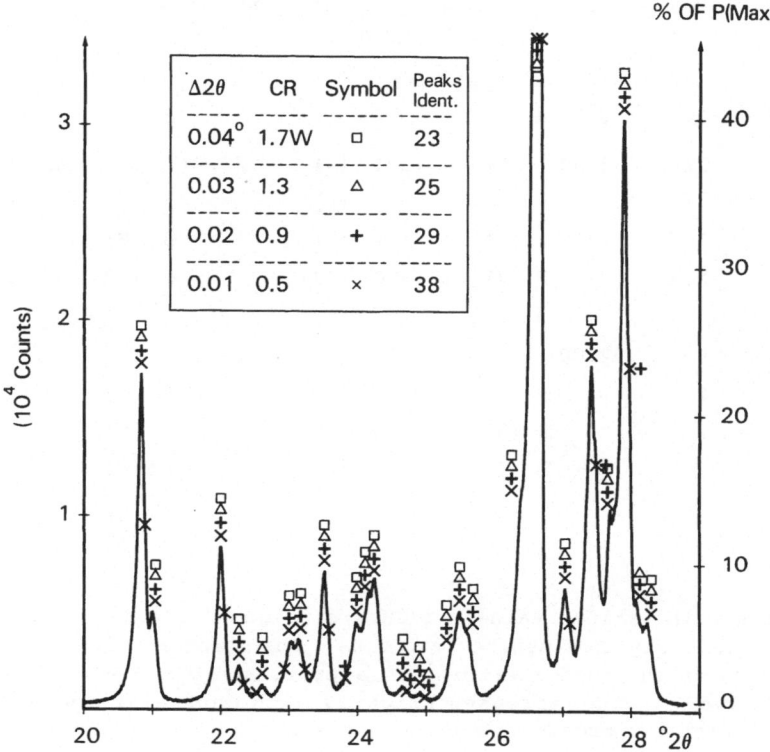

Figure 4.   Effect of step size ($\Delta 2\theta$) on peak identification.

PRECISION IN 2θ DETERMINATION

Both the counting statistics and the step size used in scanning determine
the precision in calculating the peak angles. A large number of datasets
were collected and analyzed to study the effects of these factors on the
precision.

Data were collected from the same mineral mixture with $\Delta 2\theta = 0.01°$ and
count time per step of 1 and 0.1 second. Two typical sets of these pattern
are shown in Figures 3a and 3b. Two peaks #1 at $22.00°$ and #3 at $22.28°$ were
selected to illustrate the effect of the counting statistics. The results
are summarized in the upper part of Table 1 where N is the number of profiles analyzed, $\sigma = \sqrt{P+B}$ in which P is the peak intensity above background
B. The average 2θ for N peak angle determinations is listed first and the
standard deviation is in parentheses. The P/B was 20 for peak #1 and 1.3 for
#3, and P/σ varied with count time as shown in Table 1.

Table 1.   Effect of Counting Statistics and Step Size on 2θ
Determination

			Peak #1				Peak #3			
N	$\Delta 2\theta$	CR	P/σ=28		9		8		2.5	
20	0.02	2W	22.005	(2)	22.003	(4)	22.275	(5)	22.273	(15)
20	2	1.5	6	(3)	3	(6)	7	(6)	80	(14)
20	2	1	3	(3)	2	(8)	8	(8)	76	(20)
10	1	0.5	2	(4)	21.996	(11)	76	(13)	64	(20)
10	0.01	2	22.005	(2)	22.002	(3)	22.273	(4)	22.275	(10)
20	2	2	5	(2)	3	(4)	5	(5)	73	(15)
30	3	2	5	(3)	2	(6)	5	(6)	76	(15)
40	4	2	5	(3)	2	(6)	5	(6)	73	(18)

The results with various values of CR are in good agreement with those obtained previously from synthesized data.[1] The standard deviation which is a measure of the random errors, is approximately the same for peaks with the same P/σ and increases slightly with decreasing convolution range. The errors increase as P/σ decreases and are approximately 0.002° to 0.020° for the P/σ range used. It should be noted that peak search with a small CR is sensitive to counting noise and is more likely to generate false peaks or to miss true weak peaks when counting statistics are poor. Because a small CR is advantageous for achieving the best resolution it may be necessary to analyze the data with different values of CR to obtain an optimum set of results.

The effect of step size $\Delta 2\theta$ on the precision is summarized in the lower portion of Table 1 where a convolution range of 2W was used for all profiles. The average 2θs are about the same for each peak with the standard deviations increasing slightly with increasing step size and are greater with the low P/σ profile.

The effect of step size ($\Delta 2\theta$) on the precision in peak location (2θ) is given in Table 1. Results analyzed using CR=2W on patterns with Δ = 0.01° to 0.04° are included. A large number of experimental datasets were used to obtain reliable statistical analysis (i.e., 10, 20, 30 and 40 datasets for cases with $\Delta 2\theta$ =0.01°, 0.02°, 0.03° and 0.04°, respectively). As in Table 1, the average $2\theta_{PS}$s and the standard deviations (number in the parentheses) are listed. The average $2\theta_{PS}$s remain approximately the same for the same peak (i.e., Peak #1 or 3). The maximum difference is 0.003° for profiles with poor counting statistics P/σ=2.5. Standard deviations for those having the same P/σ increase slightly as $\Delta 2\theta$ increases. In other words, the precision in peak location reduces or random error associated with $2\theta_{PS}$ increases for patterns with larger $\Delta 2\theta$.

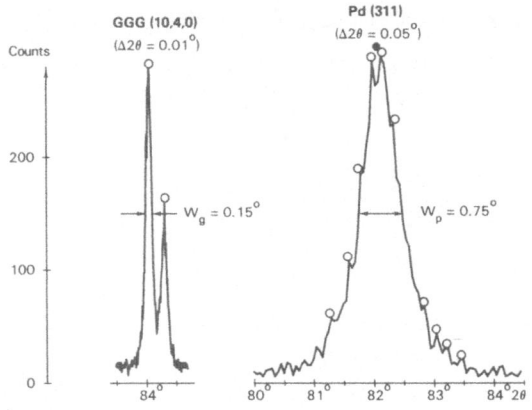

Figure 5.
Peak Search Analysis
of a broad peak.
o for CR=$1.5W_g$,
● for CR=$1.5W_p$.

ANALYSIS OF A BROAD PEAK

The peak search analysis of broad peaks, especially those with poor
statistics, often shows false noise peaks. In the present peak search
system the convolution range is made proportional to the peak width and
this problem is eliminated. This is illustrated in Figure 5 for two peaks
with approximately the same 2θ and intensity but different widths. The
narrow GGG garnet (10,4,0) reflection has W_g = 0.15° and the palladium
(311) reflection is broadened by small particle size and strain with W_p =
0.75°. Analysis with CR = $1.5W_g$ = 0.23° gave the correct α_1 and α_2 peak
positions for the GGG (10,4,0) reflection, but a number of false noise
peaks for the palladium reflection (open circles). By increasing CR to
$1.5W_p$ = 1.125°, the extraneous peaks falsely identified above were eliminated and the correct peak located (solid circle).

COMPARISON WITH THE PROFILE FITTING METHOD

A cluster of three overlapped $K\alpha_{1,2}$ doublet reflections of forsterite
(222), (402) and (231), was chosen to illustrate the different characteristics of the peak search and the profile fitting methods.[4] The peaks have
approximately the same separation of 1W and were obtained with $\Delta 2\theta$ =
0.01° and 4.3 second count time per step as shown in Figure 6.

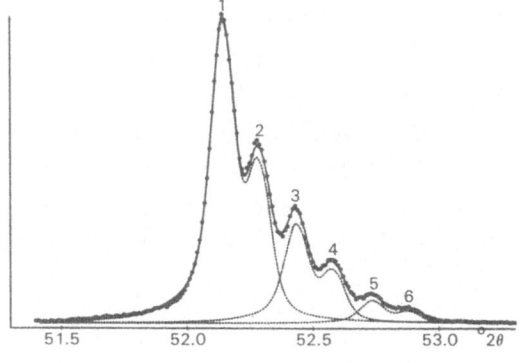

Figure 6.
Diffraction pattern of
the (222), (402) and (231)
fosterite $K\alpha_{1,2}$ reflections.

Table 2. Peak Search (PS) and Profile Fitting (PF) Results

#	2θ (°)		Peak Intensity (counts)		
	PS	PF	PS	PF	Δ(%)
1	52.144	52.144	13774	13550	2
2	52.282		8127		
3	52.431	52.435	5216	4477	17
4	52.574		2815		
5	52.731	52.735	1308	1033	27
6	52.864		680		

The experimental points are shown as dots and the solid curve was determined by profile fitting with R_{PF} = 2% The individual reflections resolved by profile fitting are shown as dotted curves.

The results are summarized in Table 2. The peak search values of 2θ were determined by CR = 1W and the peak intensities by fitting a Lorentzian curve to the upper points. The profile fitting method for determining the angles and intensities has been described elsewhere.[4] The major difference between the two methods is that the profile fitting method can resolve the individual reflections and gives results as if there was no overlapping, but the peak search results can not correct for the effect of the overlaps. Consequently the peak intensities obtained by profile fitting will be lower than those from peak search by an amount depending on the degree of overlap. The difference of intensity $\Delta = (I_{PS}-I_{PF})/I_{PF}$ increases with increasing overlap as shown in the last column of Table 2 where peak #5 has a 27% difference. The peak angles also show differences. For example, peak #1 has the same 2θ-angle for both methods because it is relatively free of overlap. Peaks #3 and #5 have peak search values 0.004° smaller than the profile fitting values due to the overlapping $K\alpha_2$ tails.

REFERENCES

1. T. C. Huang and W. Parrish, A Combined Derivative Method for Peak Search Analysis, Adv. X-Ray Anal. 27 (preceding paper) (1984).

2. W. Parrish, G. L. Ayers and T. C. Huang, A Minicomputer and Methodology for X-Ray Analysis, Adv. X-ray Anal. 23:313 (1980).

3. A. Savitzky and J. E. Golay, Smoothing and Differentiation of Data by Simplified Least Squares Procedures, Anal. Chem. 36:1627 (1964).

4. W. Parrish and T. C. Huang, Accuracy of Profile Fitting Method for X-Ray Polycrystalline Diffraction, Proc. Symp. on Accuracy in Powder diffraction, NBS Special Publ. 567:95 (1980).

EVALUATION OF EXISTING X-RAY POWDER DIFFRACTION STANDARDS FOR PHOSPHATE MINERALS

Frank N. Blanchard

Department of Geology
University of Florida
Gainesville, FL   32611

INTRODUCTION

Of the roughly 3,000 known mineral species over 290 or 10 percent are phosphates. This report deals with a recently initiated project, the purpose of which is to produce better X-ray powder diffraction standards than currently exist for selected phosphate minerals. The project involves evaluation of existing X-ray diffraction standards for phosphate minerals, and collection of new data from species for which current data are judged unsatisfactory.

ACKNOWLEDGMENTS

I am grateful to the JCPDS-International Centre for Diffraction Data for partial support by repeated grants-in-aid, and to the University of Florida Division of Sponsored Research for a one-semester, part-time student assistant.

EVALUATION

Published X-ray powder diffraction data for a substantial number of phosphate minerals are not entirely satisfactory. In order to make prudent decisions on which phosphate minerals should be restudied, an evaluation of nearly all of the phosphate mineral patterns in sets 1 through 31 of the PDF was initiated. For each of 275 phosphate minerals, the d-values, lattice parameters, and space group from the PDF card were used along with other appropriate information as input to the computer program of Appleman and Evans (1973) for indexing and refinement of lattice parameters. Refined lattice parameters were used in the

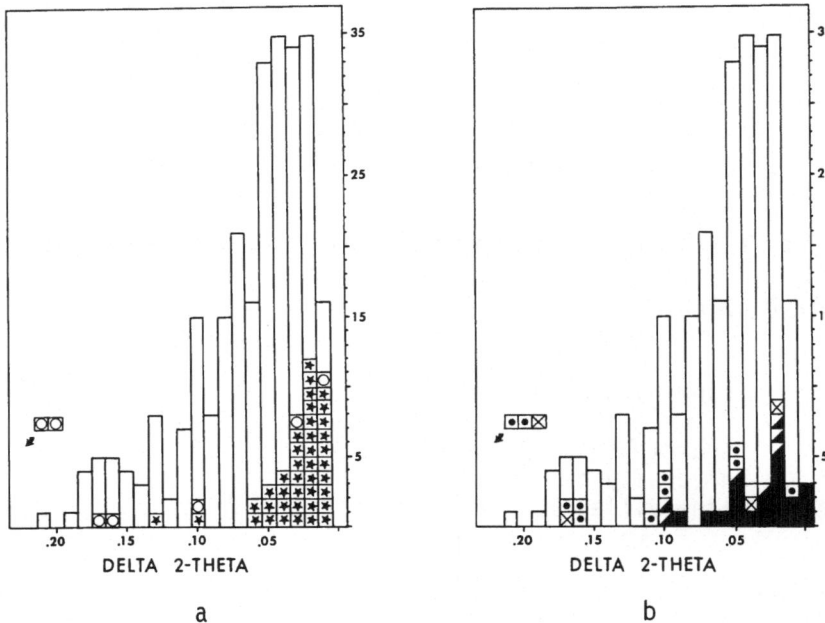

Fig. 1 (a) Histogram showing accuracy of diffraction patterns for 275 phosphate minerals. (b) Same showing results of attempt at computer identification of 44 phosphate minerals.

same program to compute the ideal 2-theta angle for each reflection, and the differences between calculated 2-theta angles and corresponding observed 2-theta angles (hereafter referred to as delta 2-theta) were used to determine the figure of merit (as defined by Smith and Snyder, 1979). The figure of merit determined for a pattern of wardite, recently prepared by the author, is expressed as F(30)=48.7(.018,35), and describes (1) the accuracy and (2) the completeness of the pattern. The 30 indicates that the first 30 observed lines of the pattern were used for the calculation. The .018 is the average delta 2-theta, that is, it is the average absolute difference between observed and calculated 2-theta angles, for the first 30 observed lines of the pattern. The 35 indicates that there are theoretically 35 possible lines from the first possible to the 30th observed line. The value 48.7 is the overall measure of accuracy and completeness of the pattern. Larger values for this figure of merit indicate higher quality.

Quality evaluations of the 275 patterns of phosphate minerals in sets 1 through 31 of the PDF maybe illustrated by histograms. Figure 1a shows the frequency distribution for the average delta 2-theta of each mineral. Seven points are too far

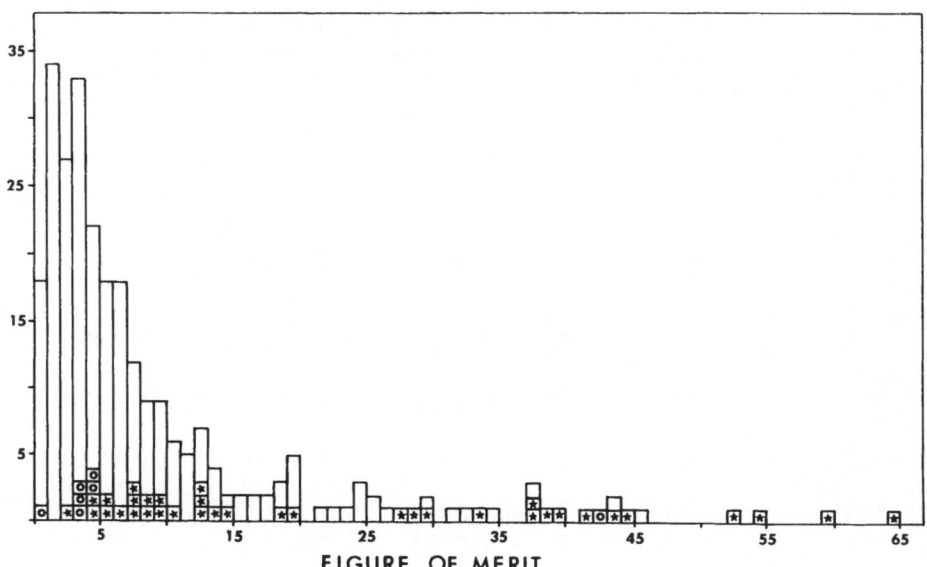

Fig. 2   Histogram showing figure of merit of 275
phosphate minerals.

to the left to be shown. Greater accuracy of the patterns is
represented to the right in the figure. For the 275 patterns
delta 2-theta ranges from 0.007 to 0.957 and the mean value is
0.071. Starred cards from the PDF are indicated and as expected
are located generally to the right. The delta 2-theta for starred
cards ranges from .009 to .129 and the mean is .030. Low quality
data (a circle in the upper right corner of the PDF card) are
indicated in Figure 1, and for these delta 2-theta ranges from
0.010 to 0.957 and the mean value is 0.234. For "i" quality cards
delta 2-theta ranges from 0.010 to 0.476 and the mean is 0.068.

Figure 2 shows the frequency distribution of values of
$F(30)$, the overall quality designation (higher quality to the
right). The range of values is from 0.295 to 117 and the mean
value is 10.6. On the figure, 4 points are too far to the right
to be shown. Most of the high values for $F(30)$ are associated
with starred cards, for which the mean $F(30)$ is 29.7, and the
range is from 2.14 to 117. Low quality cards are noted in this
figure and are mostly to the left. For these the $F(30)$ values
range from 0.549 to 42.2 and the mean is 8.65. For "i" quality
cards $F(30)$ ranges from 0.343 to 68.9 and averages 8.44.

Patterns for 13 phosphate minerals recently prepared by the
author and accepted for inclusion in the PDF yield average delta
2-thetas which range from 0.010 to 0.028 while the $F(30)$ values
range from 14.3 to 78.8 (Table 1). The average value of the figure

Table 1. Evaluation of data for 13 phosphate minerals
currently in the PDF and new data submitted by Blanchard,
1981, 1983, report to the JCPDS

MINERAL	FIGURE OF MERIT	
CARD NUMBER	FOR CURRENT PDF CARD	FOR NEW DATA
ANAPAITE		
15-583	F30 = 5.98(.167,063)	F30 = 49.9(.013,045)
ARROJADITE		
24-66	F29 = 5.59(.041,126)	F30 = 14.5(.015,142)
AUGELITE		
14-380	F30 = 6.96(.046,094)	F30 = 40.2(.014,054)
CRANDALLITE		
25-119	F30 = 6.64(.087,052)	
25-1457	F13 = 29.0(.020,022)	F30 = 34.4(.0198,44)
GOYAZITE		
11-194	F21 = 2.47(.164,052)	F30 = 78.8(.010,037)
HYDROXYL-HERDERITE		
29-1408	F30 = 25.2(.017,071)	F30 = 56.6(.011,049)
METAVARISCITE		
15-311	F30 = 1.57(.246,078)	F30 = 29.3(.0214,048)
PHOSPHOSIDERITE		
15-390	F30 = 1.98(.103,147)	F30 = 14.3(.0283,074)
ROCKBRIDGEITE		
8-159	F30 = 7.12(.042,101)	
22-356	F30 = 6.96(.052,083)	F30 = 28.8(.017,061)
SICKLERITE		
13-338	F30 = .549(.957,056)	F30 = 53.8(.0136,41)
STRENGITE		
15-513	F30 = 6.09(.046,108)	F30 = 25.7(.0191,061)
VARISCITE		
25-19	F30 = 45.9(.016,042)	F30 = 44.1(.0162,042)
WARDITE		
13-403	F30 = 11.1(.055,049)	F30 = 48.7(.0176,035)

of merit for these 13 patterns is $F(30)=39.9(.017,xx)$. Comparison
of these quality descriptions with the corresponding values for
current data for these 13 minerals indicates very substantial
improvements. Among the 275 phosphate minerals evaluated only 26
have delta 2-theta values as good or better than this average, and
only 13 have a figure of merit as good or better than this
average. Thus the quality of these 13 patterns averages in the
top 5 to 10 percent of the phosphates currently in the PDF.

IDENTIFICATION BY COMPUTER

In addition to the calculation of a figure of merit for each phosphate mineral, evaluation of patterns in the PDF is being augmented by testing the ability of a computer search-and-match program (Philips Electronics' SANDMAN) to identify phosphate minerals. At present, 44 different phosphate species, selected on an availability basis, have been tested. The results of the attempt at computer identification are shown on the histogram illustrating accuracy of the patterns for phosphate minerals (Figure 1b). Black squares indicate unequivocal identification of the mineral; that is, the correct mineral was the only one which scored high enough to be marked as "identified". This resulted in 44 percent of the cases, and 7 percent of these were from calculated patterns in the PDF. Half-black squares indicate that the mineral was listed as "identified", but as part of a mixture of two or more phases which were not actually present. This happened in 16 percent of the cases. An X indicates that the mineral was listed as a possible phase, but with very low probability (11 percent of the cases), and a dot indicates a complete failure in the attempted identification. This happened in 30 percent of the cases. As expected the correct identifications are mostly associated with more accurate patterns in the PDF and the failures are mostly associated with less accurate patterns.

Figure 3 shows the results of the attempt at computer identification for the 44 phosphate species plotted on the figure of merit diagram for 275 phosphate minerals. There is a general relationship between the quality of a pattern in the data base (as evaluated by its figure of merit) and the ability of the computer program to identify the phase. However, the relationship is not as well defined as the one relating to the accuracy of the pattern.

Even though unequivocal identification may be achieved for a particular phase, the result is not necessarily entirely satisfactory. For instance, using the 1981 database, SANDMAN correctly identified strengite; however, out of 52 common d-range lines, 19 are not listed on the current PDF card.

In some instances a phase may be correctly identified, but as a mixture with another phase which is actually absent. A new pattern for wardite which will appear in a future set of the PDF was identified by SANDMAN (using the 1981 data base) as a mixture of wardite and lautite.

In other instances an attempt to identify a single pure phase may result in the computer program finding the phase, but listing it with very low probability (hence not "identified"). Using the 1981 database, SANDMAN incorrectly identified a pattern for anapaite, which will appear in a future set of the PDF, as a mixture of albite and coloradoite. Anapaite was found as a possible phase, but was listed with a very low probability.

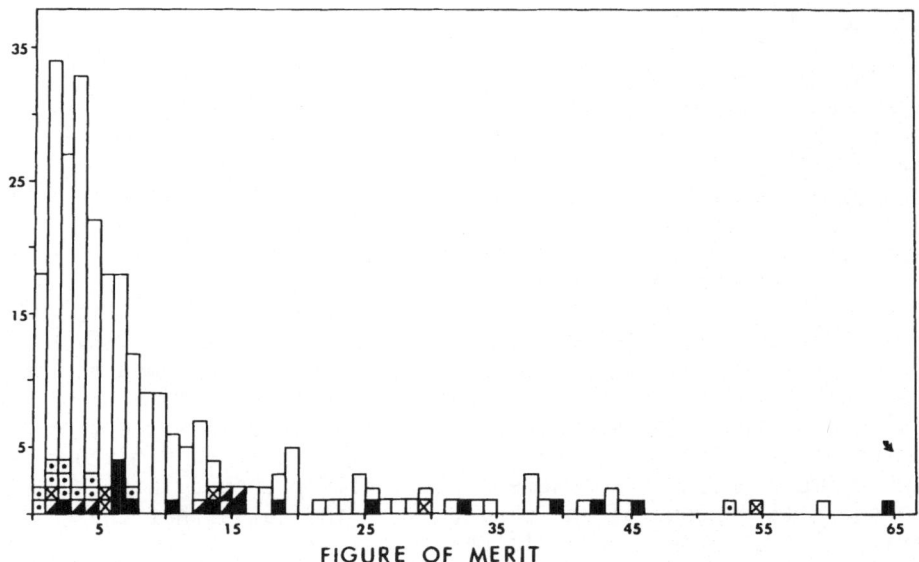

Fig. 3 Same histogram as Fig. 2, except showing the results
of attempt at computer identification of 44 phosphate
minerals.

        An attempt to identify sicklerite illustrates a complete
failure. Using the 1981 database SANDMAN failed to identify or
find the PDF pattern for sicklerite when supplied with the new
pattern which will appear in set 33 of the PDF. A mixture of
chalcopyrite and schafarzikite was incorrectly identified.

CONCLUSION

        The objective evaluation of X-ray data presented in this
report indicates that new standard patterns are needed for many of
the phosphate minerals. With the current database, moderate
success (at best) in computer identification of phosphate minerals
may be expected for no more than 60 percent of the species.

REFERENCES

Appleman, D. E., and Evans, Jr., H. T., 1973, Indexing and
least-squares refinement of powder diffraction data. Geol.
Survey Computer Contribution, No. 20, 62 p.

Smith, G. S. and Snyder, R. L., 1979, F(n): a criterion for
rating powder diffraction patterns and evaluating the reliability
of powder pattern indexing. J. Appl. Cryst. 12, 60-65.

DATA SETS FOR EVALUATION OF POWDER DIFFRACTION

SEARCH/MATCH ALGORITHMS

Diane E. Pfoertsch[a], Gregory P. Hamill[b], and
Wilson H. De Camp[c]

JCPDS-International Centre for Diffraction Data
1601 Park Lane, Swarthmore, PA 19081

INTRODUCTION

Among powder diffractionists, there is a concern about establishing criteria for testing existing and new computer methods for searching the Powder Diffraction File (PDF) data base. A large number of papers on this subject have been presented at recent Denver Conferences and at other meetings, such as the American Crystallographic Association.[1-12]

The PDF now exceeds 42,000 patterns, with an average of about thirty lines per pattern. Thus, it contains approximately 2.5 million numerical values. At its current rate of growth, the PDF will double in less than twenty years.

Exhaustive searching of the PDF is impractical, and the effective use of various pre-screening algorithms is an essential part of a rapid search/match program. However, there is no consensus about how to evaluate the efficiency, accuracy, and reliability of the many algorithms presently in use. The identification of components of a multiphase mixture presents special problems. A user selects a search/match program and the parameters to be used in its execution on the basis of many criteria. Among these are the ease of use (as perceived by the user), as well as performance related factors. Other program

----

[a]   Pennsylvania State University, University Park, PA 16802
[b]   GTE Laboratories, Inc., Waltham, MA 02254
[c]   Food and Drug Administration, Washington, DC 20204

restrictions based on chemistry, lattice types, or known phases are also important.

Comparisons of program performance are often quoted in terms of CPU time, connect time, or apparent time to the user. However, such statements are of little value in predicting program performance because they are generally system specific. Therefore, establishing a more appropriate basis for search/match performance comparisons was considered.

THE GOALS OF THE PROJECT

In response to the needs expressed by users of the PDF, the JCPDS established a Task Group on Search/Match Test Data Sets. The charge of this task group was the collection, evaluation and dissemination of a larger variety of test data sets than were currently available. The availability of these test data sets will allow meaningful comparisons of search/match programs.

In general, complete tables of experimental d's and I/I_1's for complex mixtures are not published in papers describing the application of powder diffraction methods. The Task Group obtained patterns of such mixtures from many sources. They have been evaluated with respect to accuracy of the data and its interpretation, representation of typical analytical problems, and coverage of a variety of areas of study. The precision of the data is appropriate to a variety of workers ranging from the expert to the relatively inexperienced.

Many presentations in recent years at the Denver Conference and elsewhere have described novel approaches to the "search/match problem".[5-12] However, program performance is generally evaluated only against some arbitrarily chosen standard mixture, using data obtained on a specific instrument. Test data sets will provide a method to evaluate new search/match techniques. Objective data can then be obtained to evaluate and compare programs.

In addition to program comparisons, an educational benefit can be derived from these data sets. Applications in diffraction courses are evident. Most search/match programs incorporate a large number of parameters (e.g., d-spacing error windows, minimum number of lines to be matched, etc.) which can be varied to control the program action.[1] The choice of values for these parameters is generally accepted from the programmer's default selections and only sometimes superseded by choices based on the user's experience. The values chosen have a significant effect upon the selection and sequence of the matches obtained. In the worst cases, the default conditions written into the program are accepted uncritically. Such

parameters may not be optimized for the problem at hand.
Standardized problems will provide a means whereby testing the
performance of search/match software with a particular choice of
parameters can be accomplished. Such comparisons provide a
self-teaching mechanism for the diffractionist new to computer
searching.

THE SCOPE OF THE PROBLEMS

    The Task Group began its efforts in 1981. Approximately 125
different sets of powder diffraction data were considered and
evaluated as candidates for test data sets. Most of the submissions
were inorganic problems, including many minerals or mineral
compounds. Approximately one-third were selected as being suitable
for distribution. Two examples are given in Table 1.

    In its final form, the collection of tables will be published
by the JCPDS as one part of its Methods and Practices Manual. The
full compilation of the data sets may also be obtained separately
from the JCPDS-International Centre for Diffraction Data (1601 Park
Lane, Swarthmore, PA 19081).

    Experimental data present natural conflicts which are
unavoidable. Instrument dependent factors cannot be separated from
the data, and algorithms may be computer system dependent. Both
factors have an influence on the search/match program parameters,
but these effects are precisely the ones which must be evaluated by
being tested on all kinds of data. Program performance of a
search/match procedure is a convolution of instrumental factors,
program parameters, and algorithm design which may be impossible to
unravel. Real tests require real data.

    Synthetic data have proven to be of exceptional value in the
assessment of the accuracy of computer programs used in the analysis
of single crystal X-ray data.[13] A similar approach was used in a
JCPDS round robin test,[14,15] in which powder diffraction data were
simulated by the addition of patterns from the PDF, which were then
modified by appropriate weighting, error, and resolution functions.
Program performance with such test sets is independent of
experimental variables. Such data sets have a known source and
history, and therefore an answer rather than a result. However, the
synthetic data sets proved to be too close to ideal data to provide
significant tests, and requests have been made for JCPDS to
distribute "standard mixtures" as identification tests.

    Meaningful tests of search/match methods require testing
experimental data against an experimental data base. As a result,
we elected to exclude synthetic data from the data sets, with two

minor exceptions: two pairs of data sets for three-phase mixtures include the same experimental $\underline{d}$'s, with the I/I_1 values artificially altered to correspond to compositions of approximately 1:1:1 and 6:3:1.

Identification of the phases in an experimental sample is complicated by our lack of knowledge of its true composition. An example may be found in problem 5.1 in the PDF Workbook.[16] The data correspond to α-quartz (SiO_2; PDF 5-490). Workshop students have often found that a reasonable match of these data is possible with a variety of divalent metal phosphates (e.g., $GaPO_4$, PDF 8-497). The effects of such approximate isomorphism on search/match procedures is, of course, not surprising.

The data sets which were selected cover a range of complexity, including some single phases and binary mixtures and some with as many as seven phases. L. K. Frevel's "Dow mixture" is probably the best-known "real" problem. This sample was first identified as containing five components, then later revised to seven.[17,18] This problem has been included in the test data sets. A few other complex mineral problems involving mixtures of four or more phases were included as well.

THE FORMAT OF THE DATA

The Manual to be published by the JCPDS consists of tables of $\underline{d}$-spacings and I/I_1 values. The high $\underline{d}$ cutoff is explicitly stated for each data set. The low $\underline{d}$ cutoff must be deduced from the tables. The contributed data were not truncated in any way. Supplementary information, such as chemistry, and "results" for the real problems or "answers" for the synthetic problems are provided for the tables in separate Appendices. The authors would greatly appreciate constructive comments and suggestions for further efforts from the user community.

THE QUALITY OF THE DATA

The $\underline{d}$ values are reported to the precision claimed by the submitter, and have not been rounded. Intensities have been similarly treated except that, where necessary, the data were placed on the usual scale of $I_1=100$, and rounded to the nearest integer. Data of widely varying precision are included. Although most of the data were obtained by diffractometric measurements, selected data sets obtained using Debye-Scherrer and Guinier cameras are also included. Any special techniques leading to unusual precision or accuracy are noted.

The JCPDS does not claim that the numerical values published in the Test Data Sets represent a "standard" in any sense of the word, nor that experimental measurements on similar physical samples will lead to similar results. Attempts to duplicate the results from such new data sets may not lead to similar search/match results.

CHEMICAL AND OTHER INFORMATION

It is reasonable for the powder diffractionist to expect at least limited chemical information about any sample. The value of such information in the evaluation of search/match algorithms is left to the computer programmer to determine. Such chemical data in the Test Data Sets are generally limited to a statement of the source of the experimental sample, accompanied by statements of positive (e.g., chloride is present) or negative (e.g., carbonate is absent) chemistry, that might be obtained from X-ray fluorescence (XRF) or a few rapid qualitative tests. The chemical information corresponds to physical or chemical tests which might reasonably be performed in the laboratory. The usual experimental limitations must be recognized, e.g., XRF does not provide information about the presence of elements below Na. For certain data sets, appropriate statements of sample history or other qualitative observations are included. For teaching purposes, it may be desirable to provide a student with this additional information separately.

Some of the Test Data Sets are derived from mineralogical specimens. The trained geologist will often make use of a wide variety of additional, non-chemical information in the identification of the phases present in a sample. Where appropriate, such additional information is also given. The development of algorithms for the applications of these data as pre-screens is left as a challenge to the programmer.

EXAMPLES OF THE PROBLEMS

Table 1 shows the data for the first two problems in the Test Data Sets, arbitrarily designated "JCPDS 1" and "JCPDS 2".

JCPDS 1 is an experimental pattern obtained by the Debye-Scherrer method on a white crystalline solid suspected of being a controlled substance. The d values should be considered to have only 0.05 Å precision at best. The high d cutoff is 15Å (6° for CuKα radiation). The material was soluble in water. Microscopic examination revealed that the mixture contained at least three crystal forms. Physical separation of the crystals, followed by solubility tests, showed that two forms were insoluble and one form soluble in nonpolar solvents, suggesting that the soluble form

Table 1. Examples of the test data sets

JCPDS 1		JCPDS 2	
$\underline{d}$	I/I_1	$\underline{d}$	I/I_1
9.5	40	3.702	3
7.5	5	3.591	25
5.5	30	3.013	100
5.2	20	2.791	67
4.65	50 (broad)	2.610	10
4.40	50	2.462	6
4.25	20	2.346	14
4.10	100	2.233	8
3.80	40 (broad)	2.213	7
3.70	30	2.133	23
3.60	80	2.079	5
3.40	10	2.047	7
3.30	80	1.9652	13
3.20	100	1.9061	7
3.10	100	1.8457	25
2.95	100	1.7951	9
2.84	50	1.7323	26
2.76	20	1.6946	8
2.50	20 (broad)	1.5742	15
2.35	60	1.5058	13
2.27	40	1.4261	8
plus many additional lines		1.3537	5
		1.3288	5
		1.2822	13
		1.1971	5

was organic. Upon the addition of Ag^+ to test for halides, a black deposit formed immediately.[19]

JCPDS 2 is diffractometric data for a naturally-occurring mineral association from an ore deposit. The data were corrected using $\alpha-SiO_2$ as an external standard. The high $\underline{d}$ cutoff is 20Å (4° for CuKα radiation). XRF and chemical tests confirmed the presence of S, Cu, Fe, Sb, and Zn as the major elements. Optical examination showed that at least two phases were present.

The results which should be found from these two sets of data are given in the Appendix. The effect of the use of subfiles on the search/match process may be tested by referring to the Forensic Subfile for JCPDS 1 and the Mineral Subfile for JCPDS 2.

CONCLUSIONS

The development of new algorithms for searching, as well as the improvement of existing ones, requires the availability of well-established test cases. Evaluation criteria for algorithms include not only the computer time involved and the ease of use, but also the sensitivity of the results to the parameters on which the search is based.

We wish to acknowledge the assistance of our colleagues who participated in the work of the Task Group: Donald Beard (Siemens-Allis Corp.), Prof. Dennis Canfield (Univ. of Southern Mississippi), Dr. Tim Fawcett (Dow Chemical Co.), Prof. Deane Smith (Pennsylvania State Univ.), Prof. Robert Snyder (State University of New York, Alfred) and Peter Wallace (Lawrence Livermore National Laboratory), as well as the numerous powder diffractionists who willingly and anonymously contributed problems from their laboratories.

REFERENCES

1. Nichols, M. C. and Johnson, Q. (1980). Adv. X-Ray Anal. 23, 273-278.
2. Cherukuri, S. C., Snyder, R. L. and Beard, D. W. (1983). Adv. X-Ray Anal. 26, 99-104.
3. Pyrros, N. P. and Hubbard, C. R. (1983). Adv. X-Ray Anal. 26, 63-72.
4. Goehner, R. P. and Garbauskas, M. F. (1983). Adv. X-Ray Anal. 26, 81-86.
5. Huang, T. C., Parrish, W. and Post, B. (1983). Adv. X-Ray Anal. 26, 93-98.
6. Huang, T. C., Parrish, W. and Post, B. (1982). American Cryst. Assn. Abstracts, Ser. 2, 10, 39; Winter Meeting, Gaithersburg, MD, Abstract M4.
7. Frevel, L. K. (1982). Anal. Chem. 54, 691-697.
8. Toyohisa, S., Fujiwara, I., Ui, T. and Asada, E. (1983). Adv. X-Ray Anal. 26, 89-92.
9. Goehner, R. P. and Garbauskas, M. F. (1984). This volume.
10. Smith, G. S. and Nichols, M. C. (1984). This volume.
11. Blaffert, T. (1984). This volume.
12. Winder, D. and Loftis, B. (1984). This volume.
13. Ahmed, F. R., Cruickshank, D. W. J., Larson, A. C. and Stewart, J. M. (1972). Acta Cryst. A28, 365-393.
14. Jenkins, R. (1977). Adv. X-Ray Anal. 20, 125-137.
15. Jenkins, R. and Hubbard, C. R. (1979). Adv. X-Ray Anal. 22, 133-142.
16. Hubbard, C. R., McCarthy, G. J., and Foris, C. M. (1980). PDF Workbook. JCPDS-International Centre for Diffraction Data, 1601 Park Lane, Swarthmore, PA 19081.

17.   Frevel, L. K.   (1965).   Anal. Chem. 37, 471–482.
18.   Frevel, L. K.   (1966).   Anal. Chem. 38, 1914–1920.
19.   Camp, M. J.   (1978).   Adv. X-Ray Anal. 22, 13–17.

APPENDIX: SEARCH/MATCH RESULTS

JCPDS 1
Result: hydroquinone, PDF 8–697 or PDF 22–1758; $K_2S_2O_5$,
            PDF 11–658; KBr, PDF 4–531; this mixture is a common
            formulation of photographic developers!
JCPDS 2
Result: siderite ($FeCO_3$), PDF 29–696; tetrahedrite
            ($Cu_{12}Sb_4S_{13}$), PDF 24–1318 or 11–101; a satisfactory
            match may also be obtained for the nearly isostructural
            mineral, freibergite [$(Cu,Ag,Zn)_{12}Sb_{4.4}S_{12.6}$],
            PDF 27–190; the sample is a high-yield silver ore; Ag
            forms solid solutions with both tetrahedrite and
            freibergite.

COMPUTER SIMULATION OF POWDER PATTERNS

W. Parrish, T. C. Huang and G. L. Ayers

IBM Research Laboratory

5600 Cottle Road, San Jose, CA. 95193

ABSTRACT

A method for computer simulation of X-ray powder diffraction patterns which are identical to those obtained experimentally is described. The calculated pattern is generated directly from the d's (or 2θs) and intensities of the phase(s) and is based on a profile fitting algorithm which uses the instrument function to form the profile shapes at all reflection angles. Examples of simulated patterns of mixtures, line broadening, linear and amorphous backgrounds, and counting noise are given.

INTRODUCTION

Computer simulated powder diffraction patterns are useful for many aspects of powder diffraction analysis. For example, patterns generated from known parameters can be used for the absolute test of data reduction algorithms including the determination of lattice spacings (d's) or peak angles (2θs) and intensities (I's) by peak search[1,2] and profile fitting, and the calculation of component concentrations in a mixture by various quantitative analysis methods. They can also be used for direct comparison with the observed patterns to assist phase identification[3,4], for the study of crystal chemistry of iso-structural or isomorphous compounds, for planning or selection of optimum experimental parameters to obtain observed data in the desired form, and as an education tool to illustrate various diffraction effects.

The basic parameters required for powder pattern simulation are the 2θs (or d's) and I's. One of the methods to obtain these data is by the calculation from crystal structure data including the atomic position parameters, atomic scattering factors, space group information, unit cell dimensions, absorption coefficient and other X-ray constants as done in the POWD program.[5] A simpler way is to utilize the existing value of d's and I's directly. This approach eliminates the problem of searching the literature for crystal structure information required by the POWD program, and makes it possible to analyze compounds with no crystal structure data yet available. The d's and I's of about 40,000 crystalline compounds are contained in the Powder Diffraction File (PDF) published by the JCPDS - International Centre for Diffraction Data.

NEW IBM SAN JOSE XRD GRAPHIC SYSTEM

A new interactive computer method was developed for the simulation of X-ray diffraction (XRD) patterns and it is currently run on an IBM 3083 central processing unit under time sharing operation. The procedure is outlined in Figure 1a. A color graphic station (IBM 3279) is used for interactive input/output making it possible to employ several colors in the display but only black and white reproductions are used in the paper.

Parameter Entry

All parameters required for the calculation are entered through a single menu at the color display station. An example for obtaining a quartz pattern is given in Figure 1b with entries shown in bold face. Parameters related to the graphics are entered in the upper portion of the menu. Text entered as COMMENT is printed at the top of the calculated pattern. The wavelength and instrument functions shown are Cu Kα X-rays and W*G dataset name of CU21014. The MAX. INTENSITY entry is the number of counts (not count-rate) of the highest peak in the pattern. If a value is entered the counting statistical noise is added to every datapoint in the simulated pattern. If the entry is left blank no counting noise will be included. The STEP SIZE should be entered only when it is desired to reproduce the same step increment $\Delta2\theta$ used in an experimental pattern. The 2-THETA values (LOW and HIGH) are required to plot a shorter angular range than that of the complete standard file pattern. If left blank the entire pattern will be generated. The BACKGROUND entries are required if a linear background is to be added. The number of counts at the LOW and HIGH 2θs are entered and they may be different to simulate a sloping linear background. Non-linear or amorphous background may be stored as a standard in the user's file and selected in the COMPOUND PARAMETERS section.

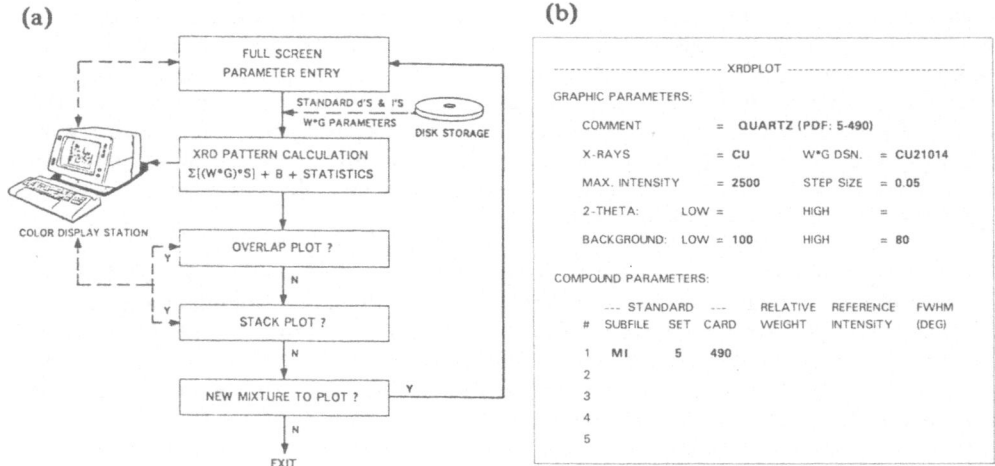

Figure 1.   (a) Procedure for the IBM San Jose XRD Graphic System;
(b) Full screen parameter entry for pattern simulation.

The lower portion of the menu COMPOUND PARAMETERS provides for selection of individual phases under STANDARD, SUBFILE, SET and CARD. Five phases can be selected on the menu shown but the number can be easily increased. The next two columns permit entries of the RELATIVE WEIGHT (in weight percent or any relative form), and the REFERENCE INTENSITY (e.g., compared to Al_2O_3 in the JCPDS file). These two values are used for plotting patterns of mixtures in the correct relative intensities (see below). When left blank the default values are set to 1. The last column FWHM (full width at half maximum) allows the inclusion of a selected value of broadening to be added to the W*G standard profiles. For a pattern of a single phase of quartz with no extra line broadening, the entries required are: SUBFILE = MI, SET = 5 and CARD = 490 as shown in Figure 1b.

Pattern Calculation

After completing the full screen entry of parameters, the selected subfile(s) stored on disk are searched to obtain the values of the d's and I's of the set and card numbers entered. The profile shape of each reflection is determined from the convolution $(W*G)*S$.[6] W*G is the instrument function previously determined using a set of standard specimens containing no specimen broadening. Each W*G profile is represented by a sum of seven Lorentzians and the parameters defining the shapes are stored in the computer. W*G profiles at intervals of $5°$ to $10°2\theta$ are stored and linear extrapolation of the parameters of a pair of W*G profiles straddling the reflection determines the shapes at any intermediate 2θ. The contribution of the reflection i of the specimen $S_i(2\theta_i, I_i, W_j)$ is represented by a Lorentzian in which $2\theta_i$ is calculated from the value of d_i and the Bragg's law, I_i directly from the standard subfile, and W_j from the FWHM of Compound j entered by the user. For no extra broadening, $W=0$. The diffraction profiles of all reflections i and compounds j in the specimen are then added together with the background to obtain a calculated pattern with its strongest intensity and background scaled to the values selected by the user. Counting statistical noise can be added by applying a normally distributed random number with a standard deviation scaled to the square root of the calculated intensity. The use of the experimental instrument function to define the profile shape makes it possible to simulate the pattern in exactly the same form obtained on the diffractometer. The entire calculation is almost instantaneous.

Graphics Displays and Results

When the calculation is completed, the results are displayed at the color station as shown in Figure 2a. In this case, all the reflections of quartz listed in Card #5-490 of the Powder Diffraction File (PDF) were plotted. Proper levels of linear background and counting noise have been added to simulate experimental conditions. A section of the pattern may also be obtained with any desired intensity and 2θ scales. For example, a small section between $63°$ to $70°2\theta$ was enlarged to fill the entire screen as shown in Figure 2b. The identification of peaks with $P/\sigma = 1$ and 2 (P = the net peak intensity above background B, and $\sigma = \sqrt{P+B}$) becomes uncertain due to poor statistics. Peaks with $P/\sigma \geq 3$ begin to emerge from the background.

Patterns of crystalline materials on amorphous substrate can also be simulated, as for example, crystalline quartz on SiO_2 glass substrate, Figure 2c. Because the noise is higher fewer quartz peaks become distinguishable from the amorphous background.

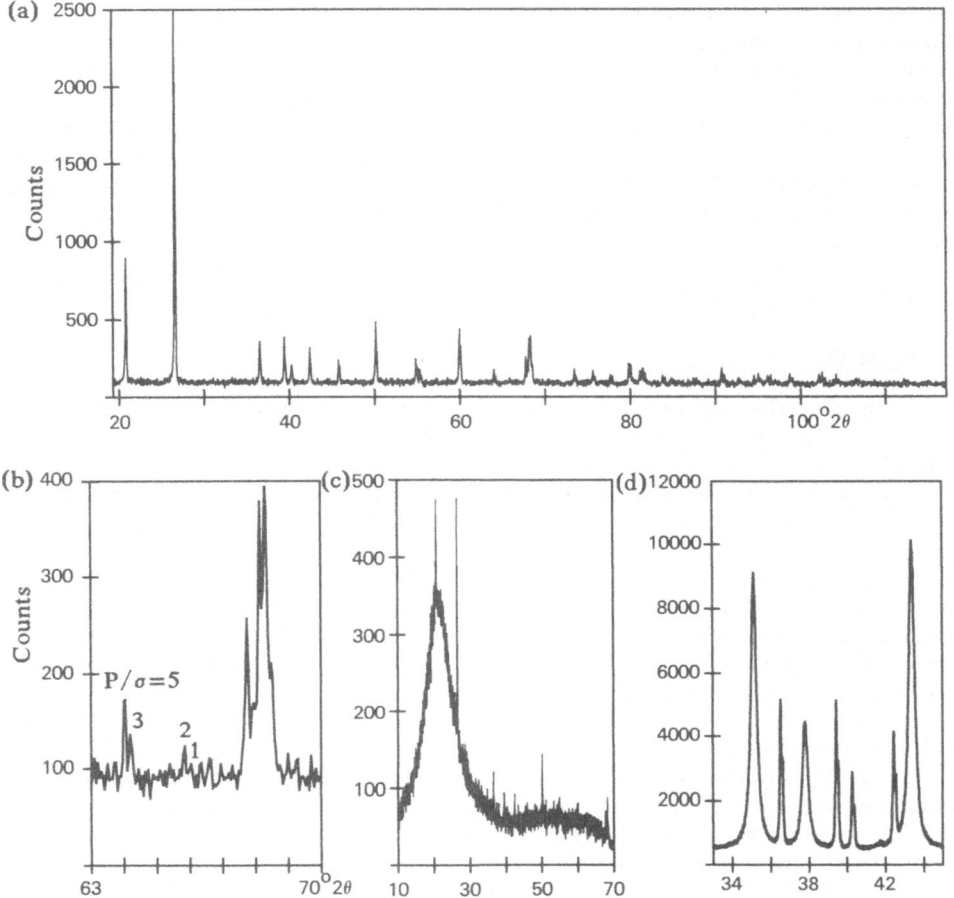

Figure 2.   **Simulated XRD Patterns: (a) Quartz from card 5-490;**
            (b) Selected section of (a) showing statistical factors;
            (c) Section of quartz on glass substrate;
            (d) Mixture of quartz (no broadening) and corundum (FWHM=0.5°2θ).

The program may also be used to illustrate the line broadening effects in a mixture of compounds with different profile widths. For example, a section of the diffraction pattern of a mixture of equal weights of SiO_2 (~10μm particles) and Al_2O_3 (~350Å particles) is plotted in Figure 2d. No extra line broadening for SiO_2, and FWHM=0.5° was entered for Al_2O_3. Peaks of these two phases are noticeably different and easily distinguished.

Overlay Plots

Overlays to form composite patterns, and selected components may also be obtained. Each individual pattern may be plotted with a different color. The overlay of three benzoic acid XRD patterns from 22° to 26°2θ is illustrated in Figure 3 (each pattern is plotted in different line type instead of color) and the composite as the upper solid line.

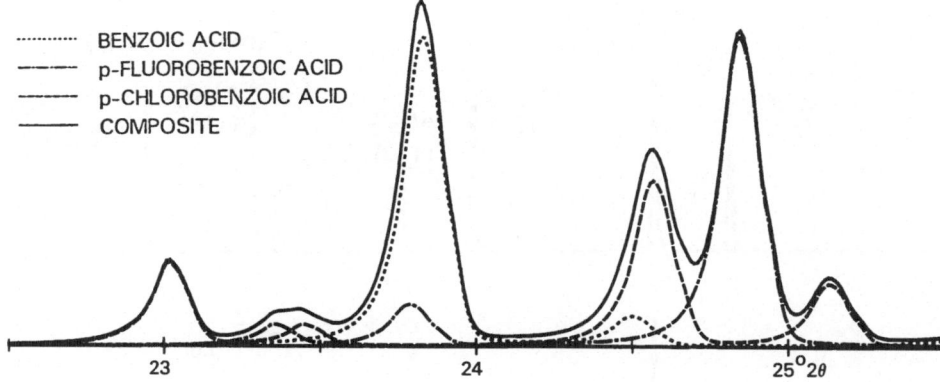

BENZOIC ACID
p-FLUOROBENZOIC ACID
p-CHLOROBENZOIC ACID
COMPOSITE

Figure 3. Overlay plot for a mixture of three benzoic acids.

Stack Plots

Another graphic display available in this system is the stack plot, frequently used to compare patterns. This option is illustrated in Figure 4 to study the effect of isomorphous replacement on lattice dimensions. The calculated patterns of three isostructural spinels show the shift of reflections to smaller $2\theta s$ as the radius of the replacing cation increases. When the Mn^{2+} ion (radius~0.80Å) replaces Zn^{2+} (0.74Å) the unit cell a increases from 8.085Å to 8.204Å; Similarly the replacement of Al^{3+} (~ 0.50Å) by Fe^{3+} (~ 0.64Å) increases a to 8.441Å.

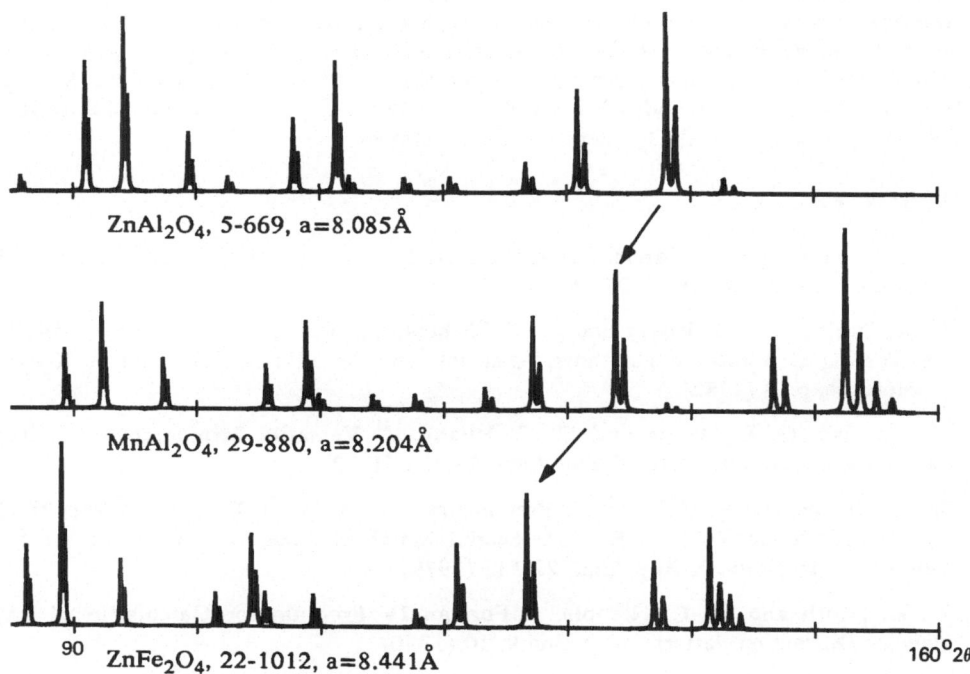

ZnAl$_2$O$_4$, 5-669, a=8.085Å

MnAl$_2$O$_4$, 29-880, a=8.204Å

90   ZnFe$_2$O$_4$, 22-1012, a=8.441Å   160°2θ

Figure 4. Stack plot of XRD patterns for isostructural spinels.

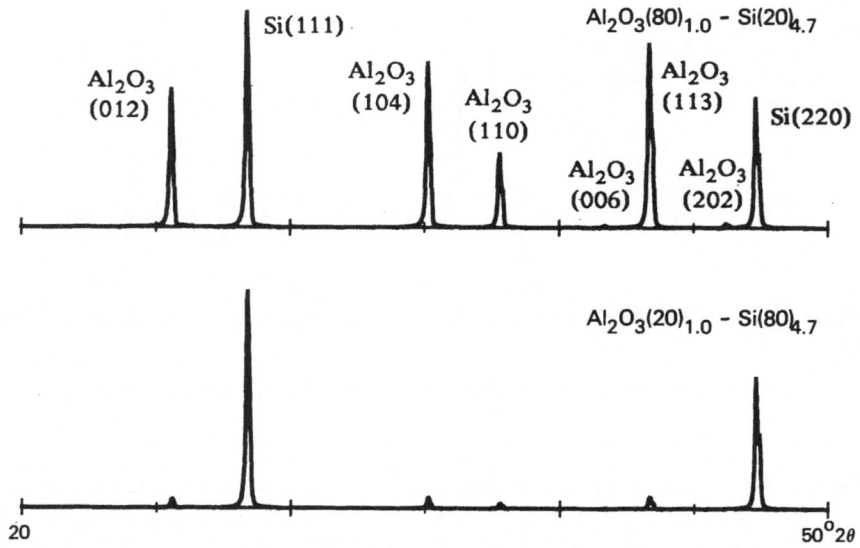

Figure 5. Quantitative synthesis of Al_2O_3 and Si mixtures.

Quantitative Syntheses

The system has also been used for quantitative synthesis studies by generating patterns of the mixtures. The product of the relative weights and the reference intensities of the individual phases are entered to scale the relative intensities of the composite pattern, as shown in Figure 5. The upper portion is a mixture of 80% Al_2O_3 and 20% Si with reference intensities of 1.0 and 4.7, respectively. The lower portion reverses the weight percentages reducing the Al_2O_3 intensities by a factor of 16.

REFERENCES

1. T. C. Huang and W. Parrish, A Combined Derivative Method for Peak Search Analysis, Adv. X-Ray Anal. 27 (1984).

2. D. K. Smith, M. E. Zolensky and M. C. Nichols, Powder Pattern Simulation, Models for Testing Computer Algorithms, Abstracts for the 31st Annual Denver X-Ray Conference, p.19 (1982).

3. W. Parrish, G. L. Ayers and T. C. Huang, A Versatile Minicomputer X-Ray Search/Match System, Adv. X-Ray Anal. 25:221 (1982).

4. R. Jenkins and C. R. Hubbard, A Preliminary Report on the Design and Results of the Second Round Robin to Evaluate Search/Match Methods for Qualitative Powder Diffractometry, Adv. X-Ray Anal. 22:133 (1979).

5. D. K. Smith and M. C. Nichols, A Fortran IV Program for Calculating X-Ray Powder Diffraction Patterns - Version 9/10 (1980).

6. W. Parrish and T. C. Huang, Accuracy of the Profile Fitting Method for X-Ray polycrystalline Diffractometry, Symp. on Accuracy in Powder Diffraction, NBS Spec. Pub. 457:95 (1980).

A USEFUL GUIDE FOR X-RAY STRESS EVALUATION (XSE)

Viktor M. Hauk and Eckard Macherauch

Institut für Werkstoffkunde, RWTH Aachen, D-5100 Aachen
Institut für Werkstoffkunde I, Universität Karlsruhe
D-7500 Karlsruhe, F.R.G.

ABSTRACT

This paper summarizes experiences available for the measurement of lattice strains in different materials with different wavelengths to evaluate stresses by means of X-rays. The recommendations given are based on previous statements[1,2]. Some principles of fundamentals of X-ray physics for the recording of interference lines with Ω- and ψ-diffractometers are dealt with. Methods applicable for the determination of the peak position of the interference lines, the assessment of linear and non-linear lattice strain distributions, and the calculation of stresses are outlined. For iron, aluminium, copper, nickel and titanium the constants for practical X-ray stress evaluation (XSE) and the parameters of measurement are tabled.

BASIC PRINCIPLES

X-ray stress analysis is based on the measurement of lattice strains dD/D_0 of special sets of {hkl}-planes in specially oriented grains of polycrystals. D_0 is the lattice spacing of the appropriate {hkl}-plane in the stress-free state. Lattice strains cause a shift

$$d(2\theta) = -2\frac{dD}{D_0} \tan\theta_0 \qquad (1)$$

of the interference lines proportional to the lattice strains. Because line shifts increase with increasing Bragg angle θ_0 in XSE interference lines with large θ-values are normally used. Table 1 summarizes appropriate X-ray wavelengths for XSE and filter materials for the selective attenuation of the $K_{\beta 1}$-radiation.

Table 1. Common target elements with appertaining wavelengths of
         eigenradiations in μm and ß-filters

Target-element	$K_{\alpha 1}$	$K_{\alpha 2}$	$K_{\beta 1}$	ß-filter
Cr	0.2289649	0.2293531	0.2084789	V
Mn	0.2101747	0.2105735	0.1910051	Cr
Fe	0.1935979	0.1939923	0.1756554	Mn
Co	0.1788893	0.1792801	0.1620703	Fe
Cu	0.1540501	0.1544345	0.1392156	Ni
Mo	0.0709261	0.0713543	0.0632253	Zr

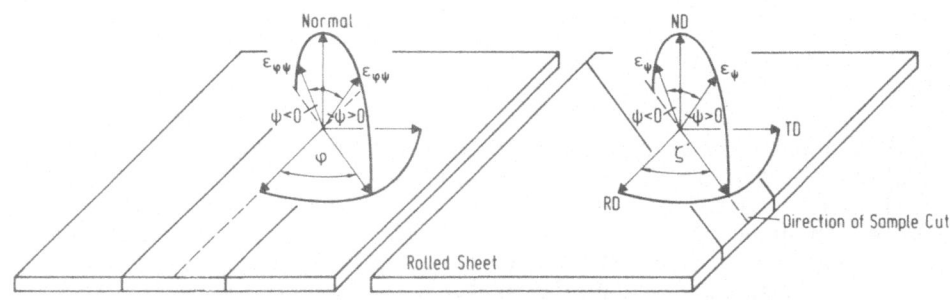

Fig. 1. Measuring system and definition of the angles φ, ψ, ζ.

Fig. 1 illustrates the coordinate system used in XSE. As has already been proved, in any case it suits to measure lattice strain dis-
tributions $\varepsilon_{\varphi, \psi} = (dD/D_0)_{\varphi, \psi}$ i.e., changes in Bragg angles $\Delta(2\theta)_{\varphi, \psi}$
in different directions ψ_i with φ = const. Alignment- and measuring
uncertainties and counting statistics influence the peak positions.[2a]

It is recommended to state the standard deviation of multiple
measurements (parameters: $\psi \lessgtr 0, \varphi$, counting time, $\Delta 2\theta$ -steps;
number of specimens, loading stresses). Furthermore, the
changes of strain or lattice parameters should be marked by an
error bar corresponding to $\Delta(2\theta) = \pm 0.01°$ in the appropriate
units of the figures. Using different {hkl}-planes and radia-
tions $D_{\varphi, \psi}$ {hkl} should be converted into $D_{\varphi, \psi}$ (100) to exhibit
possible errors of alignment and measurement at $\psi = 0$.

The spacing $\Delta(2\theta)K_{\alpha 1}K_{\alpha 2}$ of the $K_{\alpha 1}K_{\alpha 2}$-doublet of an interference
line of a stress-free material is approximated by

$$\Delta(2\theta)_{K_{\alpha 1}K_{\alpha 2}} \approx 2 \frac{\Delta\lambda}{\lambda} \tan \theta_0 \tag{2}$$

The doublet splitting increases with increasing Bragg angle Θ_0 and decreasing wavelength λ. Increasing inhomogeneous strains and/or decreasing sizes of the crystallites contributing to the X-ray lines broaden the doublet. Subtracting the underground from the intensity distribution versus 2Θ the width of the $K_{\alpha 1}K_{\alpha 2}$ peak at half height of its maximum intensity is defined as the full width at half maximum (FWHM). A doublet splitting of about 1° in 2Θ of a stress-free material is blurred even by relatively small inhomogeneous strains of the grains measured. In materials with strong inhomogeneous strains doublet splittings only occur at high Bragg angles if $\Delta(2\Theta)K_{\alpha 1}K_{\alpha 2}$ is larger than 2°.

It is recommended to determine the FWHM in the direction $\psi = 0$
as a characteristic feature of the material state measured.

For the determination of lattice strain distributions stationary or mobile diffractometers are used. Two different types are applied, the Ω- and the Ψ-diffractometer, it depends on whether the specimen axis for the ψ-tilt is vertical or parallel to the diffractometer plane. The principles of both types are illustrated in Fig. 2. Compared with Ω-diffractometers, Ψ-diffractometers are of considerable advantage[3] and are therefore increasingly applied. Both diffractometer arrangements are usually based on the Bragg-Brentano focusing principle. This means that the angular velocity of the detector is twice that of the sample around the common Θ-axis.
In order to achieve accurate results the diffractometers have to be carefully aligned by mechanical means and calibrated using stress-free powders of suitable materials.

It is recommended to align the diffractometers in the complete
range of $\psi \lessgtr 0$ such that the peak positions of the $\{hkl\}$-lines
of the calibration powder do not differ by more than $\pm$ 0.01°
in 2Θ.

Accurate lattice strain determinations can only be achieved if

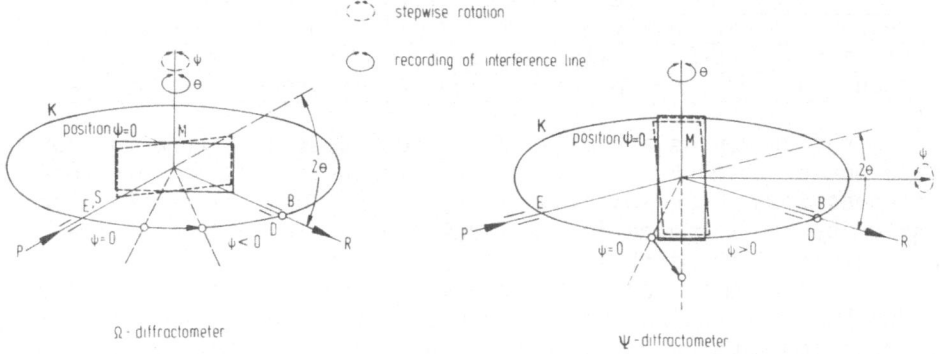

Fig. 2. Measuring arrangement with Ω- and Ψ-diffractometer.

in the specimen area irradiated by the incident X-ray beam a sufficient number of grains of the material phase of interest is randomly oriented. This presupposition is not fulfiled in coarse-grained
materials or in materials with prefered orientations of the grains.
In such cases the interference lines show different intensities in
various measuring directions $\pm \psi$. In coarse-grained materials the
number of crystallites measured can be increased by a relative motion
of the incident X-ray beam with respect to the sample's surface if
otherwise constant measuring conditions are given, by a slight pen-
dulous motion of the object surface around an axis through the center
of gravity of the irradiated area, by an enlargement of the irradiated surface area, or by using X-rays with deeper penetration. The
attenuation of X-rays depends on the wavelength, the measuring direction and the material under investigation. The linear attenuation
coefficients μ quantitatively determine the penetration depths. Characteristic values are summarized in Table 2.

Table 2. Attenuation coefficients μ in cm^{-1} of different metals for
distinct eigenradiations[4]

Metal	Density in g/cm^3	CrK_α	MnK_α	FeK_α	CoK_α	CuK_α	MoK_α
Fe	7.87	905.1	715.4	572.9	468.3	2549.9	301.4
Al	2.70	402.3	315.9	250.6	198.2	131.5	14.3
Cu	8.93	1375.2	1098.4	882.3	712.6	470.6	443.8
Ni	8.90	1290.5	1032.4	828.6	668.4	438.8	421.9
Ti	4.51	2719.5	2142.3	1700.3	1371.0	920.0	106.9

It is recommended to check the interference line intensity a-
vailable either by exposing films in front of the detector or
by successive counter registrations of the peak maxima in the
measuring range. Unless with Ω-diffractometers a continuous
decrease of intensity or with Ψ-diffractometers a constant intensity is observed with increasing ψ, the same measurements
should be repeated in neighboring sample areas, if possible.

The penetration depths are defined as those distances from the very
surface out of which 63 % or 1/e of the intensity of the interference
lines comes from. Corresponding values for Ω - and Ψ-diffractometers
are given by the formulae in the first line of Table 3. For some
materials Fig. 3 shows these penetration depths vs. $\sin^2\psi$ if distinct
{hkl}-planes are measured with specified K_α-radiations. Mean penetration depths can be calculated for $\sin^2\psi = 0.3$ i. e. $\psi = 33.21°$. Also
values are sometimes used corresponding to depths which contribute
to about 95 % or 99 % to the intensities obtained.

Table 3. Formulae of penetration depths and PLA-factors valid
for measurements with Ω- and Ψ-diffractometers

	Ω-diffractometer	Ψ-diffractometer
Penetration depths	$\dfrac{1}{2\mu}\dfrac{\sin^2\Theta - \sin^2\psi}{\sin\Theta \cdot \cos\psi}$	$\dfrac{1}{2\mu}\sin\Theta\cos\psi$
PLA	$\dfrac{1+\cos^2 2\Theta}{\sin^2\Theta}(1-\tan\psi\cot\Theta)$	$\dfrac{1+\cos^2 2\Theta}{\sin^2\Theta}$

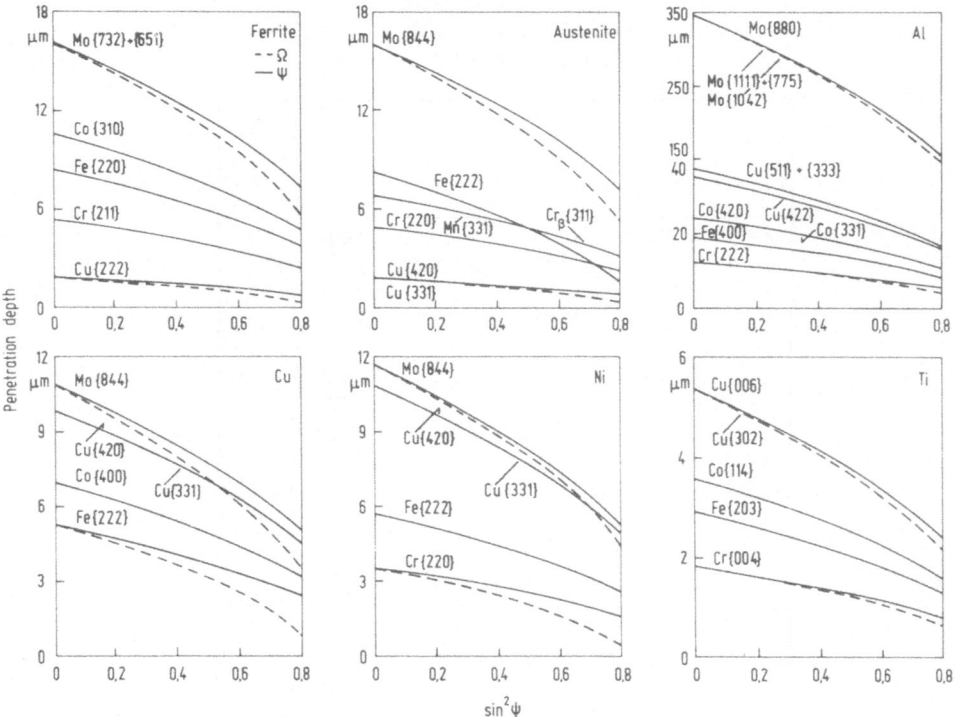

Fig. 3. Penetration depths vs. $\sin^2\psi$ of different metals
and radiations

The peak intensity, J, of an interference line is determined by

$$J \sim H\,S^2\,PLA \qquad (3)$$

where H is multiplicity factor, S structural factor, P polarization
factor, L Lorentz factor, and A absorption factor. For measurements

with Ω- or Ψ-diffractometers the PLA product is given by the equations in the second line of Table 3. The PLA values increase with increasing 2Θ. Using Ω-diffractometers PLA depends on the measuring directions ψ according to the data collected in Table 4. For measurements with Ψ-diffractometers PLA is independent of ψ.

It is recommended to apply PLA corrections on interference lines only if highest accuracy is required.

Table 4. PLA-data for measurements with Ω - diffractometer[5]

2Θ \ ψ	$0°$	$\pm\,15°$	$\pm\,30°$	$\pm\,45°$	$\pm\,60°$
$140°$	1,7970	1,6218	1,4194	1,1430	0,6642
		1,9723	2,1747	2,4511	2,9299
$142°$	1,8131	1,6459	1,4527	1,1888	0,7318
		1,9804	2,1736	2,4375	2,8945
$144°$	1,8292	1,6699	1,4860	1,2348	0,7998
		1,9884	2,1723	2,4235	2,8586
$146°$	1,8450	1,6939	1,5193	1,2809	0,8680
		1,9962	2,1707	2,4091	2,8220
$148°$	1,8605	1,7176	1,5525	1,3270	0,9365
		2,0035	2,1686	2,3940	2,7846
$150°$	1,8756	1,7410	1,5855	1,3731	1,0052
		2,0103	2,1658	2,3782	2,7461
$152°$	1,8902	1,7639	1,6181	1,4189	1,0739
		2,0165	2,1623	2,3615	2,7065
$154°$	1,9042	1,7864	1,6504	1,4646	1,1427
		2,0220	2,1580	2,3438	2,6656
$156°$	1,9175	1,8082	1,6821	1,5099	1,2115
		2,0267	2,1528	2,3250	2,6234
$158°$	1,9299	1,8294	1,7133	1,5548	1,2802
		2,0305	2,1465	2,3051	2,5797
$160°$	1,9416	1,8498	1,7439	1,5992	1,3486
		2,0333	2,1392	2,2839	2,5345
$162°$	1,9523	1,8694	1,7738	1,6431	1,4167
		2,0351	2,1308	2,2615	2,4879
$164°$	1,9620	1,8881	1,8028	1,6863	1,4844
		2,0359	2,1212	2,2378	2,4396
$166°$	1,9707	1,9059	1,8310	1,7288	1,5516
		2,0356	2,1104	2,2127	2,3899

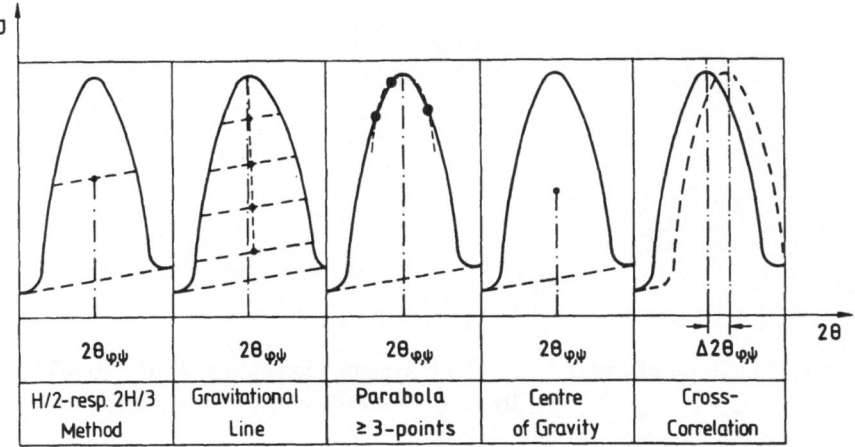

Fig. 4. Different methods to evaluate the position
of an interference line.

MEASURING PEAK POSITIONS

The most frequently used methods for the determination of the
positions of interference lines are schematically shown in Fig. 4.
In particular, there are five methods: The half resp. two third width
method, the gravitational line method, the parabola method, the center of gravity method, and the cross correlation method[6]. As in most
cases the peaks are not exactly symmetric it is suitable to use a
double slit[7] or a computer[8] to symmetrize the interference lines.
The latter method can also be used for the separation of the symmet-
ric $K_{\alpha 1}$-line from the $K_{\alpha 1}K_{\alpha 2}$ doublet. Fig. 5 shows a computer-aided
separation and symmetrization of the {211}-interference line of a
steel specimen.

It is recommended to apply symmetrizing methods for practical
measurements of peak positions except in the case of large doublet splitting.

Another method[9] to determine the position of interference lines
works with two detectors in a fixed distance according to the FWHM.
Fig. 6 illustrates the principle applied: Changing the ψ-position in

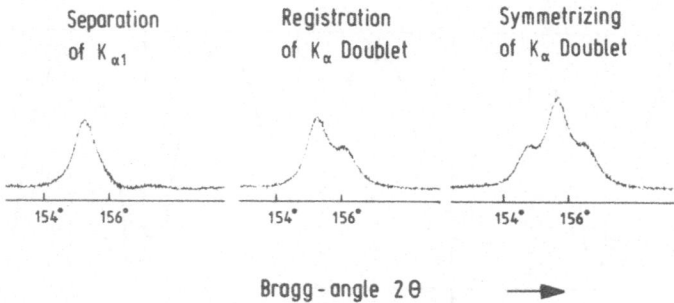

Fig. 5. Separation of $K_{\alpha 1}$ and symmetrizing of a K_α-doublet by a computer[8].

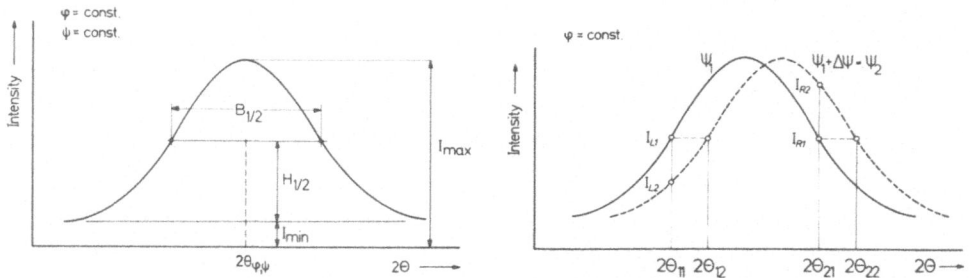

Fig. 6. Principle of automatic peak shift measurement using two detectors[9].

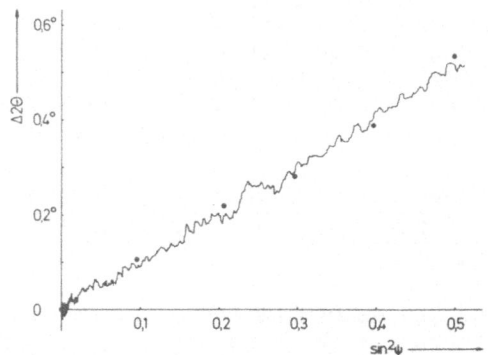

Fig. 7. Δ (2θ) versus $\sin^2\psi$ of a shot peened steel specimen quenched and tempered[9].•Δ(2θ) values determined conventionally.

a Ψ-diffractometer, the detectors automatically follow the shift
of the interference line till the intensities in both detectors are
equal again. Consequently, continuous and quick registration of $\Delta(2\Theta)$
for different values of ψ or of $\sin^2\psi$ at constant azimuth φ can be
achieved. An example is shown in Fig. 7.

BASIC TYPES OF LATTICE STRAIN DISTRIBUTIONS

From the measured peak positions $2\Theta_{\varphi,\psi}$ of the interference lines
the lattice strains in directions φ,ψ can be derived according to
equ. 1 as

$$\varepsilon_{\varphi,\psi} = -\frac{\cot \Theta_0}{2}(2\Theta_{\varphi,\psi} - 2\Theta_0) = \frac{D_{\varphi,\psi} - D_0}{D_0} \qquad (4)$$

Depending on the state of stress in the area under investigation
the $2\Theta_{\varphi,\psi}$- or $D_{\varphi,\psi}$-values corresponding to lattice strains $\varepsilon_{\varphi,\psi}$
can either be distributed linearly or non-linearly as a function
of $\sin^2\psi$ if they are determined in a macroscopic plane φ = const.
in directions $\psi \gtrless 0$. Also different $D_{\varphi,\psi}$ vs. $\sin^2\psi$ distributions
can be recorded for directions $\psi > 0$ and $\psi < 0$. This phenomenon is
called ψ-splitting. Four basic types of $D_{\varphi\psi},\sin^2\psi$- or $\varepsilon_{\varphi,\psi},\sin^2\psi$-
distributions occur. Fig. 8a shows a linear lattice strain distrib-
ution, Fig. 8b gives a ψ-split   one. A distribution with extreme
values and turning points is illustrated in Fig. 8c. Fig. 8d reveals
a simply curved lattice strain distribution. Combinations of these
basic lattice strain distributions are possible and experimentally
verified. To determine stresses from such distributions distinct
evaluation methods have to be applied.

It is recommended, independent of the stress problem to be solved,
to measure lattice strain distributions always with at least
four $\varepsilon_{\varphi,\psi}$ values in different ψ-directions in the same azimuth
plane . Additional measurements in $+\psi$- or $-\psi$-dirctions are
always of advantage

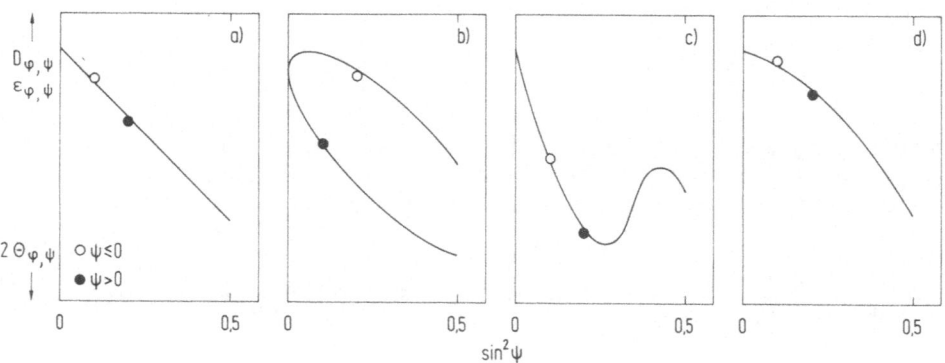

Fig. 8. Basic types of lattice strain distributions vs. $\sin^2\psi$.

If systematic deviations from linear $\varepsilon_{\varphi,\psi}$-, $D_{\varphi,\psi}$- or $2\theta_{\varphi,\psi}$ vs. $\sin^2\psi$ distributions occur as shown in Figs. 8b-d it has to be clarified whether they are due to the state of the material or to the measuring technique applied. In such cases, first the mechanical adjustment of the diffractometer should always be checked with a suitable cali- bration powder and, if necessary, be improved. As a rule the material should then be measured in more positive and negative ψ-directions than in the case of linear lattice strain distributions.

It is recommended to carry out at least 11 measurements in the ψ-range $-45° \leq \psi \leq +45°$ if non-linear lattice strain distribu- tions exist. These measurements should be repeated at several spots of the material under investigation.

If measurements with $\sin^2\psi > 0.5$ are inevitable it should be re- membered that the slightest misalignment yields large errors. It is agreed that - for fine-grained materials - deviations from linear lattice strain distributions smaller than the standard deviation (ob- tained from multiple measurements) indicate no nonlinearity. Finally, in cases where interference lines strongly differ in their intensi- ties with different ψ-angles it should be determined whether the dif- ferences are due to coarse grains or to prefered orientation of the grains.

EVALUATION OF STRESS COMPONENTS

According to the theory of elasticity the strain $\varepsilon_{\varphi,\psi}$ is given as a function of the strain components ε_{ij} (i,j = 1,2,3) by

$$\varepsilon_{\varphi,\psi} = \frac{D_{\varphi,\psi} - D_o}{D_o} = \varepsilon_{11}\cos^2\varphi\sin^2\psi + \varepsilon_{12}\sin2\varphi\sin^2\psi + \varepsilon_{13}\cos\varphi\sin2\psi + \quad (5)$$
$$\varepsilon_{22}\sin^2\varphi\sin^2\psi + \varepsilon_{23}\sin\varphi\sin2\psi + \varepsilon_{33}\cos^2\psi$$

If the shear components ε_{ij} (i $\neq$ j) are zero this equation reduces to

$$\varepsilon_{\varphi,\psi} = \frac{D_{\varphi,\psi} - D_o}{D_o} = (\varepsilon_1 \cos^2\varphi + \varepsilon_2 \sin^2\varphi - \varepsilon_3) \sin^2\psi + \varepsilon_3 \quad (6)$$

where ε_1, ε_2 and ε_3 are principle strains. Fig. 9 shows the geometri- cal presentation of this formula for the azimuth planes $\varphi = 0°$ and $\varphi = 90°$. With respect to Hooke's law equ. 6 can be written as

$$\varepsilon_{\varphi,\psi} = \frac{D_{\varphi,\psi} - D_o}{D_o} = \frac{1}{2} s_2 \left[(\sigma_\varphi - \sigma_3) \sin^2\psi + \sigma_3\right] + s_1 (\sigma_1 + \sigma_2 + \sigma_3) \quad (7)$$

with

$$\frac{\delta \varepsilon_{\varphi,\psi}}{\delta \sin^2 \psi} = \frac{1}{D_0} \frac{\delta D_{\varphi,\psi}}{\delta \sin^2 \psi} = - \frac{\cot \theta_0}{2} \frac{\delta(2\theta_{\varphi,\psi})}{\delta \sin^2 \psi} = \frac{1}{2} s_2 (\sigma_\varphi - \sigma_3) \quad \text{and} \quad (8)$$

$$\varepsilon_{\varphi,\psi=0} = \frac{D_{\varphi,\psi=0} - D_0}{D_0} = \frac{1}{2} s_2 \sigma_3 + s_1 (\sigma_1 + \sigma_2 + \sigma_3) \tag{9}$$

For $\varphi = 0$ and $\varphi = 90°$ Fig. 10 shows the interpretation of these basic equations of the $\sin^2 \psi$-method of XSE. σ_φ, the stress component in the direction $\varphi, \psi = 90°$, is given by

$$\sigma_\varphi = \sigma_1 \cos^2 \varphi + \sigma_2 \sin^2 \varphi \tag{10}$$

The quantitites s_1 and $1/2$ s_2 are the X-ray elastic constants (XEC) which depend on the $\{hkl\}$-plane involved. In the case of mechanical measurement they are given by

$$s_1 = - \frac{\nu}{E}, \quad \frac{1}{2} s_2 = \frac{1 + \nu}{E} \tag{11}$$

with E: Young's modulus and ν: Poisson's ratio.

It is recommended to apply equ. 7-11 for the evaluation of stress components if linear lattice strain distributions vs. $\sin^2 \psi$ are observed.

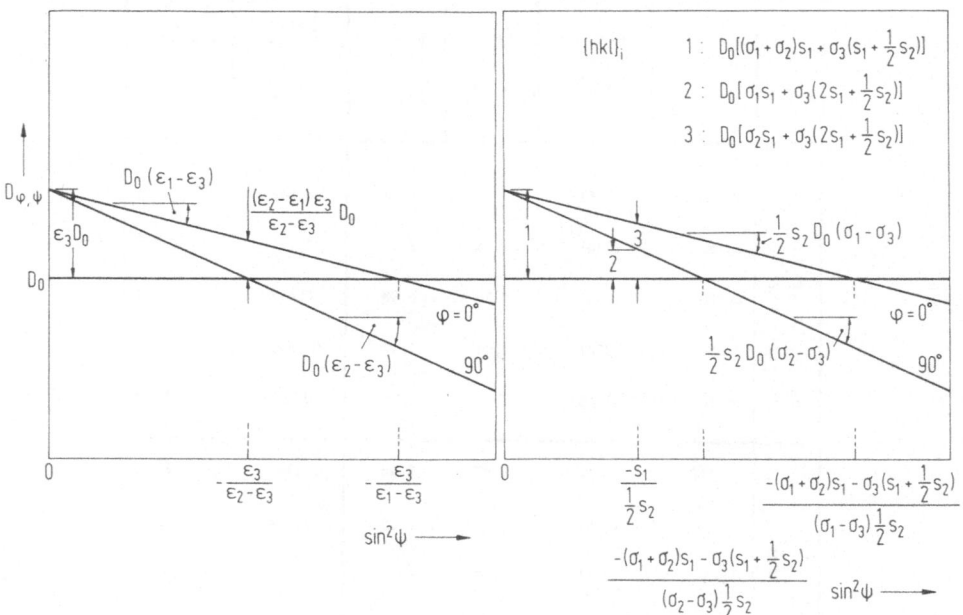

Fig. 9. $D_{\varphi,\psi}$ vs. $\sin^2 \psi$ dependences for $\varphi = 0°$ and $90°$ of a triaxial strain state with the principle strains ε_1 and ε_2 parallel to the surface of the specimen.

Fig. 10. $D_{\varphi,\psi}$ vs. $\sin^2 \psi$ for $\varphi = 0°$ and $90°$ of a triaxial stress state with the principle stresses σ_1 and σ_2 parallel to the surface of the specimen.

In most cases, however, linear lattice strain distributions exist due to two-axial surface stress states. Then equ. 7-9 hold with $\sigma_3 = 0$[10,11].

Table 5 summarizes the evaluation constants necessary for stress determinations from linear lattice strain distributions. The XEC s_1 and $1/2$ s_2 are calculated according to the modified model of Eshelby-Kröner[12,13]. The data collected are valid exactly for α-iron, aluminium, copper, nickel, and titanium, having specified single crystal elastic constants and the lattice constants "a_0" given in Table 6. Significant deviations from the data collected can occur as a consequence of strong deformations, textures, coarse grain sizes, alloying elements, and second phases. Some of these influences on XEC can be estimated[13,14].

It is recommended to apply the evaluation constants summarized in Table 5 for the materials mentiond. It should be specified if other constants are used. Experimental evaluations of XEC in a uniaxial tensile test[15] may or should be prefered.

Table 5. Constants for the evaluation of linear
lattice strain distributions

Materials	Radiation	Inter-ference {hkl}	$-\text{s}_1$ [$10^{-6}\,\text{mm}^2/\text{N}$]	$\frac{1}{2}\text{s}_2$ [$10^{-6}\,\text{mm}^2/\text{N}$]	$\dfrac{-\frac{1}{2}\cot\theta_0}{\text{s}_1}$ [$10^5\,\text{N/mm}^2$]	$\dfrac{\frac{1}{2}\cot\theta_0}{\frac{1}{2}\text{s}_2}$ [$10^4\,\text{N/mm}^2$]
Iron, ferritic - perlitic iron - basis materials	Cr-Kα	{211}	1,25	5,76	0,849	1,839
	Fe-Kα	{220}	1,25	5,76	1,242	2,691
	Co-Kα	{310}	1,66	6,98	0,497	1,178
	Cu-Kα	{222}	1,04	5,12	1,895	3,833
	Mo-Kα	{732}+{651}	1,34	6,05	0,863	1,918
Retained austenite, austenitic iron - basis materials	Cr - Kβ	{311}	1,87	6,98	0,747	2,003
	Cr - Kα	{220}	1,56	6,05	1,534	3,957
	Mn-Kα	{311}	1,87	6,98	0,659	1,767
	Fe-Kα	{222}	1,28	5,21	1,494	3,668
	Cu-Kα	{420}	1,86	6,95	0,788	2,111
		{331}	1,48	5,81	1,281	3,260
	Mo-Kα	{844}	1,56	6,05	0,833	2,147

Table 5 (continued)

Materials	Radiation	Inter-ference {hkl}	$-\frac{1}{2}s_1$ [10^{-6} mm²/N]	$\frac{1}{2}s_2$ [10^{-6} mm²/N]	$\dfrac{-\frac{1}{2}\cot\theta_0}{s_1}$ [10^5 N/mm²]	$\dfrac{\frac{1}{2}\cot\theta_0}{\frac{1}{2}s_2}$ [10^4 N/mm²]
Aluminium, Al - basis materials	Cr-Kα	{222}	4,79	18,56	0,215	0,555
	Fe-Kα	{400}	5,51	20,60	0,278	0,742
	Co-Kα	{420}	5,17	19,62	0,152	0,400
		{331}	4,92	18,93	0,285	0,741
	Cu-Kα	{511}+{333}	5,22	19,77	0,147	0,388
		{422}	4,97	19,07	0,391	1,020
		{880}	4,97	19,07	0,137	0,356
	Mo-Kα	{1111}+{775}	5,15	19,57	0,238	0,625
		{1042}	5,20	19,75	0,283	0,744
Copper, Cu - basis materials	Fe-Kα	{222}	2,02	8,52	0,995	2,360
	Co-Kα	{400}	3,75	13,71	0,190	0,521
		{420}	2,92	11,22	0,543	1,415
	Cu-Kα	{331}	2,33	9,44	0,856	2,111
	Mo-Kα	{844}	2,45	9,82	0,583	1,458
Nickel, Ni - basis materials	Cr-Kα	{220}	1,24	5,88	1,735	3,650
	Fe-Kα	{222}	1,01	5,20	1,599	3,105
		{420}	1,48	6,61	0,727	1,630
	Cu-Kα	{331}	1,17	5,68	1,361	2,804
	Mo-Kα	{844}	1,24	5,88	0,682	1,435
Titanium, Ti - basis materials	Cr-Kα	{004}	2,60	10,80	0,413	0,992
	Fe-Kα	{203}	2,82	11,50	0,369	0,904
	Co-Kα	{114}	2,72	11,24	0,416	1,008
	Cu-Kα	{006}	2,60	10,80	0,316	0,760
		{302}	2,96	12,09	0,477	1,169

If ψ-splitted lattice strain distributions occur, the stress tensor in the measured area is tilted against the reference coordinate system. Then the shear strain components ε_{ij} ($i \neq j$) influence $\varepsilon_{\varphi,\psi}$ and $D_{\varphi,\psi}$ vs. $\sin^2\psi$. In such cases the general equ. 5 has to be applied on measured values $\varepsilon_{\varphi,\psi}$ and $D_{\varphi,\psi}$ in the directions $\psi \gtrless 0$ in the azimuth planes $\varphi = 0°$, $45°$, and $90°$[16,17,18]. Fig. 11 illustrates the evaluation steps which have to be applied in practice. From the measured $D_{\varphi,\psi}$,$\sin^2\psi$-distributions (strong curves in Figs. 11a-c) linear relationships

$$\frac{1}{2} (D_{\varphi,\psi > 0} + D_{\varphi,\psi < 0}) \text{ vs. } \sin^2\psi$$

(dashed lines in Figs. 11a-11c) and

$$\frac{1}{2} (D_{\varphi,\psi > 0} - D_{\varphi,\psi < 0}) \text{ vs. } \sin|2\psi|$$

(dashed lines in Figs. 11d-11f) are expected. The slopes of the straight lines provide sufficient data for determining the components ε_{ij} of the strain tensor. Applying the generalized Hooke's law with these data the components of the stress tensor can be calculated. As an example Fig. 12 shows the strain and the stress tensor of the surface layers of a ground steel specimen measured with CrK_α radiation.

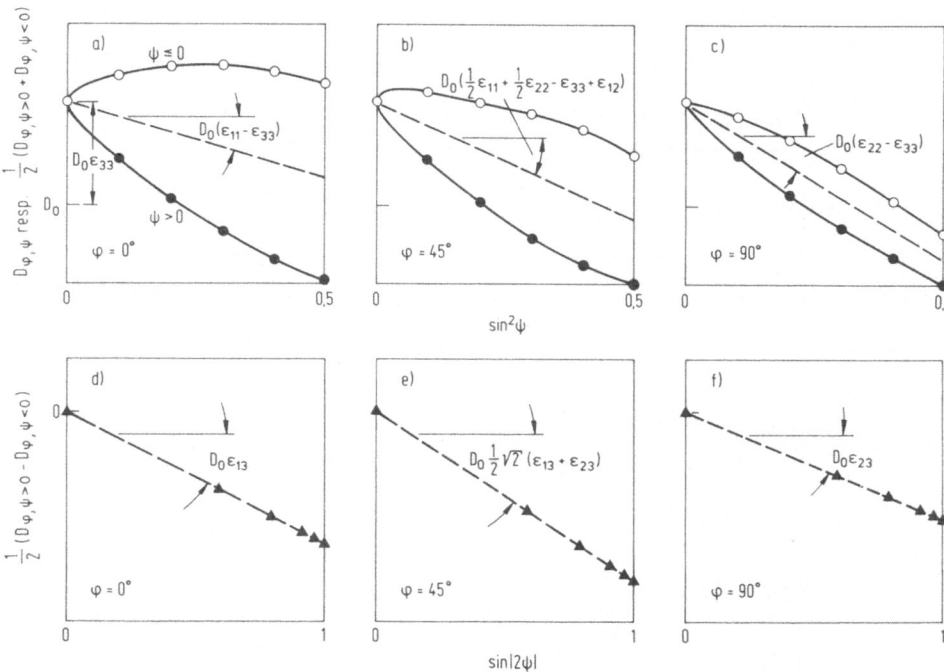

Fig. 11. Evaluation of strain tensor according to the method of Dölle and Hauk[16,17,18].

$$\begin{pmatrix} \varepsilon_{11} & 0 & \varepsilon_{13} \\ 0 & \varepsilon_{22} & 0 \\ \varepsilon_{31} & 0 & \varepsilon_{33} \end{pmatrix} \quad \begin{pmatrix} 0,95 \pm 0,2 & 0 & -4,30 \pm 0,2 \\ 0 & -6,62 \pm 0,4 & 0 \\ -4,30 \pm 0,2 & 0 & 2,80 \pm 0,06 \end{pmatrix}$$

$$\sigma_{ij} = \frac{1}{\frac{1}{2}s_2(hkl)} \left[\varepsilon_{ij} - \delta_{ij} \frac{s_1(hkl)}{\frac{1}{2}s_2(hkl) + 3s_1(hkl)} (\varepsilon_{11} + \varepsilon_{22} + \varepsilon_{33}) \right] \quad \delta_{ij} = \begin{matrix} 1\,(i=j) \\ 0\,(i \neq j) \end{matrix}$$

Figure 12. Strain and stress tensor of a ground steel specimen[19].

$$\begin{pmatrix} \sigma_{11} & 0 & \sigma_{13} \\ 0 & \sigma_{22} & 0 \\ \sigma_{31} & 0 & \sigma_{33} \end{pmatrix} \quad \begin{pmatrix} -15 \pm 7 & 0 & -74 \pm 4 \\ 0 & -146 \pm 11 & 0 \\ -74 \pm 4 & 0 & 17 \pm 5 \end{pmatrix}$$

It is recommended to apply the procedure described in Fig. 11 if ψ-splitting occurs. This method is of advantage because always linear relationships are to be assessed.

In addition to the shear stress σ_{13} (σ_{23} accordingly) the angle of tilt $\Delta\psi$ may be noted; $\Delta\psi$ is given by

$$\Delta\psi = \frac{1}{2} \arctan \left(\frac{-2\varepsilon_{13}}{\varepsilon_{11} - \varepsilon_{33}} \right) = \frac{1}{2} \arctan \left(\frac{-2\sigma_{13}}{\sigma_{11} - \sigma_{33}} \right) \tag{12}$$

If reproduceable oscillating deviations from linearity of the lattice strain distributions occur (see Fig. 8c) or lattice strain distributions that are unidirectionally curved but not ψ-splitted (see Fig. 8d), special considerations and extended measurements have to be performed. The state of-the-art in this field is reviewed by Hauk[20] in a special contribution to this conference. In many cases it is desirable if not required to determine the lattice strain distributions for several azimuths. Moreover, $D_{\varphi,\psi}$ vs. $\sin^2\psi$-distributions with a turning point and especially with additional ψ-splitting should be determined by measuring different peaks using different radiations. In this way clues can be brought to special lattice plane and penetration depth effects which may be useful for the interpretation of the state of stresses considered. Along with lattice strain measurements texture determinations have to be made, either as polefigures or at least as intensity distributions vs. $\sin^2\psi$. Additional metalphysical tests should accomplish such investigations. If principal examinations are made, besides of surface measurements always lattice strain distributions in various distances from the surface should also be determined. For this purpose surface layers have to be etched carefully. The number and thickness of the etching steps have to be coordinated with the stress gradients expected beneath the surface.

USEFUL DATA FOR XSE ON METALS AND ALLOYS

The first and most important step in XSE is to define the type of X-ray to be used. For a given material several possibilities exist as to different filter materials, different reflecting planes {hkl} with different multiplicity-factors H, different line positions 2Θ with different doublet separations ($2\Theta_{\alpha 2} - 2\Theta_{\alpha 1}$) in the stress-free state and different penetration depths. Also the calibration powders suitable for optimum diffractometer adjustments depend on the X-ray wavelengths selected. Most of the stress determinations in materials science and engineering are performed with iron-, aluminium-, copper-, nickel- and titanium-alloys. Consequently, the X-ray data necessary for measurements on pure iron, aluminium, copper, nickel and titanium should be available for orientation purposes. Corresponding data[22-24] are grouped in Table 6. The lattice constants of the calibration materials used are as follows in nm: Au 0.40786, Ag 0.40865, Cr 0.28844, Ge 0.56575, and W 0.31650.

Table 6. Data for XSE on different materials

Radiation	Wave length [nm]	Filter	Inter-ference {hkl}	Peak position $2\Theta_0$	$2\Theta_{\alpha 2} - 2\Theta_{\alpha 1}$	Penetration depth [μm], $\sin^2\psi = 0,3$		Calibration material	Inter-ference {hkl}$_C$	Peak position $2\Theta_C$
						Ω	ψ			
α- iron, ferrite, martensite;							$a_0 = 0,28665$ nm			
Cr - Kα	0,2289649	V	{211}	156,07	0,93	4,4	4,5	Cr	{211}	152,92
Fe - Kα	0,1935979	Mn	{220}	145,54	0,76	6,7	7,0	Au	{400}	143,37
Co - Kα	0,1788893	Fe	{310}	161,32	1,59	8,7	8,8	Au	{420}	157,48
Cu - Kα	0,1540501	Mono-chro-mator	{222}	137,13	0,73	1,4	1,5	Au	{422}	135,39
Mo - Kα	0,0709261	Zr	{732} +{651}	153,88	3,17	13,2	13,5	Cr[x]	{732} +{651}	150,97
Retained austenite, austenitic iron - basis materials;							$a_0 = 0,359 \pm 0,001$ nm			
Cr - Kβ	0,2084789	-	{311}	148,74	-	5,6	5,8	Au	{222}$_\alpha$	152,99[xx]
Cr - Kα	0,2289649	V	{220}	128,84	0,41	3,8	4,2	Au	{222}$_\beta$	124,59
Mn - Kα	0,2101747	Cr	{311}	152,26	0,89	5,5	5,7	W	{220}	139,81
Fe - Kα	0,1935979	Mn	{222}	138,15	0,61	6,4	6,8	Au / Cr	{400} / {220}	143,37[xx] / 143,32[xx]
Cu - Kα	0,1540501	Mono-chro-mator	{420} / {331}	147,28 / 138,53	0,99 / 0,76	1,5 / 1,4	1,6 / 1,5	Ge / Ge	{711} + {551} / {444}	152,96[xx] / 141,21[xx]
Mo - Kα	0,0709261	Zr	{844}	150,87	2,79	13,1	13,4	W[x]	{831} + {750} + {743}	149,10

Table 6 (continued)

Radiation	Wave length [nm]	Filter	Interference {hkl}	Peak position $2\Theta_0$	$2\Theta_{\alpha 2} - 2\Theta_{\alpha 1}$	Penetration depth [μm], $\sin^2\psi = 0,3$ Ω	ψ	Calibration material	Interference {hkl}$_C$	Peak position $2\Theta_C$
Aluminium, Al - basis materials;						$a_0 = 0,40491$ nm				
Cr-Kα	0,2289649	V	{222}	156,71	0,96	10,0	10,2	Au	{222}	152,99
Fe-Kα	0,1935979	Mn	{400}	145,98	0,77	15,3	16,0	Au	{400}	143,37
Co-Kα	0,1788893	Fe	{420}	162,15	1,67	20,6	20,9	Au	{420}	157,48
			{331}	148,68	0,91	19,6	20,3	Au	{331}	145,85
Cu-Kα	0,1540501	Ni	{511}+{333}	162,57	1,98	31,1	31,4	Au	{511}+{333}	157,81
			{422}	137,47	0,74	27,7	29,7	Au	{422}	135,39
Mo-Kα	0,0709261	Zr	{880}	164,51	6,41	287,6	289,9	Au[X]	{880}	159,29
			{1111}+{775}	152,50	2,99	276,9	284,2	Au[X]	{1111}+{775}	149,29
			{1042}	147,24	2,44	270,3	280,7	Au[X]	{1042}	144,53
Copper, Cu - basis materials;						$a_0 = 0,36141$ nm				
Fe-Kα	0,1935979	Mn	{222}	136,19	0,58	4,1	4,4	Au	{400}	143,37[XX]
Co-Kα	0,1788893	Fe	{400}	163,74	1,86	5,8	5,8	Au	{420}	157,48
Cu-Kα	0,1540501	Ni	{420}	144,77	0,91	8,1	8,5	Au	{422}	135,39
			{331}	136,55	0,72	7,7	8,3	Au	{422}	135,39
Mo-Kα	0,0709261	Zr	{844}	148,06	2,51	8,7	9,1	Cr[X]	{732}+{651}	150,97[XX]
Nickel, Ni - basis materials;						$a_0 = 0,35238$ nm				
Cr-Kα	0,2289649	V	{220}	133,53	0,45	2,7	3,0	Au	{311}	137,17[XX]
Fe-Kα	0,1935979	Mn	{222}	144,20	0,73	4,6	4,8	W	{310}	150,55[XX]
Cu-Kα	0,1540501	Ni	{420}	155,67	1,36	9,1	9,3	Au / Ge	{511}+{333} / {711}+{551}	157,81[XX] / 152,96
			{331}	144,65	0,91	8,7	9,1	Au	{422}	135,39
Mo-Kα	0,0709261	Zr	{844}	160,84	4,66	9,7	9,8	W[X]	{662}	155,27
Titanium, Ti - basis materials;						$a_0 = 0,29505$ nm $c_0/a_0 = 1,5873$				
Cr-Kα	0,2289649	V	{004}	155,80	0,92	1,5	1,5	Au	{222}	152,99
Fe-Kα	0,1935979	Mn	{203}	156,50	1,15	2,4	2,4	W	{310}	150,55
Co-Kα	0,1788893	Fe	{114}	154,48	1,13	2,9	3,0	W / Au	{222} / {420}	156,46[XX] / 157,48[XX]
Cu-Kα	0,1540501	Ni	{006}	161,36	7,06	4,4	4,5	Au	{511}+{333}	157,81
			{302}	148,43	1,31	4,2	4,4	W	{400}	153,54[XX]

REFERENCES

1. G. Faninger, V. Hauk, E. Macherauch and U. Wolfstieg, Recommendations for the practical use of the X-ray stress evaluation
   method (for iron base materials) (in German), Härterei-Techn.
   Mitt. 31: No. 1+2/76 109 (1976).
2. V. Hauk and E. Macherauch, The suitable performance of X-ray
   stress evaluations (XSE) (in German), in "Eigenspannungen und
   Lastspannungen, Moderne Ermittlung - Ergebnisse - Bewertung",
   Edited by V. Hauk and E. Macherauch, Härterei-Techn. Mitt. -
   Beiheft Carl Hanser Verlag München Wien: 1 (1982).
2a. M.R. James, J.B. Cohen, Study of the precision of X-ray stress
    analysis, Adv. X-Ray Anal. 20:291 (1977).
3. E. Macherauch and U. Wolfstieg, A modified diffractometer for
   X-ray stress measurement, Adv. X-Ray Anal. 20:369 (1977).
4. K. Sagel, Tables for X-ray structure analysis (in German),
   Springer-Verlag Berlin-Göttingen-Heidelberg: 123 (1956).
5. Residual stress measurement by X-ray diffraction, SAE J 784a:
   (1971).
6. H.K. Tönshoff, E. Brinksmeier and H.H. Nölke, Application of
   the cross correlation method in the X-ray residual stress
   measurement (in German), Z. Metallkde. 72: 349 (1981).
7. U. Wolfstieg, The symmetrizing of non symmetrical interference
   lines using special slits (in German), Härterei-Techn.Mitt. 31:
   No. 1 + 2/76   23 (1976).
8. V. Hauk and W.K. Krug, Computerized separation and symmetrizing
   of K_α-doublets in X-ray stress measurements (in German), Ma-
   terialprüf. 25: 241 (1983).
9. J. Hoffmann and E. Macherauch, A new method for strain determination by means of X-rays with continuous registration of
   line peaks vs. $\sin^2\psi$ (in German), in 2.: 25 (1982).
10. E. Macherauch and P. Müller, The $\sin^2\psi$-method of X-ray stress
    measurement (in German), Z. angew. Physik 13: 305 (1961).
11. V. Hauk, Residual stresses after plastic strain by tension
    (in German), Z. Metallkde. 46: 33 (1955).
12. F. Bollenrath, V. Hauk and E.H. Müller, Calculation of poly-
    crystalline elastic constants from data of single crystals (in
    German), Z. Metallkde. 58: 76 (1967).
13. V. Hauk and H. Kockelmann, X-ray elastic constants for stress
    evaluation (in German), "Eigenspannungen", ISBN 3-88355-027-2
    Deutsche Ges. f. Metallkde. Oberursel: 241 (1980).
14. V. Hauk, X-ray elastic constants (XEC) (in German), in 2.:
    49 (1982).
15. E. Macherauch and P. Müller, Evaluation of X-ray elastic con-
    stants of cold-strained Armco-iron and CrMo-steel (in German),
    Arch. Eisenhüttenwes. 29: 257 (1958).
16. P.D. Evenschor and V. Hauk, On nonlinear distributions of lat-
    tice spacings in X-ray strain measurements (in German), Z.
    Metallkde. 66: 167 (1975).
17. H. Dölle and V. Hauk, X-ray stress evaluation of residual stress
    systems having general orientation (in German), Härterei-Techn.
    Mitt. 31: 165 (1976).

18. V. Hauk, W.K. Krug, G. Vaessen and H. Weisshaupt, The residual strain-/residual stress-state after grinding (in German), Härterei-Techn. Mitt. 35: 144 (1980).

19. V.M. Hauk, R.W.M. Oudelhoven and G.J.H. Vaessen, The state of residual stress in the near surface region of homogeneous and heterogeneous materials after grinding, Metallurg. Trans. 13A: 1239 (1982).

20. V. Hauk, Stress evaluation on materials having non-linear lattice strain distributions, Adv. X-Ray Anal. 27: in press (1984).

21. R. Glocker, Materialprüfung mit Röntgenstrahlen, Springer-Verlag Berlin-Heidelberg-New York, 5. Aufl.: 139 and 434 (1971).

22. W.B. Pearson, A handbook of lattice spacings and structures of metals and alloys, Pergamon Press 2: (1967).

23. Landolt-Börnstein, Data and functions (in German), Springer-Verlag Berlin-Göttingen-Heidelberg 1: (1950) and (1955).

STRESS EVALUATION ON MATERIALS HAVING NON-LINEAR LATTICE STRAIN

DISTRIBUTIONS

Viktor M. Hauk

Institut für Werkstoffkunde, Rhein. Westf.
Technische Hochschule, D-5100 Aachen
Bundesrepublik Deutschland

ABSTRACT

The state of the art of stress evaluation on materials having non-linear lattice strain distributions is presented.

New results on heterogeneous materials with measurements conducted on both phases of the material show compensation of the shear stress components σ_{13} in ground surface layers of $(\alpha+\beta)$ brass. There is only the compensation of normal components σ_{11} of $(\alpha+\gamma)$ steel after plastic straining.

The fundamental aspects and the evaluation of macro- and micro-residual stresses on materials having preferred orientation are broadened. The use of Mo-K_α-radiation shows linear lattice strain distributions, as a result of minimizing the influence of micro residual stresses causing oscillations. The interplanar distance- or strain-polefigure shows similarities with the intensity polefigure.

The knowledge of the theoretical influence of stress distribution with depth from the surface of the material is extended. The experimental procedure should use either different radiations having different penetration depths or a low-penetrating radiation in combination with removal of surface layers.

INTRODUCTION

The X-ray method of measuring residual stresses in its modern state allows measurement even on sophisticated materials of many phases, with preferred orientation of the crystals, after plastic deformations and surface treatments. This progress was made due to the

improved measuring techniques using fully automatic diffractometers
and computerized evaluation of the measuring results. The theoretical
aspects have been broadened.

The basic types of lattice strain distributions are shown in
fig. 1. The measured Bragg'angle 2Θ is converted into the interplanar distance D (100), that depends on the direction to the sur-
face of the testpiece (φ,ψ); φ is the azimuth and $-45^0 \leqslant \psi \leqslant 45^0$.
The $D_{\varphi,\psi}$ vs. $\sin^2\psi$-dependence is in most of the actual cases linear,
may have ψ-splitting, i.e. two branches of an ellipse for $\psi \lessgtr 0$, may
have oscillations for materials with texture and shows up a curvature
if stress or micro structure gradients are present in the near sur-
face layers [1,2]. Combinations of these basic types of $D_{\varphi,\psi}$ vs. $\sin^2\psi$
dependences are possible. An example is the lattice strain distribu-
tion, in a direction $0<\varphi<90^0$ to the rolling direction of a material
having strong preferred orientation. The definitions of the angles
φ,ψ,ζ and the cut of the test piece are demonstrated in fig. 2.

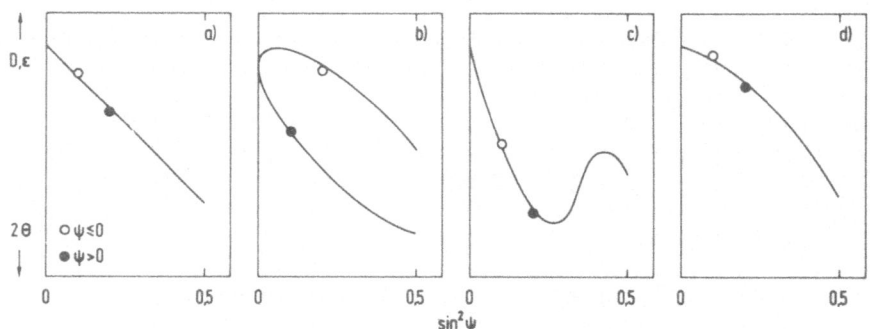

Figure 1. Basic types of lattice strain distributions.

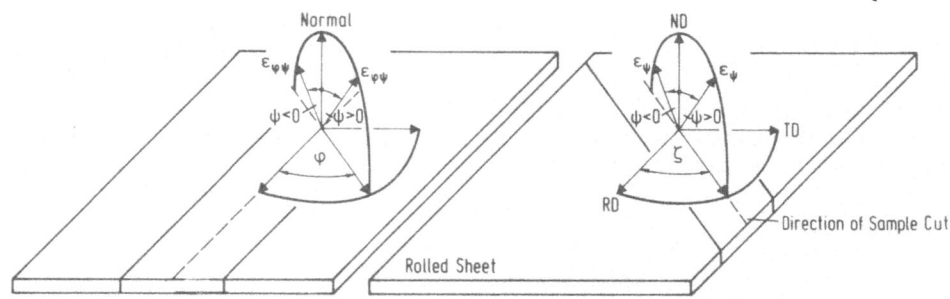

Figure 2. Definitions of specimen and measurement systems and of
          the angles φ,ψ,ζ.

The classic formula, written in terms of strain components ε_{ij}, as the result of measurement and the geometrical interpretation of the formula were given by [3,4,5]. Under the assumption of Hooke's law the stress tensor can be calculated from the strain tensor by using the X-ray elastic constants (XEC) s_1 and $\frac{1}{2} s_2$ [6,7]. The details are in the paper by Hauk and Macherauch in this conference proceedings.

One problem should be pointed out, that is the existence of components σ_{i3} (i = 1, 2, 3) in the surface-near region of a material having residual stresses. The existence and the value of σ_{33} can only be determined by very accurate measurement of $D_{\varphi,\psi}$ with a diffractometer that is aligned mechanically and calibrated with regard to the absolute value. The formulae for the dependence of D_0 on σ_{33} and for the evaluation of D_0 are shown in fig. 3; D_0 is the lattice constant in the stress free condition and ψ^* is the strainfree direction.

$$\sin^2\psi^* = \frac{-s_1}{\frac{1}{2}s_2}\left[1 + \frac{\sigma_{22} - \sigma_{33}}{\sigma_{11} - \sigma_{33}} + \frac{\left(3 + \frac{\frac{1}{2}s_2}{s_1}\right)\sigma_{33}}{\sigma_{11} - \sigma_{33}}\right] \quad ; \quad \varphi = 0°$$

$$\sin^2\psi^* \approx \frac{-s_1}{\frac{1}{2}s_2}\left(1 + \frac{\sigma_{22} - \sigma_{33}}{\sigma_{11} - \sigma_{33}}\right) \text{ when } |\sigma_{11} - \sigma_{33} + \sigma_{22} - \sigma_{33}| \gg 2|\sigma_{33}|$$

$$D_0 = \frac{D_{\varphi,\psi = 0}}{1 + s_1(\sigma_{11} + \sigma_{22} + \sigma_{33}) + \frac{1}{2}s_2\sigma_{33}}$$

Figure 3. Formulae to calculate D_0, the lattice constant of the stress free state.

An example for the evaluation of the dependence of the stress tensor with the distance from the surface is demonstrated in the fig. 4, 5 and 6. The surface preparation of the steel sample was rolling and pressing [8,9]. The lattice strain distributions in the azimuths $\varphi = 0$, 45 and 90° show ψ-splitting and linearity; fig. 4. The stress distribution with the distance from surface is demonstrated in fig. 5, the data are not corrected due to the removal of surface layers. The shear components, different in sign in the azimuths $\varphi = 0$ and 90°, decrease rapidly. The normal components σ_{11} and σ_{22} follow the results expected from theory of Hertz' pression. As fig. 6 points out, there is no dependence of the lattice constant on the depth, that means also that σ_{33} is zero. The line breadth has a minimum in the depth where the stresses are zero.

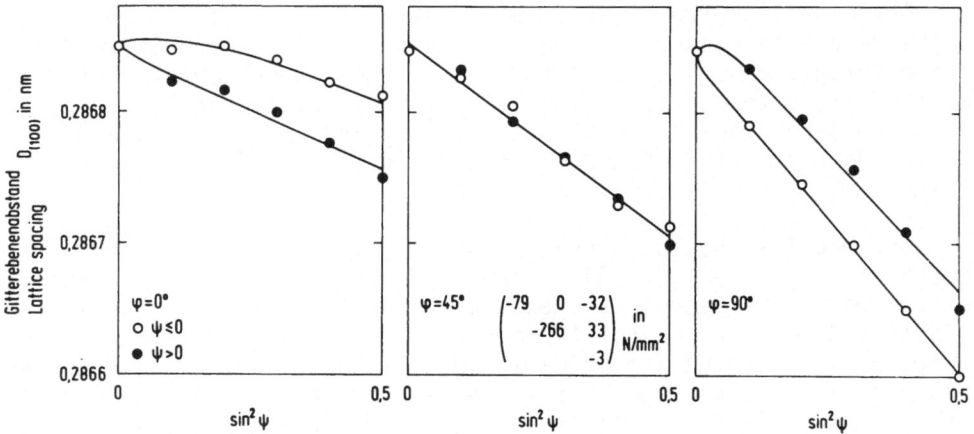

Figure 4. $D_{\varphi,\psi}$ vs. $\sin^2\psi$ distribution of a steel specimen over-rolled and pressed; Cr-Kα-radiation, $\{211\}$ peak [9].

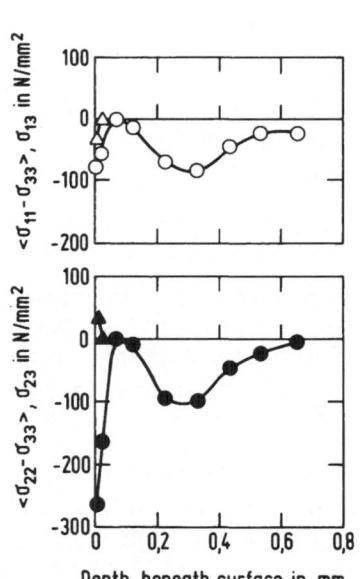

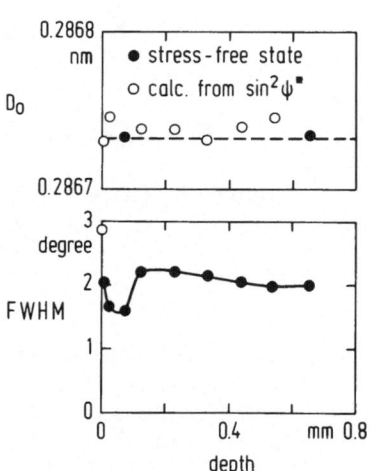

Figure 6. D_O and Full Width H. Max. distribution with depth from surface of a steel specimen overrolled and pressed [9].

Figure 5. Stress distribution with depth from surface of a steel specimen overrolled and pressed [9].

ψ-SPLITTING IN HETEROGENEOUS MATERIAL

Grinding of the surface results in ψ-splitting and this can be related to shear stress having opposite sign to the deforming tangential force [5,10,11]. A wide program is under way to study the amount, sign and depth distribution of the shear stresses in different phases of heterogeneous materials with and without preferred orientation in one or both phases to be investigated. Grinding is used as stress raiser and not as industrially used surface treatment.

One example deals with α- and β-phases of a brass sheet. Fig. 7 shows the polefigures of the α- and β-phases as delivered and fig. 8 after grinding [8,12]. The texture as rolled has been destroyed by grinding in the surface layers. The grinding results in ψ-splitting in both phases (fig. 9·and 10), the shear components and the normal components σ_{11} and σ_{22} have different values and signs in the different phases and groups of crystallites.

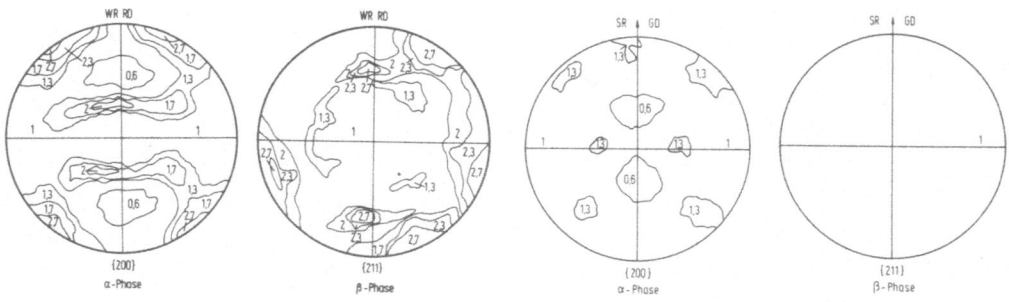

Figure 7. Polefigures of the α- and β-phase of a rolled brass sheet [8,12].

Figure 8. Polefigures of the α- and β-phase of a rolled and ground brass sheet [8,12].

The stress compensation formulae are given in fig. 11 [13,8]. The symbols are as follows:

q   volume percent of the α-phase

P   specimen thickness

σ^{I} macro residual stress

σ^{II} micro phase specific residual stress (due to peak shift)

z   depth beneath the surface

The normal stresses σ_{11} and σ_{22} compensate within the cross-section of the specimen or within the two phases. The shear components and σ_{33} should compensate in each depth. Taking into account the amount of the α- and β-phase the compensation for the two pairs of {hkl} peaks is fairly well fulfilled [43]. There exists also the compensation of the shear components in the phases Fe and Fe_3C [14,15].

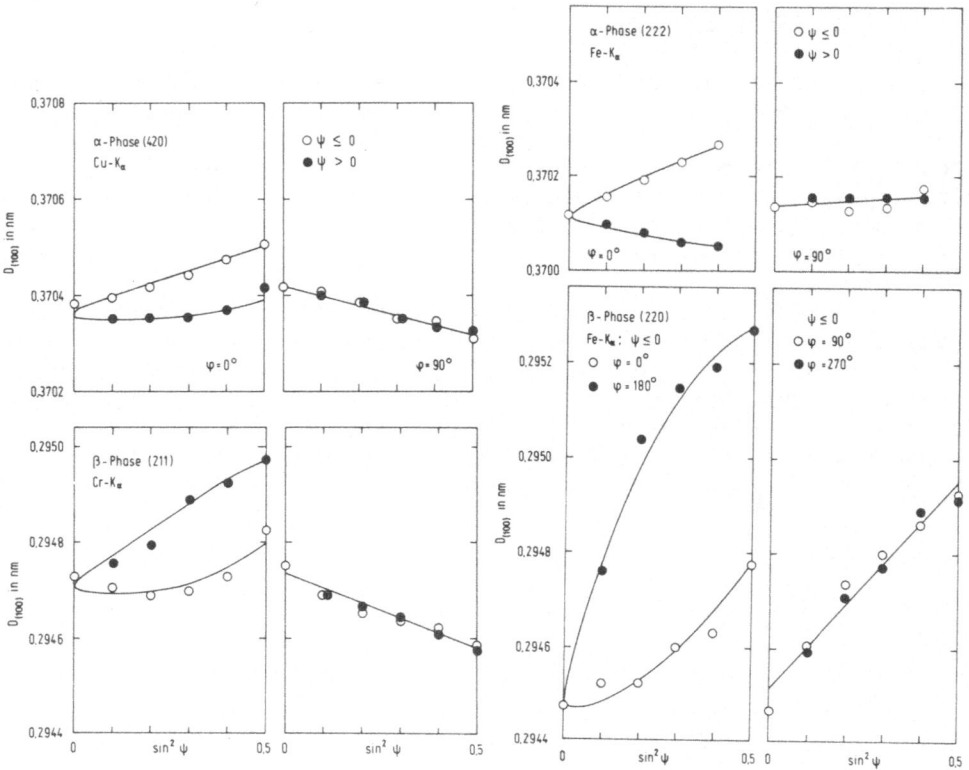

Figure 9. $D_{\varphi\psi}$ vs. $\sin^2\psi$ distribution of the α- and β-phase of a rolled and ground brass sheet [8,12].

Figure 10. $D_{\varphi\psi}$ vs. $\sin^2\psi$ distribution of the α- and β-phase of a rolled and ground brass sheet [8,12].

Normal components:

$$q\sigma_\alpha^I + (1-q)\sigma_\beta^I = \sigma^I$$

$$q\sigma_\alpha^{II} + (1-q)\sigma_\beta^{II} = 0$$

$$q\int_0^P \sigma_\alpha^{I+II} dz + (1-q)\int_0^P \sigma_\beta^{I+II} dz = 0$$

Shear components and σ_{33} in each depth:

$$q\sigma_{i3,\alpha}^{II} + (1-q)\sigma_{i3,\beta}^{II} = 0 \quad (i=1,2,3)$$

Figure 11. The stress compensation conditions of normal- and shear components [13,8].

The two-phase steel, 50 per cent α-Fe Ferrite and 50 per cent γ-Fe Austenite was tested in a similar manner, but with the addition of plastic strain [16,12]. The ground test pieces show no preferred orientation. The D vs. $\sin^2\psi$ dependences of the Ferrite and Austenite point out the following: high tensile stress components in both phases and in both azimuths parallel and perpendicular to the grinding direction, fig. 12. There is no compensation of the shear components since they have the same sign in both phases in the parallel-to-grinding direction.

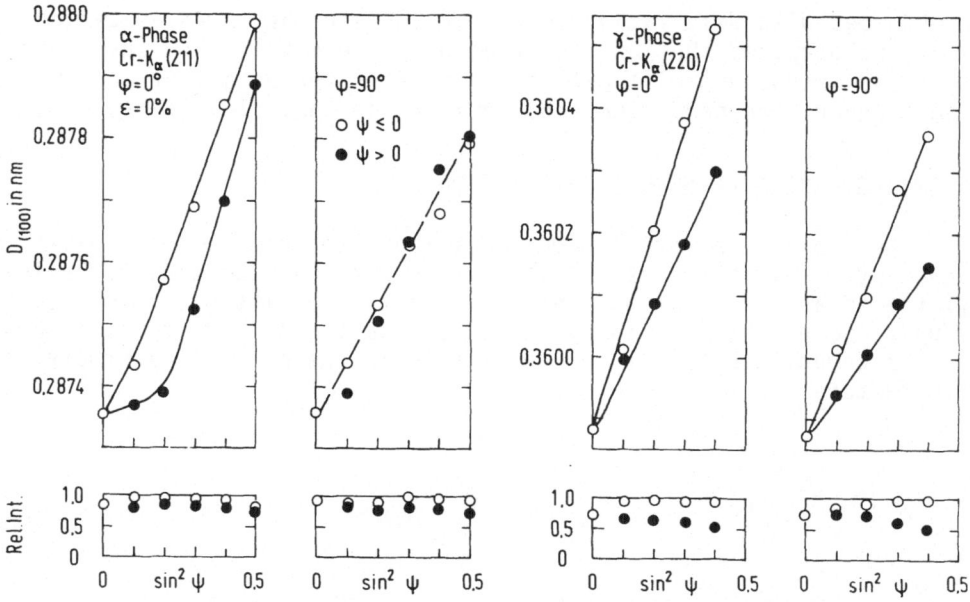

Figure 12. $D_{\varphi,\psi}$ vs. $\sin^2\psi$ distribution of the α- and γ-phase of a ground two phase steel specimen [16,12].

New is the appearance of a ψ-splitting in the direction perpendicular to grinding.

The question was the study of superposition of this stress state and the stresses caused by plastic strain (micro residual stresses due to deformation) [12]. The fig. 13 and 14 show the effect of plastic strain: the shear components vanish, the normal stresses get smaller and, in one case, change sign, but the σ_{11} stress of Austenite increases. The amount of the tensile stress in Austenite is higher than the tensile strength of the ($\alpha+\gamma$) steel. The result of the total investigation is shown in fig. 15. The deformation residual stresses [17] caused by plastic uniaxial strain compensates itself within the Ferrite- and Austenite-phases and the phase specific stresses parallel to the direction of the plastic strain superimpose to the residual stresses after grinding. The shear components and the normal components perpendicular to the grinding direction decrease with plastic strain.

The study of the state of stress in the near surface region of heterogeneous material that has undergone surface treatment and plastic deformations needs a great expenditure of measurement. The information of preferred orientation (pole-figures), line breadth, D_0 and σ_{33}, strain- and stress-tensors of at least two phases and their

dependences with deformation and depth beneath the surface must necessarily be obtained. But on the other hand the results and their evaluation and assessment will help to understand the behaviour of work
pieces under load conditions and to increase their life time.

MATERIALS HAVING PREFERRED ORIENTATION

There exists a lot of papers dealing with this subject, published over several years [2,18,19]. It is not the intention to review
all the ideas, calculations and experimental results put forward. It
is quite sure, that the X-ray stress evaluation method is not always
non ambiguous, successful and applicable in the case of materials
with texture.

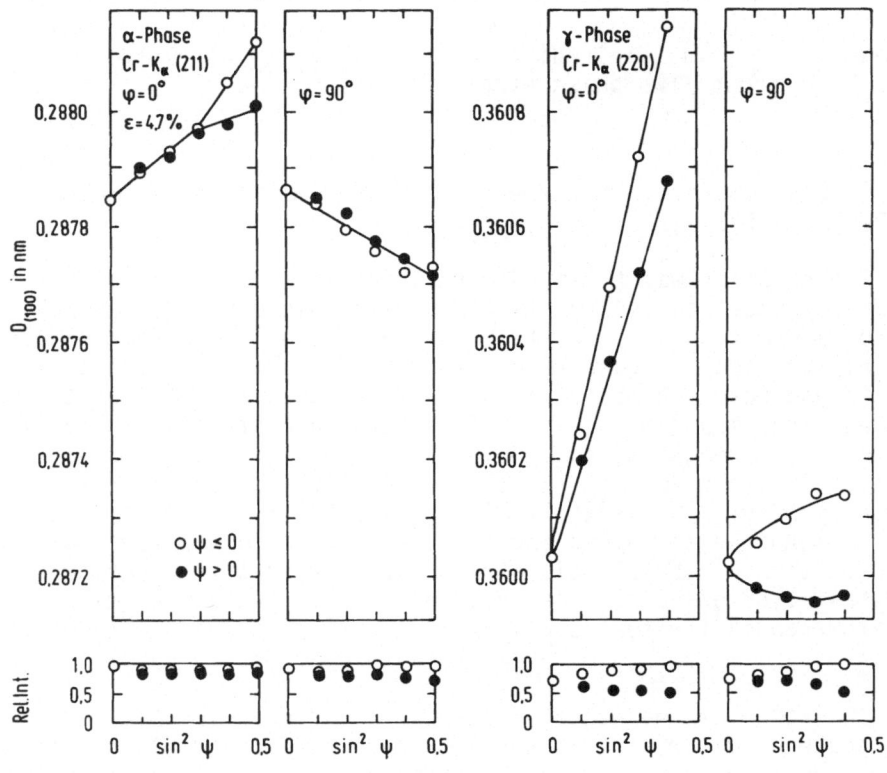

Figure 13. Same as fig. 12, but additionally 4.7 per cent
elongated [16,12].

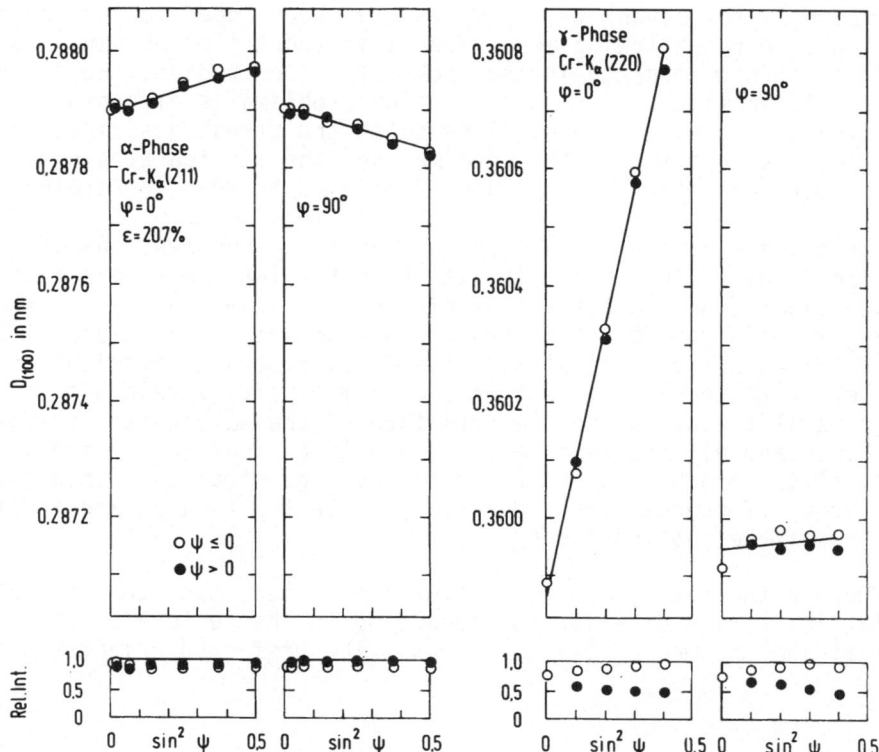

Figure 14. Same as fig. 12, but additionally 20.7 per cent elongated [16,12].

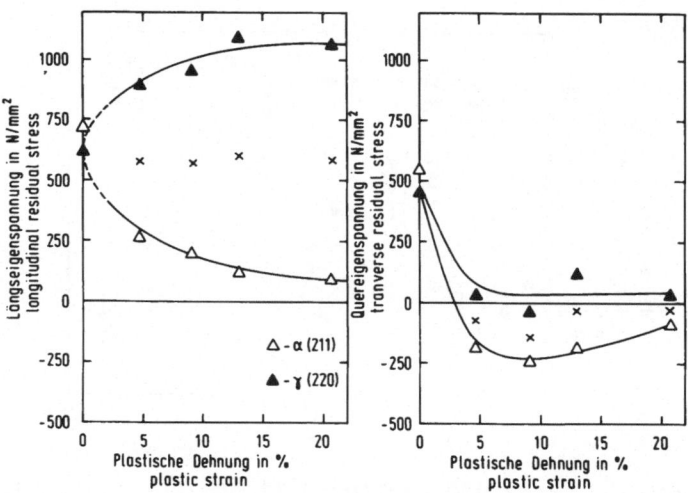

Fig. 15. Stress distribution of a ground and plastically strained two phase steel specimen [16,12].

Recently the synchrotron sources have been added to the experimental procedures allowing measurements in short time on many planes {hkl} with great accuracy in the back reflection region. The use of neutron diffraction on stress evaluation problems is also new. According to the higher penetration depth the stress distribution over the cross section of the specimen and the lattice strain distribution over the entire $\sin^2\psi$-range from 0 to 1 can be determined.

An example of the comparison between X-ray and neutron-diffraction is given in fig. 16 [20]. It shows the lattice strain distribution measured on the {211} peak of the textured steel St 52 after rolling with 75 per cent thickness reduction and an additional 4 per cent longitudinal plastic strain. The lattice strain distribution parallel to the rolling direction show compressive strain with significant oscillations. Since the variation of the wavelength is easy to achieve and all the texture poles are taken over the entire $\sin^2\psi$-range, it is possible to determine the average stresses within every {hkl} group of crystallites. Our study is in the final stage and the results will be published soon.

During the last years many efforts have been made to calculate the lattice strain distribution according to the texture of the material and the macro residual stress. The preferred orientation

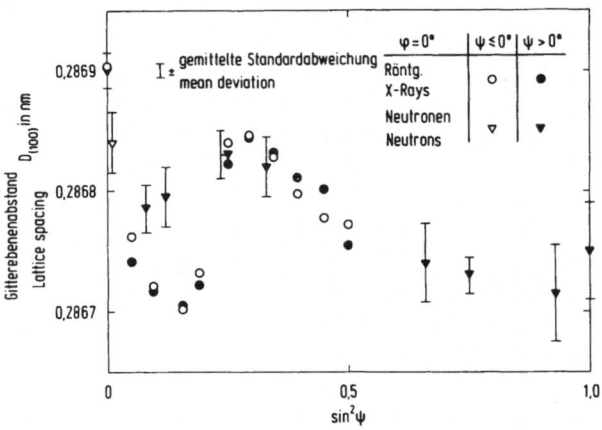

Figure 16. X-ray and neutron-diffraction of a rolled St 52 specimen (75 per cent wall thickness reduction) and additionally 4 per cent plastically strained [20].

can be described by the ideal orientations, by the inverse polefigure,
by the orientation distribution function (ODF) and the elastic behaviour can be explained by the orientation dependent X-ray elastic
constants (XEC) $_{\varphi,\psi,\zeta}$ [21-26]. All these methods do not deal with the
effect caused by micro residual stresses [19]. But there is a great
variety of oscillations due to micro residual stresses of different
origin [27-31]. The effect of micro residual stresses is obvious if the
D vs. $\sin^2\psi$-dependences are not linear for peaks {h00} and {hhh}
which is asked for by theory [32,21]. The task of today must therefore
be: to minimize the influence of micro residual stresses to measure
the macro residual stress or to evaluate the micro residual stresses
in the absence of macro residual stresses. The micro residual phase
specific stresses must also be taken into account.

One method to minimize the influence of the micro residual
stresses on the lattice strain distribution is to use the peaks of
as much as possible crystallites. In practice, to use a deep penetrating wave length, a peak with a high multiplicity factor and two
or even three superimposing planes. The following gives the effective
combination of radiation, plane and material: Mo-K$_\alpha$, {732+651},
iron and steels; Cu-K$_\alpha$, {511+333} [21], aluminum alloys; Mo-K$_\alpha$,
{911+753}, nickel alloys [33].

The following figures demonstrate impressively the use of
Mo-K$_\alpha$-radiation to get linear lattice strain distribution with
steel specimens [34]. The specimens were cut 0, 60 and 90° to the
rolling direction. Fig. 17 shows the results of measurements with
Cr- and with Mo-K$_\alpha$-radiation. The remarkable large oscillations on
{211} peak vanish when using the {732+651} peak. The texture of the
material was strong, therefore no intensity in the neighborhood of
$\psi=0$ could be detected. Fig. 18 and 19 show also linearity when
superimposing elastic strain to the textured steel specimens. A
calculation of the XEC for {732+651} peak using the inverse polefigure of the textured steel pointed out a linear behaviour as a
quasi isotropic case.

Fig. 20 demonstrates and summarizes a proposed method to evaluate residual stresses referring to the different peaks of steels to
be measured [19]. First the case of measuring the {732+651} peak where
linearity should be observed. The value of lattice strain caused by
micro residual stresses could be measured by using a small thin
specimen where macro residual stresses could not be present. Measuring {hoo} and {hhh} peaks and observing no linearity, then micro
residual stresses are present. Again the evaluation of the lattice
strain distribution of a small thin specimen helps to clarify the
amount of micro- and macro-residual stresses. In the general case of
measuring on a {hkl} peak, the procedure is not simple and the result
of measurement has to be fitted by calculation of the strain distribution taking into account the real state of texture and the macro
residual stress as parameter.

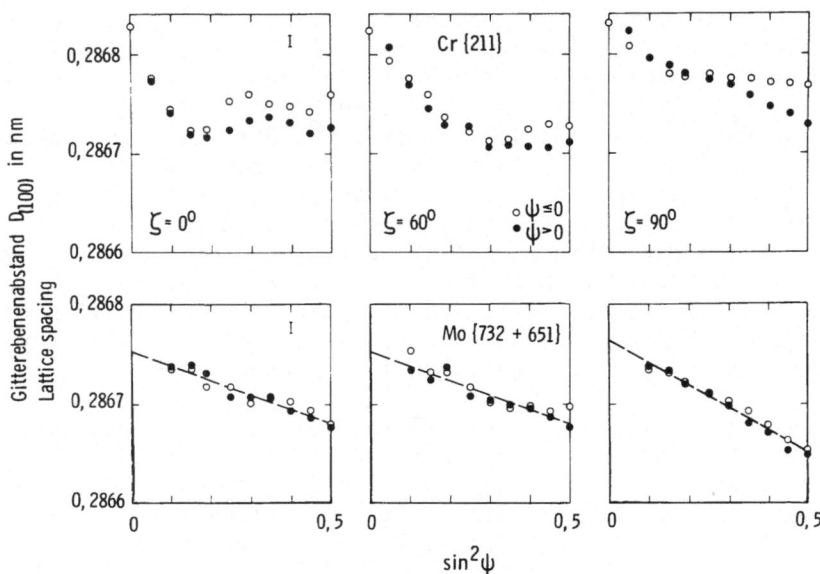

Figure 17. $D_{\varphi,\psi}$ vs. $\sin^2 \psi$ distributions of Cr $\{211\}$ and Mo $\{732+651\}$ peaks of a rolled St 52 specimen (75 per cent wall thickness reduction) [34].

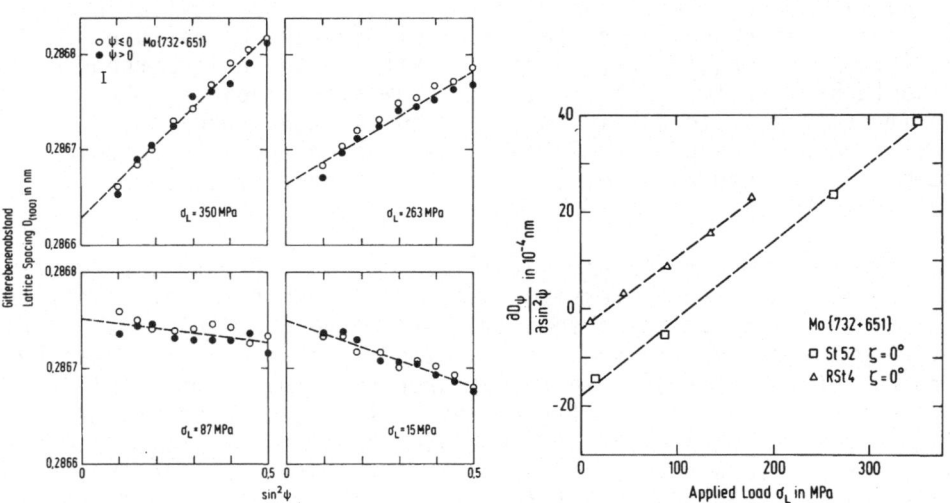

Figure 18. $D_{\varphi,\psi}$ vs. $\sin^2\psi$ distributions of Mo $\{732+651\}$ peaks of a specimen as fig. 17 but elastically strained [34].

Figure 19. Evaluation of X-ray elastic constant on rolled steel specimens [34].

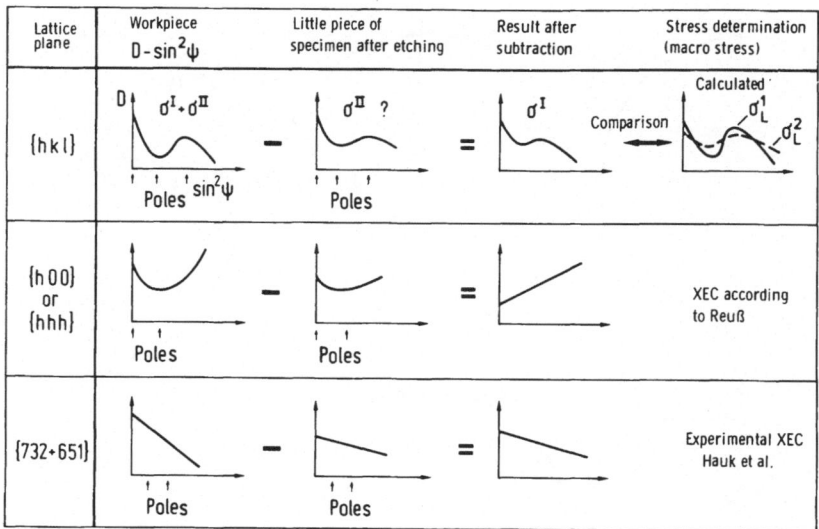

Figure 20. Evaluation of macro- and micro-residual stresses on steels having preferred orientation[19].

From the foregoing it is obvious that the state of residual macro- and micro-stresses cannot be fully described by biaxial surface stress components or by triaxial tensors. As pointed out three years ago [35] the polefigure of the lattice parameter or of the strain allover the entire orientation φ, ψ evaluated on different peaks offers the possibility to correlate the state of macro- and micro-residual strain with the physical and strength properties of the material [19]. The same procedure has taken place years ago with the usual intensity polefigures.

The fig. 21 shows polefigures [34] determined on the peaks $\{211\}$ and $\{220\}$ of the 75 per cent rolled St 52 steel by measuring $D_{\varphi, \psi \leq 45^o}$ calculating thereof ε (with $D_0 = 0,28674$ nm) and the intensity of the peaks. It is interesting to note, that the intensity and the strain polefigure look the same. That may have an importance on the formation of textures.

Summarizing the chapter on textured materials: Allthough the method to minimize or to average the influence of oscillations caused by micro stresses is the first choice to evaluate macro residual stresses in materials having preferred orientation, methods to evaluate the state of stresses in differently orientated groups of crystallites should be studied further. The strain polefigure should be used to describe the state of strain in the material. The assessment

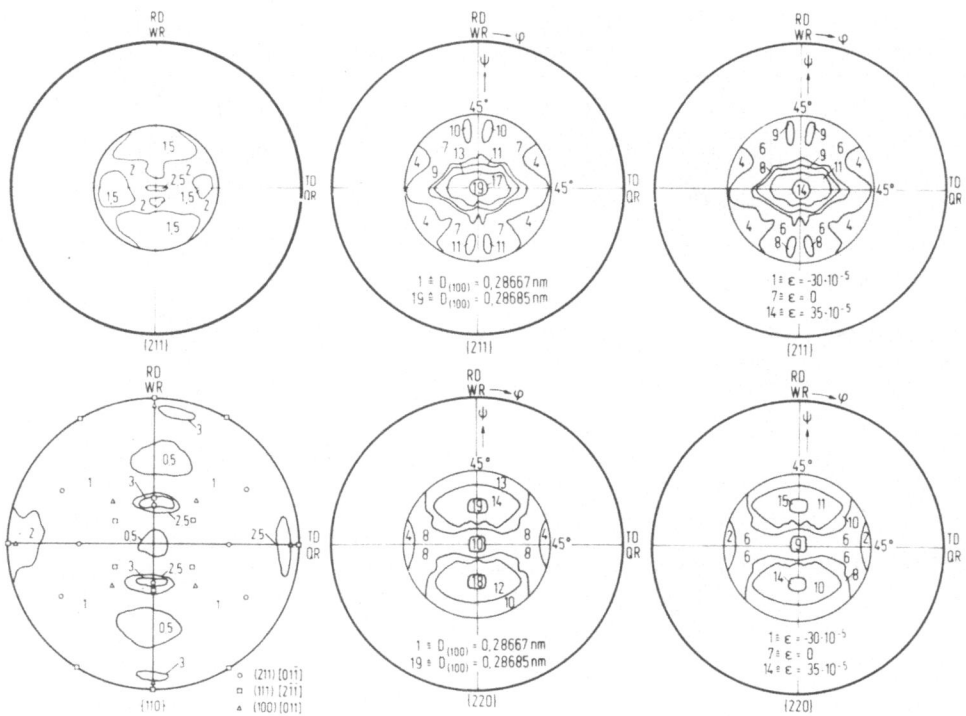

Figure 21. Intensity-, lattice constant- and strain-polefigures of
          a rolled St 52 specimen [34].

of micro residual stresses with respect to the load capability of
materials asks for detailed results.

EFFECT AND MEASUREMENT OF STRESS GRADIENTS

During the recent years steep stress gradients due to surface
machining have been investigated both theoretically and experimentally. Originally ψ-splitting and curvature resulting from gradients had not been clearly distinguished by other investigators.
Furthermore, the experimental measuring scatter has not taken duly
into consideration as regards the assessment of gradients by calculation.

The influence of multiaxial depth dependent states of residual
stress on X-ray stress evaluation was given by Dölle and Hauk [36]
taking into account the physical boundary conditions. The influence
of gradients of the components σ_{11} and σ_{22} as well as the components
perpendicular to the specimen surface σ_{33} and the shear stress σ_{13}
have been investigated. The result being that even the steepest

gradients cannot be assessed by one D vs. $\sin^2 \psi$-dependence due to the variability and inadequate measuring technique. The deviation of the measured mean stress from the surface stress $\sigma_{z=0}$ has been investigated basically in recent years but the deviation has become particularly evident for the steep gradients which are known today.

As examples are given in fig. 22 the D vs. $\sin^2 \psi$-dependences for various gradients under the assumption of a surface stress of -500 N/nm^2 and for the other data given in the figure on the left-hand side for iron, the right-hand side for aluminum. The calculation is valid for a ψ-diffractometer and Cr-K$_\alpha$-radiation. Even the steepest gradients do not show measurable curvature in the usual range of $\sin^2 \psi \leq 0.5$ (0.6).

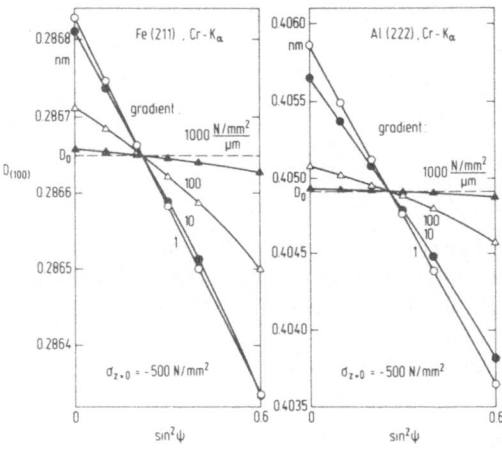

Figure 22. $D_{(100)}$ vs. $\sin^2\psi$ distributions of the stress -500 N/nm^2 on the surface and different gradients perpendicular to the surface [39].

Comments are in order on two papers that were published recently. Noyan [37] is engaged in answering the question, how the surface stress can be exactly calculated from one D vs. $\sin^2 \psi$-dependence if there are gradients of σ_1, σ_2 and σ_3. The influence of a σ_3 gradient is emphasized, although the value of σ_3 seems to be fairly high. In the paper by Sprauel and others [38] however, σ_{33} is taken as zero, and the derivative of the stress tensor and thus the gradient, is evaluated experimentally by using three radiations. Furthermore the values of the shear components on surfaces machined with tangential forces are assumed to be constant, which, however, is not true.

As many experimental results show that the decrease of the residual stress starts at a certain depth z=r, D vs. $\sin^2 \psi$-distributions were calculated with the stress gradient in σ_1 and r as parameters. In fig. 23 the results for z=1, 5, 10 and 50 µm are given, for iron as well as for aluminum. However the small curvature cannot be measured. The difference of the D-value of the D vs. $\sin^2 \psi$-dependence at $\sin^2 \psi$ = 0.3 from a straight line through the points at $\sin^2 \psi$ =0 and $\sin^2\psi$=0.6 respectively, is shown in fig. 24 in dependence of r and the gradient. Calculations were made for surface stresses of -500 and -1000 N/nm^2 for Cr- and Mo-K$_\alpha$-radiation. When the uncertainty of measurement for relatively broad peaks is taken into account, it becomes evident that only few cases can be determined by

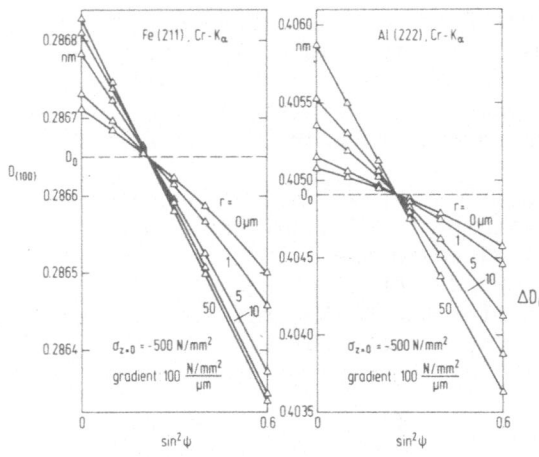

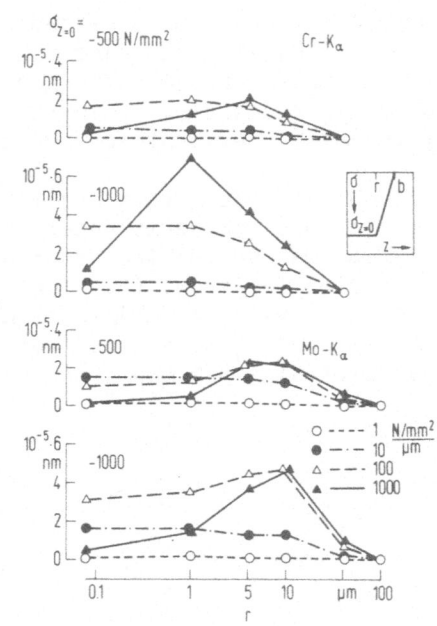

Figure 23. $D_{(100)}$ vs. $\sin^2\psi$ distributions of the stress -500 N/nm^2 on the surface a gradient of 100 N/nm^2 per µm and different depths from the surface where the stress gradient is effective [39].

Figure 24. The difference ΔD between curvature and straight line in $D_{(100)}$ vs. $\sin^2\psi$ dependence according fig. 22 and fig. 23 for Fe [39].

measurement. Besides, these cases cannot be realized without ambiguity.

It is concluded that the curvature is not influenced by the gradient alone but that the gradient and the starting point r of the stress decrease must be related to the penetration depth of the X-rays. Comparing figs. 22 and 23 it is seen that both variation in gradients and variation in r can alter the D vs. $\sin^2\psi$ distribution. Therefore it is not possible to deduce the stress on the surface, the starting point r and the stress gradient unambiguously.

Therefore, a method to determine the stress distribution as function of the depth should make use of the measuring results of the penetration depths of the various K_α-radiations that are available. The distribution of real stress with depth can be determined from the measured mean stresses by means of known mathematical methods [40,41,42]. An example of this procedure is given in fig. 25 [42]. Here the mean stresses measured with various radiations as a function of penetration depth were converted into real stress distribution and compared with

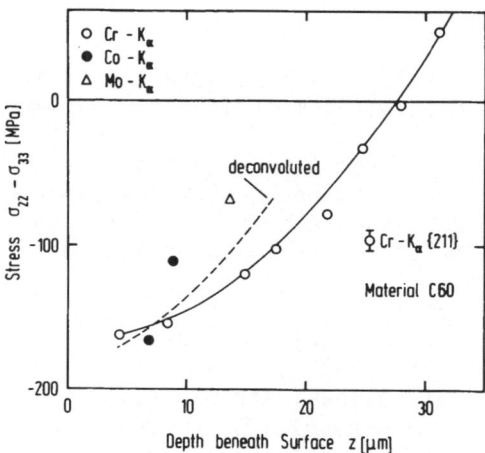

Figure 25. Stress distribution in a ground steel specimen evaluated with different radiations or with Cr-radiation and removal of surface layers [42].

the values determined after etching and the use of the relatively low-penetrating Cr-K_α-radiation. The correspondence can be improved by a correction caused by the removal of stress layers.

The procedure to evaluate stress gradients with the depth from the surface should be as follows: to measure the D vs. $\sin^2\psi$ dependence with different wavelength i.e. different penetration depths or/ and to use a low-penetrating radiation after removal of surface layers. The influence of the absorption of the X-rays should be deconvoluted. If the etching procedure is used the stress profile should be corrected according to the removal of the surface layers. Gradients are observed on normal stress components σ_{11} and σ_{22}. Gradients of the shear components σ_{13} and σ_{23} can easily be measured by the ψ-splitting in the appropriate azimuth. The existence and a gradient of σ_{33} should be checked very carefully. The lattice constant in the stress-free condition D_0 should be measured as function of the depth from surface up to really stress-free region. Different materials-physical tests have to be applied in order to decide wether a σ_{33}- or D_0-profile as a function of the depth is present. Problems due to several phases and preferred orientation are being investigated. Possible relaxation of stresses σ_{i3} after etching should be taken in consideration.

ACKNOWLEDGEMENTS

Thanks are expressed to the organizers of the conference for inviting the author to present this paper. The Deutsche Forschungsgemeinschaft has granted the support of the journey.

REFERENCES

1. H. Dölle and V. Hauk, System of possible lattice strain distri-
    butions on mechanically loaded metallic materials (in German),
    Z. Metallkde. 68: 725 (1977).
2. V. Hauk and G. Vaessen, Evaluation of non-linear lattice strain
    distribution (in German), in: "Eigenspannungen und Lastspan-
    nungen, Moderne Ermittlung-Ergebnisse-Bewertung", Edited by
    V. Hauk and E. Macherauch, HTM-Beiheft Carl Hanser Verlag
    München Wien: 38 (1982).
3. P. D. Evenschor and V. Hauk, On non-linear distributions of
    lattice interplanar spacing at X-ray strain measurements
    (in German), Z. Metallkde. 66: 167 (1975).
4. H. Dölle and V. Hauk, Evaluation of residual stress systems
    arbitrarily oriented by X-rays (in German), HTM 31: 165 (1976).
5. V. Hauk, W. K. Krug, G. Vaessen and H. Weisshaupt, The residual
    strain-/stress-condition after grinding (in German),
    HTM 35: 144 (1980).
6. V. Hauk and H. Kockelmann, X-ray elastic constants of ferritic,
    austenitic and hardened steels (in German), Arch. Eisen-
    hüttenwes. 50: 347 (1979).
7. V. Hauk, X-ray elastic constants (XEC) (in German),
    same as 2: 49 (1982).
8. V. Hauk, P. J. T. Stuitje and G. Vaessen, Presentation and
    compensation of residual stresses in machined surface layers
    of heterogeneous materials (in German), same as 2: 129 (1982).
9. E. Broszeit, V. Hauk, K. H. Kloos and P. J. T. Stuitje, to be
    published.
10. H. Dölle and J. B. Cohen, Residual stresses in ground steels,
    Metallurgical Transaction 11 A: 159 (1980).
11. H. Krause and H. Jühe, Contribution to X-ray evaluation and
    assessment of systems of residual stresses of friction
    loaded surfaces (in German), in: "Eigenspannungen, Entste-
    hung-Berechnung-Messung-Bewertung", Deutsche Ges. für Metall-
    kunde, Oberursel: 121 (1980).
12. V. Hauk and P. J. T. Stuitje, Residual stresses in the phases
    of heterogeneous materials after machining (in German),
    Symposium Karlsruhe (1983), in press.
13. V. Hauk, R. Oudelhoven and G. Vaessen, On the state of residual
    stresses after grinding (in German), HTM 36: 258 (1981).
14. T. Hanabusa and H. Fujiwara, On the relation between ψ-splitting
    and microscopic residual shear stresses in unidirectionally
    deformed surfaces, same as 2: 209 (1982).
15. H. Krause and M. Mathias, Strain measurement on the cementite-
    and ferrite-phase after different machining of surfaces (in
    German), to be published.
16. V. Hauk, E. Schneider, P. Stuitje and W. Theiner, Comparison of
    different methods to determine residual stresses nondestruc-
    tively, in "New Procedures in Nondestructive Testing", Edited
    by P. Höller, Springer-Verlag, Berlin, Heidelberg, New York:
    561 (1983).

17. V. Hauk, Residual stresses by deformation (in German),
     same as 2: 92 (1982).
18. V. Hauk and W. K. Krug, Superposition of load- and residual
     stresses, fundamental tests (in German) same as 2: 133 (1982)
19. V. Hauk, Residual stresses, their importance for science and
     technique (in German), Symposium Karlsruhe (1983), in the
     press.
20. V. Hauk, W. K. Krug and L. Pintschovius, to be published.
21. V. Hauk and H. Sesemann, Deviations from linear distributions
     of lattice interplanar spacings in cubic metals and their
     relation to stress measurement by X-rays (in German),
     Z. Metallkde. 67: 646 (1976).
22. H. Dölle and V. Hauk, Influence of mechanical anisotropy of
     polycrystalls (texture) upon the X-ray stress determination
     (in German), Z. Metallkde. 69: 410 (1978).
23. H. Dölle, V. Hauk and H. Zegers, Calculated and measured XEC
     and lattice strain distributions in textured steels (in
     German), Z. Metallkde. 69: 766 (1978).
24. H. Dölle and V. Hauk, Evaluation of residual stresses in tex-
     tured materials by X-rays (in German), Z. Metallkde. 70:
     682 (1979).
25. V. Hauk and H. Kockelmann, Calculation of the distribution of
     the intensity and the lattice strain out of inverse pole-
     figures, Z. Metallkde. 69: 16 (1978).
     V. Hauk and H. Kockelmann, Lattice strain distribution of
     plastically deformed specimens of pure and silver alloyed
     copper (in German), Z. Metallkde. 71: 303 (1980).
26. C. M. Brakman, Residual stresses in cubic materials with
     orthorhombic or monoclinic specimen symmetry: Influence
     of texture on ψ-splitting and non-linear behaviour,
     J. Appl. Cryst. 16: 325 (1983).
27. F. Bollenrath, V. Hauk and W. Weidemann, To the interpretation
     of lattice residual stresses in plastically deformed ferrite
     (in German), Arch. Eisenhüttenwes. 38: 793 (1967).
28. T. Shiraiwa and Y. Sakamoto, The X-ray stress measurement of
     the deformed steel having preferred orientation, The 13th
     Jap. Congr. on Mater. Res.-Metal. Mater.: 25 (1970).
29. V. Hauk, D. Herlach and H. Sesemann, Non linear distributions
     of lattice interplanar spacings in steels, their origin,
     calculation and their relation to stress measurement (in
     German), Z. Metallkde. 66: 734 (1975).
30. R. H. Marion and J. B. Cohen, Anomalies in measurement of
     residual stress by X-ray diffractions, Adv. X-Ray Anal. 18:
     466 (1975).
31. P. F. Willemse, B. P. Naughton and C. A. Verbraak, X-ray resi-
     dual stress measurements on cold-drawn steel wire,
     Mat. Science Engg. 56: 25 (1982).
32. P. D. Evenschor and V. Hauk, X-ray elastic constants and distri-
     butions of interplanar spacings of materials with preferred
     orientation (in German), Z. Metallkde. 66: 164 (1975).

33.   V. Hauk and G. Vaessen, to be published.
34.   V. Hauk and G. Vaessen, X-ray stress evaluation on steels
      having preferred orientation (in German), Symposium
      Karlsruhe (1983), in press.
35.   V. Hauk, X-ray methods for measuring residual stress, in
      "Residual stress and stress relaxation", Edited by E. Kula
      and V. Weiss, Plenum Press New York and London: 117 (1982).
36.   H. Dölle and V. Hauk, The theoretical influence of multiaxial
      depth-dependent residual stresses upon the stress measurement
      by X-rays (in German), HTM 34: 272 (1979).
37.   I. C. Noyan, Effect of gradients in multiaxial stress states on
      residual stress measurements with X-rays, Metallurgical
      Transactions 14 A: 249 (1983).
38.   J. M. Sprauel, M. Barral and S. Torbaty, Measurement of stress
      gradients by X-ray diffraction, Adv. X-Ray Anal., in
      press.
39.   V. Hauk and W. K. Krug, to be published.
40.   T. Shiraiwa and Y. Sakamoto, X-ray stress measurement and its
      application to steel, Sumitomo Search 7: 109 (1972).
41.   A. I. Mel'ker and V. G. Pavlova, Determination of residual
      stresses with a gradient by means of X-ray diffraction
      (Translation from Russian), Ind. Lab. 42: 376 (1976).
42.   V. Hauk, R. Oudelhoven and G. Vaessen, The state of residual
      stress in the near surface region of homogeneous and heter-
      ogeneous materials after grinding, Metallurgical Transactions
      13 A: 1239 (1982).
43.   In meantime the paper, Equilibrium conditions for the average
      stresses measured by X-rays by I. C. Noyan was published in
      Metallurgical Transactions 14 A September: 1907 (1983).

X-RAY MULTIAXIAL STRESS ANALYSIS TAKING ACCOUNT

OF STRESS GRADIENT

Toshihiko Sasaki and Makoto Kuramoto

The Institute of Vocational Training
1960 Aihara, Sagamihara, Kanagawa 229, Japan

Yasuo Yoshioka

Musashi Institute of Technology
1 Tamazutsumi, Setagaya, Tokyo 158, Japan

INTRODUCTION

Since the method of X-ray multiaxial stress analysis (ψ-splitting problem) was proposed by Dölle and Hauk[1], their method is often utilized for residual stress evaluation on a processed surface. However, as pointed out by themselves, a theoretical problem still remains on the assumption of the stress state. Namely, the effect of ψ-splitting is impossible unless stress gradients with respect to the direction of the depth in σ_{13} or σ_{23} are present, because these components and σ_{33} have to vanish at the outer surface. Actually, we often find the 2θ vs. $\sin^2\psi$ relations which do not agree with their theory. In this paper, we proposed a new method for X-ray multiaxial stress analysis in which the effect of stress gradient was considered. The basic equation of this method was solved by means of the integral method proposed by Lode and Peiter[2]. The validity of the present method was proved through a numerical simulation and an experiment.

BASIC ASSUMPTION

Stress State

The coordinate system is defined in Fig.1. We firstly assumed that stress gradient with respect to the direction of the depth z is linear. Thus, under multiaxial stress state, the stress at the depth z, $\sigma_{ij}(z)$ is expressed as:

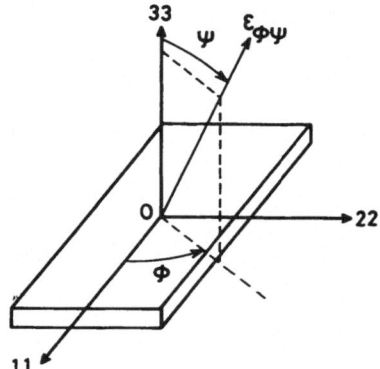

Fig.1. Definition of the coordinate system and
the angle ϕ and ψ.

$$\sigma_{ij}(z)=\sigma_{ij_0}+A_{ij}\cdot z= \begin{bmatrix} \sigma_{110} & \sigma_{120} & 0 \\ & \sigma_{220} & 0 \\ & & 0 \end{bmatrix} + \begin{bmatrix} A_{11} & A_{12} & A_{13} \\ & A_{22} & A_{23} \\ & & A_{33} \end{bmatrix}\cdot z \qquad (1)$$

Here, σ_{ij_0} is a stress at the surface (z=0) and A_{ij} is stress
gradient.

Penetration Depth of the X-rays

The penetration depth of the X-rays (T) can be expressed by the
first term of eq.(2). As this result is almost linear to $\sin^2\psi$ within
the range of $-45°\leqq\psi\leqq45°$, in this study, we used the second term which
was proposed by Lode and Peiter[2].

$$T = \frac{\sin^2\theta-\sin^2\psi}{2\mu\cdot\sin\theta\cdot\cos\psi} \simeq T_0-T'\cdot\sin^2\psi \qquad (2)$$

Here, θ is Bragg's angle, μ is linear absorption coefficient and T_0,
T' are constants.

Lattice Strain

When the stresses shown by eq.(1) are present, the lattice
strain determined by the X-ray method $<\varepsilon\phi\psi>$ is expressed as:

$$\begin{aligned}
<\varepsilon_{\phi\psi}> &= \int_0^{z_1} \varepsilon_{\phi\psi}(z)\cdot\exp(-z/T)\ dz \ / \int_0^{z_1} \exp(-z/T)\ dz \\
&= \frac{1+\nu}{E}(\sigma_{110}\cos^2\phi + \sigma_{120}\sin2\phi + \sigma_{220}\sin^2\phi)\cdot\sin^2\psi \\
&\quad - \frac{\nu}{E}(\sigma_{110} + \sigma_{220})
\end{aligned}$$

$$+ W \cdot T \cdot \{ \frac{1+\nu}{E}(A_{11}\cos^2\phi + A_{12}\sin2\phi + A_{22}\sin^2\phi) \cdot \sin^2\psi$$

$$+ \frac{1+\nu}{E}A_{33}\cos^2\psi - \frac{\nu}{E}(A_{11}+A_{22}+A_{33})$$

$$+ \frac{1+\nu}{E}(A_{13}\cos\phi + A_{23}\sin\phi) \cdot \sin2\psi \} \tag{3}$$

Here, W is a weighting coefficient. If $z_1=T$ we obtain $W=0.42$, and if $z_1=\infty$ we obtain $W=1$.

STRESS EVALUATION

To solve σ_{ij_0} and A_{ij} from eq.(3), we used the integral method. Eq.(3) can be rewritten as:

$$<\varepsilon_{\phi\psi}> = B_0 + B_2\cos2\psi + B_4\cos4\psi + A_2\sin2\psi + A_4\sin4\psi \tag{4}$$

Thus, we obtain:

$$
\begin{bmatrix} B_0 \\ B_2 \\ B_4 \\ A_2 \\ A_4 \end{bmatrix} =
\begin{bmatrix}
a_{11} & a_{12} & a_{13} & a_{14} & a_{15} & 0 & a_{17} & 0 & a_{19} \\
a_{21} & a_{22} & a_{23} & a_{24} & a_{25} & 0 & a_{27} & 0 & a_{29} \\
0 & 0 & 0 & a_{34} & a_{35} & 0 & a_{37} & 0 & a_{39} \\
0 & 0 & 0 & 0 & 0 & a_{46} & 0 & a_{47} & 0 \\
0 & 0 & 0 & 0 & 0 & a_{56} & 0 & a_{57} & 0
\end{bmatrix}
\begin{bmatrix} \sigma_{110} \\ \sigma_{120} \\ \sigma_{220} \\ A_{11} \\ A_{12} \\ A_{13} \\ A_{22} \\ A_{23} \\ A_{33} \end{bmatrix} \tag{5}
$$

Multiplying eq.(4) by 1, $\cos2\psi$, $\cos4\psi$, $\sin2\psi$ and $\sin4\psi$, respectively, and then integrating each equation from $\psi=-45°$ to $\psi=45°$, we obtain the following equations.

$$C_0 = \int_{-45°}^{45°}<\varepsilon_{\phi\psi}>d\psi = \frac{\pi}{2}B_0 + B_2$$

$$C_2 = \int_{-45°}^{45°}<\varepsilon_{\phi\psi}>\cos2\psi d\psi = B_0 + \frac{\pi}{4}B_2 + \frac{1}{3}B_4$$

$$C_4 = \int_{-45°}^{45°}<\varepsilon_{\phi\psi}>\cos4\psi d\psi = \frac{1}{3}B_2 + \frac{\pi}{4}B_4 \tag{6}$$

$$S_2 = \int_{-45°}^{45°}<\varepsilon_{\phi\psi}>\sin2\psi d\psi = \frac{\pi}{4}A_2 + \frac{2}{3}A_4$$

$$S_4 = \int_{-45°}^{45°}<\varepsilon_{\phi\psi}>\sin4\psi d\psi = \frac{2}{3}A_2 + \frac{\pi}{4}A_4$$

Thus, we obtain:

$$B_0 = 2/\pi \cdot (C_0 - B_2)$$

$$B_2 = (\pi/4 \cdot C_2 - 1/2 \cdot C_0 - 1/3 \cdot C_4)/(\pi^2/16 - 11/18)$$

$$B_4 = 4/\pi \cdot (C_4 - 1/3 \cdot B_2) \tag{7}$$

$$A_2 = 4/\pi \cdot (S_2 - 2/3 \cdot A_4)$$

$$A_4 = (\pi/4 \cdot S_4 - 2/3 \cdot S_2)/(\pi^2/16 - 4/9)$$

By using the Simpson's formula, we can obtain the values of C_0, C_2, C_4, S_2 and S_4 from the experimental data. And by substituting these into eq.(7), the values of B_0, B_2, B_4, A_2 and A_4 can be found.

On the other hand, all the components of the coefficient matrix of eq.(5) can be found if the condition of the measurement(T_0, T' and ϕ), the X-ray elastic constants($-\nu/E$ and $(1+\nu)/E$) and the weighting coefficient(W) are known. Therefore, if we carry out the X-ray measurement under the condition of $\phi=0°$, $45°$ and $90°$, we can obtain the fifteen equations from eq.(5). These include eight equations which are independent to each other. Then, by using the equation(8) which is derived by substituting $\psi=0°$ into eq.(3), together with these eight equations, all the stresses can be solved.

$$<\varepsilon_{\phi\psi}>_{\psi=0°} = -\frac{\nu}{E}(\sigma_{110}+\sigma_{220}) - \frac{\nu}{E}(A_{11}+A_{22})\cdot W \cdot T_0 + \frac{1}{E}\cdot W \cdot T_0 \cdot A_{33} \tag{8}$$

Although there are sixteen equations in all, we chose the equations for $B_2(0°)$, $B_4(0°)$, $A_2(0°)$, $B_0(45°)$, $B_4(45°)$, $B_2(90°)$, $B_4(90°)$, $A_2(90°)$ and $<\varepsilon\phi\psi>_{\psi=0°}$ respectively.

NUMERICAL SIMULATION

The following stress state was assumed.

$$\sigma_{ij}(z) = \begin{bmatrix} 100 & -50 & 0 \\ & -100 & 0 \\ & & 0 \end{bmatrix} + \begin{bmatrix} -40 & 10 & 20 \\ & 50 & -2 \\ & & -40 \end{bmatrix} \cdot z \tag{9}$$

Here the unit of σ_{ij_0} is MPa and that of A_{ij} is MPa/μm. The simulation was performed on the αFe(211) diffraction plane by means of Cr-Kα X-rays. Both the opened and the closed circles in Fig.2 were obtained by substituting eq.(9) into eq.(3) and converting $<\varepsilon\phi\psi>$ into $<2\theta\phi\psi>$. The constants used for the calculation are summarized in Table 1. The stress evaluation was carried out according to the present method by using the relations symbolized by the opened circles in Fig.2. Namely, the integrals of eq.(6) were performed with $5°$-interval ranged from $\psi=-45°$ to $\psi=45°$. The number of data used for the stress evaluation, 19x3=57, is almost as same as that for the Dölle-Hauk method.

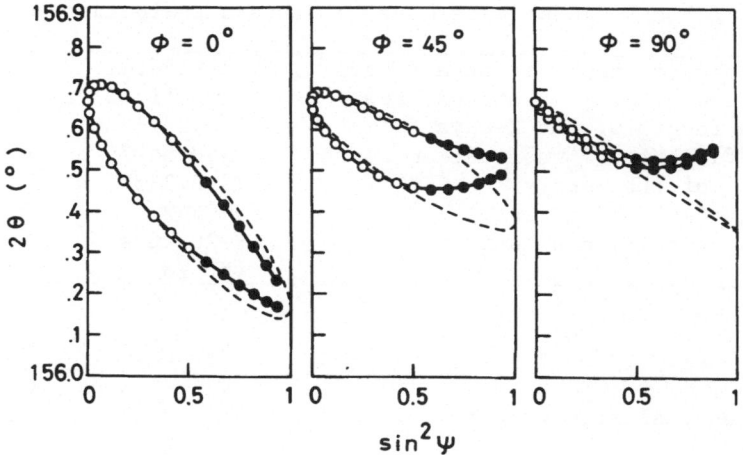

Fig.2. The 2θ vs. $\sin^2\psi$ diagrams (Numerical simulation).
(O,●) Calculated by substituting the assumed
stresses(eq.(9)) into eq.(3). (O) Used for the stress
evaluation. (----) Obtained from the Dölle-Hauk method.
(———) Obtained from the present method.

The result was as follows:

$$\sigma_{ij}(z) = \begin{bmatrix} 103 & -51 & 0 \\ & -103 & 0 \\ & & 0 \end{bmatrix} + \begin{bmatrix} -42 & 10 & 20 \\ & 52 & -2 \\ & & -40 \end{bmatrix} \cdot z \tag{10}$$

The accuracy of the method is sufficient for practical use.

To know a sensitiveness of the method to experimental error,
10% error was added to $<\varepsilon\phi\psi>$ corresponding to $\phi=0°$ and $\psi=-25°$.
Namely, about $0.03°$ was added to 2θ. The stresses obtained were as
follows.

$$\sigma_{ij}(z) = \begin{bmatrix} 126 & -60 & 0 \\ & -99 & 0 \\ & & 0 \end{bmatrix} + \begin{bmatrix} -53 & 16 & 20 \\ & 52 & -2 \\ & & -40 \end{bmatrix} \cdot z \tag{11}$$

The influence of the error added is seen mainly in σ_{110}, σ_{120}, A_{11}
and A_{12}. The accuracy would be increased by smoothing data.

The present method was compared to the Dölle-Hauk method. The
same data as the above simulation were used for the Dölle-Hauk method.

Table 1. Constants used for stress evaluation

Diffraction angle at stress free	$2\theta = 156.41°$
Lattice spacing at stress free	$d_0 = 0.117021$ nm
Wave length of the X-rays	$\lambda = 0.2291$ nm
Coefficients of penetration depth	$T_0 = 0.54079$ nm
of the X-rays	$T' = 0.33595$ nm
Weighting coefficient	$W = 0.42$
X-ray elastic constants	$E = 206$ GPa
	$\nu = 0.28$

The stresses evaluated were as follows:

$$\sigma_{ij} = \begin{bmatrix} 79 & -34 & 37 \\ & 24 & -4 \\ & & -69 \end{bmatrix} \tag{12}$$

The dotted lines in Fig.2 are the 2θ vs. $\sin^2\psi$ relations obtained by substituting eq.(12) into the basic equation adopted in the Dölle-Hauk method. As eq.(12) is considered to be the stresses averaged over the penetration depth of the X-rays, the comparison was done by using the weighted averaging stress described as:

$$\langle\sigma_{ij}\rangle = \int_0^{z_1} \sigma_{ij}(z)\cdot\exp(-z/T)\ dz \ / \int_0^{z_1} \exp(-z/T)\ dz$$

$$= \sigma_{ij_0} + A_{ij}\cdot W\cdot T. \tag{13}$$

The following were obtained from the assumed stresses(eq.(8)) and the analyzed stresses(eq.(9)) according to the above equation(13).

$$\langle\sigma_{ij}\rangle = \begin{bmatrix} 9 & -27 & 45 \\ & 14 & -5 \\ & & -91 \end{bmatrix} \quad \text{from eq.(9)} \tag{14}$$

$$\langle\sigma_{ij}\rangle = \begin{bmatrix} 8 & -27 & 45 \\ & 15 & -5 \\ & & -91 \end{bmatrix} \quad \text{from eq.(10)} \tag{15}$$

Here $T = T_0$ was used for the calculation of eq.(13). Though there is a good agreement between eq.(14) and eq.(15), both of them do not agree with the result of the Dölle-Hauk method. The disagreement is caused by the assumption in their method that the 2θ vs. $\sin^2\psi$ relation is exactly like an ellipse. Namely, their method is a kind of approximate solution when the 2θ vs. $\sin^2\psi$ relation becomes non-ellipse such as Fig.2. It is also pointed out that the weighted averaging stress varies with the value of T used in eq.(13).

EXPERIMENT

An experiment was performed on a ground steel by means of Cr–Kα X-rays. The experimental conditions are summarized in Table 2. Fig. 3 shows the 2θ vs. $\sin^2\psi$ relations obtained. The shape of each diagram is different from an ellipse. After the experimental data were smoothed into a five-order-polynomial, the stress evaluation was performed. The result was as follows.

$$\sigma_{ij}(z) = \begin{bmatrix} -247 & -30 & 0 \\ & -373 & 0 \\ & & 0 \end{bmatrix} + \begin{bmatrix} 57 & 19 & -11 \\ & 43 & -1 \\ & & 8 \end{bmatrix} \cdot z \qquad (16)$$

$$<\sigma_{ij}> = \begin{bmatrix} -118 & 13 & -25 \\ & -276 & -2 \\ & & 17 \end{bmatrix} , \ (T=T_0) \qquad (17)$$

The real lines in Fig.3 are the 2θ vs. $\sin^2\psi$ relations obtained by substituting eq.(16) into eq.(3). They are passing nearly through the data points.

The experimental data were also analyzed according to the Dölle-Hauk method. The result was as follows.

$$\sigma_{ij} = \begin{bmatrix} -137 & -1 & -17 \\ & -287 & -1 \\ & & 9 \end{bmatrix} \qquad (18)$$

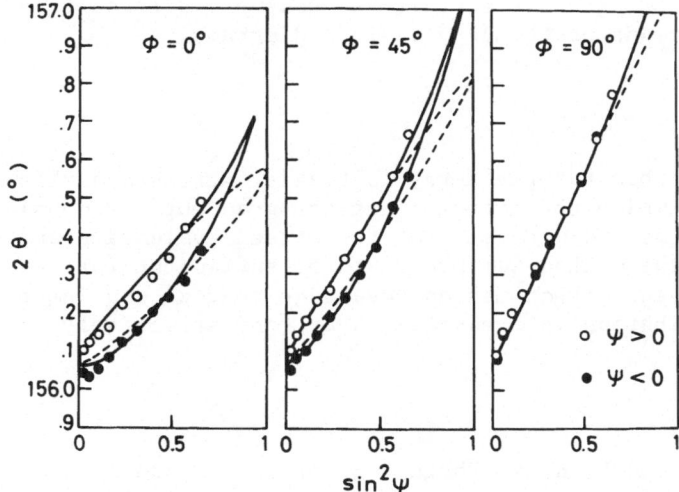

Fig.3. The 2θ vs. $\sin^2\psi$ diagrams measured at the 211 reflection of a ground steel. (– – – –) Obtained from the Dölle-Hauk method. (———) Obtained from the present method.

Table 2.   Experimental conditions

Grinding conditions :	
Process	Up grinding (4 strokes)
Grinding wheel	White fused almina, vitrified wheel. (size:ϕ180x16, mm)
Grinding fluid	solution x 50
Work speed	50 mm/sec
Spindle speed	2800 rpm
Actual depth of cut	5.7 μm
Average roughness	0.18 μm
X-ray conditions :	
Characteristic X-ray	Cr-Kα
Diffractin plane	αFe(211)
Scanning method	ψ-constant method
Determination of peak position	Half value breadth method
Irradiated area	4x4 mm^2
Tube voltage and current	30 kV, 10 mA
Range of measurement	ϕ=0°,45° and 90° −45°$\leq\psi_0\leq$45° (5°-interval)
Instrument used	Rigaku MSF-1M type

The dotted lines in Fig.3 are the 2θ vs. $\sin^2\psi$ relations obtained by the Dölle-Hauk method. Compared to the result of the numerical simulation, the difference is little except σ_{12} between the stresses of the Dölle-Hauk method and those of the present method(eq.(17)). Thus, the relation between the stresses obtained from the two methods vary due to the stress state. It is, however, difficult to know the relation by means of the 2θ vs. $\sin^2\psi$ diagrams.

CONCLUSION

    A new method for the X-ray multiaxial stress analysis was proposed, in which the effect of stress gradient was considered. The integral method was used for the stress evaluation and the accuracy of the method was proved to be sufficient for practical use. The present method has an advantage that we can inspect the stress distribution in a surface layer nondestructively.

REFERENCES

1. H. Dölle and V. Hauk, Röntgenographische Spannungsermittlung
     für Eigenspannungssysteme allgemeiner Orientierung,
     Härterei-Tech. Mitt., 31, 165:168 (1976).
2. W. Lode and A. Peiter, Numeric röntgenographischer
     Eigenspannungsanalysen oberflächennaher Schichten,
     Härterei-Tech. Mitt., 32, 235:240 (1977).

DETERMINING STRESSES IN THE PRESENCE OF NONLINEARITIES IN

INTERPLANAR SPACING VS. $\text{SIN}^2\psi$

I. C. Noyan and J. B. Cohen

I. C. Noyan is research assistant and graduate student,
J. B. Cohen is Frank C.Engelhart Professor of Materials
Science and Engineering and Technological Institute
Professor

Northwestern University, Dept. of Mat. Sci. and Eng.,
Technological Institute, Evanston, Il. 60201

ABSTRACT

The physical meaning of non-linearities in "d" vs.
$\sin^2\psi$ lines, encountered in X-ray measurements of surface residual stresses in polycrystalline materials is investigated. It is
shown that when oscillations are present in any one reflection,
switching to another reflection to obtain a straight line in "d"
vs. $\sin^2\psi$ is feasible only under very special conditions. We
also discuss the effect of "quasi-homogeneous" strain distributions and investigate the effects of ψ-range on the accuracy of
X-ray residual stress measurements when "ψ-splitting" is present.
A new geometric error is also discussed that can not be detected
by the "annealed powder" method often used for alignment.

Introduction

In the past few years the basic equations used for the determination of surface residual stresses in polycrystalline ma-

terials by X-rays have undergone considerable modification and
expansion.   Traditional methods assume a bi-axial stress state
that is uniform in the surface layers penetrated by the X-ray
beam[1],[2] .   This assumption is based on the concept that this
penetration depth is too shallow to be affected by the stresses
in the third dimension. Methods based on this assumption predict
a linear variation of interplanar spacing "d" with $\sin^2\psi$, with a
slope that is proportional to the stress in the measurement direction σ_ϕ.   The angles Φ and Ψ between the normal of the diffracting planes L_3 and the surface normal of the specimen S_3, are
shown in figure $\overline{1}$. If the components of the (assumed) bi-axial
stress tensor exhibit steep gradients in the volume sampled by
the X-ray beam , curvature occurs in the "d" vs. $\sin^2\psi$ plot.
The slope of a least squares fit to such data yields the stress
in the $\underline{S}_{\underline{\Phi}}$ direction, averaged over the depth of penetration[3].

     The assumption in these traditional methods, that stress
components in the direction of the surface normal are negligible
in the volume sampled by the X-ray beam was shown to be invalid
in several recent studies[4],[5],[6]. Stress components of appreciable
magnitude in this direction have been detected in the surface layers of materials prepared in certain ways. Recent theory as
well as experiment show that, when only the normal stress σ_{33} is
present in the direction of the surface normal, curvature occurs
in the "d" vs. $\sin^2\psi$ data[7],[9]. The degree of such curvature depends on the steepness of the gradient in σ_{33}. It has also been
shown that even for very small curvature in "d" vs. $\sin^2\psi$ data,
analysis with the bi-axial assumption causes appreciable error in
the calculated surface stress, and that a high ψ -range
($\sin^2\psi \geq$ .4) should be used even when tri-axial analyses, (taking
into account the presence of σ_{33}) are employed . When the shear
stresses σ_{13}, σ_{23} are present in the measurement volume, splitting
of the "d" vs. $\sin^2\psi$ data results; that is, "d" vs. $\sin^2\psi$ plots
have opposite curvature for negative and positive ψ. Analysis of
such data has been described in detail by Dölle.   (see also
references 4,5 for the use of this analysis). We present here
the first study of the effect of ψ-range on the accuracy of this
analysis.

     Both the traditional methods and the methods that have been
expanded to include the effect of the stresses in the direction
of the surface normal predict a smooth variation in the "d" vs.
$\sin^2\psi$ curves with no large oscillations. However such oscillations are often encountered in practice and various explanations
have   been   given for their cause[6],[10],[11] .   The latest
study attributes the cause of oscillations to elastic anisotropy
(i.e change in the elastic constants of the material with each
ψ-tilt), and suggests the use of h00 and hhh type reflections for
X-ray residual stress analysis when oscillations are present in

the hkl type reflections, since calculations indicate that the elastic constants associated with such reflections are isotropic.

In this paper we will investigate the effects of the ψ-range on the accuracy of residual stress results determined by X-rays when ψ-splitting is present. It will be shown that the ψ-range is very important and must be chosen according to the stress gradient(s) existing in σ_{13} and/or σ_{23}. Information about the gradient can be obtained from a plot of "a_2" vs. $\sin|2\psi|$ (where $a_2 = \frac{1}{2}(d_{\Phi\psi+} - d_{\Phi\psi-})$). Furthermore a geometric error will be discussed which causes ψ-splitting of the same shape as that caused by the presence of σ_{13} and/or σ_{23}. This error can not be detected by the current "annealed- powder" alignment technique used to minimize geometric contributions to the observed "ψ-splitting". We will also examine the presence of oscillations in "d" vs. $\sin^2\psi$ and show that oscillations are expected in h00 and hhh reflections when they are present in the hkl reflections. Computer simulation and experimental data are presented to test the above ideas.

The Basic Equations

The orthogonal coordinate systems used in the following discussion are shown in figure 1. The specimen axes are defined such that $\underline{S}_1$, $\underline{S}_2$ are in the plane of the specimen surface. The laboratory system, in which diffraction is occurring, is defined such that $\underline{L}_3$ is in the direction of the normal to the family of planes (hkl) whose "d" spacings are being measured by X-rays. $\underline{L}_2$ is in the plane defined by $\underline{S}_1$ and $\underline{S}_2$ and makes an angle Φ with $\underline{S}_1$. In what follows, primed tensor quantities refer to the laboratory system $\underline{L}_i$ and unprimed quantities to the sample system $\underline{S}_i$, following the convention established by Dölle[6].

If the unstressed lattice spacing "d_o" for the material under examination is known, strains in the $\underline{L}_3$ direction may be obtained from the formula

$$(\varepsilon'_{33}) = (d_{\Phi\psi} - d_o)/d_o . \qquad (1)$$

this strain may be expressed in terms of the strains in the $\underline{S}_i$ coordinate system using the tensor transformation

$$(\varepsilon'_{33})_{\Phi\psi} = a_{3k}a_{3l}\varepsilon_{kl} \qquad (2\text{-}a)$$

where a_{3k}, a_{3l} are the direction cosines between the axes $\underline{L}_3$ and $\underline{S}_k$ or $\underline{S}_l$ respectively. For arbitrary angles Φ and ψ, (2-a)

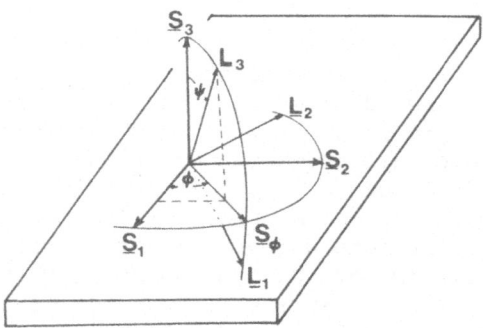

FIGURE 1: Definition of the angles and orientation of the
laboratory system $\underline{L}_i$ with respect to the sample system $\underline{S}_i$.

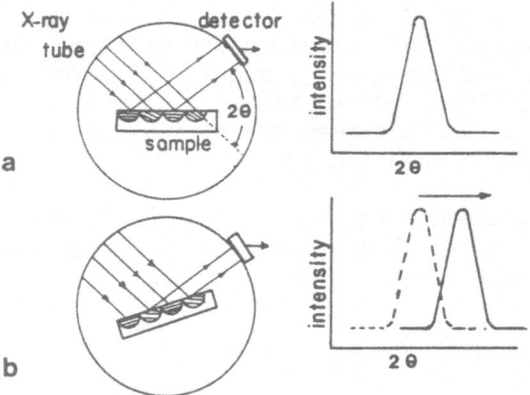

Figure 2: Schematic of a diffractometer in a stress measurement.
a) Certain atomic planes satisfy Bragg's law and diffract X-rays
at a   value which depends on the spacing of the planes.   This
spacing is affected by the stresses existing in the surface layers. b) After the specimen is tilted, diffraction occurs from
other grains but from the same set of planes. Since the normal
stress component on these is different than in (a), the plane
spacing will be different, as will the diffraction angle.

becomes

$$(\epsilon'_{33})_{\Phi\psi} = \{\epsilon_{11}\cos^2\Phi + \epsilon_{12}\sin2\Phi + \epsilon_{22}\sin^2\Phi - \epsilon_{33}\}\sin^2\psi - \epsilon_{33}$$

$$+ (\epsilon_{13}\cos\Phi + \epsilon_{23}\sin\Phi)\sin 2\psi \qquad\qquad (2\text{-}b)$$

Once the strains in $\underline{S}_i$ are obtained (by techniques described in reference 6, or by least squares solution), one can obtain the stresses in the sample coordinate system from:

$$\sigma_{ij} = C_{ijkl}\epsilon_{kl} \qquad\qquad (3)$$

Where C_{ijkl} are the stiffness coefficients of the material, referred to the $\underline{S}_i$ coordinate system.

Presence of Oscillations

It must be noted that the above development makes no assumptions about whether the material under investigation is polycrystalline or is a single crystal. It is valid for both isotropic and anisotropic materials, and predicts a linear "d" vs. $\sin^2\psi$ behaviour (for ϵ_{13}, $\epsilon_{23} = 0$) or regular "ψ-splitting" (ϵ_{13} and/or $\epsilon_{23} \neq 0$). There are however two implicit assumptions in the above treatment[1,2]:
(i) The strain tensor for the material under examination is a symmetrical second rank tensor.
(ii) The strains ϵ_{kl} in the specimen coordinate system $\underline{S}_i$ are homogeneous in the total diffraction volume sampled during the experiment.

The second assumption is very important for X-ray residual stress determination in polycrystalline specimens. It is well known that the grains that contribute to the diffraction profile (from which "$d_{\Phi\psi}$" is determined) are different for every ψ-tilt (figure 2). Thus, when we use the $(\epsilon'_{33})_{\Phi\psi}$, determined from these mutually exclusive subsets of the total irradiated volume to calculate the strain tensor for the surface coordinate system $\underline{S}_i$, we are assuming, a-priori, that ϵ_{kl} in the coordinate system $\underline{S}_i$ is invariant (i.e. homogeneous) no matter where the origin of $\underline{S}_i$ is located within the total irradiated volume. An experimentally determined linear "$d_{\Phi\psi}$" vs. $\sin^2\psi$ plot means that this assumption is justified and equation 2 is valid for the volume of material examined. On the other hand, oscillations in "$d_{\Phi\psi}$" vs. $\sin^2\psi$ mean that an inhomogeneous strain distribution exists at the surface, with each diffracting subset having their unique strain tensor (ϵ_{kl}) for the coordinate system $\underline{S}_i$. In such a case the strain along a given direction in the sample coordinate system will be different at different points and this will cause non-uniform strain gradients in the irradiated volume whose variation increases with increasing oscillations in "$d_{\Phi\psi}$" vs.

$\sin^2\psi$. It is rather improbable to have certain grains in the
irradiated volume that are free from the effect of these gradients, which is the assumption made when switching to an hOO and
hhh reflection (without moving the beam position) after observing oscillations in an hkl reflection. Consider figure 3;   When
oscillations are present in "$d_{\phi\psi}$" vs. $\sin^2\psi$ for the hkl reflection, ε_{kl} are different at points 1,3,5. In order for "$d_{\phi\psi}$" vs.
$\sin^2\psi$ to be linear for the hhh reflection, ε_{kl} should be constant
at points 2,4,6. When grain size is small and the gradients are
steep (i.e, large oscillations), this is probable if the inhomogeneous regions have a very small mean size and volume fraction.
Otherwise all points will have large, non-uniform gradients in
order to keep the displacements constant across grain boundaries
in the measurement volume. Thus one would expect oscillations in
all reflections when large oscillations are present in any one
reflection.

        In figure 4 the "$d_{\phi\psi}$" vs. $\sin^2\psi$ plots from a cold-rolled
(90%) α-brass specimen, which was then polished to .05 microns
(diamond paste), are shown for various elastic loads applied
in-situ on a diffractometer. Also shown, as a measure of texture
are the relative intensities of the X-ray peak for each ψ-tilt.
It is seen that the material has texture, and oscillations occur
for both (222) and (311) reflections. This result can not be explained by the "elastic anisotropy" concepts developed in references [6,7]. In a previous paper Dölle and Cohen concluded oth-
erwise but the hOO reflection used was at low 2θ angle ($\sim105^\circ 2\theta$)
where any oscillations would be harder to detect   (since
$\Delta d/d = -\cot\theta \cdot \Delta 2\theta/2$).

        The origin of the non-uniform strain gradients   (and hence
the reason for the oscillations) is not important for the arguments presented above. Such gradients may arise from inhomogeneous stress gradients[11], inhomogeneous plastic flow in the volume
under examination[10,15,16,17], or changes in the elastic constants
[6,8] of the material in various subsets of the total irradiated
volume. This last effect may be due to large grains oriented
differently, texture[6], or the presence of other phases with different elastic constants[8]. When an elastic stress is applied at
the boundaries of such materials it will cause a reaction stress
field which acts to keep the displacements constant across phase
(or grain) boundaries.   Such a reaction stress field will be inhomogeneous[19], and thus, cause oscillations. For a given applied
stress the inhomogeneity will be more pronounced (and the oscillations larger) in the phases that have a smaller volume fraction
and small grain sizes.

        Different methods have been proposed so far to obtain the
stress state in the surface layers for some of the cases listed
above. However so far no combined method that is rigorously

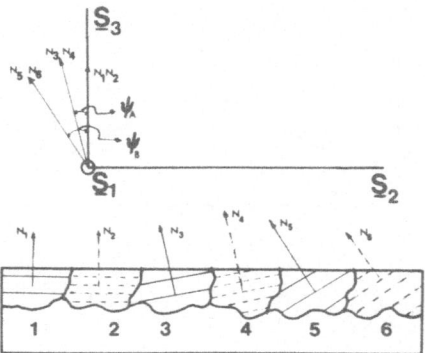

Figure 3: Hypothetical specimen whose grains 1, 3, 5 and 2, 4, 6 will diffract at $\psi = 0$, ψ_A, ψ_B, for hkl and hhh reflections respectively.

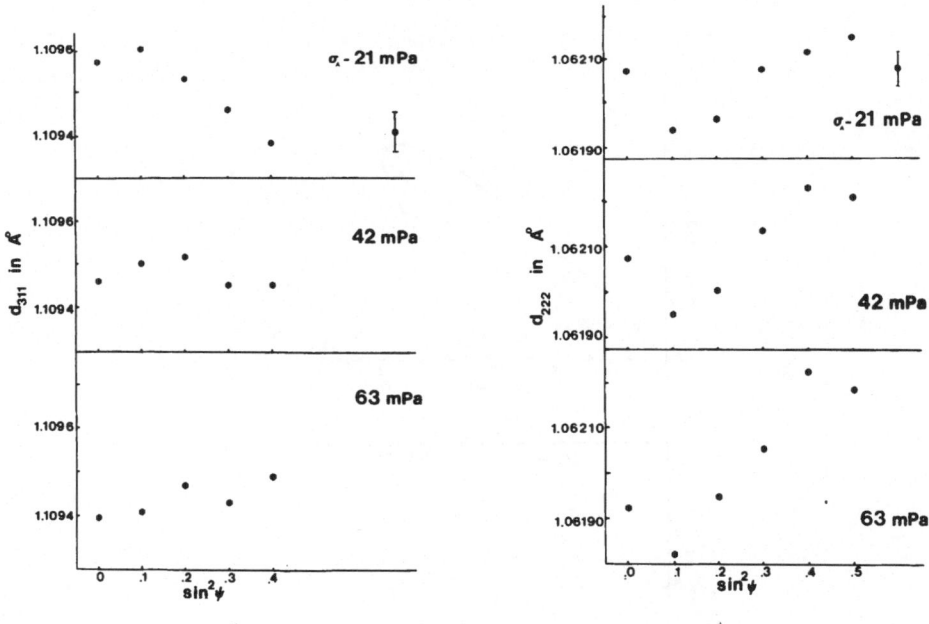

a   b

Figure 4: Interplanar spacing "d" vs. $\sin^2\psi$ lines obtained from a cold-rolled alpha-brass specimen (using filtered Fe-K radiation and a solid-state detector to reduce the background due to fluorescence). The specimen was loaded to a different elastic load for each measurement.
a) "d" vs. $\sin^2\psi$ data for 311 reflection (121° 2θ) at loads 21, 42, 63 mPa.
b) "d" vs. $\sin^2\psi$ data for 222 reflection (131° 2θ) at the same loads.

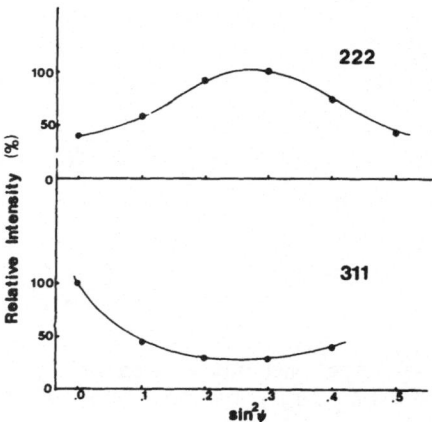

4c) Relative intensities for both reflections at each ψ-tilt   show
the presence of texture.

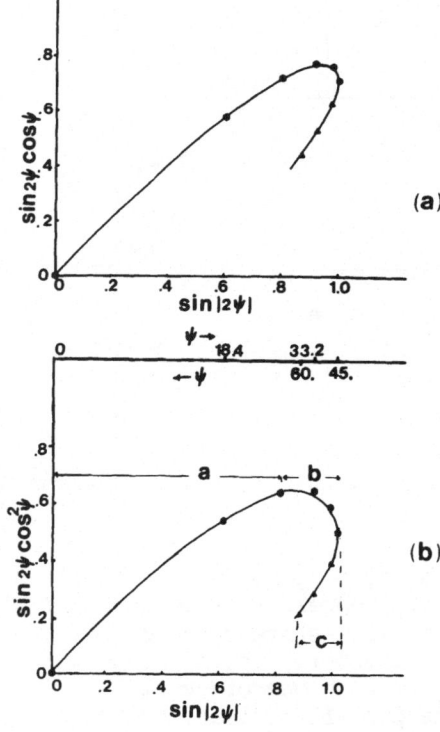

Figure 5:   Variation of   the   function $\sin 2\psi (\cos\psi)^{n_{13}}$ over
$\psi=(0,60)$ for $n_{13}=1$ (a) and $n_{13}=2$ (b).

based on elasticity theory is available. Such a method is under investigation at Northwestern University and will be reported at a later date.

ψ-Splitting In Isotropic Materials with Homogeneous Strain Fields

For isotropic materials equation (3) becomes[12]:

$$\sigma_{ij}=[E/(1+\nu)]\epsilon_{ij}+\delta_{ij}\Big[\nu E/[(1+\nu)(1-2\nu)]\Big]\epsilon_{kk} \qquad (4\text{-}a)$$

or, inverting the equation and expressing strains in terms of stresses:

$$\epsilon_{ij}=[(1+\nu)/E]\sigma_{ij}-(\nu/E)\sigma_{kk}\delta_{ij} \qquad (4\text{-}b)$$

In the equations (4), ν is Poisson's ratio, E is Young's modulus δ_{ij} is Kronecker's delta and summation over repeated indices is assumed. Substituting equation 4-b in equation 2-b:

$$(\epsilon'_{33})_{\Phi\psi}=[(1+\nu)/E]\{\sigma_{11}\cos^2\Phi+\sigma_{22}\sin^2\Phi+\sigma_{12}\sin 2\Phi-\langle\sigma_{33}\rangle_\psi\}\sin^2\psi+[(1+\nu)/E].$$

$$\langle\sigma_{33}\rangle_\psi-(\nu/E)\sigma_{kk}+[(1+\nu)/E]\{\langle\sigma_{13}\rangle_\psi\cos\Phi+\langle\sigma_{23}\rangle_\psi\sin\Phi\}\sin 2\psi \qquad (5)$$

Where carats imply averages over the penetration depth at a particular ψ-angle.

The solution to this equation is given by Dölle[6] where the terms a_1 and a_2 are formed such that:

$$a_1=\tfrac{1}{2}\{(\epsilon'_{33})_{\Phi\psi+}+(\epsilon'_{33})_{\Phi\psi-}\}=(1+\nu)/E\{\sigma_{11}\cos^2\Phi+\sigma_{22}\sin^2\Phi+\sigma_{12}\sin 2\Phi$$

$$-\langle\sigma_{33}\rangle_\psi\}\sin^2\psi-(\nu/E)\sigma_{kk}+(\tfrac{1+\nu}{E})\langle\sigma_{33}\rangle_\psi \qquad (6\text{-}a)$$

$$a_2=\tfrac{1}{2}\{(\epsilon'_{33})_{\Phi\psi+}-(\epsilon'_{33})_{\Phi\psi-}\}=[(1+\nu)/E]\{\langle\sigma_{13}\rangle_\psi\cos\Phi+\langle\sigma_{23}\rangle_\psi\sin\Phi\}\cdot$$
$$\sin|2\psi| \qquad . \qquad (6\text{-}b)$$

The stresses σ_{11}, σ_{22}, σ_{33}, σ_{12} are then obtained from the slopes and intercepts of a_1 vs. $\sin^2\psi$ for Φ=0,90,45, respectively. The stresses σ_{13}, σ_{23} are obtained from the slopes of least-squares lines for a_2 vs. $\sin|2\psi|$ for ψ =0,90.

Equations (6) contain terms (σ_{ij}, i,j=3) which can only exist as gradients in the near surface layers, being, by definition, zero at the surface. Thus, in a strict sense, they violate the condition of homogeneity, required for the use of equations (2). However, assuming that the variation of these stresses is only along $\underline{S}_3$, and is a single valued function with depth within the penetration distance of X-rays, and that the gradient within the irradiated volume is small, and also assuming that no gradients along $\underline{S}_1$ and $\underline{S}_2$ are present for any element of the stress tensor, it can be shown that the deviation of experimental points from that predicted by equation (2-b) will be small[7,8]. We will call the state described by the above assumptions "quasi-homogeneity" and all the discussion that follows will be based upon these assumptions.

Consider equation (6-b). Since X-rays penetrate to different depths at each ψ-tilt, a different region in the gradient for σ_{13}, σ_{23} will be sampled. This effect when coupled with the fact that $\sin|2\psi|$ is a multi valued function over $\psi=0,60$ (the accessible ψ- range in most experiments), causes non-linearity in a_2 vs. $\sin|2\psi|$. This non-linearity and its effect on the stresses determined from the slope of a_2 vs. $\sin|2\psi|$ lines are now examined.

Assume that the stresses σ_{3j} can be expressed by a power-law in the measurement volume:

$$\sigma_{3j} = a_{3j} \cdot Z^{n_{3j}} \qquad (7)$$

where Z is the direction coordinate along $\underline{S}_3$, measured into the material and a_{3j} , n_{3j} are constants over the depth of penetration. The average stress value observed by X-rays at any ψ-angle will then be[8]:

$$\langle \sigma_{3j} \rangle_\psi = k_{3j} \tau_\psi^{n_{3j}} \qquad (8)$$

where $\tau_\psi = \sin 2\theta \cdot \cos \psi / 2\mu$ is the penetration depth of X-rays at any angle for ψ-goniometer geometry.

Substituting τ_ψ into (8) we have:

$$\langle \sigma_{3j} \rangle_\psi = k'_{13} (\cos\psi)^{n_{3j}} . \qquad (9)$$

For $\phi=0$, substituting (10) into (6-b) we obtain:

$$(a_2)_\psi = [(1+\nu)/E] k'_{13} (\cos\psi)^{n_{13}} \cdot \sin|2\psi| . \qquad (10)$$

Now let us examine the behaviour of the function $\cos(\psi)^{n_{13}} \sin|2\psi|$. In figure (5-a,b) this behaviour is plotted for $n_{13}=1,2$ respectively. It is seen that:
(i) a_2 vs. $\sin|2\psi|$ is non-linear.
(ii) over a range of $\psi=0,60$ it exhibits splitting. This effect is due to the different regions of the stress gradient sampled at each ψ-tilt. The steeper the gradient, the greater the splitting.
(iii) Three major regions of the a_2 vs. $\sin|2\psi|$ curve may be de-

fined:
a) The low angle region ($\psi < 33.21^{\circ}$ for the cases treated).
b) The apex region ($39.23^{\circ} < \psi < 45^{\circ}$).
c) The high angle region ($\psi > 50^{\circ}$).

It can be seen from the curves that the slope determined from the low-angle region is smaller than that determined from the high angle region. The slope determined from the apex region is of opposite sign to the slopes determined from the regions a and c.

In order to obtain some magnitudes for these slopes, and the σ_{13} obtained thereof, computer simulation was used. The "d" vs. $\sin^2\psi$ curves were synthesized for a given stress profile, and the results analyzed by least-squares. The following conditions were assumed:

$$\langle\sigma_{13}\rangle_\psi = k_{13}\tau_\psi^2, \langle\sigma_{33}\rangle_\psi = k_{33}\tau_\psi^2, (\sigma_{11})_z = (\sigma_{22})_z = -400\text{mPa}$$

The assumed stress profiles are shown in figure 6. The corresponding "d" vs. $\sin^2\psi$ curve is shown in figure 7, with the terms a_1 vs. $\sin^2\psi$, a_2 vs. $\sin|2\psi|$ shown in figures 8-a,8-b respectively. The results of the analysis are summarised in tables I,II.

As was reported earlier[9], the normal stresses σ_{11},σ_{33} are determined accurately over the high ψ-range. However the stress σ_{13} shown is a totally unrealistic value compared to the actual profile for this region. By comparison, the total range or the lower ψ-range (table I) yield values that are more representative of the average value. However, when the a_2 vs. $\sin|2\psi|$ is divided into the three regions previously defined and analyzed (table II) we see that in comparison to the input profile, the most accurate value is obtained from the low angle region. This is due to the fact that this region best approximates a fit to our starting parameters. In practice however, the angular boundaries of these regions may change depending on the steepness and the shape of the gradient. It is suggested that the complete a_2 vs. $\sin|2\psi|$ plot for $0^{\circ} \le \psi \le 60^{\circ}$ for the case at hand be determined before a decision is made over the region of the best linear fit. Alternately, it may be possible to use the shape of the a_2 vs. $\sin|2\psi|$ curve to determine the type of stress gradient in the shear stresses σ_{13},σ_{23} by determining the best fitting function to the plot using a least squares method[11].

Alignment Errors That Produce ψ-Splitting

It is well known that positioning of the sample over the center of the diffractometer is critical for the accuracy in the parafocusing method of X-ray residual stress measurement and

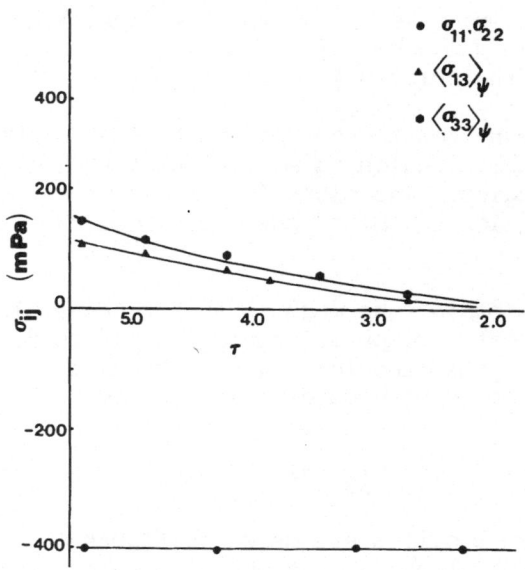

Figure 6: Variation of the (assumed) stresses within the penetration volume in the hypothetical sample.

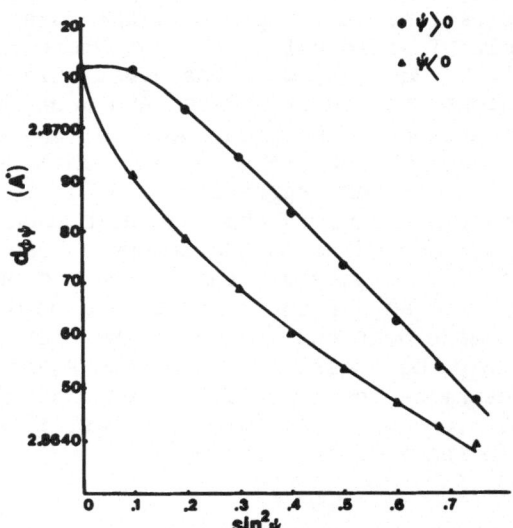

Figure 7: "d " vs. sin²ψ profile for a hypothetical steel specimen which has the stress profiles shown in figure 6 in the volume
irradiated by X-rays.

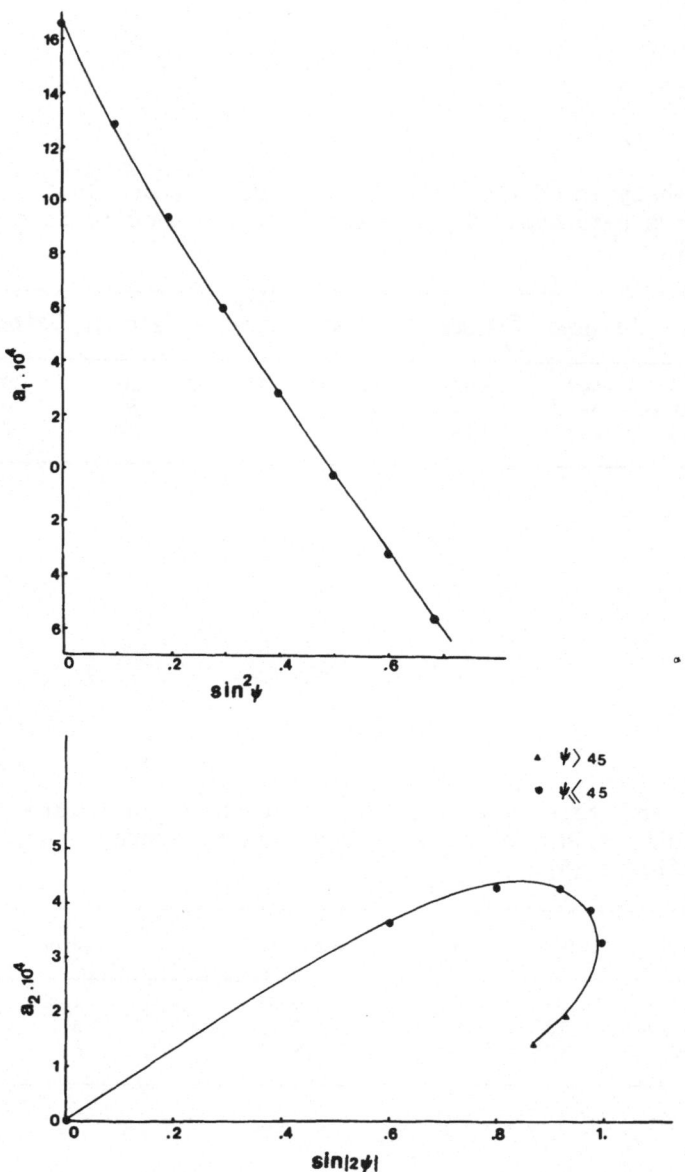

Figure 8: a_1 vs. $\sin^2\psi$ (a) and a_2 vs. $\sin|2\psi|$ (b) profiles calcu-lated from the data shown in figure 7.

Table I: Analysis of the "d" vs. $\sin^2\psi$ curves shown in figure (7) over various ψ-ranges. The (assumed) stress profile is shown in figure (6).

ψ-range	σ_{11}(MPa)	σ_{22}(MPa)	σ_{33}(MPa)	Error in σ_{11}(MPa)	σ_{13}(MPa)	σ_{23}(MPa)
0-60	-442	-442	107	42	50	0
0-33.21	-562	-562	55	162	86	0
39.23-60	-412	-412	81	12	278	0

Table II: Analysis of a_2 vs. $\sin |2\psi|$ curve shown in figure (8-b) over the different ψ-ranges. The assumed stress profile is given in figure (6).

ψ-range	σ_{13}(MPa)	σ_{23}	Corr. Coeff.	Region
0-26-57	96	0	.9952	a
33.21-45.00	-182	0	-.913	b
50.84-60	183	0	.992	c

that any displacement from the correct position may produce ψ-splitting. Among the various methods used to check this the most widely used is the " stress-free powder " technique in which a (stress-free) powder is dusted on the sample surface and then the sample surface adjusted until no ψ-splitting occurs in the "d" spacings obtained from the powder. It has also been shown that the ψ-splitting shape caused by sample misalignment*is not similar to the shape caused by σ_{13}, σ_{23}. In the following, we will treat an error that can not be detected by the stress-free powder method and gives the same shape for ψ-splitting as equation (2-b):

Assume that the zero position for (ψ-motion), read from the odometer of a diffractometer is different from the geometric definition of zero by Δ degrees, i.e:

$$\psi_{obsv} = \psi_{true} + \Delta \qquad (11)$$

where Δ is defined in figure 9. For a well-aligned diffractometer Δ will be negligible. If, however, the $\psi=0$ point is not adjusted with the actual (flat) specimen in question, Δ up to $1°$ is possible. With portable residual stress devices coming into use, higher Δ are possible, especially if measurements are being made near fillets, bends etc. Substituting (11) into (5) we obtain: (for σ_{13}, $\sigma_{23} = 0$)

$$(\epsilon'_{33})_{\Phi\psi} = k_{\Phi}(\cos^2\Delta - \sin^2\Delta)\sin^2\psi + (k_{\Phi}/2)\sin 2\Delta \sin 2\psi + k_{\Phi}\sin^2\Delta$$

$$+ [(1+\nu)/E']\sigma_{33} - \nu/E \; \sigma_{kk}$$

$$\text{where } k_{\Phi} = [(1+\nu)/E]\{\sigma_{11}\cos^2\Phi + \sigma_{12}\sin 2\Phi + \sigma_{22}\sin^2\Phi - \sigma_{33}\} \quad (12)$$

From equation (12) it is seen that an error Δ from the true ψ position will cause splitting proportional to $k_{\Phi}\sin 2\Delta$ for positive and negative ψ. The shape of the split "d" vs. $\sin^2\psi$ curve will be the same shape as predicted by equation (2) with $\Delta = 0$ and σ_{13} and/or σ_{23} finite since in both cases the effect is caused by the argument " $\sin 2\psi$".

When all the stress components in equation 12 are zero, as in the case of an annealed powder,

$$(\epsilon'_{33})_{\Phi\psi} = d_{\Phi\psi} - d_0/d_0 = 0, \text{ thus, } d_{\Phi\psi} = d_0$$

---

*The equation for the difference in peak shift for a sample displacement ΔX from the center of the diffractometer between an angle ψ and $\psi = 0$ is[18]:

$$\delta(\Delta 2\theta)_{0,\psi} = (360/\pi)\Delta X \cos\theta \{(1/R_{GC}) - \sin\theta/(R_P^1\sin\theta + \psi)\}$$

where ΔX is the sample displacement, θ is the Bragg angle, R_{GC} is the goniometer radius and R_P is given by:

$$R_P^1 = R_{GC} \cdot (\cos(\psi + [90-\theta]))/\{\cos[\psi - (90+\theta)]\}$$

thus, the amount of split between $+\psi$ and $-\psi$ will be: $\Delta\delta(\Delta 2\theta) = \delta(\Delta 2\theta)_{0,\psi+} - \delta(\Delta 2\theta)_{0,\psi-}$. The difference is not a function of $\sin 2\psi$.

for all Φ,ψ independent of Δ. Thus the stress free powder method
of alignment is inadequate to detect the effects due to this
error. Verification of the true-zero for ψ movement presents no
problem on a diffractometer if a flat sample is used. For curved
samples or when a portable unit is used for measurements, a thin
sample with known stresses, but no σ_{13} ,σ_{23} , that can be placed
over the actual measurement surface is needed. An alternative is
to rotate Φ 180°. and repeat the $\pm\psi$ measurements. It can be seen
from equations (13) and (5) that while for true ψ-splitting the
$d_{\Phi\psi}+,d_{\Phi\psi}-$values will be interchanged, no such change will occur
for the missetting case, as:

$$k_{\Phi+180} = k_{\Phi} \qquad (\Phi = 0^{\circ}, 90^{\circ})$$

Magnitude of Errors Caused by an Error in True ψ Position

Computer simulation was used in this case also. For a given
error Δ , the "d" vs. $\sin^2\psi$ plots were synthesized and then analyzed using equations (6-a,b). In figure 10 the "d" vs.
$\sin^2\psi$ curve for $\Delta=3^{\circ}$ is shown. The following parameters are used
in the data analysis:

$$(\sigma_{11})_z = (\sigma_{22})_z = -400 \text{ mPa}, \langle \sigma_{33} \rangle_\psi = k_{33}\tau^2_\psi , \text{ all other } \sigma_{ij}=0$$

The σ_{33} gradient is the same as shown in figure 6

The analysis results for figure 10 are shown in table III .
It is seen that, as in the first case ($\Delta =0^{\circ}$, $\sigma_{13} > 0$),   the
most accurate results for σ_{ii} are obtained from the high
ψ-range. The error in σ_{13} (the results in this case are only due
to Δ) are highest for this range. This is again due to the in-
teraction of the $\sin|2\psi|$ term with a stress gradient, in this case
σ_{33} through k_{Φ} in equation (13).It is also evident that for
this case also, a_2 vs. $\sin|2\psi|$ will exhibit splitting over $0<\psi<60^{\circ}$.

in figure 11 the errors in σ_{ij} are plotted as a function of
Δ when the high ψ-range is used in the analysis. Also plotted, is
the apparent σ_{13}'values caused by the presence of Δ (versus Δ) for
both high and low ψ-ranges. It is seen that use of a low ψ-range
to evaluate the slope of a_2 vs. $\sin|2\psi|$ line will keep the appar-
ent σ_{13} value below 20 mPa ($\Delta<2^{\circ}$) which is generally within experimental error.

Conclusions

1) Oscillations in experimentally obtained "d " vs. $\sin^2\psi$ data
indicate presence of an inhomogeneous strain state in the surface
layers of the sample.

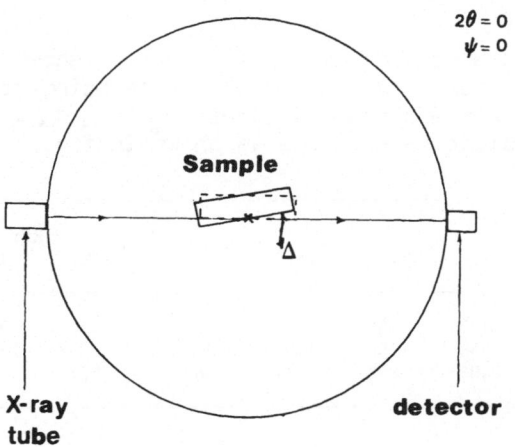

Figure 9: Definition of the "ψ-missetting" for a diffractometer.

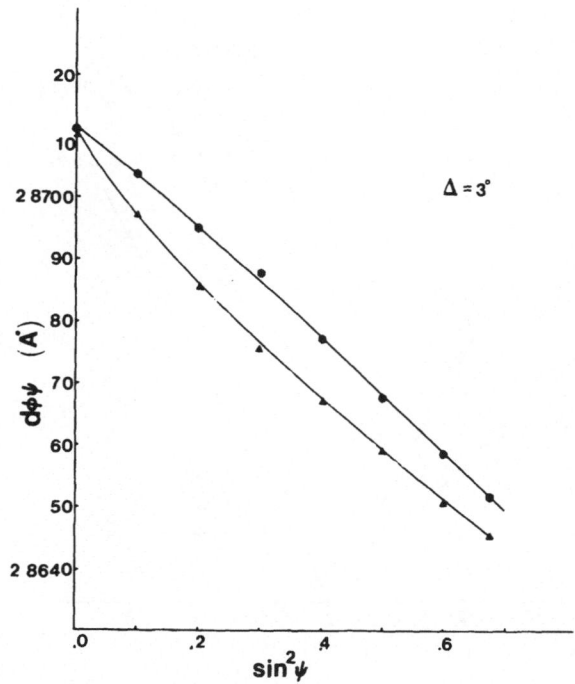

Figure 10: "d" vs. sin²ψ for a hypothetical specimen analyzed
in a diffractometer with a ψ-missetting of 3 .

Table III : Analysis of "d" vs. $\sin^2\psi$ curve   shown   in   figure
(10) .   (The   stressσ_{13} is due to the ψ-missetting error in this
case.) For this case $\sigma_{11} = \sigma_{22} = -400.$mPa, $\sigma_{33} = k_{33}\tau_x^2$,   all   other
$\sigma_{ij} = 0$.   The σ_{33} gradient is the same as shown in figure (6).

ψ-range	σ_{11}	σ_{22}	σ_{33}	σ_{13}	σ_{23}	Error in σ_{11}
0-60	-438	-438	107	26	26	38
0-33.21	-556	-556	56	30.5	30.5	156
39.23-60	-408	-408	82	56	56	9

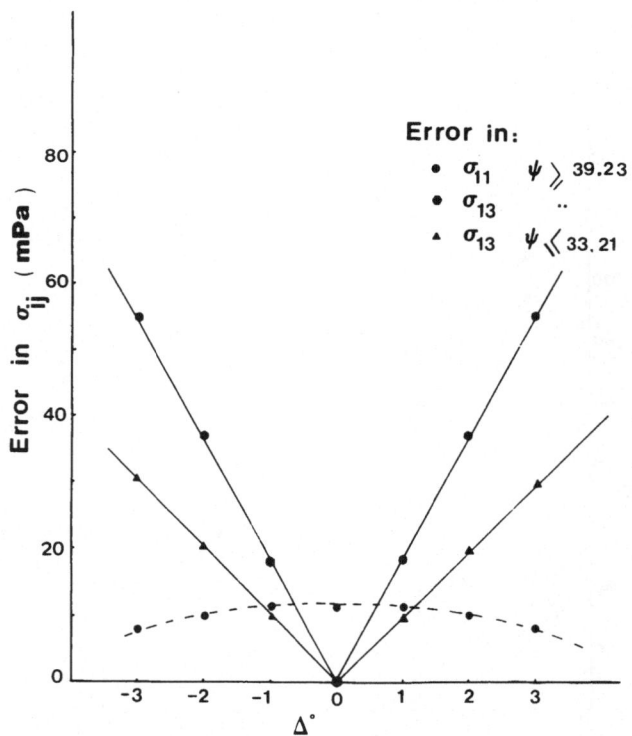

Figure 11:   Variation of errors in σ_{11}, σ_{13} with ψ-missetting.

2) The degree of inhomogeneity increases with the magnitude of oscillations and such inhomogeneity causes strain gradients in the surface layers.

3) When large oscillations are present in any one reflection obtained from a given irradiated volume, oscillations can also be present from any other reflection, obtained from the same volume, unless the volume fraction of inhomogeneous regions is small.

4) When strain gradients are only a function of depth in an irradiated volume, a "quasi-homogeneous" strain state may exist if certain other conditions, examined in the text, are fullfilled. In such a case, experimentally determined "d " vs. $\sin^2\psi$ lines deviate slightly from the curves predicted by the equations applicable to homogeneous strain distributions.

5) Even for quasi-homogeneous strain states, application of solutions developed for homogeneous strain states may cause large errors in the stress results if the methods are not appropriately modified for the true strain distribution in the surface layers.

6) Qualitative ideas about the actual strain distribution may be obtained from the deviation of a_1 vs. $\sin^2\psi$ and a_2 vs. $\sin|2\psi|$ plots from linearity. Our examinations here and previously published[9] show that when the quasi-homogeneous gradients in the irradiated volume obey a power law, the stress results obtained from the analysis described by Dolle will be more accurate if:

i) a_1 vs. $\sin^2\psi$ is analyzed over a high ψ-range,

ii) a_2 vs. $\sin|2\psi|$ is analyzed over a low ψ-range.

The linearity of a_2 vs. $\sin|2\psi|$ must be checked for each experiment before the linear range is decided upon.

7) A missetting Δ from the ψ true-zero position of a diffractometer will cause an apparent ψ-splitting which will have the same shape as a ψ-split caused by the shear stresses σ_{13}, σ_{23}. This error can not be detected by the stress-free powder technique currently in use.

8) The missetting error can be detected by rotating the specimen $180°$ in Φ and repeating the measurements for $\mp\psi$. if $(d_{\Phi\psi^+} = d_{\Phi\psi^-})_{\Phi=\Phi+180°}$ then the error is due to misalignment in ψ.

ACKNOWLEDGEMENTS

The financial support for this research was provided by ONR under contract No. N00014-C-0116. The X-ray measurements were made in the diffraction facility of Northwestern University's Materials Research Center, supported in part by the MRL-NSF Program under grant No. DMR-8216972. One of us (I.C. Noyan) thanks the Turkish Scientific Research Foundation (TUBITAK) for supplying a NATO grant, enabling him to study in the U.S.A. Also, we thank Mr. Paul Rudnik for help with the x-ray measurements.

REFERENCES

1. B. D. Cullity, Elements of X-ray diffraction, 2nd ed., Addison-Wesley, Reading MA., pp. 447-479 (1978)
2. C. S. Barret, T. B. Massalski, Structure of Metals, 3rd ed., McGraw-Hill, New York, NY, pp. 465-485 (1966)
3. T. Shiraiwa, Y. Sakamoto, X-ray Stress Measurement and Its Application to Steel,Sumito Search, 7:159-169 (1972)
Metall. Trans. A, 11 A: 159 (1980)
4. H. Dölle, J. B. Cohen, Residual Stresses in Ground Steels, Metall. Trans. A, 11 A: 159 (1980)
5. J. W. Ho, I. C. Noyan, J. B. Cohen, V. D. Khanna, Z. Eliezer: Residual Stresses and Sliding Wear, Wear, 84: 183 (1983)
6. H. Dölle, Influence of Multi-Axial Stress States, Stress Gradients and Elastic Anisotropy on the Evaluation of Residual Stresses by X-rays, J. Appl. Cryst. 12: 489 (1979)
7. J. B. Cohen, H. Dölle, M. R. James, Stress Analysis From Powder Diffraction Patterns,National Bureau of Standards Special Publication 567, pp 453-77 (1980)
8. I. C. Noyan, Equilibrium Equations for the Average Stresses measured by X-rays: Met. Trans. A., in press
9. I. C. Noyan, Effect of Gradients in Multi-Axial Stress States on Residual Stress Measurements with X-Rays.Met. Trans. A, 14-A:249(1982)
10. R. H. Marion, J. B. Cohen, Anomalies in Measurement of Residual Stress by X-ray Diffraction,in "Adv. in X-ray Analysis vol. 18" eds. W. 1. Pickles, C. S. Barret, J. B. Newkirk, C. O. Ruud, Plenum, New York, N.Y. (1978) 18: 466 (1975)
11. W. Lode, A. Peiter, Numerik Rontgenographischer Eigenspannungsanalysen Oberflaschennaher Schichten, Harterei Tech. Mitt. 32: 235 (1977)
12. J. F. Nye, Physical Properties of Crystals, Oxford,pp. 3-30.(1976)
13. H. Dölle and J. B. Cohen, Evaluation of (Residual) Stresses in Textured Cubic Materials, Met. Trans. A, 11 A: 831 (1980)
14. S. R. McEwen, J. Faber, A. P. L. Turner, The use of Time-of-flight Neutron Diffraction to the Study of Grain Interaction Stresses,Acta. Met., vol. 31: 657 (1983)
15. K. Hashimoto, H. Margolin, The Role of Elastic Intraction Stresses on the Onset of Slip in Polycrystalline Alpha Brass, Acta. Met. 31: pp. 773-785 (1983)
16. Ibid, , pp. 787-800
17. G. B. Greenough, Residual Stresses in Plastically Deformed Polycryst alline Metal Aggregates, Proc. Roy. Soc., 167-A:556 (1949)
18. M. R. James, Ph. D. Thesis, Northwestern Uni, Evanston, Ill., pp.74 (1977)
19. T. Mura, Micromechanics of Defects in Solids, Martinus Nijhoff, Hague (1982)

ON THE USE OF SYNCHROTRON RADIATION FOR THE STUDY

OF THE MECHANICAL BEHAVIOUR OF MATERIALS

Marc Barral, Jean-Michel Sprauel, Jean-Lou Lebrun,
Gérard Maeder, and Stephan Megtert

Ecole Nationale Supérieure d'Arts et Métiers
151, Boulevard de l'Hôpital
75640 Paris Cedex 13, France
Lure, Bat 209 C, UPS, 91405 Orsay, France

INTRODUCTION

The study of the mechanical behaviour of materials by X-ray measurements in a classical laboratory is limited by the possibilities of the X-ray tubes used. Some crystallographic planes are not conducive to good diffraction conditions and the radiation characteristics may not be optimum. The use of synchrotron radiation resolves many of these problems by providing a continuously variable wavelength which allows measurements of stress and stress gradients to be carried out in very good conditions. The high intensity and perfectly monochromated radiation with a small beam divergence are very helpful for microstrain measurements.

This paper presents the results obtained with a view to testing, on various diffraction planes, the REUSS hypothesis adopted in our calculations of the anisotropic elastic constants of a textured material using the Orientation Distribution Function. The different penetration depths obtainable with various diffracting planes allowed stress gradients to be determined on a ground steel. Also, the recording of $\{110\}$ and $\{220\}$ diffraction peaks on low carbon steel enabled us to use the WARREN-AVERBACH analysis for calculating the probability distribution function for the columns length of the coherent domains and for the mean square of microstrains.

LURE main facilities

The LURE laboratory (ORSAY, FRANCE) uses a synchrotron radiation emitted by the ORSAY electron storage ring which, under optimum conditions, works with an energy of 1.8 GeV and an electron current of 500 mA. The beryllium window used to separate the storage ring from the utilization line limits the emitted radiation to 4 Angstroms. At the energy level of 1.8 GeV and a wavelength of 1.6 Angstroms, 90% of the radiation passes through a cone of 0.36 mrad vertical aperture. This characterizes a very small beam divergence[1]. The installation used provides a continuously variable wavelength from 1.19 to 1.98 Angstroms, which is perfectly monochromated by a germanium curved crystal. The center of the goniometer is located at a distance of 10 meters from the monochromator.

In stress measurements, the use of this radiation allows us to work at a constant value of the diffraction angle 2θ, in an Ω-geometry. For best accuracy in the stress determination, it is necessary to have the diffraction angle 2θ as large as possible. Using a position sensitive detector (PSD), a constant angle of $2\theta = 156°$ was chosen for all the diffraction planes.

STRESS MEASUREMENTS ON TEXTURED MATERIAL

Principle of the method [2,3]

In a general way, the measured strains are linked to the macroscopic stresses by the relation already presented[4]:

$$\varepsilon_{\phi\psi} = F_{ij} \cdot \sigma_{ij}$$

where the terms F_{ij} are not tensor components, but mere coefficients. For quasi-isotropic materials, for instance, these F_{ij} coefficients depend only on the classical X-ray elastic constants $1/2\,S_2$ and S_1 and the sample orientation angles Φ and Ψ (Figure 1). Thus,

$$F_{ij} = 1/2\,S_2(hkl) \cdot \rho_{3i} \cdot \rho_{3j} + S_1(hkl) \cdot \delta_{ij}$$

with[2]

$$\rho_{ij} = \Phi_{ik} \cdot \Psi_{kj} \quad (\rho_{ij}, \Phi_{ij}, \Psi_{ij}: \text{transformation matrix components})$$

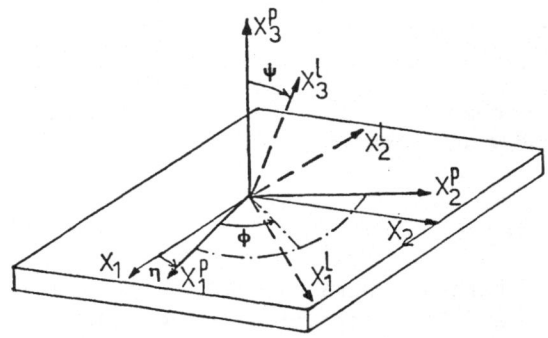

Figure 1   Definition of the measurement angles Φ and Ψ

For a textured material, on the other hand, the mechanical behaviour cannot be characterized by only two elastic constants and all the terms F_{ij} must be calculated.

To carry out calculations of the anisotropic elastic constants for a textured material, a program based on the REUSS hypothesis has been developped at ENSAM. This program uses the Orientation Distribution Function (ODF) for the quantitative characterization of the texture.

X-ray measurements

To determine the experimental curves of F_{11} vs $\sin^2\psi$, a series of stresses in the range 0 to 200 MPa have been applied to a specimen held in a four-point bending machine. The values of F_{11} are estimated from the following expression :

$$F_{11} \text{ meas.} = \frac{\partial \varepsilon_{\phi\psi}}{\partial \sigma_{11} \text{appl.}}$$

It is worth noting that F_{11} is the term most representative of the anisotropic mechanical behaviour.

Different diffracting planes were chosen for the value of their multiplicity factor and orientation factor $\Gamma(hkl)$ (Table 1).

Table 1 : LURE experimental conditions for stress measurements on a textured material

RADIATION (Ans.)	1.98	1.77	1.50	1.40
PLANES	{220}	{310}	{321}	{400}
MULTIPLICITY	12	24	48	6
ORIENTATION FACTOR	0.25	0.09	0.25	0.00
GEOMETRY	AUTOMATED GONIOMETER AND MONOCHROMATOR OMEGA GEOMETRY FOUR-POINT BENDING MACHINE ALUMINIUM BACK-FILTERED			
DETECTION	POSITION SENSITIVE DETECTOR (PSD)			
PHI ANGLE	$0°$			
PSI ANGLES	10 PSI : FROM $-42.13°$ TO $+39.23°$			

The sample used is a low carbon steel, cold-rolled 58%, annealed at 680 °C during 10 hours, uniaxially deformed in the rolling direction to 30% strain; stress-relief annealed at 250 °C and presents a main texture orientation {111}⟨112⟩.

For the $\{220\}$ planes (Figure 2), there is very good agreement between the measured and the calculated values of F_{11}, with some divergence for $\sin^2\psi < 0.1$ which at present we cannot explain. For the $\{310\}$ planes (Figure 3), the agreement is even better.

For the $\{400\}$ planes (Figure 4), however, considered as a texture independent plane in the REUSS hypothesis [5,6], demonstrated by the theoretical straight line, the measured term F_{11} shows an oscillation around that straight line and, so, in an absolute

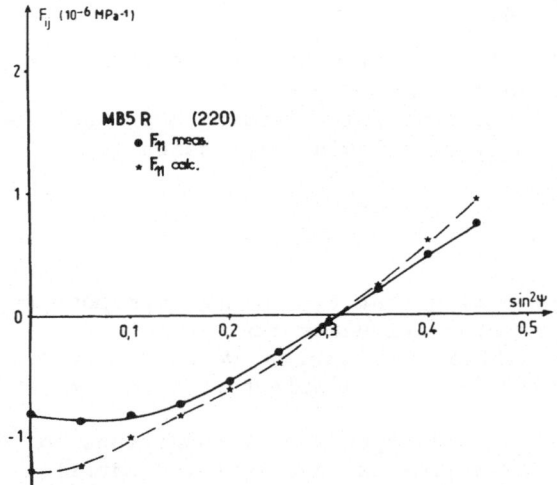

Figure 2   Comparison of the measured and calculated values of F_{11} for the $\{220\}$ planes

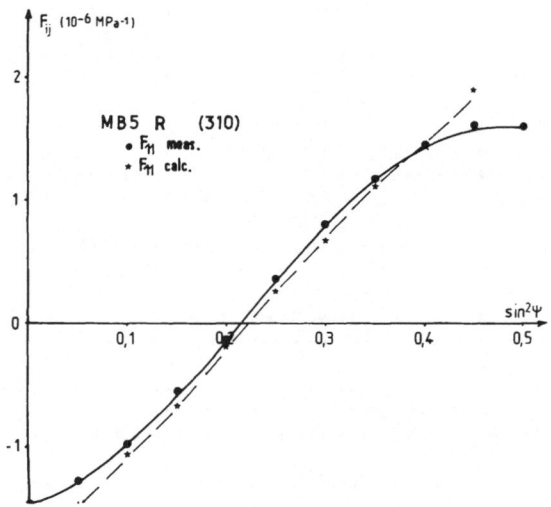

Figure 3   Comparison of the measured and calculated values of F_{11} for the $\{310\}$ planes

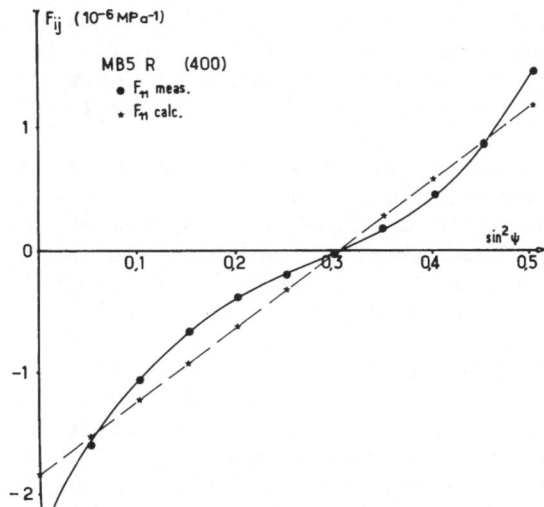

Figure 4   Comparison of the measured and calculated values of F_{11}
           for the {400} planes

sense, the {H00} planes are not texture independent. This result
also demonstrates the limitations of the REUSS hypothesis. These
oscillations cannot be explained by the evolution of the peak broadening.

Stress measurements

      The values of the applied and measured stresses are compared
using either the classical isotropic curve-fitting or the anisotropic one with the anisotropic elastic constants F_{ij} calculated.
This comparison for the {310} and {220} planes is shown in Table 2.

Table 2 : Comparison between the anisotropic fitting and
          the isotropic one for stress determination

PLANE	{310}		{220}	
APPLIED STRESSES (MPa)	MEASURED STRESSES (MPa)			
	ANISOTROPIC FITTING	ISOTROPIC FITTING	ANISOTROPIC FITTING	ISOTROPIC FITTING
0	1	1	5	5
35	40	42	39	35
70	72	75	76	67
105	112	118	98	85
140	149	158	137	121
175	183	193	174	152

For each plane, the anisotropic fitting leads to more accurate results particularly at higher stresses. It may be remarked also that isotropically determined stresses are higher than the applied ones for the {310} reflection and lower. for the {220} reflection.

This result may be linked to the value of the orientation factor for each one of these planes. Other results obtained for other planes allow us to conclude that for planes with an orientation factor $\Gamma(hkl)$ lower than 0.2, which is the value of the mechanical orientation factor Γ_0, isotropically determined stresses will be overestimated and vice-versa[2].

MEASUREMENTS OF STRESS GRADIENTS ON A GROUND STEEL

A theory underlying the method of measuring stress gradients by X-ray diffraction was presented by Sprauel et al[7] at the 31st DENVER Conference. The main point of this method is to carry out measurements for various X-ray penetration depths determined by the wavelength of the radiation used and the incidence angle Ψ. The utilization of various wavelengths implies diffraction by different planes for which the X-ray elastic constants vary with the crystallographic anisotropy of each one of these planes. The continuously variable wavelength from LURE radiation allows us to choose many planes with a better distribution of penetration depth from 2 μm to 12 μm and measurements may be carried out more rapidly than in a classical laboratory.

Table 3 : LURE experimental conditions for the authors'
          measurements of stress gradient on ground steel

RADIATION (Ans.)	(@)			
	1.50	2.29	1.98	1.77
PLANES	{321}	{211}	{220}	{310}
PENETRATION DEPTH (m)	2.2	5.8	8.7	12.1
ORIENTATION FACTOR	0.25	0.25	0.25	0.09
GEOMETRY	AUTOMATED GONIOMETER AND MONOCHROMATOR OMEGA GEOMETRY ALUMINIUM BACK-FILTERED			
DETECTION	POSITION SENSITIVE DETECTOR (PSD)			
PHI ANGLES	8 PHI: 0°, 45°, 90°, 135°, 180°, 225°, 270°, 315°			
PSI ANGLES	9 PSI : FROM −39.23° TO +39.23°			

(@) : OBTAINED IN A CLASSICAL LABORATORY WITH λ Cr Kα

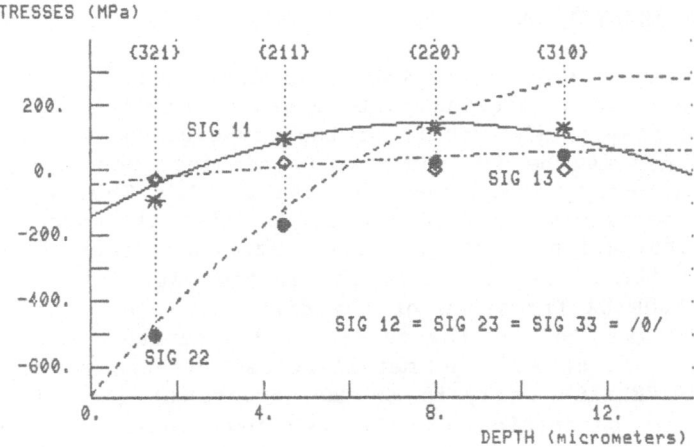

Figure 5   Stresses for the {321}, {211}, {220} and {310} planes as
a function of depth.   (Lines : Sprauel et al's theory[7],
◇,✳,●: Classical triaxial method of analysis.)

The sample used in these measurements was a ground alloy steel
(0.35% C, 1% Cr, 0.6% Mo).   Table 3 shows the experimental
conditions.

Figure 5 summerizes the results obtained and shows the variation of the different stresses as a function of the depth below
the surface.   The stress σ_{11} is parallel to the direction of lay.
Its value at the extreme surface is −150 MPa, which is characteristic of mild grinding, and shows a maxima in tension at a depth of
8 μm.   The gradient of σ_{11} near the surface is 70 MPa/μm.   The
stress σ_{22} is perpendicular to the direction of lay.   It is highly
compressive (−700 MPa) at the extreme surface with a maxima in
tension at a depth of 12 μm.   The gradient of σ_{22} near the surface
is 150 MPa/μm.   The stress σ_{13} is essentially zero, thus we have a
biaxial state of stresses and their gradients.

It is clear from these results that the first order approximation of the classical, triaxial methods is sufficient for stresses
to be determined at the mean depth of penetration, particularly for
low values of the latter.   Where there is a strong stress gradient
(e.g. σ_{22}) near the surface, the predictions of the two methods of
analysis at high penetration depth are significantly different.

MICROSTRAIN ANALYSIS ON A COLD ROLLED STEEL

Analysis of the broadening of a diffraction peak allows the microstructure of a material to be characterized[8,9]. The shape of a diffraction peak is the result of the convolution of three factors:
The first is the instrumental broadening, depending on the X-ray beam geometry, which must be minimized. The second and the third are metallurgical effects to be distinguished as follows: (i) the distribution of the particle size characterized by its mean value, (ii) the distribution of the lattice strains in each particle. The FOURIER transform of the diffraction peak obtained on an annealed specimen and on the cold-rolled specimen studied allow the instrumental effect and the metallurgical effects to be separated. The WARREN-AVERBACH analysis on two orders of the same reflexion leads to the separation of the two metallurgical effects which characterize the state of plastic deformation of the material. LURE facilities are very helpful for this type of analysis because of the high intensity of its radiation which provides a good statistical definition of the diffraction peak. The optimum geometry of the X-ray beam, with a very small divergence minimizes the instrumental broadening, thus allowing a very good deconvolution of the metallurgical effects from the instrumental one to be made. The perfectly monochromated wavelength obtainable from the LURE facility gives a very low background.

For a wavelength of 1.98 Å, peak acquisition was carried out on the $\{110\}$ planes at $2\theta = 58.5°$ and on the $\{220\}$ planes at $2\theta = 156°$. The storage of data needed about 30 min and processing 15 min. The experiments were conducted on a low carbon steel, cold-rolled at 70% for the specimen studied and annealed at 575 °C for the reference specimen.
The diffraction peaks obtained for the $\{220\}$ reflection are shown in Figure 6. There is a very low background as well as a high profile broadening induced by plastic deformation.

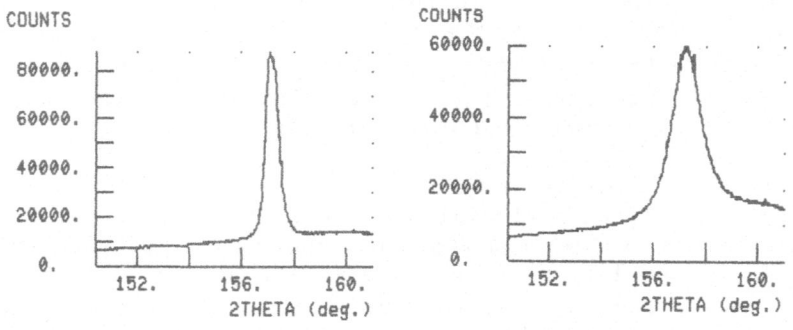

a) annealed   b) cold-rolled
Figure 6   Diffraction peaks for the $\{220\}$ planes

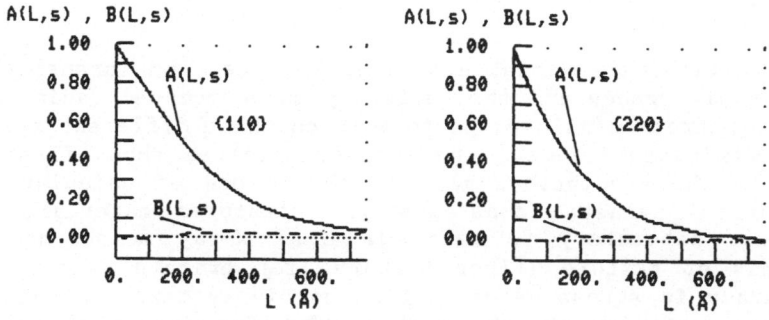

Figure 7   FOURIER coefficients for the cold-rolled specimen.

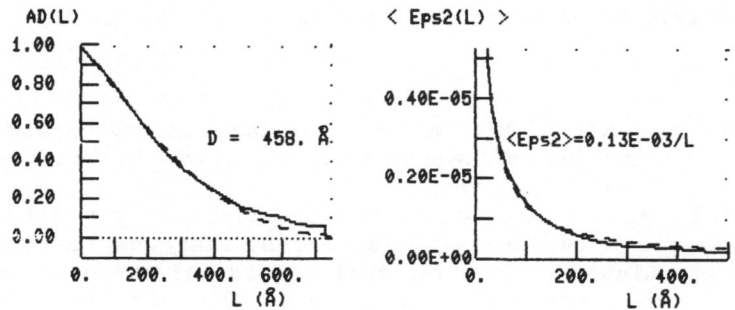

Figure 8   a) Function AD(L), b) Function $\varepsilon^2(L)$

The variation of the FOURIER transform with the columns length of the coherent domains can be seen in Figure 7. The coefficients are obtained after deconvolution from the instrumental effect and using the {110} and {220} reflections on the cold-rolled specimen.

Thus, A(L,s) is the real part of the FOURIER transform. It is representative of the particle size and of the distorsion induced by dislocations. B(L,s) is the imaginary part of the FOURIER transform. It is representative of stacking faults, and ought to be zero in cubic centered materials. Its value, here, is less than 0.04. From the FOURIER coefficients obtained for 2 orders of reflection, WARREN-AVERBACH analysis provides the function AD(L), representative of particle size effect and the function $\varepsilon^2(L)$ representative of distorsion effect (Figure 8). The curve AD(L) vs L (columns length) allows the determination of the mean value of the particle size, here D = 458 Å. It may be noticed that there is no hook or crochet effect in this analysis, which demonstrates the quality of the data obtained. The curve $\varepsilon^2(L)$ vs $\underset{-3}{L}$ is hyperbolic and its constant can be determined, here : 0.132×10.

CONCLUSIONS

The continuously variable wavelength, the high intensity, the small beam divergence and the perfectly monochromated radiation of the LURE synchrotron allowed us to work on three different areas of X-ray stress measurements. The results obtained show: first, the validity of the assumptions made in the method of calculation of the anisotropic elastic constants of a textured material, which enables our laboratory to make classical X-ray measurements on other kinds of texture (fiber texture, for example) with a very good accuracy in stress determination on highly textured material; second, the validity of the method used for determining stress gradients; third, the quality of the data which allows a micro-strain analysis to be carried out with confidence. These results will be used as reference in the further development of the single profile analysis being currently developed.

REFERENCES

1- P. DAGNEAUX et al, L'utilisation du rayonnement synchrotron en France (The use of synchrotron radiation in France), Ann. Phys., 9, 1975, pp. 9-65
2- M. BARRAL, Mesures de contraintes résiduelles par diffraction des rayons X sur des matériaux présentant une texture cristallographique (X-ray residual stress measurements on materials with a cristallographic texture), Doct.-Ing. thesis, Université PARIS VI, 1983
3- C. M. BRACKMAN, Residual stresses in cubic materials with orthorhombic or monoclinic specimen symmetry: influence of texture on Ψ splitting and non-linear behaviour, J. Appl. Cryst., 16, 1983, pp. 325-340
4- H. DOLLE, V. HAUK, Rontgenographische Ermittlung von Eigenspannungen in texturierten Werkstoffen (X-ray determination of residual stresses on textured materials), Z. Metallkde, 70, 1979, pp. 682-685
5- H. DOLLE, V. HAUK, Einfluss der mechanischen Anisotropie des Vielkristalls(Textur) auf die rontgenographische Spannungsermittlung (Influence of the mechanical anisotropy of polycrystals (texture) on the X-ray determination of stresses), Z. Metallkde, 69, 1978, pp. 410-417
6- H. DOLLE, J.B. COHEN, Evaluation of (residual) stresses in textured cubic metals, Met. Trans., 11A, 1980, pp. 831-836
7- J.M. SPRAUEL, M. BARRAL, S. TORBATY, Measurements of stress gradients by X-ray diffraction, Adv. in X-ray Anal., 26, 1982, pp. 217-224
8- B. E. WARREN, "X-ray diffraction", Addison Wesley P.C., USA, 1969
9- C. N. J. WAGNER, in "Local atomic arrangements studied by X-ray diffraction", ed. J. B. COHEN and J. E. HILLIARD, Gordon and Breach, NEW-YORK, 1965, pp. 219-270

THE MEASUREMENT OF ELASTIC CONSTANTS FOR THE DETERMINATION OF STRESSES BY X-RAYS

K.Perry, I.C.Noyan, P.J.Rudnik, and J.B.Cohen

Department of Materials Science, The Technological
Institute, Northwestern University, Evanston,
Illinois,60201

INTRODUCTION

Residual and applied stresses (σ_{ij}) are often measured via X-ray diffraction, by calculating the resultant elastic strains (ε_{ij}) from the measured change in interplanar spacing ("d "). This method is non-destructive, reasonably reproducible (typically ±14 MPa), can be carried out in the field[1], and is readily automated to give values to an operator-specified precision[2]. Let L_i represent the axes of the measuring system with L_3 normal to the diffracting planes, and P_i represent the sample axes. These axes are illustrated in Figure 1. In what follows, primed stresses and strains are in the laboratory system, while unprimed values are in the sample system. The strains in the direction L_3 are referenced with the angles Φ and ψ in Figure 1, and can be written in terms of the stresses in the sample[3]:

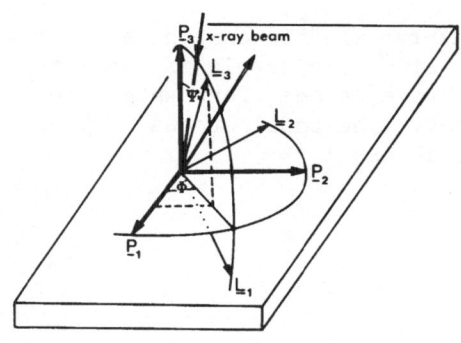

Fig. 1: P_i: Sample Axial System.

L_i: Measuring System.

$$(\epsilon'_{33})_{\Phi\psi} = \frac{d_{\Phi\psi}-d_o}{d_o} = \tfrac{1}{2}S_2\{\sigma_{11}\cos^2\Phi+\sigma_{12}\sin2\Phi+\sigma_{22}\sin^2\Phi-\sigma_{33}\}\sin^2\psi$$

$$+\tfrac{1}{2}S_2\sigma_{33}+S_1(\sigma_{11}+\sigma_{22}+\sigma_{33})+\tfrac{1}{2}S_2(\sigma_{13}\cos\Phi+\sigma_{23}\sin\Phi)\sin2\psi. \tag{1}$$

Here d_o is the d-spacing in a stress-free material, S_1 and $\tfrac{1}{2}S_2$ are the so-called X-ray elastic constants and the first term in parentheses on the right hand side of Equation 1 will be called $\sigma_{\Phi\psi}-\sigma_{33}$. The first three terms are the stress in the surface at an angle Φ to the σ_{11} axis.

For an isotropic material the X-ray elastic constants can be written in terms of Poisson's ratio (ν) and Young's modulus (E),

$$S_1 = -\nu/E \tag{2a}$$
$$\tfrac{1}{2}S_2 = (1+\nu)/E \tag{2b}$$

Other estimates for S_1 and $\tfrac{1}{2}S_2$, such as those by Neerfield[6] and Kröner[7], are also available. For an anisotropic material these values depend on texture and method of processing and must be uniquely measured.

The normal components σ_{33}, σ_{13}, and σ_{23} are zero at the surface, but the X-ray beam penetrates a sufficient depth so that their contribution can be detected[4,5]. Their presence leads to curvature in $d_{\Phi\psi}$ vs. $\sin^2\psi$, which for the shear terms is opposite in sense for $+\psi$ and $-\psi$. The presence of texture and/or the variation in stress from point to point under the X-ray beam can lead to large oscillations in this relationship[8,9]. If both effects are absent, $d_{\Phi\psi}$ vs. $\sin^2\psi$ is linear and from the slope $\sigma_{\Phi\psi}$ is obtained. This is the common practice, and in such a case measurements at only one Φ and two ψ tilts are sometimes employed. However, the absence of these effects must be verified before such a simple procedure is applied. Other procedures are available for more complex situations[5,9]. In any case, the measured X-ray elastic constants are required.

The simplest way to measure the X-ray elastic constants is to apply a uniaxial elastic load, say σ_{11}^{APP}, to a sample of the same material under the same conditions as the piece for which the strains to be measured will be used. The total stress is then $\sigma^{APP}+\sigma^{RES}$, where σ^{RES} is the residual stress :

$$\sigma^{APP}+\sigma^{RES}= \left[\frac{\partial(\epsilon'_{33})_{\Phi\psi}}{\partial\sin^2\psi}\right]\bigg/(\tfrac{1}{2}S_2) \quad . \tag{3}$$

When $\Phi = 0$, $\sigma^{RES} = \sigma_{11}^{RES}, \sigma_{33}^{RES}$,

$$\sigma^{APP} + \sigma^{RES} = \frac{\partial(\epsilon'_{33})_{\Phi\psi}}{\partial\sin^2\psi} \Big/ (\tfrac{1}{2}S_2)$$

$$= \frac{\partial d_{\Phi\psi}}{\partial\sin^2\psi} \Big/ \left(d_o \cdot \tfrac{1}{2}S_2\right) = m' \Big/ \left(d_o \cdot \tfrac{1}{2}S_2\right) . \tag{4}$$

Thus $\tfrac{1}{2}S_2$ can be obtained from the slopes (m'') of several plots of "d" vs. $\sin^2\psi$ at different σ_{11}^{APP}, that is :

$$\tfrac{1}{2}S_2 = m''/d_o , \tag{5a}$$

where:

$$m'' = \frac{\partial m'}{\partial\sigma_{11}^{APP}} . \tag{5b}$$

Similarly :

$$S_1 = m'''/d_o , \tag{6a}$$

where :

$$m''' = \frac{\partial d_{\Phi,\psi=0}}{\partial\sigma_{11}^{APP}} . \tag{6b}$$

Errors in the results result from both counting statistics and geometric errors. Consider first the statistical errors. James and Cohen have derived an equation for the variance (V) of m' (which is in terms of the variance of the peak location 2θ).

Assume that one has a straight line: $m' = m'' \cdot \sigma_{11}^{APP} + b$. Then using the equation[10]:

$$m'' = \frac{\sum_i(\sigma^{APP} - \overline{\sigma^{APP}})(m'_i - \overline{m}')}{\sum_i(\sigma_i^{APP} - \overline{\sigma_i^{APP}})^2} . \tag{7}$$

If $X = f(x_1, x_2, x_3, \ldots)$, V is given by the following[10]:

$$V(X) = \left(\frac{dX}{dx_1}\right)^2 V(x_1) + \left(\frac{dX}{dx_2}\right)^2 V(x_2) + \text{---} . \tag{8}$$

Applying this to Equation (7) :

$$V(m'') = \frac{\sum_i\left(\sigma_i^{APP} - \overline{\sigma^{APP}}\right)^2 V(m')}{\left[\sum_i\left(\sigma_i^{APP} - \overline{\sigma^{APP}}\right)^2\right]^2} \tag{9}$$

Therefore, combining this with Equation (5a) leads to :

$$V(\tfrac{1}{2}S_2) = V(m'')/d_o^2 . \tag{10}$$

Following the same procedure for S_1,

$$V(S_1) = V(m''')/d_o^2 \quad . \tag{11}$$

The principal instrumental errors are those due to sample displacement,   axis missetting, and horizontal X-ray beam divergence[11]. Formulae for the variance in 2θ due to these effects can be found in this reference, and the error propagated into S_1 and $\frac{1}{2}S_2$ using the above equations. The two variances can then be added. (It can be shown that for S_1, the instrumental factors for the stationary slit method are zero.)

To apply these equations requires a nearly linear "d" vs. $\sin^2\psi$ plot. It is unclear from a survey of the literature on X-ray elastic constants[12,13] that this has always been the case. Also, errors have usually been estimated after repeating the measurement only once. Proper evaluation of the errors by the methods described here has never been done. There are reports of large effects of plastic deformation on the elastic constants[14,15]. These may be valid, or could arise from large curvature or oscillations in "d" vs. $\sin^2\psi$. There are also reports of different stresses obtained from different peaks[15]. A new systematic determination of constants for the various reflections of practical interest is sorely needed. In this paper we describe an automated system for this purpose, by which the constants can be obtained to an operator specified precision.

The paucity of carefully determined X-ray elastic constants is not surprising. If six different ψ values and five stress levels are employed, the thirty measurements can take 18 to 24 hours with a normal detector. Automation is needed; also the use of a position sensitive detector can reduce the time by an order of magnitude[16].

HARDWARE

Our miniature tensile device is shown in Figure 2, mounted on a diffractometer. The specimen (A) is held in place by two grips (B), which have been precisely machined to minimize bending. One of the grips is attached to a gear assembly (C) to which a high torque Slo-Syn stepping motor is attached (D). There are 200 steps per revolution and movement is directed by a Motorola 8080 type microprocessor so that the specimen can be loaded and unloaded automatically.

The other grip is attached to a load cell (E), Model 41, manufactured by Sensotec Inc. of Columbus, Ohio. The load cell is bolted to a 0.5 inch thick circular metal plate which is attached

Fig. 2: Tensile device mounted on X-ray unit for measurement of
         elastic constants. A - tensile specimen; B - grips;
         C - gears; D - stepping motor; E - load cell; F - device
         holder; G - track; H - micrometer adjustment.

to the body of the load cell. The force on the sample is
transmitted through the grip via a threaded screw which runs
through the center of the cell. The Model 41 senses the deflec-
tion between the outer rim bolt holes and a threaded inner hub.
The cell used was designed for loads up to 5000 pounds.

The output of this cell is read with a 450-D single channel
amplifier, also manufactured by Sensotec,which provides a signal
conditioner and digital indicator. The mechanical strain on the
sample can be measured by either cementing a thin foil strain
gauge to the back of the sample, or attaching a clip-on extensome-
ter. This strain is read by a Model 4412 Voltmeter manufactured
by Data Technology Corp. The outputs of both the 450-D and the
4412 were modified so that they could be interfaced with the microprocessor.

The tensile device is mounted onto a sample holder (F), designed so that the tensile device can be moved horizontally, vertically, and rotated normal to the specimen surface.   This holder
is mounted onto a track (G) and can be moved along the track by
means of an attached micrometer (H), allowing for accurate specimen positioning.   All 2θ and ψ movements were made by the Slo-Syn
motors, via computer control, while the counting was recorded by
the microcomputer.

SOFTWARE

The computer package is written in XYBASIC, a computer
language copyrighted by the Mark Williams Chemical Company of Chicago, Illinois and designed especially for process control, data
acquisition, and real time applications with 8080-based computers.
Our package for elastic constant determination contains the following features :

a. A separate alignment program for determination
   of sample displacement. (This is determined
   from the slope of the lattice parameter a_0 vs.
   the Nelson-Riley function for three or more
   peaks.)
b. On-line peak location using a least-squares
   parabolic fit to the top of a peak.
c. Determination of elastic constants to an operator specified accuracy, or using a preset number of counts.
d. Operator specification of stress values to be
   used in measurement.
e. Operator choice of psi tilts to be used in
   measurement.
f. Operator choice of number of data points to be
   used for parabolic fit to a peak.
g. Option of scattering factor correction.
h. Operator choice of preliminary and final scan
   steps and counts.
i. Optional linear background subtraction.
j. Optional sample oscillation through $\pm\theta$.
k. Optional peakshift correction. (This is due to
   the effect of $K\alpha_2$ on the $K\alpha_1$ position, which
   varies with the peak shape.)
l. Calculation of statistical error with the
   optional calculation of geometric error, due
   to divergence and effects of sample and/or
   psi axis displacement.
m. Calculation of Young's modulus using an attached
   mechanical strain gauge, and the corresponding
   stress-strain plot.

n. Plots of d vs. $\sin^2 \psi$ for all stress values;
   also, plots for m' vs. stress and for $d_{\psi=0}$ vs.
   stress.
o. Use of any detector.
p. Storage of data on a separate flexible disk for
   use with a separate data manipulation program, if
   changes in various terms are desired.

A multiple scan procedure is employed for peak location and
to make an estimate of the time required to achieve a desired precision. This is described in reference 2. It is accomplished by
multiplying the time needed for a single peak by the number of
ψ and σ^{APP} values to be employed. This allows the operator an opportunity to choose a larger error if the time is excessive. Tests
of the device are described below.

EXPERIMENTAL DETAILS

The materials examined and their preparation are described in
Table I. Flat tensile specimens were cut to dimensions of 2.75
inches long by 0.4 inch wide and had reduced sections which were
1.75 inches long by 0.25 inch wide. Typical operation conditions
are given in Table II . It is to be emphasized that oscillations
of the sample on the diffractometer can considerably reduce oscillations in d vs. $\sin^2 \psi$. Although it was not done here, it is
also sometimes helpful to shot peen or grit blast a sample. This
minimizes texture in the surface and can also reduce oscillations.

Table I. Sample Preparation

Specimen	Starting Thickness	Treatment	Final Thickness
1100 Al	.45"	cold rolled to a 90% reduction	.045"
70-30 α-Brass	.247"	cold rolled to a 90% reduction	.024"
304 stainless steel	.059"	cold rolled as received	.059"
1075 steel	.032"	cold rolled as received	.032"
Ni	.031"	cold rolled as received	.031"

Table II. Operating Conditions

Beam Size on Sample   .15" x .45"
Divergent Slit   1°
Receiving Slit   .18°

Tube Voltage   Cu - 40kV   Fe - 40kV
Tube Current   Cu - 20mA   Fe - 15mA
Goniometer Radius   8.125"
Six ψ Tilts
Seven Point Parabolic Fit
2° Oscillation to Reduce Oscillations in d vs. $\sin^2\psi$
10,000 - 15,000 Counts/Point
Background Subtraction
No Scattering Factor Correction
No Peak Shift Correction

RESULTS

Replicate measurements with nickel are given in Table III. The columns labelled "stat" give errors which are estimated from Equations 10 and 11. It can be seen that these are somewhat less than the actual variation. A similar set of data for a brass sample with a preset error of about 20 percent of the S value (rather than the 5 percent error used with the nickel specimen) gave good agreement with the calculated error. Therefore, unless the error is set very low, Equation 10 does give an estimate of the error in $\frac{1}{2}S_2$ with only a single measurement. Errors in S_1 are often larger than the statistical estimates. This is probably due to the fact that any oscillations or curvature in d vs. $\sin^2\psi$ violates the initial assumption of linearity.

An attempt was made to see if any other factors might affect the results. A dial gauge placed on the sample indicated that some displacement occurred during and after loading. For the most part, the displacement was 2×10^{-3} inch or less. Occasionally displacements as large as 5×10^{-3} inch were found. Calculations indicated that the largest change in $\frac{1}{2}S_2$ due to this effect would be 3 percent. The constant S_1 is unaffected by this when the stationary slit method is used.

Some stress relaxation occurred during measurements at a given load. For aluminum, this could change $\frac{1}{2}S_2$ by as much as 6 percent for a 400 reflection, and 1.5 percent for the 422. For softer materials the change was much less (0.1 percent for nickel).

A comparision of the nickel results with other data is given in Table IV. Results for other materials tested are shown in Tables V and VI. Included are some h00 and hhh reflections;

Table III. Results of 10 Replicate Measurements of
Elastic Constants Using the 313 Reflection
of Nickel

Run #	XREC* $S_2/2$	Stat.	Error Instr.	Total	XREC* S_1	Stat.
1	4.740	.216	.028	.218	-.757	.102
2	3.655	.227	.030	.229	-.411	.064
3	4.116	.195	.029	.197	-.739	.059
4	4.004	.221	.030	.223	-.587	.069
5	4.000	.197	.029	.199	-.606	.063
6	4.210	.210	.029	.212	-.742	.062
7	4.128	.199	.029	.201	-.776	.063
8	3.593	.185	.028	.187	-.635	.050
9	3.763	.211	.029	.213	-.518	.082
10	4.330	.187	.029	.189	-.611	.057
Mean	4.054	.204	.029	.206	-.638	.067
St. Dev.	.340				.117	

*Units of 10^{-8} psi^{-1}

Table IV. Elastic Constants of Ni: 313 Reflection
(in 10^{-8} psi^{-1})

Method	$S_2/2$	S_1
This work	4.05 ± .34	-.64 ± .11
Mechanical Measurement**	4.49	-1.06
X-Ray Experimental Calibration*	3.83 ± .14	-.83 ± .04
Voight (Constant Strain)*	3.81	-.84
Reuss (Constant Stress)*	3.66	-.79
Neerfield (Average of Voight and Reuss)*	3.73	-.82
Kröner***	3.58	-.77
Calculated from Handbook****	4.37	-1.03

*   Reference 13
**   E. Macherauch, Experimental Mechanics, 6, (1966), pp. 140-153.
*** Calculated from single crystal data.
**** Metals Handbook ASM, Metals Park, Ohio.

Table V. Experimental and Theoretically Calculated
Values of $S_2/2$ (in 10^{-8} psi^{-1})

Material	λ	hkl	$S_2/2$	Total Error	Voight	Reuss	Neerfield[6]	Kröner[7]
Al	Cu	422	12.19	.27	13.13	12.84	12.99	13.01
	Fe	400	10.49	.28	13.13	14.97	14.05	13.96
			11.28	.25				
α-Brass	Cu	331	6.94	1.22	6.85	7.23	7.04	6.98
	Fe	222	4.22	.82	6.85	4.83	5.84	6.14
			4.36	.83				
304 stainless steel	Cu	331	4.48	.20	4.01	3.82	3.92	3.93
	Fe	222	3.75	.55	4.01	3.09	3.55	3.63
			3.51	.38				
1075 steel	Fe	220	4.17	.17	4.01	4.12	4.07	4.06
	Cu	222	3.05	.24	4.01	3.09	3.55	3.63
			2.41	.25				
Ni	Fe	311	4.04	.35	3.64	4.98	4.31	4.19
	Fe	222	3.12	.25	3.64	2.76	3.20	3.28
			3.57	.24				

Table VI. Experimental and Theoretically Calculated
Values of S_1 (in 10^{-8} psi^{-1})

Material	λ	hkl	S_1	Total Error	Voight	Reuss	Neerfield	Kröner
Al	Cu	422	-3.81	.08	-3.38	-3.29	-3.34	-3.35
	Fe	400	-3.03	.11	-3.39	-4.00	-3.70	-3.67
			-3.20	.09				
α-Brass	Cu	331	-1.31	.39	-1.64	-1.77	-1.71	-1.69
304 stainless steel	Cu	331	-.94	.05	-.81	-.74	-.78	-.78
1075 steel	Fe	220	-1.05	.06	-.81	-.85	-.83	-.83
	Cu	222	-.77	.06	-.81	-.50	-.66	-.68
			-.76	.06				
Ni	Fe	311	-.61	.10	-.78	-1.23	-1.01	-.97
	Fe	222	-.28	.08	-.79	-.50	-.73	-.67
			-.21	.08				

ignoring grain interaction stresses, theory indicates that oscillations in d vs. $\sin^2\psi$ due to texture should be eliminated. However, this does not always occur. For α-brass and nickel, there
was some reduction in such oscillations for the peaks shown in
Table V. In both cases the hhh reflection is at the same or higher
2θ value as the hkl reflection; thus any oscillations should be
equally clear since the peakshift $\Delta 2\theta$ is proportional to tan θ.

If the oscillations are not large, two ψ tilts are sufficient.
Recalculating the elastic constants in Table III for $\psi = 0^\circ$ and 45°
changes $\frac{1}{2}S_2$ by only 3 percent.

In summary, software and hardware for the fully automated determination of X-ray elastic constants have been demonstrated with
several materials. Equations have been developed and tested to
allow estimates of the error in these constants without repeating
the measurement, regardless of whether or not automation was used.
It is hoped that future reports on these constants will include
such error estimates.

ACKNOWLEDGEMENTS

The financial support for this research was provided by ONR
under contract number N00014-C-0116. The authors thank Mr. J.
Hahn for the construction of the tensile device, Mr. P. Weiss
for his assistance in interfacing the stress and strain measurements, and Mr. G. Raykhtsaum for assistance with the programming
and in setting up the diffractometer. The measurements were made
in the X-ray diffraction facility of Northwestern University's Materials Research Center, supported in part by the MRL-NSF program
under grant number DMR-8216972. This paper constitutes a portion
of a thesis submitted (by K. Perry) in partial fulfillment of the
requirements for the M.S. degree at Northwestern University.

REFERENCES

1. M.R. James and J.B. Cohen, "PARS"- A Portable
   X-ray Analyzer for Residual Stresses, J. of
   Testing and Evaluation, 6:91 (1978).
2. M.R. James and J.B. Cohen, Study of the Precision of X-Ray Stress Analysis, Adv. in X-Ray
   Anal., 20:291 (1977).
3. H Dölle, The Influence of Multiaxial Stress
   States, Stress Gradients, and Elastic Anisotropy on the Evaluation of (Residual) Stress-
   by X-Rays, J. Appl. Crystall., 12:489 (1979).

4. H.Dölle and V. Hauk, Einfluss der mechanischen
    Anisotropie des Vielkristalls (Textur) auf die
    rontgenographische Spannungsermittlung, $\underline{Z. Met}$-
    $\underline{allkunde}$, 69:410 (1978).

5. H. Dölle and J. B. Cohen, Residual Stresses in
    Ground Steels, $\underline{Met. Trans. A}$, 11A:159 (1980).

6. H. Neerfeld, The Calculation of Stress from
    X-Ray Elongation Measurements, $\underline{Mitt. KWI Eis}$-
    $\underline{enforsch.}$, 24:61 (1942).

7. E. Kröner, Berechnung der elastischen Konstan-
    ten des Vielkristalls aus den Konstanten des
    Einkristalls, $\underline{Z. Physik}$, 151:504 (1958).

8. C.M. Brakman, Residual Stresses in Cubic Mater-
    ials with Orthorhombic or Monoclinic Specimen
    Symmetry: Influence of Texture on ψ-Splitting
    and Non-linear Behaviour, $\underline{J. Appl. Crystall.}$,
    16:325 (1983).

9. I.C. Noyan and J.B. Cohen, Determining
    Stresses in the Presence of Non-linearities
    in Interplanar Spacing vs. $\sin^2\psi$, this volume.

10. W. Volk,"Applied Statistics for Engineers",
    McGraw-Hill Book Co., New York City (1958).

11. M.R. James and J.B. Cohen, The Measurement of
    Residual Stresses by X-Ray Diffraction Tech-
    niques, $\underline{Treatise on Materials Science and}$
    $\underline{Technology}$, 19A:1 (1980).

12. Society of Automotive Engineers, "Residual
    Stress Measurement by X-Ray Diffraction", SAE
    Handbook J784a, 2nd. ed., Soc. Auto. Eng.,
    Inc., New York City (1971).

13. K. Kolb and E. Macherauch, Der Anisotropie-
    einfluss rontgenographische Gitterdehnungs-
    messungen an Nickel, $\underline{Z. Physik}$, 162:119
    (1961).

14. A.L. Esquivel, X-Ray Diffraction Study of the
    Effects of Uniaxial Plastic Deformation on
    Residual Stress Measurements, $\underline{Adv. in X-Ray}$
    $\underline{Anal.}$, 12:269 (1969).

15. R.H. Marion and J.B. Cohen, The Need for Ex-
    perimentally Determined X-Ray Elastic Con-
    stants, $\underline{Adv. in X-Ray Anal.}$, 20:355 (1977).

16. M.R. James and J.B. Cohen, The Application of
    a Position Sensitive X-Ray Detector to the
    Measurement of Residual Stresses, $\underline{Adv. in}$
    $\underline{X-Ray Anal.}$, 19:695 (1976).

X-RAY STRAIN MEASUREMENTS IN IV-VI-SEMICONDUCTOR SUPER-LATTICES AT LOW TEMPERATURE

E.J. Fantner, H. Clemens, and G. Bauer

Institut für Physik, Montanuniversität Leoben

A-8700 Leoben, Austria

ABSTRACT

Multilayers composed of many thin films of PbTe and $Pb_{1-x}Sn_xTe$ on BaF_2 substrates were grown epitaxially by hot-wall-vapor deposition. In order to investigate the fraction of the total misfit (2.5×10^{-3} at x=0,12) accommodated by misfit strain we have performed strain measurements on these superlattices by two different X-ray diffractometer techniques. We also report on substrate induced strain due to different thermal expansion coefficients of films and substrate. For film thicknesses smaller than 300 nm there is clear evidence for almost complete accommodation of lattice mismatch by misfit strain. Below room temperature the substrate induces a tensile strain which is comparable to that of the misfit strain.

INTRODUCTION

The accommodation of lattice mismatch across the interface between epitaxial films has been treated by van der Merwe[1]. Below a critical film thickness it is energetically favourable for misfit to be accommodated by uniform elastic strain. With increasing film thickness he proposes an accommodation shared between dislocations and strain. This behaviour of lattice mismatch accommodation offers the opportunity of growing superlattices of high crystalline quality besides structures grown by lattice matched constituents. For sufficiently thin layers and a lattice mismatch of the unstrained SL-con-

171

stituents not too high, the lattice mismatch is totally
accommodated by strain, so that no misfit defects are
generated at the interfaces[2,3]. For example, $GaP/GaAs_{0.4}$
$P_{0.6}$ strained-layer superlattices were grown at a misfit of 1.5×10^{-2} up to individual layer thicknesses of
about 60 Å without generation of misfit defects[4].

For IV-VI-heterojunctions it is well known that the
interface recombination velocity increases with increasing lattice mismatch. To minimize the number of misfit
dislocations at the interface, ternary and quarternary
lattice-matched heterostructure lasers were grown with
very low threshold current densities[5]. This condition
of latticematch restricts the variety of combination of
layer materials for the use of the wide spectral range
of IV-VI-lasers. The growth of dislocation free strained
layer IV-VI-superlattices probably overcomes these limitations as was demonstrated by the continuous 300 K laser
operation of strained $GaAs-In_xGa_{1-x}As$ superlattices for
III-V-compounds[6].

Although various methods to measure strain in single
crystalline films have been reported[7-10], to our knowledge no direct strain measurements on multiple hetero-
structures or superlattices have been performed. It is
the purpose of this paper to report for the first time
on low temperature strain measurements which yield information on the strain of the individual superlattice
constituents.

EXPERIMENTAL

PbTe- and $Pb_{0.88}Sn_{0.12}Te$ films were deposited by a
modified hot-wall technique[11] from PbTe and PbSnTe
sources after initially depositing a 300-500 nm PbTe-
buffer layer on cleaved {111}-BaF_2. Various samples with
individual layer thicknesses of 30-300 nm and up to 50
layers were grown. The abruptness of the interfaces was
tested by Auger depth profile measurements[12].

To determine the strain in the PbTe- and the
$Pb_{1-x}Sn_xTe$-layers as well as the substrate, the lattice
constants of various lattice planes parallel and inclined to the interface were measured. Using a standard
Ω-diffractometer the $CuK\alpha$ - and $MoK\alpha$ - reflections were
measured at the highest order possible. The net accura-
cy of the technique was better than 2×10^{-4} in all cases.
The lattice parameter of unstrained single crystalline
BaF_2 (6.2000 Å) served as internal standard.

For the experiments below 300 K a special X-ray cryostat connected to a two stage closed-cycle refrigerator
was used (Fig.1). The sample holder was attached to the

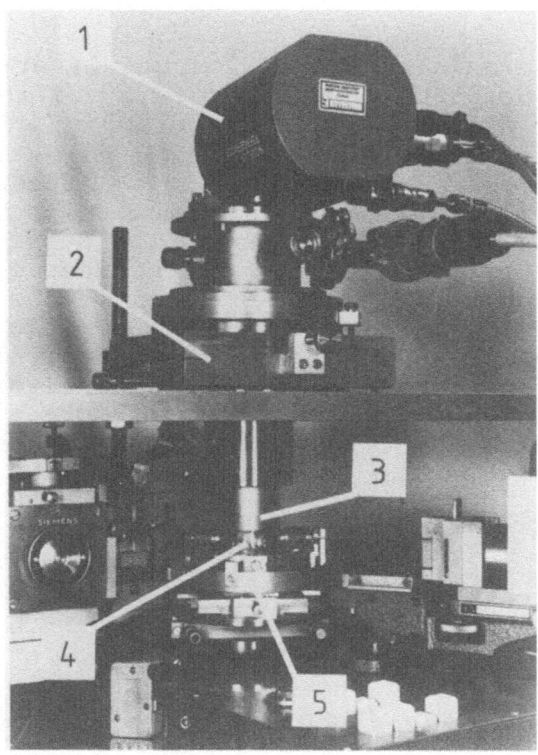

Fig.1   Set up of the low temperature X-ray diffractometer
        (1) 2-stage closed cycle cold head, (2) adjustable
        cold head support (3 translation, 1 rotation fa-
        cilities), (3) X-ray cryostat with mylar windows
        and flexible bellows, (4) sample, (5) fine-adjust-
        ment attachment.

second stage of the cold head by a OFHC heat wick to
provide good thermal conductance. The sample is mechanically connected with the fine-adjust attachment by two
reinforced teflon mounting rods, which are fed through
the two flexible bellows. Thus, the cryostat and the
sample adjustment are mechanically decoupled. This allows
a precise orientation of the sample even at low temperatures by rotation (two perpendicular rotation axes in
the diffractometer plane) and translation in the direction normal to the sample surface. This low temperature
X-ray cryostat with its fine-adjustment equipment fits
to the standard Siemens goniometer and is easily adapted
for any horizontal or vertical diffractometer.

RESULTS

Elastic strain in the individual layers - compressive
or tensile - ought to be manifested by different lattice
constants for different crystallographic orientations
with respect to the interface normal direction. For a
tensile strain one expects a maximum lattice constant
in the interface plane and a minimum value in the perpendicular direction. For compressive strain the opposite situation occurs. Measured differences of the lattice
constants for various lattice planes of the <211>-zone
of the superlattice constitutents (PbTe, $Pb_{0.88}Sn_{0.12}Te$)
and the unstrained BaF_2 are shown in Fig.2 for a sample
of SL period of 200 nm. The 300 K data clearly show that
the PbTe-layers are subjected to a compressive strain,
the PbSnTe-layers to a tensile strain in the film plane.
From the slopes of the lattice constants vs. $\sin^2\psi$ [13,14]

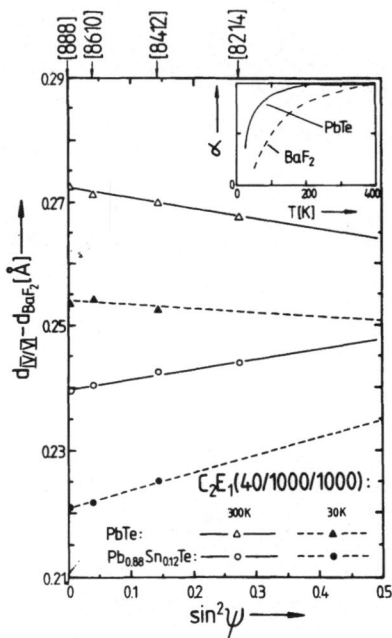

Fig.2   Experimental and calculated differences of SL-lat-
        tice constants at 300 K and 30 K for various pla-
        nes of the <211>-zone.The sample parameters are:
        40 layers of 100 nm thick PbTe- and PbSnTe-layers.
        (Insert: thermal expansion coefficients of PbTe
        and BaF_2 vs. temperature)

for PbTe and PbSnTe, which are almost equal but of opposite sign, we deduced an absolute value of strain of
$(1.3\pm0,1)\times10^{-3}$.ψ is the angle of inclination between
the lattice plane investigated and the interface plane.
At a temperature of 30 K the PbSnTe shows a strongly enhanced tensile strain, whereas the PbTe compressive
strain is reduced by the same amount as the strain in
PbSnTe is increased.

        The 300 K results of Fig.2 are confirmed by the da-
ta presented in Fig.3 for 3 samples of individual layer
thicknesses ranging from 30-300 nm. As is illustrated in the insert,a strain in the layer plane changes
the angle of inclination of the lattice planes tilted
relative to the interface direction. This means, that
the angle ψ, which describes the orientation of a certain lattice plane in an unstrained crystal, is increased for a compressive and decreased for a tensile strain.

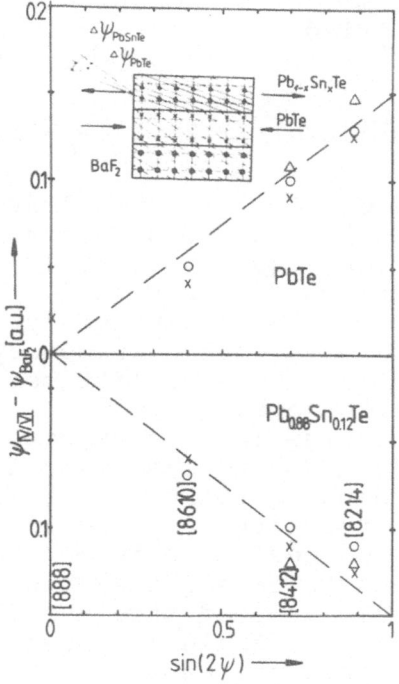

Fig.3   Measured inclination angles of planes of the <211>
        zone for 3 samples of different SL-periodicity:
        o...60 nm, x...200 nm, Δ...600 nm.

For the special symmetry in our samples - two of the
main axes of strain parallel to the film plane, the third
perpendicular to it - a linear dependence of $\Delta\psi$ on sin
(2ψ) is valid [14]. Within the experimental uncertainty the
data yield quantitatively the same strain for all the
three samples despite of the extremly different layer
thicknesses.

DISCUSSION

      The solid lines in Fig.2 are calculated values de-
rived for the assumption of the lattice mismatch of
2.5×10^{-3} to be totally accommodated by misfit strain. For
equal thicknesses and the elastic constants of the PbTe-
and the PbSnTe-layers the strain is shared between the
two constituents. The good quantitative agreement suggests that only few misfit dislocations are generated to
accommodate the lattice mismatch. From Fig.3 one can deduce that this suggestion holds for layers as thick as
300 nm. This is considerably thicker than all corresponding data known from III-V compounds. This means that the
critical thickness h where it becomes energetically favourable to create misfit dislocations is larger than
300 nm. In a simplified model for isotropic crystals proposed by Matthews [15] h is given by equ.(1):

$$h = G_A \cdot b \left(\ln(h/b) + 1 \right) / 2\pi \cdot f \cdot (G_A + G_B)(1+\nu) \qquad (1)$$

The elastic moduli G_A, G_B are approximately equal for
PbTe and PbSnTe. Using a Burgers vector b=4,59Å and
Poisson ratio ν=0.066, equ.(1) gives 100 nm for the relevant lattice mismatch f of 2.5×10^{-3}. This means that
the model of Matthews underestimates the critical thickness in IV-VI-compounds. This discrepancy is consistent
with data from PbGeTe/PbTe superlattices [16] deduced from
etch pits at the interface.
      At low temperatures BaF_2 is expected to induce addi-
tional tensile strain due to its lower thermal expansion coefficient compared to PbTe (see insert in Fig.2).
We have calculated the strain due to the difference of
the thermal expansion coefficients of the film and the
substrate.   For the PbTe- and the PbSnTe-layers the
sum of this thermally induced strain and the calculated
room temperature misfit strain is shown by the broken
lines in Fig.2. Despite the enormous strain which is observed the good quantitative agreement rules out any
glide processes between the substrate and films or the
films themselves.

ACKNOWLEDGEMENTS

Work supported by Fonds zur Förderung der wissenschaft-
lichen Forschung and by Forschungsförderungsfonds der
gewerblichen Wirtschaft.

REFERENCES

1) J.H. van der Merwe, On the stresses and energies associated with intercrystalline boundaries, Proc.Phys.
   Soc. A63: 616 (1950).
2) J.W.Matthews, and A.E.Blakeslee, Defects in epitaxial
   multilayers: Misfit dislocations, J.Cryst.Growth 27:
   118 (1974).
3) J.E.Hilliard, Artificial layer structures and their
   properties, in: "Modulated Structures", J.M.Cowley,
   J.B.Cohen, M.B.Salamon, and B.J.Wuensch, ed., American Institute of Physics, New York (1979).
4) G.C.Osbourn, R.M.Biefeld, and P.L. Gourley, A GaAs$_x$
   P$_{1-x}$/GaP strained-layer superlattice, Appl.Phys.Lett.
   41: 172 (1982).
5) Y.Horikoshi, M.Kawashima, and H.Saito, PbSnSeTe-
   PbSeTe lattice matched double-heterostructure lasers,
   Jap.J.Appl.Phys. 21: 77 (1982).
6) M.J.Ludowise, W.T.Dietze, C.R.Lewis, M.D.Camras,N.
   Holonyak, Jr., B.K.Fuller, and M.A.Nixon, Continuous
   300 K laser operation of strained superlattices, Appl.
   Phys.Lett. 42: 487 (1983).
7) E.W.Hearn, Stress measurements in thin films deposited
   on single crystal substrates through X-ray topography
   techniques, Adv. X-ray analysis, 20: 273 (1977).
8) E.Estop, A.Izrael, and M.Souvage, Double-crystal
   spectrometer measurements of lattice parameters and
   X-ray topography on heterojunctions GaAs/Al$_x$Ga$_{1-x}$As,
   Acta Cryst. A32: 627 (1976).
9) W.J.Bartels and W.Nijman, X-ray double-crystal diffractometry of Ga$_{1-x}$Al$_x$As epitaxial layers, J.Cryst.
   Growth 44: 518 (1978).
10) A.Bohg, Measurements of stresses in thin films on
    single crystalline substrates, phys.stat.sol.(a)
    46: 445 (1978).
11) H.Clemens, E.J.Fantner, and G.Bauer, Hot wall epi-
    taxy system for growth of multilayer IV-VI- com-
    pound heterostructures, Rev.Sci.Instr. 54:685(1983).
12. K.E.Ambrosch, H.Clemens, E.J.Fantner, G.Bauer, M.
    Kriechbaum, P.Kocevar, and R.J.Nicholas, Structural
    and electronic properties of PbTe/Pb$_{1-x}$Sn$_x$Te super-
    lattices, Surf.Sci. (in print).

13) E.J.Fantner, B.Ortner, W.Ruhs, and A.Lopez-Otero,
    Misfit strain in epitaxial IV-VI-semiconductor films,
    Lecture Notes in Phys. 152: 59 (1981).

14) B.Ortner, Röntgenographische Spannungsmessung an ein-
    kristallinen Proben, in "Eigenspannungen (residual
    stresses)", Deutsche Gesellschaft f.Metallkunde, ed.,
    DGM, Karlsruhe (1983).

15) J.W.Matthews, Coherent interfaces and misfit dis-
    locations, in "Epitaxial Growth" (PartB), A.H.Alper,
    J.L. Margrave, and A.S. Nowick, ed., Academic Press,
    New York (1975).

16) D.L.Partin, Growth of PbGeTe - thin film structures
    by molecular beam epitaxy, J.Vac.Sci.Techn. 21: 1
    (1982).

RECENT APPLICATIONS OF XSA IN HEAT TREATMENT AND FATIGUE OF STEELS

E. Macherauch and B. Scholtes

Institut für Werkstoffkunde I
Universität Karlsruhe (TH)
D 7500 Karlsruhe   FRG

This paper is intended to give an exemplary review of recent investigations performed in the X-ray laboratory of the Institut für Werkstoffkunde I of the Universität Karlsruhe, FRG, concerning particular problems of residual stresses of heat-treated and fatigued steels. The experimental work was mainly performed with computer-controlled Karlsruhe type ψ-diffractometers[1]. If linear distributions of residual lattice strains occurred the $\sin^2\psi$-method was applied to determine residual stresses. The experiments were performed with plain carbon steels of 0.22 and 0.45 wt.-% carbon (German grade Ck 22 and Ck 45) and some low alloyed steels.

CALCULATION AND MEASUREMENT OF RESIDUAL STRESSES OF QUENCHED STEELS

Basic studies about the generation of residual stresses in cylinders due to quenching have been performed[2,3]. A finite element program was developed taking into account the dependence on temperature and microstructure of all thermal and mechanical properties of the materials under consideration, such as thermal conductivity, heat capacity, heat transfer coefficient, thermal expansion coefficient , Young's modulus, Poisson's ratio, yield strength and stress-strain curve[4]. Besides the transient temperature distributions also the elasto-plastic deformations and stresses occurring during the cooling period of the specimen with the finite element mesh shown in Fig. 1 were computed. The nonlinear solution was obtained with the initial stress method[5]. The influences of strain hardening and Bauschinger effect were considered using the kinematic hardening rule[6]. The v. Mises yield criterion was applied. In any case the totally accumulated local plastic deformations determined the residual stresses after cooling. As an example, in Fig. 2 results for a Ck 45 steel cylinder (150 mm length, 50 mm diameter) quenched from

600ºC to 0ºC are shown. The time dependent generation of thermal
stresses during cooling and the formation of quenching residual stresses
in the midlength of the cylinder are plotted vs. the radius. The final residual stress profiles 60 s after quenching are given by thick
solid lines. Perspective views of the stress distributions for a quarter
of the cylinder are plotted in Fig. 3. Perpendicular to the z,r-plane
tensile stresses are plotted upwards and compressive stresses down-
wards. The total residual stress state with axial, radial, tangential
and shear stress components is shown. Residual shear stresses only
occur near the end planes of the cylinder. Therefore, only in the midlength area of the cylinder the axial, radial, and tangential stresses
are principle stresses.

Fig. 1. Finite element structure.

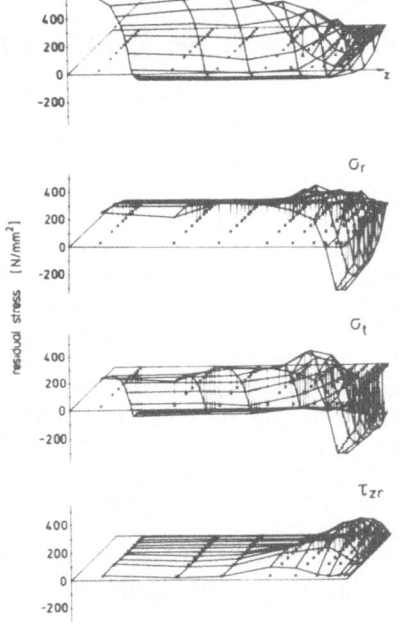

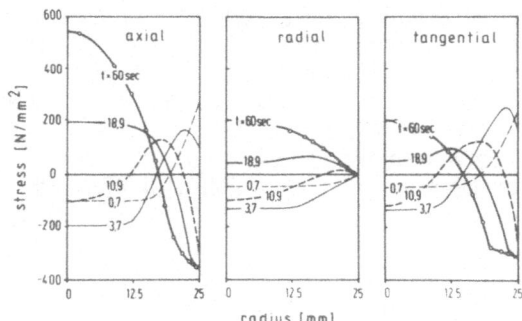

Fig. 2. Thermal stresses and residual stresses during quenching.

Fig. 3. Residual
stress distribution
over the cross-section.

Predicted and experimentally determined residual stresses showed
good agreement for transformation-free quenched as well as for hardened
cylinders. This is exemplarily shown in Fig. 4 for a cylinder (40 mm
diameter, 100 mm length) quenched from 830º C to 20º C in water. On the
left axial and tangential surface residual stresses are plotted versus
z. Along the cylinder axis computed and measured stresses agree well.
However, the measured axial stresses in the midlength area are a little
smaller than the calculated ones. On the right, the radial and tangential stresses of an endplane of the cylinder are plotted versus r.
Apart from some discrepancies near the center of the endplane again
calculated and measured stresses agree quite well. Obviously, X-ray
stress measurements are an important tool for the improvement of residual

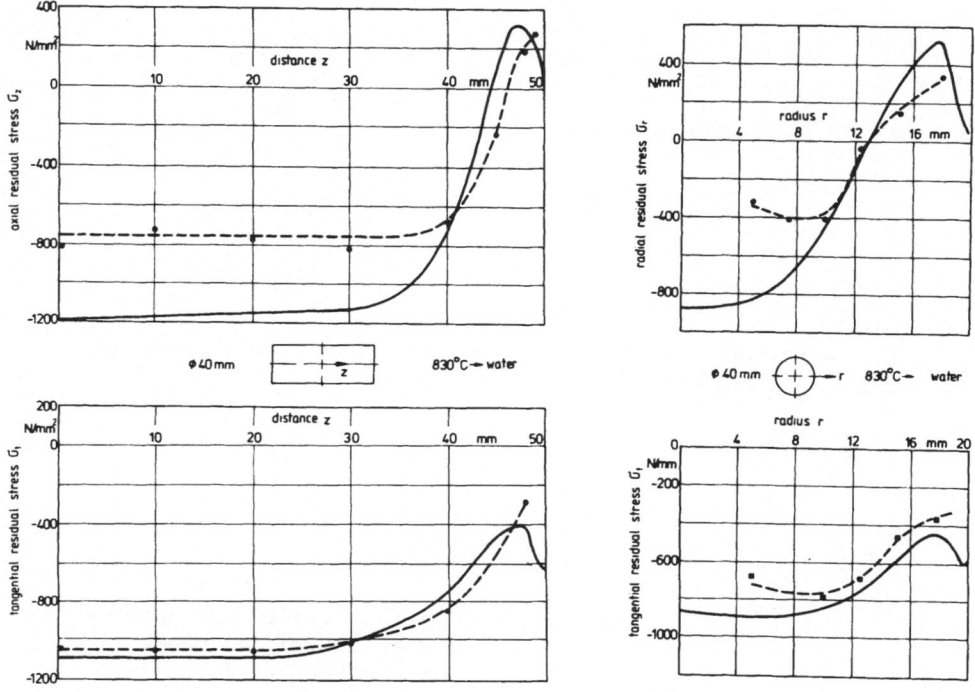

Fig. 4.   Comparison between calculated and measured residual stresses.

stress calculation methods and allow a relatively simple control of parameters influencing the calculated residual stress patterns.

In further investigations the development of residual stress distributions of end-quenched cylinders (Jominy tests) has been examined[7]. Typical results are illustrated in Fig. 5 for the steels Ck 45 and 42 CrMo 4. On the left hardness HV 10 is plotted vs. the distance from the quenched end. As expected both steels show the same hardness at the endplane but different hardness profiles along the cylinder length. On the right the axial residual surface stress profiles of the steels are plotted vs. the distance from the quenched end. The alloyed steel 42 CrMo 4 always shows tensile residual stresses. This is due to the dominating transformational residual stresses. For the plane carbon steel Ck 45 only near the quenched end tensile residual stresses are measured. In distances $z > 5$ mm compressive residual stresses occur. Obviously, besides changing the hardenability alloying elements also change the residual stress distributions in end-quenched cylinders. It was shown that a correlation exists between the maximum residual stress values and the ideal critical diameter Φ of different steels.

Also attempts of modelling the Jominy test with the above mentioned finite element method have been successful (see Fig. 6).

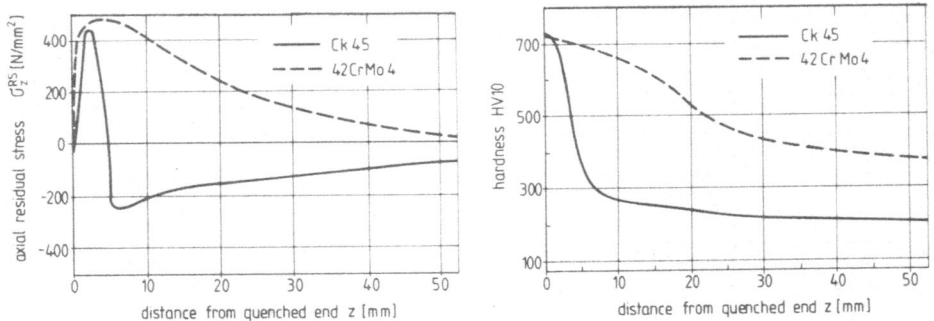

Fig. 5. Axial residual stresses and hardness of end-quenched cylinders ($T_a = 840^o$ C).

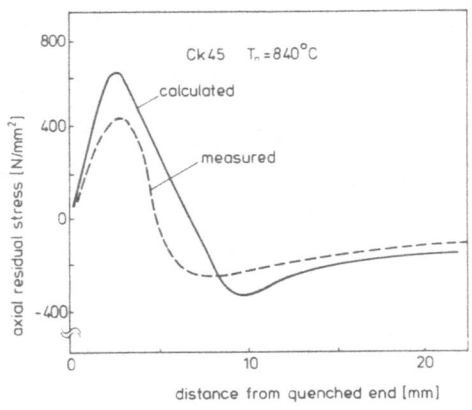

Fig. 6. Calculated and measured residual stresses of an end-quenched Ck 45-cylinder.

CONSEQUENCES OF MACHINING RESIDUAL STRESSES ON THE FATIGUE STRENGTH OF DIFFERENTLY HEAT TREATED STEELS

It is of high practical importance to know the effects of machining residual stresses on the fatigue behavior of differently heat-treated parts. Usually, if at all, the fatigue strength is generally thought to be increased by 1st kind compressive and decreased by tensile residual stresses[8,9]. However, besides macro residual stresses neither the creation of micro residual stresses nor the formation of characteristic surface topographies can be avoided during machining. Therefore, doubtless assessments of the effects of 1st kind residual stresses on the fatigue strength demand comparisons of Woehler-curves determined from different batches of machined specimens with different magnitudes of 1st kind residual stresses but identical micro residual stresses and topographies. In the left part of Fig. 7 an example is

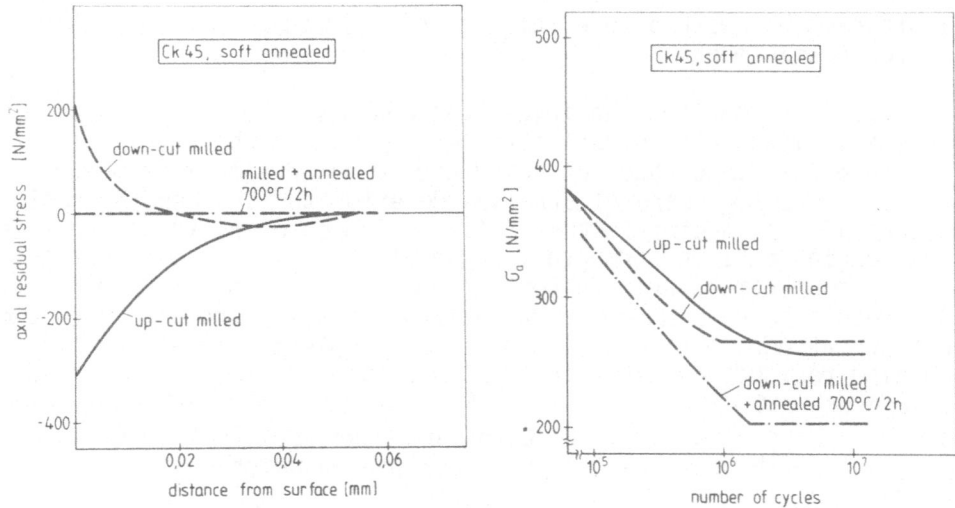

Fig. 7.   Residual stress distributions near the surface
and Woehler-curves of differently treated specimens.

shown for soft annealed Ck 45[10-12]. The residual stresses produced by
up-cut and down-cut milling are illustrated. In both cases approximately the same surface hardness, the same surface roughness, the same
surface workhardening state but surface residual stresses of -300 N/mm^2
and +210 N/mm^2 parallel to the milling direction were produced. For
comparison also the residual stress distributions after stress relief
annealing are plotted. Annealing of 2 h at 700^0 C reduces the machining
macro-stresses to zero values. In the right part of Fig. 7 appertaining Woehler-curves are shown. Inspite of a difference of about
500 N/mm^2 in the surface residual stresses the differently milled
samples yield at least equal fatigue strengths. However, the stress
relieved specimens show a markedly smaller fatigue strength than the
milled specimens. The conclusion can be drawn that machining macro residual stresses do not influence the fatigue strength of soft annealed
steel specimens whereas machining micro residual stresses are
of considerable influence. These statements are valid also for other
machining processes than milling. As an example in Fig. 8 the Woehler-
curves of planed specimens are shown in the as-planed state, after
annealing 2 h at 540^0 C and after annealing 2 h at 700^0 C. The same behavior as discussed before is observed.

On the other hand, as shown in Fig. 9, the same steel in the as-
hardened state with grinding surface residual stresses of -220 N/mm^2,
+60 N/mm^2, and +890 N/mm^2 reveals clearly separated Woehler-curves
with remarkable differences in the fatigue strengths. Obviously, in
this particular case the macro residual stresses are fully effective.
Similar experiments were conducted with batches of quenched and tempered but differently ground specimens. Also in this case Woehler-
curves split up according to sign and magnitude of the surface resid-

ual stresses but with a lower amount as observed with the hardened
and ground specimens.

Fig. 10 summarizes the investigations discussed. It is seen that
compressive 1st kind residual stresses improve the bending fatigue
strength of hardened resp. quenched and tempered specimens whereas
tensile machining residual stresses deteriorate the bending fatigue
strength of such material states. The influence of residual stresses
on the bending fatigue strength turns out to be the more distinct the
harder the specimens are. In soft annealed specimens 1st kind residual stresses have no effect on the fatigue strength. The main reason for
the observations mentioned is the almost complete relaxation of the
machining residual stresses in steel specimens of smaller hardness
than approx. 350 HV during bending fatigue loading. Similar conclusions have been drawn by W. P. Evans and coworkers from systematic
experiments with heat-treated and shot-peened notched steel
specimens [13].

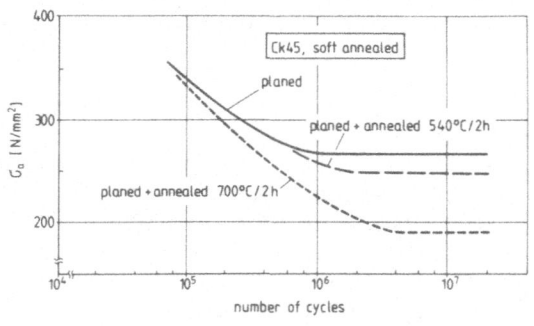

Fig. 8. Woehler-curves of
planed specimens.

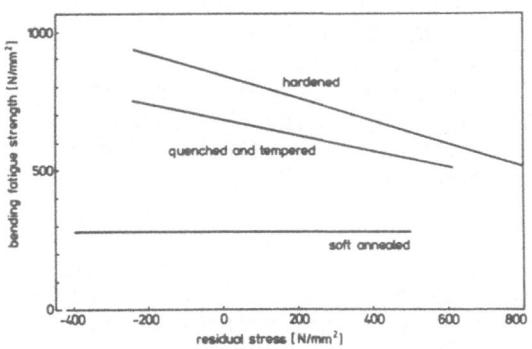

Fig. 10. Influence of macro
residual stresses on the bending fatigue strength.

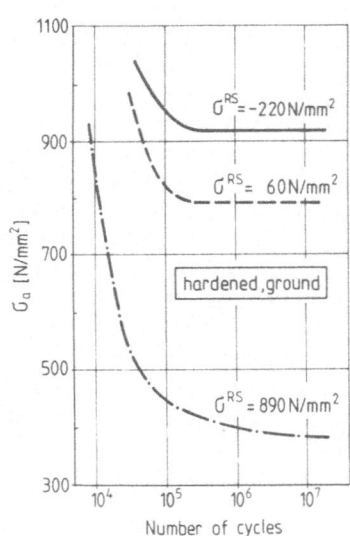

Fig. 9. Woehler-curves of
hardened and ground specimens.

INFLUENCE OF RESIDUAL STRESSES ON THE FATIGUE STRENGTH OF SMOOTH
AND NOTCHED STEEL SPECIMENS

The experiments described in the last chapter inspired similar
investigations with notched specimens of the same plain carbon steel.
Very recently, batches of notched specimens could successfully be
produced with different magnitudes and signs of surface residual
stresses but nearly identical workhardening states and topographies
in the notch root[14]. For example, in the case of soft annealed specimens with a notch factor K_t = 2.5 notch root residual stresses of
+210 N/mm^2 and -230 N/mm^2 were obtained by up-cut resp. down-cut
milling. The Woehler-curves of such specimens are shown in Fig. 11.
Despite of a difference in magnitude of the surface residual stresses of ~450 N/mm^2 no influence of the sign of the residual stresses
on the fatigue strength of the notched specimens can be seen. Similar
experiments were carried out with batches of quenched and tempered
specimens having the same notch factor K_t=2.5 but different notch root
residual stresses between -250 N/mm^2 and +380 N/mm^2 produced by milling. In this case a marked influence of residual stresses on the
fatigue strength occurs. In Fig. 12 the results of the quenched and
tempered specimens are compared with those of the soft annealed specimens. Further experiments are under way with quenched specimens of
the same type. As far as can be seen for the moment similar principles
are valid as in the case of smooth specimens. It should be emphasized,
however, that these conclusions are drawn from investigations on flat
side notched specimens.

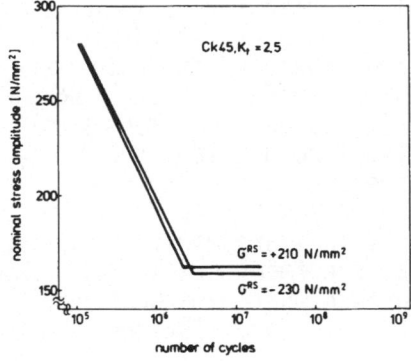

Fig. 11. Woehler-curves
of notched specimens.

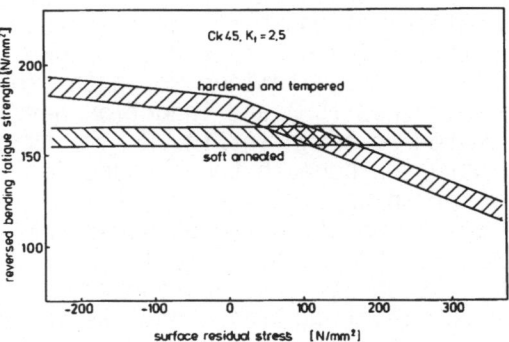

Fig. 12. Influence of the residual
stress in the notch surface on the
fatigue strength.

EFFECTS OF RESIDUAL STRESSES ON THE SUBSURFACE FATIGUE CRACK
INITIATION OF DIFFERENTLY SURFACE HARDENED STEEL SPECIMENS

Several investigations indicate subsurface fatigue crack initiation in surface hardened steel specimens which were shot peened or
thermo-chemically treated[15]. Recently, the inherent aspects of subsurface cracking were clarified[16,17]. In one of these investigations
two batches of Ck 45 specimens were hardened and after electrolytic
removal of thin surface layers shot peened with two different shot
sizes producing approximately equal surface residual stresses of
-800 N/mm^2 but different maximum residual stresses of about -1100
and -1300 N/mm^2 in different distances from the surface. Subsequently,
the specimens were fatigued in reversed bending tests. After shot
peening 41 % resp. 54 % larger fatigue strengths than in the unpeened
state were observed. Accompanying SEM-investigations (see Fig. 13)
showed that in the case of unpeened specimens fatigue cracks always
initiated at the very surface, whereas, in peened specimens, however,
always at distinct distances from the surface if not too large loading stresses were applied. Fig. 14 summarizes the experimental results. Thin straight lines indicate the loading stress profiles for
given surface stress amplitudes. The points of crack initiation observed under distinct stress amplitudes are marked by dots. Non-pro-
pagating cracks are denoted by cercles. The full line determines the
local fatigue strengths of subsurface layers calculated by a modified
Goodman relation with the assumption that local residual stresses act
like mean stresses. The diagram shows that propagating cracks only
initiate at such sites in or beneath the surface where the local loading stress amplitudes exceed the residual stress controlled local
fatigue strengths.

This idea of residual stress-controlled local fatigue strength
has also successfully been applied to steel specimens subjected to induction hardening, case hardening, surface rolling, and nitriding[8].
However, in these particular technological treatments tensile strength
and fatigue strength of surface areas are strongly effected by local
compositional and structural changes. Taking these effects into consideration local fatigue strength profiles for surface rolled (SR),
case hardened (CH) and induction hardened (IH) steel specimens were
estimated as shown in Fig. 15. The thin solid line displays the loading stresses beneath the surface for specimens with a diameter of
20 mm for an applied surface stress amplitude of 700 N/mm^2. The crack
initiation points observed under these conditions are marked by dots.
Obviously, in the differently treated steel specimens crack initiation
always occurs in such distances from the surface where for the first
time the local applied stress is larger than the local fatigue
strength.

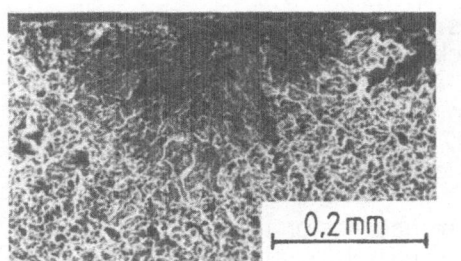 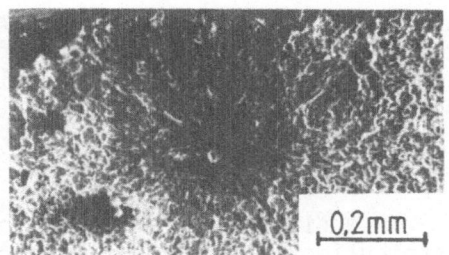

Fig. 13. Micrographs of crack initiation points.

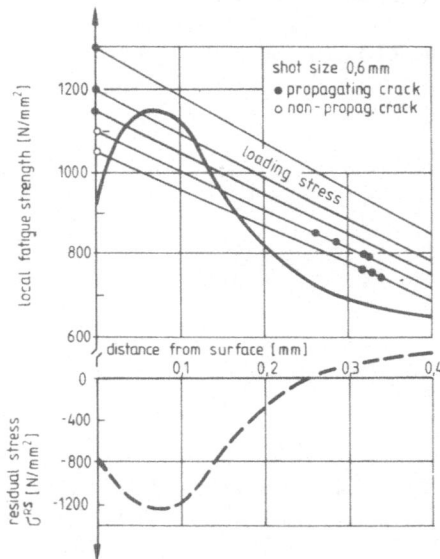
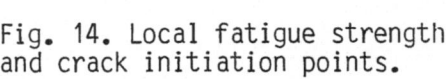

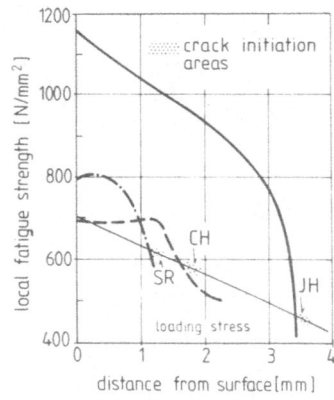

Fig. 14. Local fatigue strength
and crack initiation points.

Fig. 15. Crack initiation areas
for surface hardened specimens.

DEVELOPMENT OF RESIDUAL STRESSES DURING FATIGUE CRACK GROWTH

The creation and redistribution of residual stresses in front and behind a growing fatigue crack have been analysed recently[18]. Single
step tests as well as single overloading tests were carried out taking into account the influence of plastic deformations at maximum load
as well as plastic back deformations during unloading on the residual
stress states. Fig. 16 shows schematically the expected distributions
of residual stresses near the tip of a fatigue crack due to different

assumptions about the unloading process. The theoretical predictions taking into account the effects of crack closure and plastic deformation during unloading are in excellent agreement with recent X-ray investigations[19] on tension-to-tension loaded RCT-specimens ($\Delta\sigma$=21 N/mm2, R = 0.05). This can be seen from Fig. 17 where results are shown for a crack with an original length of 15 mm grown up to 17 mm.

In addition, the influence of overloadings on the residual stress distribution of cracked specimens was studied. Applying 10 overloadings of $\Delta\sigma$ = 43 N/mm2 to the specimen state considered in Fig. 17 increases the crack length to 17,7 mm and results in the characteristic redistribution of the residual stress profile shown in Fig. 18. Accompanying crack growth measurements show that crack growth retardation effects due to overloadings can clearly be correlated with distinct residual stress fields near the tip of propagating fatigue cracks.

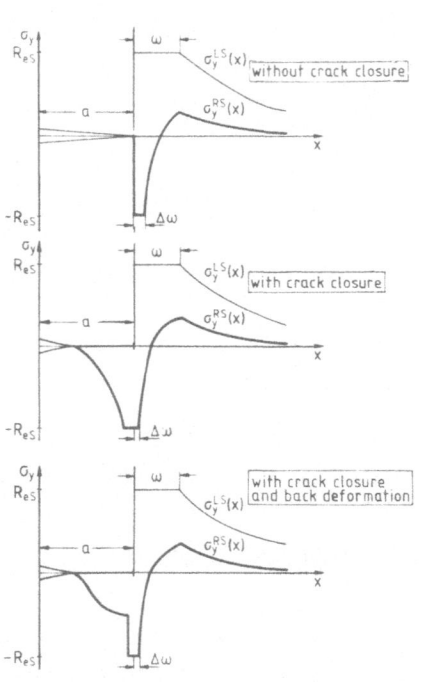

Fig. 16.　Calculated distributions of residual stresses near the crack tip

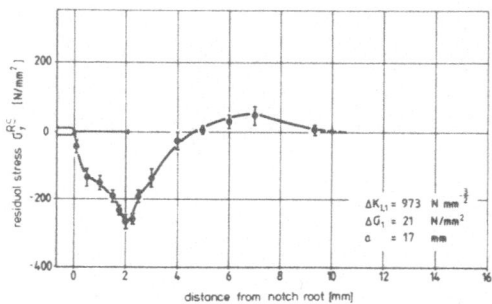

Fig. 17.　Measured residual stress distribution near the crack tip after single step loading.

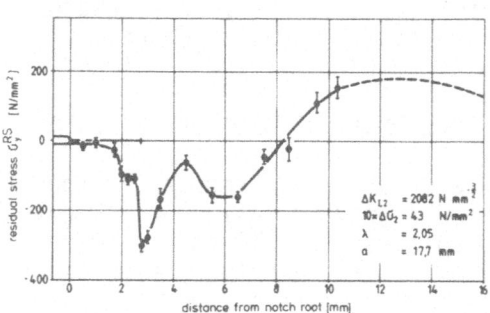

Fig. 18.　Residual stress distribution near the crack tip after applying overloads.

REFERENCES

1. B. Scholtes and E. Macherauch, Residual Stress Determination, annual meeting of TMS/AIME, Atlanta 1983, ASTM STP, to be published.
2. R. Schröder, Dr.-Ing. Diss., Universität Karlsruhe (TH) (1983).
3. R. Schröder, B. Scholtes, and E. Macherauch, Experimentelle und theoretische Analyse der Ausbildung von Eigenspannungen in abgeschreckten Stahlzylindern, in:"Eigenspannungen", V. Hauk and E. Macherauch, eds., Deutsche Gesellschaft für Metallkunde e.V., Oberursel (1983).
4. H. J. Yu, U. Wolfstieg, and E. Macherauch, Berechnung von Eigenspannungen mit Hilfe eines speziellen Finite-Element-Programmes, Arch. Eisenhüttenwes. 49:499 (1978).
5. W. Prager, in:"Probleme der Plastizitätstheorie", Birkhäuser, Basel, Stuttgart (1955).
6. G. C. Nayak and O. C. Zienkiewicz, Elasto-plastic stress analysis. A generalization for various constitutive relations including strain softening, Int. J. Num. Meth. Engng. 5:113 (1972).
7. P. Graja, H. Müller, and E. Macherauch, Eigenspannungsmessungen an Stirnabschreckproben, see 3.
8. E. Macherauch and K. H. Kloos, Bewertung von Eigenspannungen , in: "Eigenspannungen", V. Hauk and E. Macherauch, eds., Carl Hanser Verlag, München (1982).
9. Residual stresses for designers and metallurgists, J.Larry and V. Walle, eds., Proceed. Conf.9-10. April, Chicago (1980), Am. Soc. Met., Metals Park, Ohio.
10. B. Syren, Dr.-Ing. Diss.,Universität Karlsruhe (TH) (1978).
11. B. Syren, H. Wohlfahrt, and E. Macherauch, The influence of residual stresses and surface topography on bending fatigue strength of machined Ck 45 in different heat treatment conditions , Proceed. 2nd Int. Conf. on Mech. Beh. of Mat., Special Volume, Boston (1976).
12. B. Syren, H. Wohlfahrt, and E. Macherauch, Der Einfluß von Schleifbearbeitungen auf das Biegewechselverhalten von Ck 45 in verschiedenen Wärmebehandlungszuständen, Härterei-Techn. Mitt. 37:236 (1982).
13. W. P. Evans, R. E. Ricklefs and J. F. Millan, X-ray and fatigue studies of hardened and cold-worked steels, in: "Local atomic arrangements studied by X-ray diffraction," Gordon and Breach, New York (1966).
14. H. J. Böhmer, J. E. Hoffmann, D. Eifler, and E. Macherauch, Zum Einfluß bearbeitungsbedingter Eigenspannungen auf das Wechselverformungsverhalten von Kerbstäben aus Ck 45, Proceed. DVM-Tag, 9-11 Nov, Düsseldorf (1983).
15. K.-H. Kloos, VDI-Ber. Nr. 268:63 (1976).
16. P. Starker, Dr.-Ing. Diss., Universität Karlsruhe (TH) (1981).

17. P. Starker, H. Wohlfahrt, and E. Macherauch, Der Amplitudenein-
    fluß auf die Bildung von Ermüdungsanrissen in gehärteten
    und kugelgestrahlten Biegeproben aus Stahl Ck 45, Arch. Ei-
    senhüttenwes. 51:439 (1980).
18. H. Führing, Berechnung von elastisch-plastischen Beanspruchungs-
    abläufen in Dugdale-Rißscheiben mit Rißuferkontakt auf der
    Grundlage nichtlinearer Schwingbruchmechanik, Veröffentli-
    chung des Instituts für Statik und Stahlbau der TH Darmstadt
    (1977).
19. E. Welsch, B. Scholtes, D. Eifler, and E. Macherauch, Oberla-
    stungsbedingte Eigenspannungsverteilungen in rißspitzennahen
    Werkstoffbereichen und deren Einfluß auf die Ausbreitung von
    Ermüdungsrissen, see 3.

STRESS MEASUREMENT AT FATIGUE CRACK TIP USING PSPC

Yasuo Yoshioka

Musashi Institute of Technology

1 Tamazutsumi, Setagaya, Tokyo 158, Japan

INTRODUCTION

The growth of fatigue crack builds up the compressive residual stress with the plastic zone in front of the crack tip. The presence of this residual stress is one of the main causes for the crack closure [1]. The consequence of residual stress is a reduction of the crack tip stress intensity variation during a load cycle. Although this effect is well understood in a qualitative manner, it has not been completely quantified because of a lack of information about the load and residual stress distributions at the crack tip.

There are few reports on the X-ray stress measurement in the vicinity of the crack tip [2,3,4]. A main reason is an experimental difficulty in determining the stress in a small region because of the low X-ray intensity when the conventional detector such as X-ray film or 0-dimensional counter was used.

This paper reports an X-ray study of applied and residual stress measurement near the fatigue crack tip using a PSPC technique which has a high counting efficiency [5]. Discussions are performed on the effect of residual stress due to the stress intensity and the crack growth rate.

MATERIAL AND EXPERIMENTAL PROCEDURE

The material used was a 0.15% carbon steel which was annealed at 600°C for 8 hours. The specimen has the planer dimensions of an ASTM standard 1 inch(25.4 mm) thickness compact tension type, but the thickness employed was nominally 12 mm.

   Fatigue tests were conducted under sinusoidal tension-tension
load control, the stress ratio R being 0.5 and 0.05, respectively.

   X-ray stress measurements were made using a Cr-target microbeam
X-ray generator and a PSPC system with the high counting efficiency
[4]. To obtain the stress in a very small region, a 200 μm diameter
pin-hole slit was placed at the front of the X-ray window. The real
X-ray beam on the specimen was about 600 μm diameter. The specimen
was oscillated about the irradiated point as a center of oscillation through a range of ±5° in order to prevent roughness of profile. The direction of stress measured was perpendicular to the
crack growth direction. The $Sin^2\psi$ method was employed for the
stress determination ($\psi_0=0°,0°,15°,30°,40°$ and $40°$).

EXPERIMENTAL RESULTS

Crack Growth Behaviour

   Fig.1 shows the relation between the stress intensity range, ΔK,
and the crack growth rate, da/dN. The data exhibits a custmary
straight line and the influence of stress ratio R is not observed
in the experimental range.

Residual Stress Distribution

   A fatigue test was performed under the condition of Pmax of 6.9
KN(700 Kgf) and stress ratio R of 0.05. Residual stress distribu-

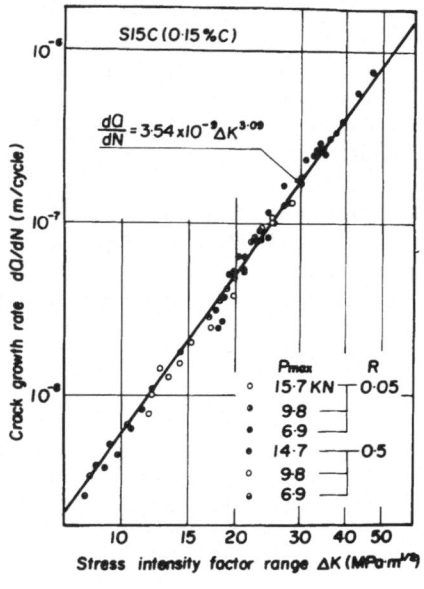

Fig.1 Relation between
crack growth rate and
stress intensity range
of 0.15% carbon steel.

tions were measured when a crack length reached 24.4 mm and 31.9 mm, respectively. The mean value of 95% confidence limit of stress by the $\text{Sin}^2\psi$ method [7] was about 25 MPa. A time required for a stress determination was about 20 minutes when a maximum count number was preset to 512.

Results are shown in Fig.2. The compressive residual stress shows a maximum value near the crack tip and it gradually decreases with distance from the tip. The value of maximum compressive residual stress depends on the crack length or the stress intensity. By Rice's analysis [6], it should be equal to the yielding stress of material used irrespective of the stress intensity. The line drawn in Fig.2 shows an analytical distribution of residual stress at 24.4 mm crack length. The experimental value is smaller than this value and it exceeds the yielding stress at 31.9 mm length. Similar results by means of X-rays were reported by other researchers [3,8].

Another result under the condition of Pmax=9.8 KN(1000 Kgf) and stress ratio R of 0.5 is shown in Fig.3. The behaviour of residual stress is similar to the former case. The maximum residual stress at each crack length was plotted against the stress intensity range as shown in Fig.4. We find parallel lines which depend on the stress ratio R, and the value for R equal to 0.5 is about twice as much as that for R equal to 0.05, at same ΔK.

Stress Distribution Under Loading

Stress distributions near a crack tip were measured at loads up to 9.8 KN statically applied on a specimen which has a crack length

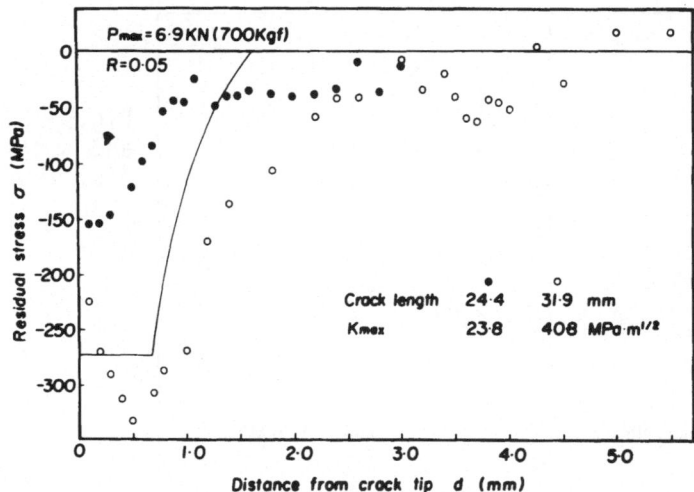

Fig.2   Residual stress distributions at crack tips.
(Pmax=6.9 KN, R=0.05)

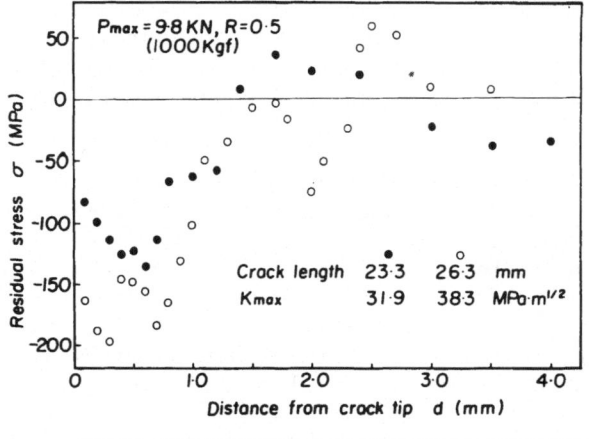

Fig.3 Residual stress distributions at crack tips (Pmax=9.8 KN, R= 0.5).

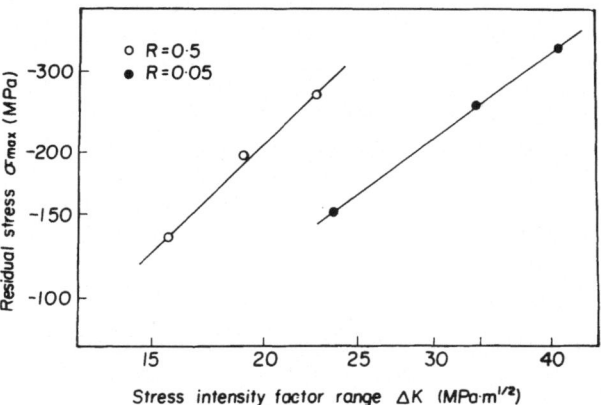

Fig.4 Relation between maximum compressive residual stress at crack tip and stress intensity range.

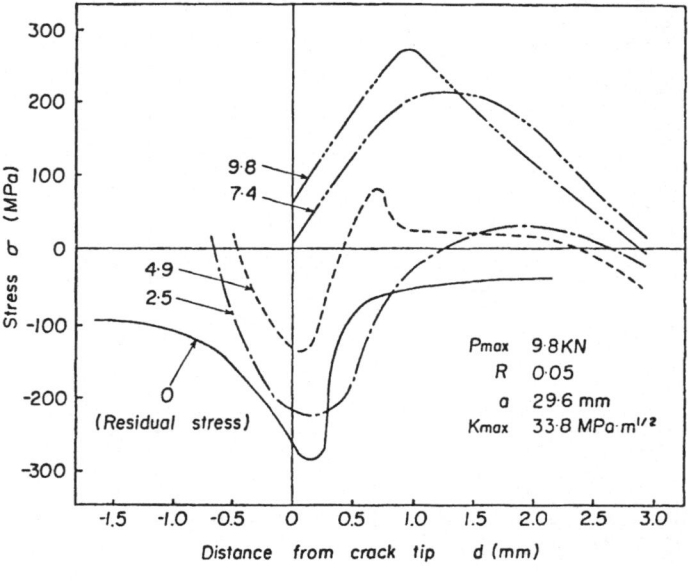

Fig.5 Residual and Applied stress distributions at crack tip.

of 29.6 mm (Pmax=9.8 KN, R=0.05). Fig.5 shows these results. Stress measurement on the crack was impossible at the load of 7.4 KN and above because of the crack opening. The maximum stress is observed on a position at a distance 0.5-1 mm apart from the tip and the compressive stress still remains near the crack tip even if half of the maximum load was applied.

DISCUSSION

The analysis by Rice is based on the assumptions of the perfectly elastoplastic material and a single stress cycle. In practice, however, the effect of stress cycles should be considered. Fig.6 shows a Debye-Scherrer pattern from a fatigue crack tip by means of microbeam X-rays. There are many spots in the ring and each spot is diffracted from each subgrain formed polygonally by the cyclic stresses. Such fine spots are not observed in a ring from a crack tip formed by a monotonic load. The spotty pattern suggests that the cyclic stresses affect the formation of residual stress at the crack tip. The yielding stress at the crack tip would increase with the crack length and the maximum residual stress would also increase in proportion to the stress intensity factor.

The crack growth behaviour is independent of the stress ratio R as shown in Fig.1. We discuss this behaviour in relation to the residual stress at the crack tip. The effective stress for the crack growth is assumed as the sum of the applied stress and the residual one. The effective stress intensity range ΔK_{eff_1} at a crack length a and the stress ratio R=0.05($\doteqdot$0) is

$$\Delta K_{eff_1} = (\sigma_{max_1} + \sigma_{r_1}) \sqrt{a} \; \Phi \tag{1}$$

where σ_{r_1} is the effective residual stress affecting the crack opening and closure and Φ is a calibration factor. If R is 0.5, ΔK_{eff_2} can be written as

$$\Delta K_{eff_2} = (\sigma_{max_2} - \sigma_{min_2} + \sigma_{r_2}) \sqrt{a} \; \Phi = (\sigma_{max_1} + \sigma_{r_2}) \sqrt{a} \; \Phi \tag{2}$$

$$\therefore \; \sigma_{max_2} = 2\sigma_{max_1}, \quad \sigma_{min_2} = \sigma_{max_1}$$

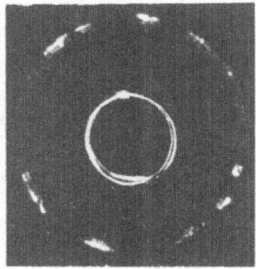

Fig.6 An example of Debye-Scherrer pattern diffracted from a fatigue crack tip by microbeam X-rays.(100 μm diameter pin-hole slit)

As shown in Fig. 4, the maximum compressive residual stress at R=0.5 is about two times that for R=0.05($\doteqdot$0) at the same ΔK. But this stress is not σ_{r_2}, and σ_{r_1} should be equivalent to the maximum stress near the crack tip at which $\sigma_{min_2}(=\sigma_{max_2}/2)$ was applied.   The value of this stress at σ_{min_2} is about half the maximum compressive residual stress under load-free condition as shown in Fig. 5. Hence, σ_{r_1} is equal to σ_{r_2}.   Thus, the following relation can hold:

$$\Delta K_{eff_1} \doteqdot \Delta K_{eff_2} \tag{3}$$

and it can be concluded that an R dependency is not observed for the crack growth behaviour in the material used here.

CONCLUSION

The measurement of stress distribution ahead of fatigue crack tips was attempted by means of microbeam X-rays using a PSPC system.

Based on the experimental result, it could be explained that a linear relation between the crack growth rate and the stress intensity range holds approximately regardless of the stress ratio R on the material used in this study.

This discussion is based on the assumption of the same crack length regardless of stress ratio R. An analysis under any crack length is the subject for a future study. However, stress measurement by the PSPC system[9] will clarify such problems because a local stress with high precision can be measured in a short time.

REFERENCES

1.   W.Elber, ASTM STP, 486:230(1971).
2.   S.Taira, K.Tanaka and T.Shimada, J. JSMS, 21:1142(1972).
3.   K.Honda, N.Hosokawa, T.Sarai and K.Okamoto, J. JSMS, 31:210 (1982).
4.   W.Schlosberg and J.Cohen, Met. Trans., 13A:1987(1982).
5.   Y.Yoshioka, K.Hasegawa and K.Mochiki, Adv. X-Ray Analysis, 23: 325(1980).
6.   J.Rice, ASTM STP, 415:247(1967).
7.   Y.Yoshioka, K.Hasegawa and K.Mochiki, Adv. X-Ray Analysis, 26: 209(1983).
8.   D.Stephan and K.Richter, Crystal Research and Technology, 16: 57(1981).
9.   M.James and J.Cohen, Adv. X-Ray Analysis, 19:695(1975).

DIRECT X-RAY MEASUREMENT OF RESIDUAL STRAINS IN TEXTURED STEEL

C. P. Gazzara

Army Materials & Mechanics Research Center
Watertown, Massachusetts 02172

INTRODUCTION

One of the most detrimental effects on the accuracy of an X-ray diffraction residual stress analysis, XRDRSA[1], is found in the examination of textured materials. The degree of elastic anisotropy and texture is in general agreement with the extent of the error in the residual stress. Several approaches have been made to correct for the effects of texture, particularly involving experimental techniques. Reviews of such efforts are given by H. Dölle[2], V.M. Hauk[3] and G. Maeder, J.L. Lebrun and J.M. Sprauel [4], just to mention a few.

A brief chronology of the texture corrections involved in XRDRSA follows. With isotropic materials the d spacing of a crystal lattice, d, is assumed to vary linearly with $\sin^2\psi$. With textured materials the d vs $\sin^2\psi$ relationship is nonlinear. This is due to the anisotropy of the elastic constants and their departure from a random distribution, or taking on a preferred orientation.

In 1975, R.H. Marion and J.B. Cohen[5] were the first to give serious consideration to correcting these texture induced curvatures in d, using a direct method, applying crystallite concentration data obtained from pole figures to the X-ray diffraction data. R.H. Marion and J.B. Cohen aligned the ψ, or β, angle along the rolling direction (R.D.) of α-Fe for the reflecting plane (211) with the highest diffracting Bragg angle, 2θ, (CrK) and deduced that the curvature in d is in phase with the distribution function $f(\alpha,\beta)$, i.e., the diffracted peak intensity maxima, h, should align with the d maxima.

197

The nonlinearity along R.D. of d vs $\sin^2\psi$ was treated by T. Shiraiwa and Y. Sakamoto(6) to determine the effect of elastic constants and plastic anisotropy in textured cold rolled steel specimens. The elastic constants employed were taken from single crystal data.

V. Hauk and H. Sesemann (7) found in steels, that two texture independent directions occur in which the lattice strain is isotropic or quasi-isotropic. These two ψ^{**} values were employed to set the linear relationship between the strain and $\sin^2\psi$.

A generalized treatment was developed by H. Dölle and V. Hauk (8) for relating the lattice strain to the elastic constants, and the angles ϕ and ψ, for anisotropic materials. Although this method has much potential and was further developed by H. Dölle(2) and H. Dölle and J.B. Cohen(9), the computations for the strain tensor must be performed for each system in a specific material.

An attractive experimental approach, to be tested in the future, was suggested by H. Dölle and J.B. Cohen(9) employing {h00} and {hhh} reflections, which retain the linear d vs $\sin^2\psi$ character. As an alternative, the {310} reflection was recommended in place of the {211} in steel to avoid these texture effects along the R.D. This procedure, as well as the other direct methods described in references (5)(6) and (7) assume a quasi-isotropic system whereby the elastic interaction, or coupling, between the grain and its surrounding matrix is not considered (2).

The Marion and Cohen technique(5) was applied, using parallel beam optics in the case of highly textured 4340 steel with a triplex texture (10)(11). The results of this study showed that a displacement, or "phase shift", of the f and d or 2θ, curves with ψ_0 is evident (see figure 1). This mismatch or lag in $2\theta_{min}$ (or d^o_{max}) behind the $Cr(211)K$ intensity maximum could not be accounted for with systematic errors or beam geometry. The effect of this error on the corresponding computed stress values, along with the stress values obtained by shifting the f and d curves into alignment, is shown in figure 2. It was suggested* that this effect could be due to the characteristics of the diffraction system, in that the θ and ψ motion are independent, and that a 2θ-ψ follower (12) might account for this mismatch in the f and d curves.

Another suggestion† was that the side inclining method be employed through the 2θ-ψ field.

*Prof. T. Ericson of the Linkoping Inst. of Tech., Linkoping, Sweden brought the reference (12) to the attention of the author.
†R. Chrenko, G.E. Research Labs, is credited for this suggestion.

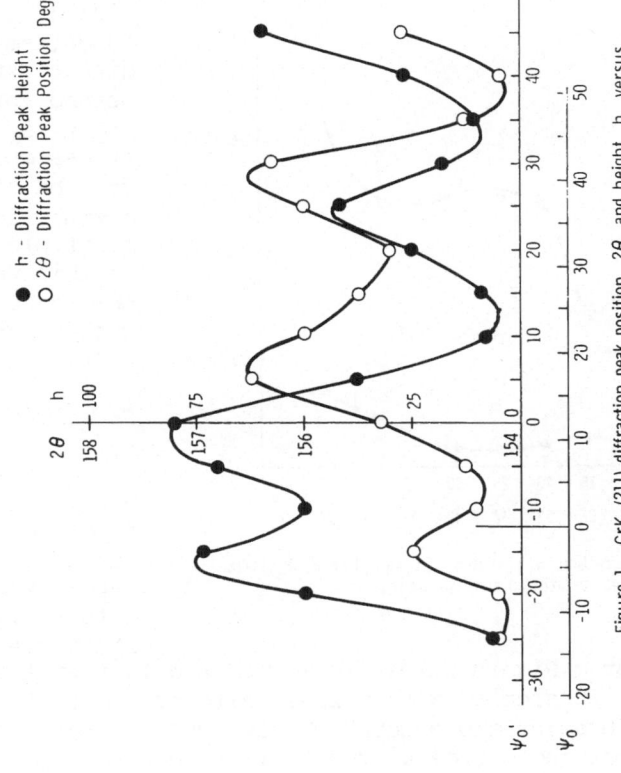
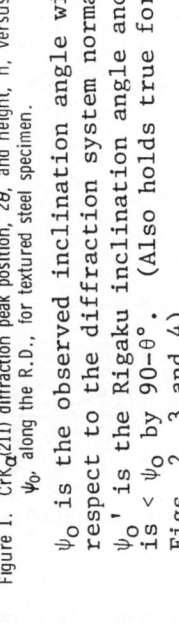

Figure 1. $CrK_\alpha(211)$ diffraction peak position, 2θ, and height, h, versus ψ_0, along the R.D., for textured steel specimen.

ψ_0 is the observed inclination angle with respect to the diffraction system normal. ψ_0' is the Rigaku inclination angle and is < ψ_0 by $90-\theta°$. (Also holds true for Figs. 2, 3 and 4).

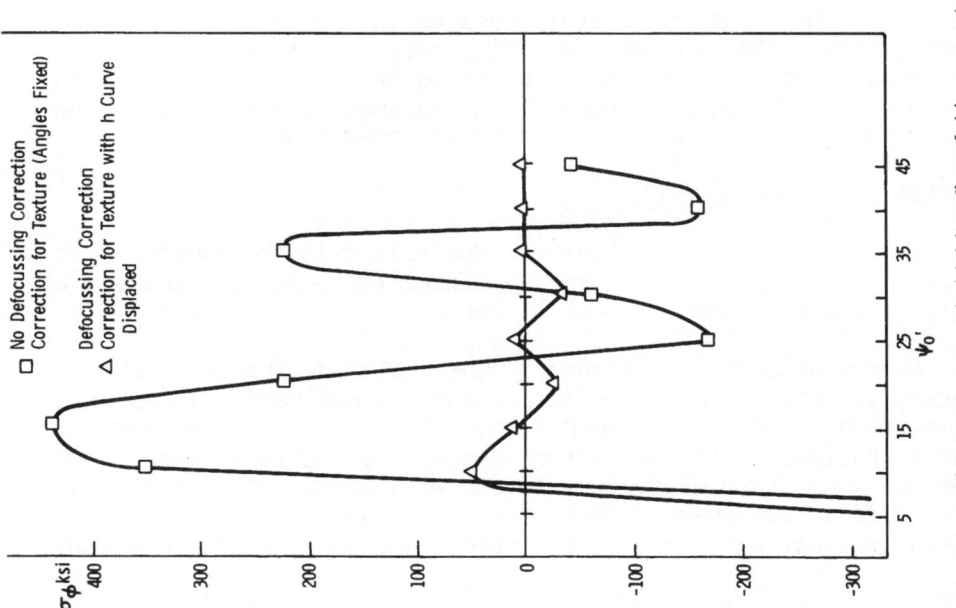

Figure 2. Residual stress, σ_ϕ, calculated from figure 1 data, versus ψ_0, as observed and with the 2θ and h curves placed in phase.

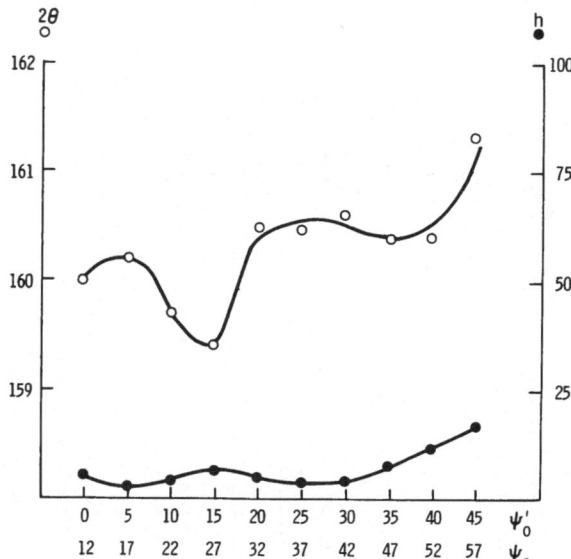

Figure 3.   CoK$_\alpha$(310) diffraction peak position, 2θ, and height, h, versus ψ_0, for a textured steel specimen.

The main objective of this work is to determine a direct experimental method for correcting X-ray residual stress measurements for the effects of texture in rolled steel plate. It was in this vein that the above mentioned suggestions were tested.

PROCEDURE

For the most part, the X-ray diffraction data presented in this report was taken with a Rigaku Strainflex MSF/PSF system. A custom Rigaku Strainflex† was used to test the case of the "phase" shift problem with a 2θ-ψ follower, or that ψ be held constant. A chromium and a cobalt X-ray tube, operated at 30 KV and 10 ma., were the source of the K$_\alpha$ filtered radiation utilized in this work. A standard 1 degree divergence beam and receiving slit with a built-in collimator provided the specified "parallel beam" optics. This system has a side inclining feature for which the inclination angle can be α, or if so desired ψ; this inclination angle is 90° with respect to the 2θ angular sweep axis. The sample examined was studied in a prior investigation (10)(11).

RESULTS

The results of this study are summarized in the figures that follow. Plots are shown of the 2θ diffracted peak position and the diffracted X-ray function f or h, versus ψ_0.

As may be seen from figure 3, CoKα (310) diffracted peak intensity is not constant with increasing ψ_0, but varies in an nonlinear manner indicating the presence of texture. The variation of d or 2θ follows a similar texture dependence. This is because this specimen has a triplex texture system with broad local strain induced diffraction peaks. Therefore, the suggestion made(2) and tested, was not pursued for this steel, and deferred at this time.

†The custom Strainflex was designed for a special application for the Babcock and Wilcox Co.

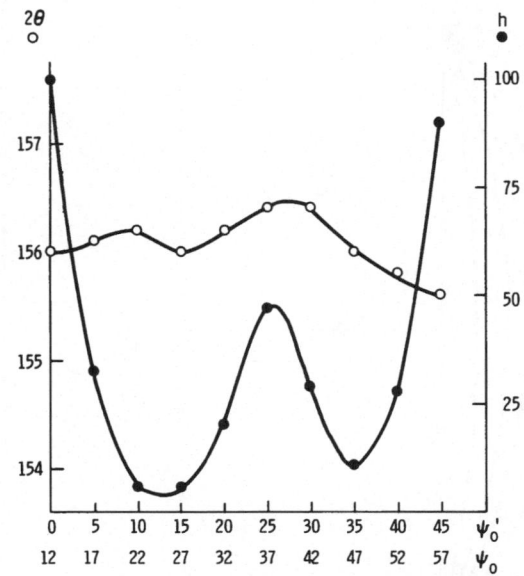

Figure 4.   CrKα(211) diffraction peak position, 2θ, and height h, versus ψ_0, employing a 2θ-ψ follower system (ψ_0 is constant for the 2θ scan).

Side Inclining	$\epsilon_{\phi\psi\alpha}$	$\psi_{\phi\alpha}$	$d_{\phi\psi\alpha}$	ϕ_s
α = 0, Normal System	$\epsilon_{\phi\psi}$	ψ_ϕ	$d_{\phi\psi}$	ϕ

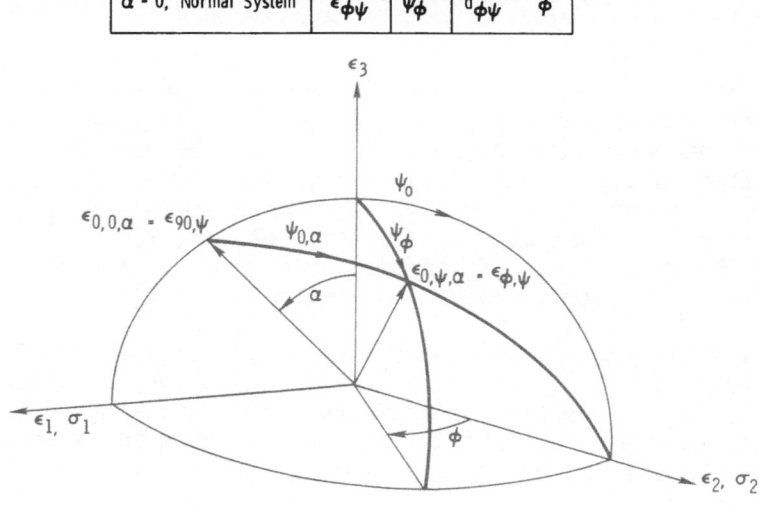

Figure 5.   A stress strain ellipsoid illustrating the parameters for the side inclining and those normally employed (α = 0).

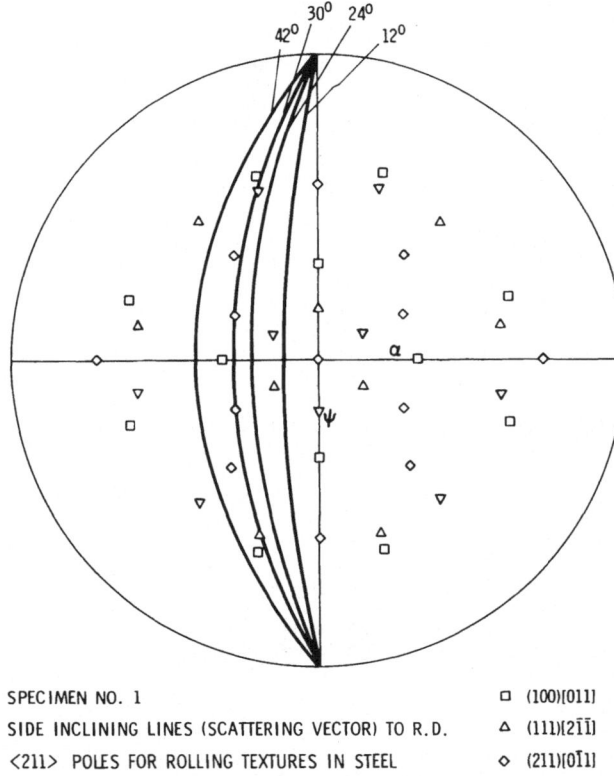

SPECIMEN NO. 1

SIDE INCLINING LINES (SCATTERING VECTOR) TO R.D.

⟨211⟩ POLES FOR ROLLING TEXTURES IN STEEL

☐ (100)[011]

△ (111)[2$\bar{1}\bar{1}$]

◇ (211)[0$\bar{1}$1]

Figure 6. A stereographic projection of the ⟨211⟩ poles for three texture modes in steel. Constant α curves are shown for the experimental conditions employed.

The data involving the $2\theta-\psi$ follower are summarized in figure 4 with results that are similar to those taken without the $2\theta-\psi$ follower as shown in figure 1. Although the results of this experiment are not conclusive, indications are that the complete removal of this "phase effect", using the $2\theta-\psi$ follower, would be improbable.

An attempt to find a texture independent path through the (211) pole figure was undertaken on the steel specimen. The notation for the parameters in the case of the side inclining system is illustrated in figure 5 and compared with the normal nomenclature. A series of h and 2θ runs were made, over the range of $\psi_{o,\alpha}$ angles, for constant α values of 0°, 12°, 24°, 30°, and 42°. A pole figure showing the paths taken may be seen in figure 6. The associated h and 2θ vs $\psi_{o,\alpha}$ plots are given in figure 7.

For the textured steel specimen tested, it is evident that a tilt along α of 42° is necessary to provide a texture free path. (The (211) K_α intensities do not vary with increasing angle β or ψ). In addition, the 2θ peak position is nearly constant with increasing $\psi_{o,42}$, but this is not the case for all other values of α tested (see figure 7).

The implications of these results are revealing. For example, the following points should be noted. (a) A texture free path need not exist for the (310) reflection in textured steel, as can be seen from the data at α=0. (see figure 3) (b) A texture free path

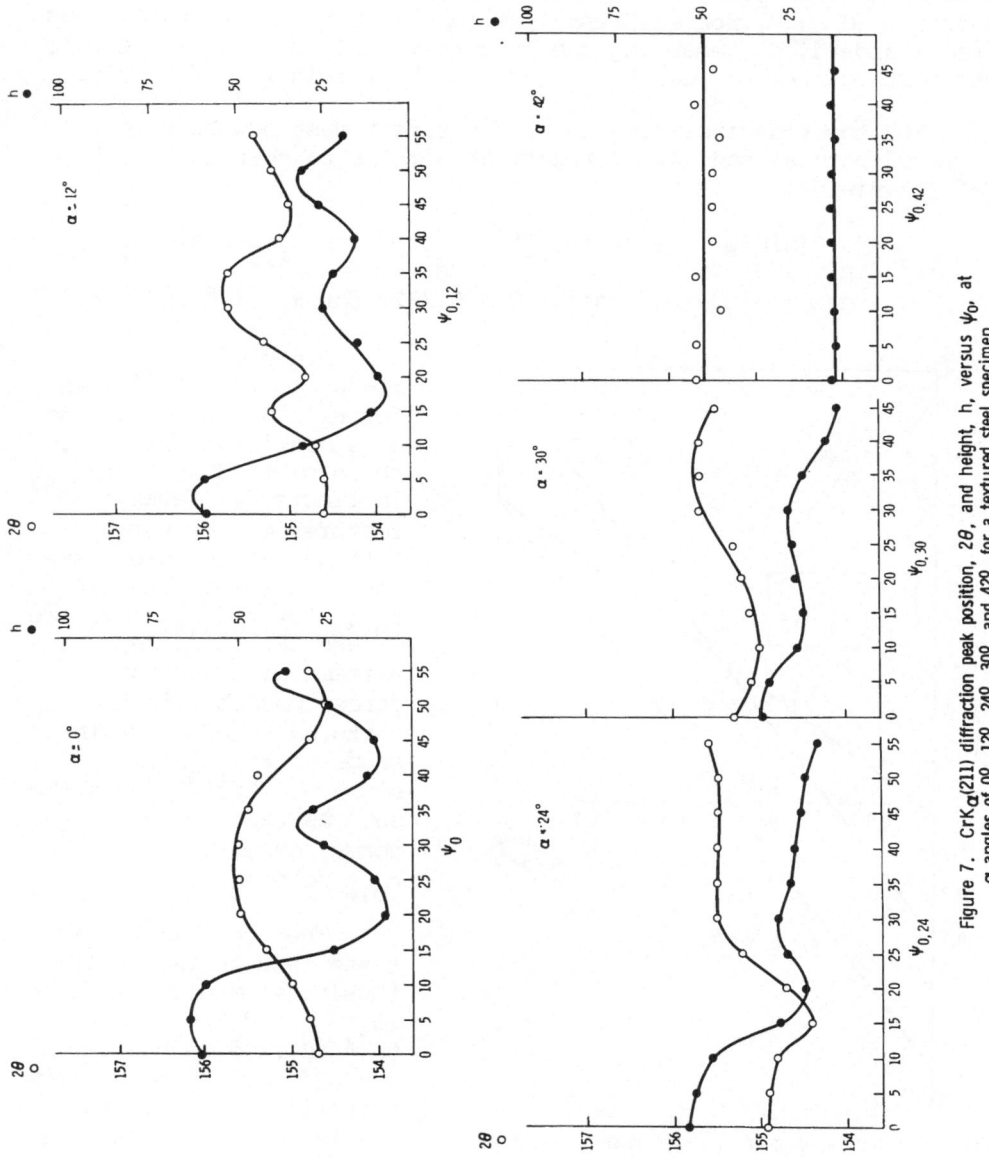

Figure 7. CrKα(211) diffraction peak position, 2θ, and height, h, versus ψ₀, at α angles of 0°, 12°, 24°, 30°, and 42°, for a textured steel specimen.

can be achieved for the side inclining case ($\alpha=42°$) in spite of the broad peak width of the (211) reflection (c) The d spacings for the data in figure 7, for $\alpha=12° - 42°$, do not include the d spacing for the surface normal, $d_\perp$. It is customary to apply $d_{\phi,\psi} - d_\perp$ to denote the d spacing as a function of ψ_ϕ.   (d) The d spacing at $\alpha=42°$, or $d_{\alpha=42}$, for a stress free sample, can be used as a stress free standard, d_o, enabling one to correct values of d_o for errors. Hence it may now be possible to obtain the strain $\varepsilon_{\phi,\psi}$ directly.

For the side inclining case, the strain must now be given in terms of both σ_1 and σ_2. A treatment similar to that conducted in ref (1) yields:

$$\varepsilon_{o,\psi,\alpha} = \frac{\sigma_1}{E} [\sin^2\psi_{\phi,\alpha} - \nu \cos^2\psi_{\phi,\alpha}] + \frac{\sigma_2}{E} [C \cos^2\psi_{\phi,\alpha} - \nu \sin^2\psi_{\phi,\alpha}]$$

where $C = \sin^2\alpha - \nu \cos^2\alpha$; $C = 0.282$, for $\alpha = 42°$ and $\nu = 0.3$

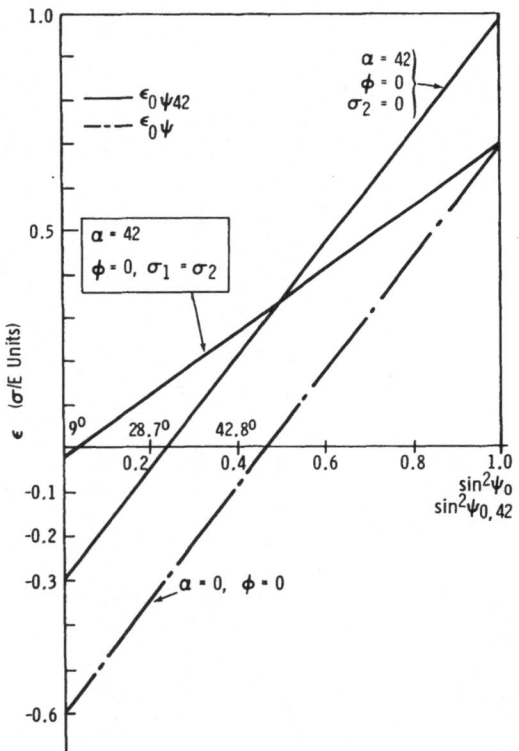

Figure 8. Strain, $\varepsilon_{0,\psi,42}$ or $\varepsilon_{0,\psi}$, versus $\sin^2\psi_{0,42}$ for the side inclining and $\sin^2\psi_0$ ($\alpha = 0$) for the "normal" case.

Plots of $\varepsilon_{o,\psi,42}$ vs $\sin^2\psi_{0,42}$ for $\sigma_1=\sigma_2$ and for $\sigma_2=0$, as in the case of a cantilever bar, are shown plotted vs $\sin^2\psi_{0,42}$ in figure 8. Several features may be worth noting: (a) The plots are linear.   (b) The values of $\psi_{0,42}$ for $\varepsilon_{0,\psi,42} = 0$, are $9°$ and $28.7°$ for the extreme distributions of stress possible in the textured steel. (c) The slope of $\varepsilon_{o,\psi,42}$ vs $\sin^2\psi_{0,42}$ for $\sigma_2=0$ is the same as that shown for the normal case, or $\varepsilon_{o,\psi} /\sin^2\psi_o$.

The application of these results may follow thusly: (a) Measure $\Delta d/d$ or $(d_{\phi,\psi,\alpha} - d_o)/d_o$ for a texture free path, i.e. at $\alpha = 42°$, obtaining d_o from a strain free bar, if available.   (b) Determine the direction of the principal strains. In this case, σ_1 is parallel to the rolling direction of the steel and $\phi = 0$. (c) From the above equation, determine σ_1 in terms of σ_2, at zero strain, and extrapolate to $\varepsilon_{o,90,\alpha}$ or σ_1/E.

Another approach may be to obtain samples of similar composition and texture and calibrate the cantilever bars under known loads. The established texture-independent path would ensure a linear behavior of the strain with $\sin^2\psi$, as in figure 8, so that extrapolation procedures would provide reasonable accuracy.

ACKNOWLEDGEMENTS

Thanks are extended to D. Lee of Babcock and Wilcox for his help with the $2\theta-\psi$ follower data, to T. Ericson and R. Chrenko for their suggestions, to A. Zarkades for the steel specimen, and to R. Hinxman for his assistance.

REFERENCES

1. "Residual Stress Measurement by X-ray Diffraction", SAEJ784a, Society of Aut. Engrs. Inc., Warrendale, PA (1971)

2. H. Dölle, The Influence of Multiaxial Stress States, Stress Gradients and Elastic Anisotropy on the Evaluation of Residual Stress by X-rays, J. Appl. Cryst, 12:489 (1979).

3. V.M Hauk, X-ray Methods for Measuring Residual Stress, Sagamore Army Materials Proc., 28:117, Plenum Press, New York (1982).

4. G. Maeder, J.L. Lebrun, and J.M Sprauel, Present Possibilities for the X-ray Diffraction Method of Stress Measurement, NDTITDS, 14:235 (1981).

5. R.H. Marion, and J.B. Cohen, Anomalies in Measurement of Residual Stress by X-ray Diffraction, Adv. X-ray Anal., 18:466 (1975).

6. T. Shiraiwa, and Y. Sakamoto, The X-ray Stress Measurement of the Deformed Steel Having Preferred Orientation, Soc. Mat. Sci., 25, Kyoto, Japan (1970).

7. V. Hauk, and H. Sesemann, Abweichungen von Linearen Gitterbenenabstandsverteilungen in Kubischen Metallen und ihre Berucksichtigung bie der Rontgenographischen Spannungermittlung, Z. Metallk. 67:646 (1976).

8. H. Dölle, and V. Hauk, Einfluss der Mechanischen Anisotropie de Vielkristalls (Textur) auf die Rontgenogriphische Spannungermittlung, Z. Metallk. 69:410 (1978).

9. H. Dölle, and J.B. Cohen, Evaluation of Residual Stresses in Textured Cubic Metals, Met. Trans. A., 11A:831 (1980).

10. C.P. Gazzara, X-ray Residual Stress Measurement Systems for
     Army Material Problems, <u>Sagamore</u> <u>Army</u> <u>Materials</u> <u>Proc.</u>, 28:369,
     Plenum Press, New York (1982).

11. C.P. Gazzara, The Measurement of X-ray Residual Stress in
     Textured Cubic Materials, <u>Proc.</u> <u>Fall</u> <u>Mtg.</u>, <u>SESA</u>, Keystone, CO,
     32 (1981).

12. B. Jaensson, A Principal Distinction Between Different Kinds
     of X-ray Equipment for Residual Stress Measurement, <u>Mat.</u> <u>Sci.</u>
     <u>and</u> <u>Engrg.</u>, 43:169 (1980).

A STUDY OF PLASTIC FLOW AND RESIDUAL STRESS DISTRIBUTION CAUSED

BY ROLLING CONTACT

Shin-ichi Nagashima,* Niritaka Tanaka,*
and Toshinori Ohtsubo**

*Department of Mechanical Engineering, Yokohama National
University, Hodogaya-ku, Yokohama, 240 Japan
**Nippon Steel Corporation, Kitakyushu, 805 Japan

INTRODUCTION

It is well known that a characteristic plastic flow occurs
when two discs are rolled together at loads exceeding the yield
point; even in the case of rolling contact without slip, a deformed layer shows marked plastic flow with respect to the direction
opposite to the rolling direction. This phenomenon was first found
by Crook[1] and Welch[2]. Later, Johnson and his co-workers explained
why the forward flow occurs under pure normal stresses[3-5].

The aim of the present study is to reproduce the various types
of plastic flow under different testing conditions of rolling
contact, such as nearly pure rolling contact, rolling contact by
friction drive, and rolling-sliding contact. After a test of
repeated rolling contact, the mode of plastic flow and the residual
stress distribution at the surface of the specimen were observed.
Further, the experimental data were compared with the result of
analytical calculation by the method of Merwin and Johnson[5].

SPECIMENS FOR EXPERIMENTS

The chemical composition and mechanical properties of the
specimen are shown in Table 1. And the shape of disc specimens and
the alignment of them for the contact rolling test are shown in
Fig.1.

Table 1. Chemical composition and mechanical properties
of the specimen

C	Si	Mn	P	S	Cu	PS	TS	El	RA
0.36	0.26	0.72	0.024	0.028	0.27	327 MPa	651 MPa	38%	65%

Table 2. The specimen numbers and the condition of experiments

Specimen Number		Diameter of discs D,mm	Slip rate s, %	Revolution x10^4	Theoretical contact breadth a, mm	Load P,kN	Max. contact stress p_O,GPa
P1	P1-1	31.42	-0.01	0.91	0.18	2.35	1.34
	P1-2	28.57	0.01	1.0			
P2	P2-1	31.43	-0.08	36.0	0.19	2.94	1.20
	P2-2	28.60	0.08	40.0			
P3	P3-1	31.43	-0.01	90	0.17	2.94	1.31
	P3-2	28.58	0.01	100			
RS	RS-1	31.43	-13.7	0.27	0.15	2.35	1.14
	RS-2	28.60	12.1	0.30			
FR	FR-1	31.42	-	36.0	0.18	2.94	1.34
	FR-2	28.41	-	40.0			

EXPERIMENTAL CONDITION

The experimental condition is shown in Table 2, where the number 1 and 2 mean driver and follower sides, respectively. In order to obtain the same velocity at the surface of the two discs, the gear ratio of the driver and follower was fixed at 60/66 and the distance of the axes was 30 mm, then the standard size of discs was fixed to D_1 = 31.43 and D_2 = 28.57 mm, respectively. This means that the gear ratio of 60/66, 56/70 and without gear give pure rolling, rolling-sliding, and friction driven rolling contact, respectively. These tests were carried out under dry conditions. The slip rate is given by the following equation:

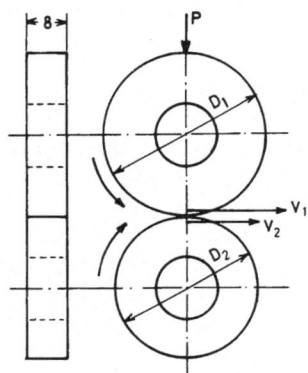

Fig.1. Alignment of disc specimen for the test.

$$s_1 = (v_1 - v_2) / v_1; \quad s_2 = (V_2 - v_1) / v_2$$

where, v_1 and v_2 are surface velocity of discs, respectively. Therefore, positive slip means that the direction of rotation of the disc and that of the tangential force are the same.

EXPERIMENTAL CONDITION FOR THE X-RAY STRESS MEASUREMENTS

The residual stresses were measured by x-ray diffraction using the method proposed by Dölle and Hauk[6], for the tested specimen after repeated rolling contact. Distributions of residual stress along the depth direction were measured by removing the surface layers successively by electrolytic polishing, and the stress relief from removing surface layers was corrected by the method

Table 3. Experimental condition of X-ray stress measurement

Method	Dölle's method, parallel beam
Characteristic radiation	Cr-Kα
Diffraction plane	211
Tube voltage and current	33 kV, 5 mA
Time constant	5 sec
Scanning speed of detector	1 deg / min
Speed of recording chart	10 deg / min
Size of irradiation mask	1 x 3 mm^2
Determination of peak angle	half-value method
Setting angle of ϕ	0, 45, 90 deg
Setting angle of ψ	0, ±15, ±26, ±36, ±45 deg

of Moore and Evans[7]. As shown in Table 3, Cr Kα radiation was used
and the spacing of (211) plane of stress free specimen was measured
as d_0 = 1.1699 Å. The stress was calculated using the value of
Young's modulus E = 206 GPa.

EXPERIMENTAL RESULTS; PLASTIC FLOW

        The typical mode of plastic flow caused by rolling contact is
shown schematically in Fig. 2, and the results are as follows:
(1) In the case of rolling contact with a small negative slip less
than 0.02%, no plastic flow was observed at the surface layer and
forward flow was observed in a deeper region. In the case of small
positive slip less than 0.04%, slight backward flow was observed at
the surface layer and forward flow was observed at a deeper region.
(2) In the case of friction driven rolling contact without gear
drive, nearly the same tendency as above was observed. (3) In the
case of rolling-sliding contact with a slip rate of more than 12
%, marked plastic flow due to a tangential force was observed even
in the surface layer. For the specimen with the slip rate of 12 %,
only backward flow was observed just under the surface layer to the
deeper region, while negative slip rate as large as -13% showed
remarkable forward plastic flow in the same region. The reason why

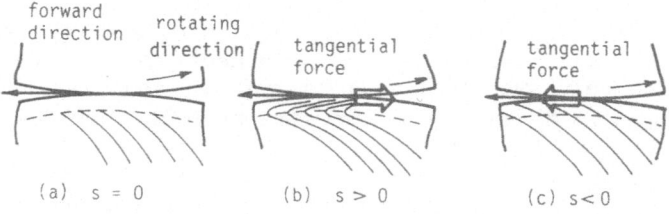

Fig.2. Pattern of plastic flow caused by rolling contact.
    (a) Pure rolling contact, (b) rolling-sliding contact
    with positive sliding, (c) rolling-sliding contact with
    negative sliding.

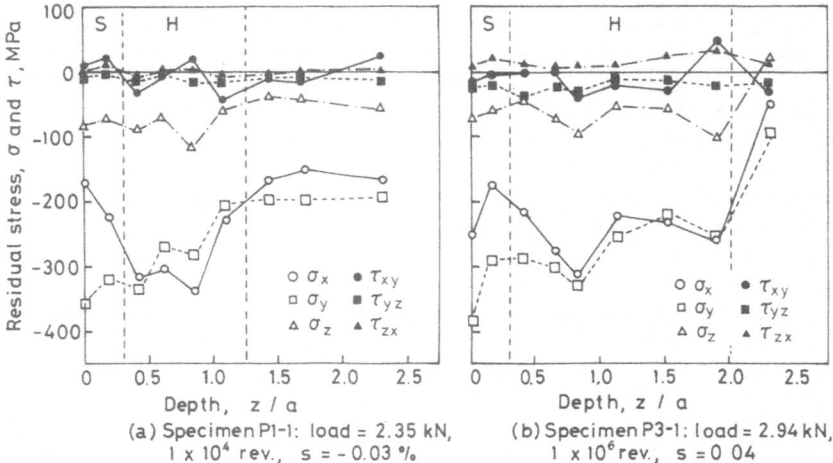

Fig.3. Residual stress distribution for the pure rolling contact.

the plastic flow turned to the reverse direction from the normal
forward flow is considered to be due to the large tangential force
applied at the surface contact layer.

EXPERIMENTAL RESULTS; RESIDUAL STRESS DISTRIBUTIONS

The distrbution of the residual stress components along the
depth direction showed nearly the same tendency for all rolling con-
tact conditions; that is, normal stress components σ_x, σ_y, σ_z are
compressive from the surface to the depth of z = 2.5 a, where a is
the semi-contact width, and the x-, y-, z-axes are tangential, axial
and radial directions of the disc, respectively. The component σ_x
and σ_y showed characteristic change with the rolling condition and
σ_z was less than 100 MPa in value and decreased gradually along the
depth direction. The shear stress components are less than ±50 MPa.
Only τ_{zx} showed as large a value as 100 MPa in the case of rolling-
sliding contact.
The detailed change in stress components for typical rolling
conditions along the depth direction is as follows:
(1) Nearly pure rolling contact: The changes in σ_x and σ_y are nearly
the same along the depth direction for both negative and positive
slip (Fig.3). The only exception is in the case of negative slip, where
the mode of change in σ_x and σ_y is opposite near the surface layer.
(2) Friction driven rolling contact: The change in stress components
along the depth direction is nearly the same as in the case of pure
rolling contact. (3) Rolling-sliding contact: As described above, if
a tangential force is applied at the surface, marked plastic flow is
formed by a severe shearing force and affects the residual stress.
The value of residual stress σ_y is the largest, σ_x is the second
and σ_z is the next, as shown in Fig.4, and they hold rather large

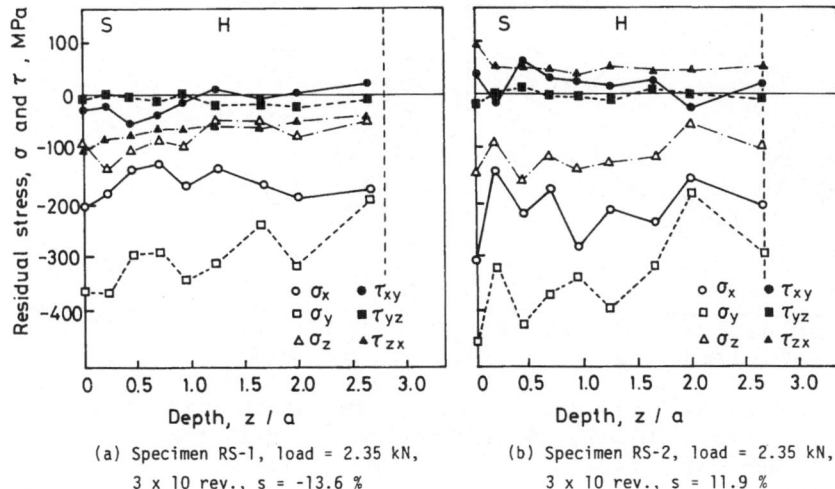

Fig.4. Residual stress distribution for rolling-sliding contact.

values even in the region as deep as 400 μm. The change in stress
distribution of the specimen with positive slip is nearly the same
as in rolling contact without slip, while in the case of negative
slip, the stress distribution is rather flat. The shear stress components are less than ±100 MPa, and are positive in the case of
positive slip, and vice versa.

COMPARISON OF EXPERIMENTAL RESULTS WITH THE CALCULATED DATA

The mode of plastic flow and the residual stress distribution
under repeated rolling contact was calculated by the method of Merwin and Johnson[3] with an approximate numerical analysis of the
elastic-plastic strain cycles. This treats the problem of two rolling cylinders to be simplified by the following model:
(1) The system is replaced by a cylinder rolling on the plane surface of a semi-infinite solid.
(2) The material of the cylinder and the plate is elastic-perfectly
plastic, i.e., no work hardening and isotropic.
(3) The plane strain criterion is assumed.
The present authors used this method, and tried to solve the
case when tangential forces are applied together with a normal
stress. The ranges of variables are as follows:
(1) The maximum contact stress p_o applied was less than 6 times
the yield stress in simple shear k, and calculated for the value
of p_o/k = 3.2 - 5.8.
(2) The values of the friction coefficient were 0 and 0.1.
(3) The ratio of the tangential and normal forces T/N were ±0.1,
±0.05 and 0.

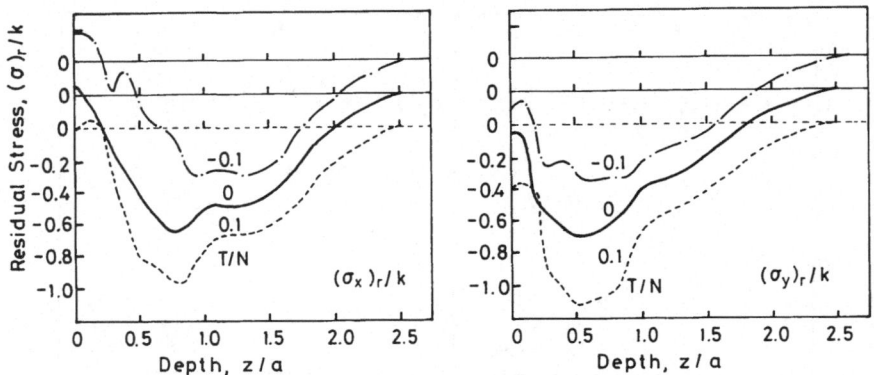

Fig.5. Change in residual stress distribution with tangential
       force; T/N = 0, ±0.1. Work-hardening is taken into
       account. p_o/k = 5.4 - 4.0, μ = 0.1.

(4) Work hardening was taken into account by the following model:

$$\sigma = F \, \epsilon^{0.25} = F\{\epsilon_e \cdot g(n)\}^{0.25},$$

$$g(n) = 1 - \{1 \, / \, (n + 1)^a\}$$

where n is the number of the rolling cycle, and ϵ_e is the total
sum of the residual plastic strain $(\epsilon_y)_r$, and a is the constant.
      The comparison of the calculated and experimental results
showed the best result for the work-hardening model. (See Fig.3
and Fig.5). Wo et al[8]. have obtained the same stress distribution
in steels shot peened (as Fig.3) and after wear (as Fig.4).

ACKNOWLEDGEMENTS

      The authors wish to express their gratitude to Prof. Shiratori
for advice on the calculation of residual stresses.

REFERENCES

1. A. W. Crook, Proc. Instn mech. Engrs. Lond. 171:187 (1957).
2. N. C. Welsh, Proc.Conf.Lubr.Wear,Instn mech.Engrs,Lond.701(1957).
3. K. L. Johnson and J. A. Jefferis, Proc.Symp.Fatigue in Rolling
   Contact, Instn mech. Engrs, Lond, 54 (1963).
4. J. E. Merwin and K. L. Johnson, ibid.,145 (1963)
5. K. L. Johnson, ibid.,155 (1963)
6. H. Dölle and V. Hauk, Härterei-Techn. Mitt., 31:165 (1976).
7. M. G. Moore and W. P. Evans, SAE Trans., 66:340 (1958).
8. J. W. Ho et al.,Wear, 84:183 (1983).

X-RAY FRACTOGRAPHY ON STRESS CORROSION CRACKING OF

HIGH STRENGTH STEEL

Yukio Hirose

Faculty of Education, Kanazawa University
1-1 Marunouchi, Kanazawa 920, Japan

Zenjiro Yajima

Faculty of Engineering, Kanazawa Institute of Technology
7-1 Oogigaoka, Nonoichi, Kanazawa 921, Japan

and

Keisuke Tanaka

Faculty of Engineering, Kyoto University
Yoshida Honmachi, Sakyo-ku, Kyoto 606, Japan

INTRODUCTION

X-ray fractography is a new method utilizing the X-ray diffraction technique to observe the fracture surface for the analysis of the micromechanisms and mechanics of fracture. The line broadening of X-ray diffraction profiles and the residual stress are two of the important X-ray parameters. Among them, the X-ray residual stress has been confirmed to be particularly useful for the fracture surfaces of high strength steels, and applied to the fracture surface of fracture toughness specimens[1] and the fatigue fracture surface.[2]

In the present paper, the distribution of residual stress beneath the fracture surface of a high strength steel (AISI 4340) made by stress corrosion cracking was measured with the X-ray diffraction technique. The fracture surface was also observed by scanning electron microscopy. Based on the results of both observations, the mechanisms of stress corrosion cracking were discussed.

EXPERIMENTAL PROCEDURE

Experimental Conditions

The material used for experiments is a high strength low alloy steel AISI 4340. The chemical compositions (wt%) of the material are as follows: 0.39C, 0.74Mn, 1.38Ni, 0.78Cr, 0.23Mo. Compact tension specimens were machined from plates by cutting slices at the right angle to the axis from hot rolled round bar with 100 mm diameter. The specimens with the plate thickness B=5.5 mm and the sharp V notch tip were used for crack growth experiment. The specimen dimensions were reported in a previous paper.[3] The specimens were normalized at 880°C for 1 hour. They were austenized at 850°C for 1 hour and then oil quenched. The tempering processes were conducted at 200°C or 400°C for 2 hours. The specimen surface was ground off by 1.2 mm to remove the decarburized surface layer after heat treatment. Finally, all the test specimens were electropolished. The mechanical properties of material are given in Table 1. The prior austenite grain size of material after heat treatment is about 12 μm.

Table 1. Mechanical properties of material.

Tempering temperature TT (°C)	Tensile strength σ_B (MPa)	Yield stress σ_Y (MPa)	Elongation E_l (%)
200	1880	1530	4.5
400	1400	1330	6.5

Stress Corrosion Cracking Tests

The stress corrosion cracking tests were conducted on a simply constructed lever arm tensile machine, in which it was posible to keep load constant or the stress intensity factor K constant during testing by changing the load. The environment is circulating 3.5% NaCl solution and the temperature of solution is kept at 16±2°C.

The growth rate of stress corrosion cracking as a function of the stress intensity factor has been reported elsewhere.[3,4] Figure 1 shows the relation for the materials used for the present fractographic study. The curves are divided into three regions: Region I near K_{ISCC}, Region III near the unstable fracture, K_{SC}, and Region II, which lies between the two. K_{ISCC} is the value of K above which the crack becomes unstable. In particular, it is interesting to note that a plateau velocity region exists for the higher-tempered specimens.

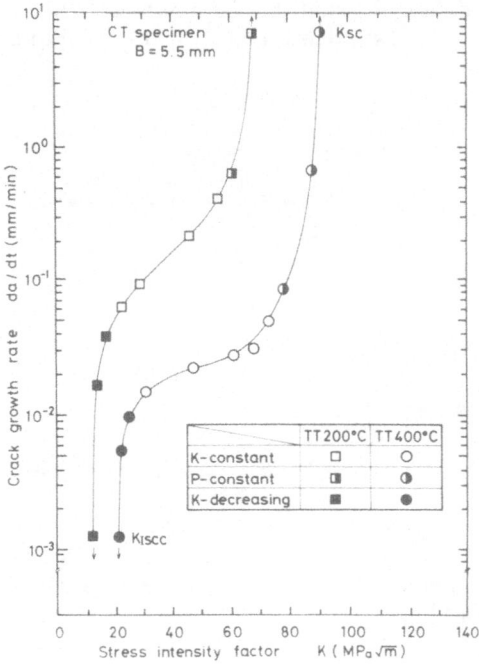

Fig. 1.   Relation between crack growth rate and stress intensity
          factor.

Inspection of Fracture Surface

The distribution of residual stress in the stress corrosion
fracture surface was measured with the X-ray diffraction method.
The method adopted the standard $\sin^2\psi$ method by using the parallel
beam of Cr-K_α X-rays as described in a previous paper.[5] The area
irradiated by X-rays was of 1 mm width and 4 mm length at the middle
of the thickness of fracture surfaces of the specimens. The residual
stress in depth was successively measured by using an electro-
polishing method. The morphology of the fracture surface was
observed by a scanning electron microscopy (SEM).

EXPERIMENTAL RESULTS

X-Ray Observation of Fracture Surface

Figure 2 shows the distribution of residual stress in the stress
corrosion fracture surface. The value of residual stress in the
vicinity of the fracture surface is in tension. It increases
gradually and then decreases with increasing distance from the
surface. The plastic zone size ω_y is defined as the distance at
which the residual stress approaches to the initial value (49 MPa).

The relation between ω_y and K/σ_Y is shown in Fig. 3, where σ_Y is the yield stress. It is noted that ω_y is proportional to the square of K/σ_Y as

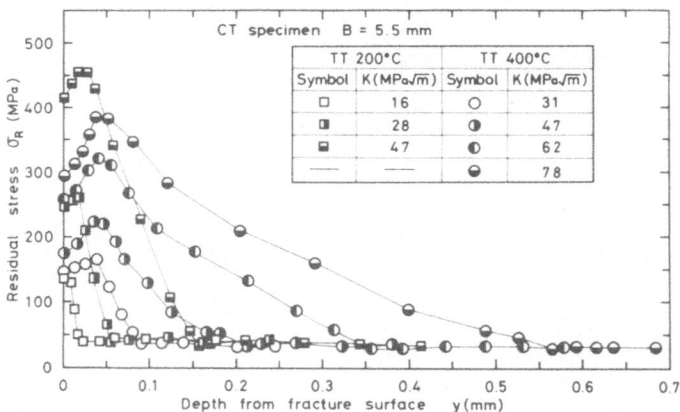

Fig. 2.   Residual stress distribution from fracture surface at 200°C and 400°C.

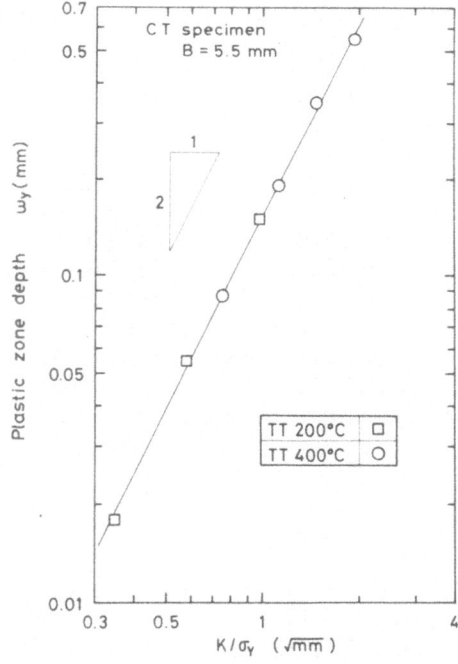

Fig. 3.   Relation between plastic zone depth and stress intensity factor divided by yield stress.

$$\omega_y = \alpha \ (\ K/\sigma_Y \ )^2 \tag{1}$$

where α is 0.16. The proportional constant α for the fracture
surface of the fracture toughness specimen of the same material was
0.13[5] which agreed with the finite element analysis for perfectly
plastic material by Levy et al.[6] Therefore, a larger value of α
for stress corrosion fracture surfaces corresponds to an increase
in the yield stress due to hydrogen.[7]

Scanning Electron Microscope Observation of Fracture Surface

Figure 4 presents the example of scanning electron micrographs
of fracture surfaces of 200°C tempered material. The fracture
surface is primarily along the prior austenite grain boundary at low
K values as seen in (a), but begins to contain more tear ridges
between grain boundary facets at higher K values as seen in (b)
and (c). Fractographic observation showed a similar tendency in

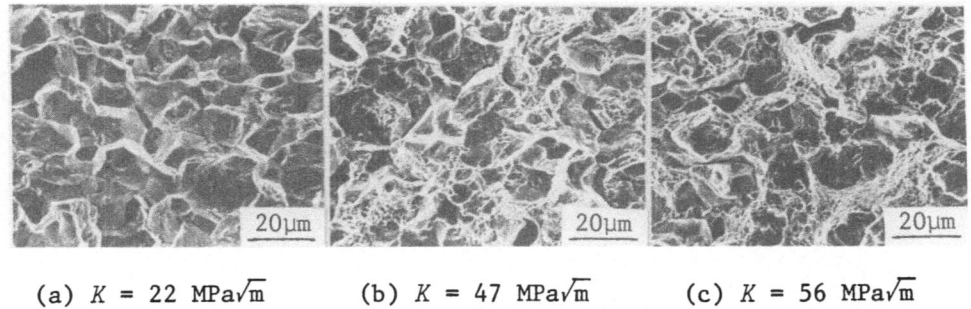

(a) K = 22 MPa$\sqrt{\mathrm{m}}$ (b) K = 47 MPa$\sqrt{\mathrm{m}}$ (c) K = 56 MPa$\sqrt{\mathrm{m}}$

Fig. 4. Scanning electron micrographs of stress corrosion fracture
surfaces of 200°C tempered steel.
(Crack growth direction is from bottom to top).

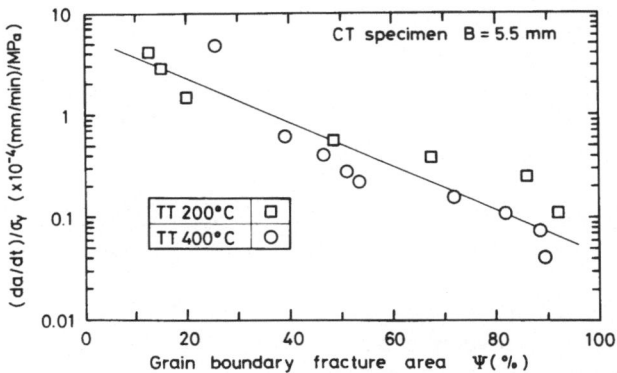

Fig. 5. Relation between crack growth rate divided by yield stress
and grain boundary fracture areal fraction.

400°C tempered material. The areal fraction of grain boundary fracture to the total fracture surface was measured for several K values. The areal fraction ψ was found to be more closely related to the rate da/dt than to K. The relation between ψ and $(da/dt)/\sigma_Y$ is shown in Fig. 5. The relation for both material is expressed by

$$da/dt = \sigma_Y \ exp \ (\ C_1 - C_2 \ \psi \ ) \hspace{3cm} (2)$$

where C_1 and C_2 are constant.

DISCUSSION

The residual stress σ_R measured on the fracture surface is plotted as a function of ψ in Fig. 6. The data indicated with the solid marks are the values measured on the unstable fracture surface in 3.5% NaCl solution. The unstable fracture surface does not show very small amount of grain boundary facets. Without respect to the tempering temperature, the relation between ψ and σ_R is

$$\sigma_R = (\ 1 - \psi \ ) \ \overline{\sigma}_R \hspace{3cm} (3)$$

and $\overline{\sigma}_R$ is nearly equal to the value measured on the unstable fracture surface in NaCl solution. The residual stress σ_R is related to da/dt as ψ is a function of da/dt (See Fig. 5).

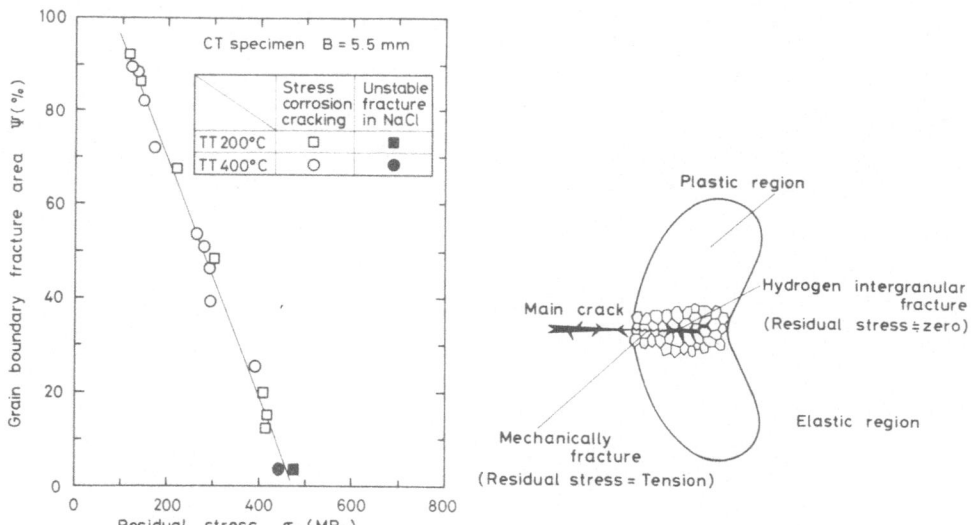

Fig. 6.   Relation between grain   Fig. 7.   Crack growth model.
          boundary fracture areal
          fraction and residual
          stress.

A strong relation between the residual stress and the grain boundary can be explained based on a model of stress corrosion cracking illustrated in Fig. 7. The growth of stress corrosion cracks in high strength steel in NaCl solution environment is generally recognized as hydrogen assisted cracking.[8] Most mechano-chemical models such as those proposed by Gerberich et al.,[9,10] Leeuwen[11] and Hirose et al.[12~14] are based upon the micromechanisms that the crack advances step-wise by connecting detached microcracks formed ahead of the main crack tip. Detached microcracks are supposed to be formed due to condensed hydrogen migrating under a gradient of negative hydrostatic pressure distribution near the crack tip. Our previous observation also supported the micromechanisms mentioned above.[4] The detached cracks were often observed ahead of the main crack tip. The crack was sometimes clearly observed to grow intermittently. Microcracks are supposed to be found along the prior austenite grain boundaries in the plastic region near the crack tip. The linking of the main crack with microcracks takes place when the mechanical instability condition is satisfied. A successive occurrence of the process results in an intermittent growth of a crack. The tear ridges observed on the fracture surface are thought to be made during mechanical fracturing. On the mechanical unstable fracture in NaCl solution, the residual stress σ_R is tensile as indicated by the data points on the abscissa in Fig. 6. The relation given by Eq. (3), which is found between the residual stress and the areal fraction of grain boundary fracture, can be explain if we assume that the residual stress is zero on the grain boundary fracture and is $\overline{\sigma}_R$ on the tear ridges.

CONCLUSIONS

The results obtained are summarized as follows:

(1) In the distribution of the residual stress, the size of the plastic zone ω_y was determined as the depth where the residual stress reached to the initial value. It is related to the stress intensity factor K and the yield stress σ_Y through

$$\omega_y = 0.16 \ (\ K/\sigma_Y \ )^2$$

A larger value of the proportional constant 0.16 indicates an increase in the yield stress due to hydrogen.

(2) The relation between the areal fraction ψ of prior austenite grain boundary fracture surface and growth rate da/dt of stress corrosion cracks was obtained as

$$da/dt = \sigma_Y \ exp \ (\ C_1 - C_2 \ \psi \ )$$

where constants C_1 and C_2 were independent of tempering temperature.

(3)   The residual stress σ_R measured on the fracture surface by the X-ray diffraction method was tensile and changed as a function of areal fraction of grain boundary fracture

$$\sigma_R = (1 - \psi) \overline{\sigma}_R$$

where $\overline{\sigma}_R$ is nearly equal to the value measured on the unstable fracture in NaCl solution.

REFERENCES

1.   Z. Yajima, Y. Hirose, and K. Tanaka, X-Ray Diffraction Observation of Fracture Surfaces of Ductile Cast Iron, Advances in X-Ray Analysis, 26:291 (1983).

2.   K. Tanaka, and N. Hatanaka, Residual Stress Near Fatigue Fracture Surfaces of High Strength and Mild Steels Measured by X-Ray Method, J. Soci. Mat. Sci. Jap., 31:215 (1982).

3.   Y. Hirose, K. Tanaka, and K. Okabayashi, Nucleation and Growth of Stress Corrosion Cracking in Notched Plates of High-Strength Low-Alloy Steel, J. Soci. Mat. Sci. Jap., 27:545 (1978).

4.   Y. Hirose, and K. Tanaka, Nucleation and Growth of Stress Corrosion Cracks in Notched Plates of High Strength Steels, ICM 3, II:409 (1979).

5.   Z. Yajima, Y. Hirose, and K. Tanaka, X-Ray Diffraction Study on Fracture Surface Made by Fracture Toughness Tests of Blunt Notched CT Specimen of High Strength Steel, J. Soci. Mat. Sci. Jap., 32:783 (1983).

6.   N. Levy, P.V. Marcal, W.J. Ostergren, and J.R. Rice, Small Scal Yielding Near a Crack in Plane Strain: A Finite Element Analysis, Int. J. Frac. Mech., 7:143 (1971).

7.   J.P. Hirth, Effects of Hydrogen on the Properties of Iron and Steel, Metall. Trans. 11:816 (1980).

8.   C.D. Beachem, A New Model for Hydrogen-Assisted Cracking (Hydrogen "Embrittlement"), Metall. Trans. 3:437 (1972).

9.   W.W. Gerberich, and Y.T. Chen, Hydrogen-Controlled Cracking-An Approach to Threshold Stress Intensity, Metall. Trans. 6A:271 (1975).

10.   W.W. Gerberich, J. Garry, and J.F. Lesser, Grain Size and Concentration Effects in Internal and External Hydrogen Embrittlement, Effect of Hydrogen on Bhavior of Materials, 70, TMS-AIME, New York (1976).

11.   H.P. Van Leeuwen, The Kinetics of Hydrogen Embrittlement a Quantitative Diffusion Model, Eng. Frac. Mech. 6:141 (1974).

12.   Y. Hirose, and K. Tanaka, Micromechanisms of Stress Corrosion Cracking in High Strength Steel, ICMC 8 I:553 (1981).

13.   Y. Hirose, and T. Mura, Nucleation Mechanisms of Stress Corrosion Cracking, Eng. Frac. Mech. 16:339 (1983).

14.   Y. Hirose, and T. Mura, Growth Mechanism of Stress Corrosion Cracking in High Strength Steel, To be published in Eng. Frac. Mech.

DETERMINATION OF RESIDUAL STRESSES IN

TRANSFORMATION-TOUGHENED CERAMICS

M. R. James, D. J. Green and F. F. Lange

Rockwell International Science Center
Thousand Oaks, CA 91360

INTRODUCTION

The strength of ceramics or glasses can be increased by placing their surfaces into compression. Techniques include ion exchange, temperature glazing, surface chemical reactions and stress-induced phase transformations. Although most of these techniques are well recognized, little effort has been expended in experimentally determining the magnitude of the compressive stress, and in particular, to use experimental evidence to identify important material and process parameters that need to be controlled. The goal of this investigation was to determine some of the factors that effect the magnitude, profile and depth of the compressive layer introduced by a structural phase transformation. X-ray residual stress measurements were used to directly determine the state of the surface residual stress.

Measurements were made on both polycrystalline Al_2O_3 and on the Al_2O_3/ZrO_2 system. The former material was used to investigate grinding stresses in a single phase system. Compressive stresses in the range of 125 to 145 MPa and extending < 15 μm deep were introduced during diamond grinding. It is believed that the compressive surface layer coincides with the plastic layer produced by the elastic/plastic interaction of the abrasive grains with the ceramic.[1]

Transformation induced residual stresses were investigated in the Al_2O_3/ZrO_2 system. This effect has been studied by several investigators (as reviewed by Green, et al[2]) and it has been shown that ceramics containing ZrO_2 can be made substantially stronger by grinding. The grinding stresses provide the energy for the martensitic transformation of tetragonal ZrO_2 to monoclinic to take place.[3] In this study, x-ray measurements confirmed that stresses as high as 1 GPa could be induced.

The magnitude of the residual surface stresses are shown to be controlled primarily by the volume fraction of tetragonal ZrO_2 that transforms to monoclinic as confirmed by varying the percent of ZrO_2 in the composite. The experimental data also indicates that the residual stress profile depends on the free energy change during the transformation (which can be modified by adding stabilizing oxides) and mildly on the grain size; the latter determines the critical applied stress required to induce the transformation. Another technique was used to increase the extent of the transformation over that caused by grinding. This was done by removing the stabilizing oxides such as Y_2O_3 from the surface to spontaneously induce the transformation. The results indicate this procedure is advantagous because it does not introduce microcracks, and the depth of the transformation extends significantly deeper than from grinding.

EXPERIMENTAL PROCEDURE

Material

The single phase material was hot pressed Al_2O_3 (1500°C/2 h) having an average grain size of 2 μm. Bar specimens (~ 0.3 by 0.6 by 1 cm) were produced and ground on four sides with a 320 grit diamond wheel removing ~ 25 μm of material with each pass.

Several series of Al_2O_3/ZrO_2 composites were used. One series was fabricated to contain different v/o ZrO_2. For this series, Y_2O_3 was used as the alloy addition; the Y_2O_3 content was increased with increasing v/o ZrO_2 in the composite. Since the elastic modulus of the composites decreases with increasing v/o ZrO_2, the m/o Y_2O_3 must be increased to retain the tetragonal structure.[5] Sintering was carried out at 1600°C/2 h.

X-Ray Conditions

The simpler x-ray diffraction techniques for measuring residual stress have received much study and are readily available in manuals[6] and textbooks on x-ray diffraction.[7,8] The theory on which these methods are based refer only to homogeneous elastic, isotropic materials in a state of biaxial stress, but the techniques have been applied to two phase materials. The technique is based on a change in the Bragg angle with tilt of the specimen, which implies the presence of surface strain. Conversion of the measured strain into surface stress is most easily accomplished by assuming isotropic elasticity. To partially compensate for anisotropic behavior along the particular (hkl) planes, the elastic constants specific to that plane are used rather than bulk values of the aggregate. Denoting the "x-ray elastic constant" as S_2, isotropic elasticity theory yields $S_2 = (1 + \nu)/E$, where ν and E

are Poisson's ratio and Young's modulus, respectively. The
conversion also assumes that the surface is in a state of biaxial
stress to the depth penetrated by the x-rays and that the principal strain axes are coplanar with the surface.

The x-ray elastic constants for the <hkl> direction can be
calculated with knowledge of the single-crystal elastic compliance
constants or determined by direct measurement. Table 1 lists the
values of $E/(1 + \nu)$ for two sets of planes obtained by the experimental method described previously,[1] and through calculations
using single crystal compliance constants.[9] Note that the calculated value of $E/(1 + \nu)$ is ~ 25% greater than the experimental
values. Discrepancies between measured and calculated elastic
constants for metal alloys have also been noted.[10,11] The cause
of these discrepancies is not clear in this case, but it is
important to accurately measure the elastic constants as described
here rather than to use calculated values.

The particular computer-automated instrumentation and technique used to determine surface stresses is similar to that
described previously.[12] Each measurement was taken using eight
tilt angles (0° to 45°). The peak of the diffraction profile at
each tilt angle was determined by the apex of a least-squares
parabola fit to the top portion of the profile. The profiles were
quite broad (> 3° 2θ full width, half maximum) and symmetric
around the apex. To ensure that grinding did not introduce errors
resulting from the principal strain components being inclined to
the surface,[9] some measurements were conducted at both positive
and negative tilt angles and gave identical results.

For the present work, both CuKα (λ = 1.542Å) and CrKα (λ =
2.291Å), radiations were used. The Al$_2$O$_3$ (602) diffraction peak
was chosen for the Cu radiation (2θ ≈ 142°), and the (119) peak
for the (119) peak for the Cr radiation. These peaks were chosen
on the basis of the diffracting angle being as high as possible,
to maximize the measured peak shift. In addition, other factors
such as intensity and interference from nearby peaks were also
considered. Table 2 lists the penetration depths (normal to
surface) for both CuKα and CrKα at 2θ = 142° and 135°, respectively. CuKα is much more penetrating than CrKα, and the penetration depth decreases with increasing v/o ZrO$_2$. The surface stress

Table 1. Values of $E/(1 + \nu)$ for Al$_2$O$_3$

Radiation	Crystal Plane	$E/(1 + \nu)$ (GPa)	
		Experimental	Calculated
Cu	(602)	212	270
Cr	(119)	246	313

Table 2.   Penetration Depth of Diffracted X-rays[†] (μm)

Radiation	2θ	Volume % ZrO₂*						
		7.5	15	20	30	40	50	60
CrKα	135°	6.5	5.5	4.9	4.1	3.5	3.0	2.6
CuKα	142°	20.8	17.2	12.6	10.6	9.1	8.0	5.2

*Alloy additions to ZrO_2 (viz Y_2O_2 or CeO_2) not taken into account.

†Depth at which 50% intensity of x-rays is diffracted was used in calculation.

determination is a mean value integrated over the depth penetrated by the radiation used (weighted by an exponential function due to absorption).   If the residual surface stress is not constant with increasing depth and/or its depth is less than the depth penetrated by diffracting x-rays, one would expect different mean stress values for different characteristic x-rays.   Also, if the stress gradient were sharp enough over the penetration depth, a curvature in the d vs $\sin^2\psi$ plot would be exhibited.[13,14]   Occasionally, slight curvature was evident, but most often the relationship was extremely linear.

RESULTS

Surface Stress in Al_2O_3

Table 3 summarizes results from pure Al_2O_3.   Compressive stresses on the ground surface are substantial, ~ 135 MPa (to 170 MPa if single-crystal results for $E/(1 + \nu)$ are used), whereas the 1 h anneal reduced the stresses to within the limits of experimental error.   Results of the deeper penetrating Cu-radiation (50% diffracted within a depth of 26 μm from surface) and those obtained by polishing the surface to remove up to 10 μm of material suggest that a majority of the compressive stresses are within a surface layer <15 μm deep.   Most of the deviations are within that indicated from statistical counting errors (± 45 MPa), indicating that the stress state is uniform over the surface.   The results of polishing suggest that even this procedure introduces a small amount of compressive residual stress.

Effect of ZrO_2 Volume Fraction

Figure 1 illustrates that the residual compressive stresses increase with increasing v/o ZrO_2 retained in its tetragonal

Table 3. Surface Stress Determinations in Al_2O_3

Specimen	Radiation	Surface Treatment	Average Compressive Surface Stress* (MPa)
1	Cu	Ground	35 ± 16 (3)†
1	Cr	Ground	125 ± 7 (2)
2	Cr	Ground	145 ± 21 (2)
2	Cr	5 μm removed	70 ± 28 (2)
2	Cr	10 μm removed	83 ± 56 (2)
3	Cr	Annealed	29 ± 49 (3)

*Calculated using experimental value of $E/(1 + \nu)$.

†Number of measurements is given in parentheses.

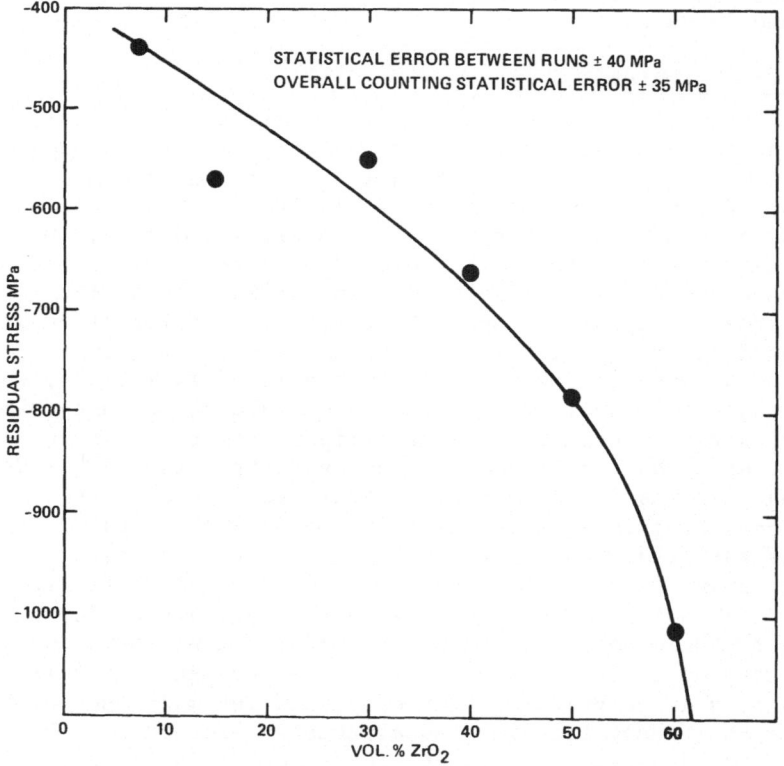

Fig. 1 Average residual surface stress measurements for a
series of Al_2O_3/ZrO_2 composites. Measurements made
with CrKα radiation.

structure in the sintered material. Namely, the compressive
stress can exceed 1 GPa. It should be noted that these results
may be somewhat modified by the fact that ZrO_2 is a strong
absorber of x-rays (see Table 1). That is, the depth sampled with
the $CrK\alpha$ radiation for 7.5 v/o ZrO_2 is about three times that for
the 60 v/o ZrO_2, which may bias the apparent average stress magnitude to higher values with increasing ZrO_2 content.

A comparison of these experimental values to theoretical
predictions has been made[2] based on the measured depth of transformation. The data show that there is a reasonable correlation
between the amount of ZrO_2 transformed and the measured stress.
On the whole the measured stress appears to be slightly higher
than the calculated values and some of this difference could be
explained by the plastic deformation that occurs during grinding
of ceramic surfaces.

By changing the amount of stabilizing oxide and the ZrO_2
grain size, we have shown that these modify the surface residual
stress in the manner predicted by Lange.[3-5] However, the changes
in residual stress were small in comparison to the effect of the
ZrO_2 volume fraction.[2]

Y_2O_3 Removal Technique

In the transformation-toughened materials used in this study,
it has been found necessary to alloy the ZrO_2 with an oxide such
as Y_2O_3 or CeO_2,[5] in order to retain the tetragonal phase.
Therefore, if a material once fabricated with such an addition is
heat-treated in ZrO_2 powder containing no alloy addition, the
alloying oxide is expected to be removed from the surface region
due to the concentration difference. Once this occurs, the
chemical free energy associated with the ZrO_2 transformation will
be increased in the surface region and the transformation is
expected to proceed.[15] This effect is confirmed in Table 4, where
the apparent fraction of ZrO_2 in the monoclinic phase (not corrected for x-ray penetration) is determined from the areas under
the (111) tetragonal and monoclinic x-ray diffraction peaks ($CuK\alpha$
radiation). It was found that after the heat treatment the amount
of monoclinic ZrO_2 is substantially increased when compared to
either the as-fired or annealed (after grinding) surface.
Moreover, it was found that a greater fraction of ZrO_2 was transformed by the heat treatment than by surface grinding. Depth
profiling of the monoclinic content showed the transformation took
place as deep as 30 μm. Residual stress measurements confirm that
the surface is in compression (Table 4), and the magnitude depends
on the length of heat treatment as expected. Green[15] has shown
that improved resistance to indentation cracking is achieved
through this Y_2O_3 removal process.

Table 4. Fraction of Monoclinic ZrO_2 for an $Al_2O_3/30$ v/o ZrO_2
Composite for Various Surface Treatments

Surface Treatment	Monoclinic Fraction	Surface Stress
As fired	<0.02	–
Ground (320 Grit Diamond)	0.11	-550 MPa
Annealed (1400°C-16 h)	<0.02	-180
Y_2O_3 Removal		
-1400°C (1 h)	0.31	-140
-1400°C (4 h)	0.60	-330
-1400°C (16 h)	0.67	-550

DISCUSSION

Green[16] has shown that the strengthening aspect of
compressive surface stress depends on the crack size and the
magnitude and, to a large degree, the depth of the compressive
zone. Maximum strengthening occurs when the crack is completely
embedded in the compressive zone. The control of the magnitude
and depth will depend on the process being used to induce the
surface compression. It is important, therefore, to understand
the process variables, particularly if the compressive stress
varies with zone depth. Moreover, it should also be remembered
that the process itself may change the surface flaw characteristics, e.g., by subcritical extension of the surface cracks.
Such effects could be incorporated into the analysis provided the
final crack sizes are known. As an alternative to increasing the
depth of the compressive zone, the strengthening could also be
increased by reducing the size of the surface cracks.

It has been shown that the most significant factor in obtaining a deep compressive zone is the extent of transformation.
While grinding suffices to produce high compressive stresses on
the surface, the depth is only $10 \sim 15$ μm. Flaw populations in
ceramics can have cracks significantly deeper than this. The
removal of stabilizing oxide on the surface was shown to provide a
mechanism to achieve a deeper transformation and thus a deeper
zone of compressive residual stress.

ACKNOWLEDGEMENTS

The authors would like to acknowledge the financial support
of the Office of Naval Research, Contract No. N00014-77-C-0441.

REFERENCES

1. F. F. Lange, M. R. James and D. J. Green, Determination of
    Residual Surface Stresses Due to Grinding in Polycrystal-
    line Al_2O_3, $\underline{J}$. $\underline{Am}$. $\underline{Ceram}$. $\underline{Soc}$., 66: C16 (1983).
2. D. J. Green, F. F. Lange and M. R. James, Factors Influencing
    the Residual Surface Stresses Due to a Stress-Induced Phase
    Transformation, $\underline{J}$. $\underline{Am}$. $\underline{Ceram}$. $\underline{Soc}$., to be published, 1983.
3. F. F. Lange, Compressive Surface Stresses Developed in
    Ceramics by an Oxidation-Induced Phase Change, $\underline{J}$. $\underline{Am}$.
    $\underline{Ceram}$. $\underline{Soc}$., 63: 38 (1980).
4. F. F. Lange, Transformation Toughening: Part 5, $\underline{J}$. $\underline{Mater}$.
    $\underline{Sci}$., 17: 255 (1982).
5. F. F. Lange, Transformation Toughening, Part 1, $\underline{J}$. $\underline{Mater}$.
    $\underline{Sci}$., 17: 225 (1982).
6. M. E. Hilley, J. A. Larson, C. F. Jatczak and R. E. Ricklefs
    (eds.), Residual Stress Measurements by X-ray Diffraction,
    SAE Information Report J784a, SAE, Warrendale, Penn. (1971).
7. H. P. Klug and L. E. Alexander, "X-ray Diffraction Procedures,"
    2nd edition, John Wiley & Sons, 755-790 (1974).
8. B. D. Cullity, "Elements of X-ray Diffraction," 2nd edition,
    Addison-Wesley Publ. Co., 447-479 (1978).
9. H. Dolle, The Influence of Multiaxial Stress States, Stress
    Gradients and Elastic Anisotropy on the Evaluation of
    Residual Stresses by X-rays, $\underline{J}$. $\underline{Appl}$. $\underline{Cryst}$., 12: 489
    (1979).
10. P. S. Prevey, A Method of Determining the Elastic Properties
    of Alloys in Selected Crystallographic Directions for X-ray
    Diffraction Residual Stress Measurements, $\underline{in}$ "Advances in
    X-ray Analysis", 20 (Ed. by H. F. McMurdie et al), Plenum
    Press, New York, 345-354 (1977).
11. R. H. Marion and J. B. Cohen, The Need for Experimentally
    Determined X-ray Elastic Constants, IBID 355-368.
12. M. R. James and J. B. Cohen, Study of the Precision of X-ray
    Stress Analysis, IBID, 291-307.
13. M. R. James and J. B. Cohen, The Measurements of Residual
    Stresses by X-Ray Diffraction Techniques, in "Treatise on
    Materials Science and Technology", 19A ed. H. Hermon,
    Academic Press, NY, 1-62 (1980).
14. I. C. Noyan, Effect of Gradients in Multi-Axial Stress States
    on Residual Stress Measurements with X-rays, $\underline{Met}$. $\underline{Trans}$.,
    14A: 249 (1983).
15. D. J. Green, A Technique for Introducing Surface Compression
    into Zirconia Ceramics, $\underline{J}$. $\underline{Am}$. $\underline{Ceram}$. $\underline{Soc}$., in press.
16. D. J. Green, Compressive Surface Strengthening of Brittle
    Materials, $\underline{J}$. $\underline{Am}$. $\underline{Ceram}$. $\underline{Soc}$., to be published.

SMALL AREA X-RAY DIFFRACTION TECHNIQUES; APPLICATIONS OF THE
MICRO-DIFFRACTOMETER TO PHASE IDENTIFICATION AND STRAIN
DETERMINATION

C.C. Goldsmith and G.A. Walker

IBM General Technology Division, East Fishkill Facility
Hopewell Junction, New York   12533

ABSTRACT

The trend towards smaller devices and packages in the semiconductor industry makes it increasingly important and
increasingly difficult to obtain useful X-ray diffraction information from the small areas employed. This study covers applications of a Rigaku[1] micro-diffractometer to measure strain and
obtain phase information from areas less than 100µm in diameter.
Examples of residual stress mappings between electrical vias only
100µm apart and phase determination on a single electrical via
120µm diameter will be shown.

INTRODUCTION

   X-ray diffraction studies, such as phase analysis and
residual strain measurements, generally require an area millimeters in size for analysis. In the electronics industry, where
the trend is towards smaller devices and packages, this large
area requirement precludes many X-ray diffraction measurements on
actual product. The analyst is therefore forced to make measurements on large area pseudo samples which have been subjected to
the same processes as the actual product. The validity of
extrapolating the results obtained from measurements on these
pseudo samples to actual product can be questioned. This is
particularly true in the case of residual stress measurements if
dimensions are reduced to a point where stress concentration
effects predominate or where interface phases and strains affect
material properties. The types of X-ray analysis necessary to

229

discover problems of this type require sophisticated instrumentation capable of accurate X-ray line analysis.[2]

This study will describe several examples in which a
microbeam X-ray diffractometer was used to measure residual
stresses and obtain phase information from areas less than
100 micrometers in diameter.

EXPERIMENTAL

The instrumentation used in this study consists of a Rigaku
micro-diffractometer connected to a 12KW rotating anode X-ray
point source. The rotating anode source is necessary to enhance
the X-ray intensity to compensate for the extreme aperturing for
the small beam sizes employed. The micro-diffractometer is
really the only unique segment of the system. Figure 1 shows the
major components of the micro-diffractometer. They consist of
two different detectors, collimator, sample holder mounted in an
alignment micrometer and an optical microscope. The micro-
diffractometer is based on Debye-Scherrer optics (Figure 2)
rather than the Bragg-Brentano optics used on most conventional
diffractometers. In this design the sample remains stationary
while the detector scans directly along the X-ray beam path.
This allows the detectors to employ a circular receiving aperture
which allows detection of a complete Debye ring enhancing
counting statistics. The scintillation detector (Figure 1) works
in transmission mode and covers the two-theta range from
5 degrees to approximately 60 degrees. This detector also has a
pin hole in the center for goniometer alignment. The ring shaped
detector, which detects approximately 270 degrees of the Debye
ring due to the specimen holder, is a sealed gas-filled detector
that works in either transmission or reflection mode. This
detector covers the two theta range from 30 degrees to
150 degrees.

The alignment of the X-ray beam onto a specific area of a
specimen is done with an optical microscope. The alignment of
the optical microscope to the X-ray beam is accomplished by
replacing the sample (Figure 1) with a special pinhole alignment
device. The pinhole is adjusted to maximize the X-ray intensity
onto the center of the scintillation detector. The cross hairs
in the optical microscope are then aligned to the pinhole in the
alignment device. This optically fixes the area where the X-ray
beam strikes. The alignment device is then replaced with a
specimen, which can be adjusted with the micrometers until the
specific area to be studied is centered on the cross hairs in the
microscope.

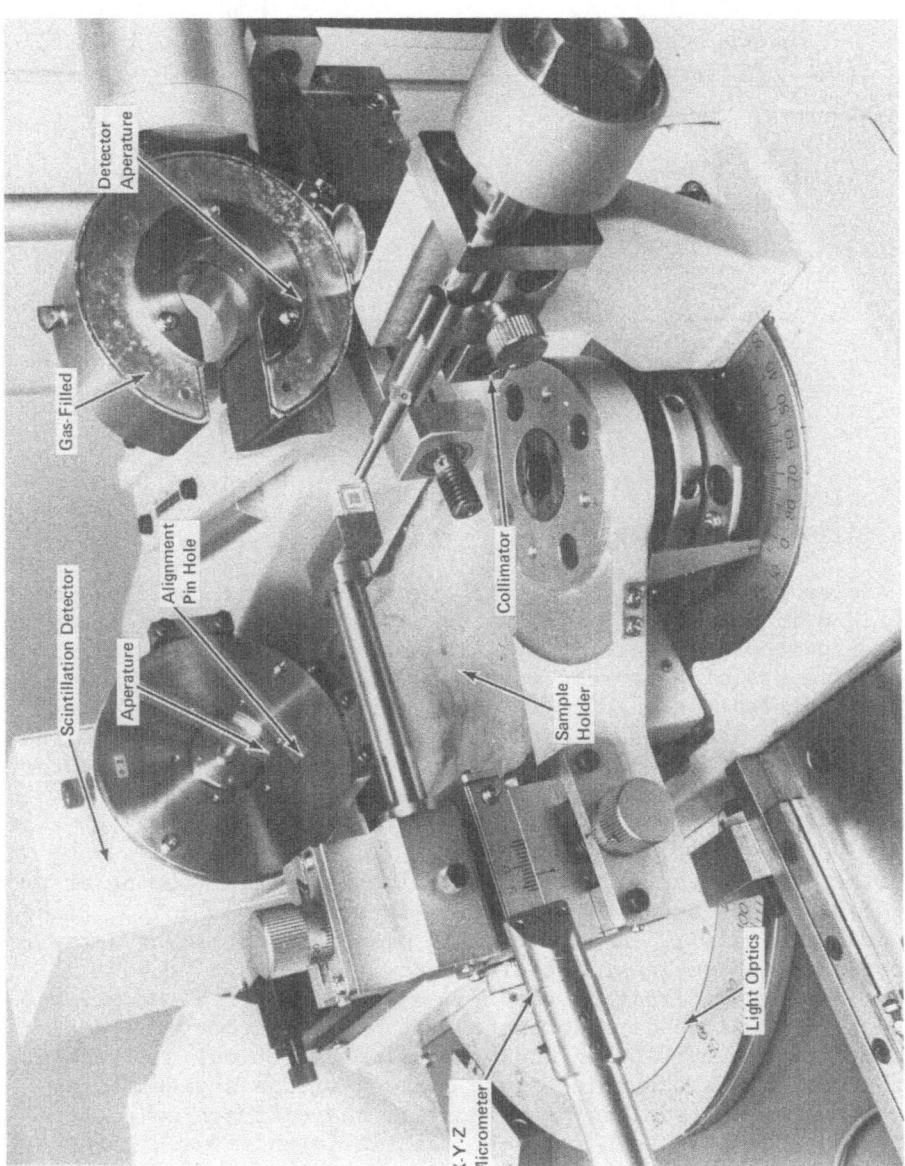

Fig. 1. Major Components of Micro-Diffractometer

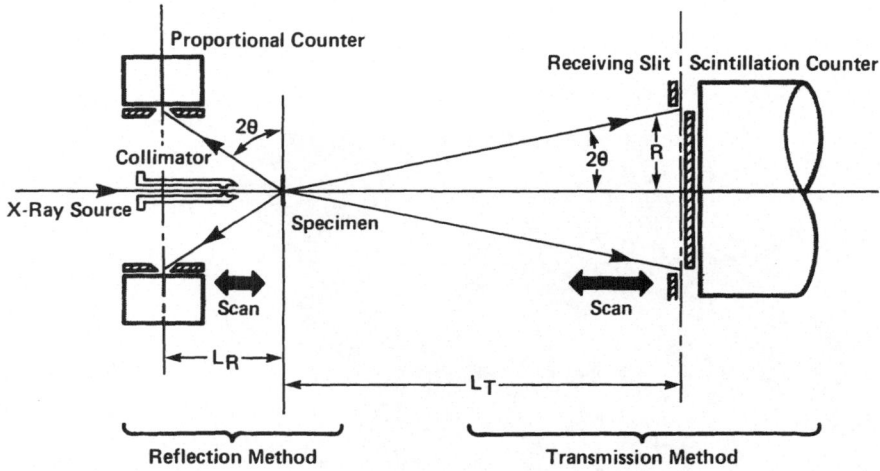

Fig. 2.   Geometrical Arrangement

APPLICATION AND RESULTS

Several examples of analysis on the microelectronic package described in Reference 3 will serve to demonstrate application of the X-ray micro-diffractometer. The package is composed of several layers of patterned molybdenum metallurgy on alumina sheets sintered together to form a monolithic structure. The exposed metal features on both surfaces of the substrate are then plated with nickel and gold. The gold is then diffused into the nickel to form a thin metallurgical solid solution of gold-nickel.

The first example described involves a "stain" on one or two of the many pads used for connection to the silicon chip. Since the pads are only 125 micrometers in diameter, obtaining X-ray patterns for phase identification becomes difficult if not impossible. A diffraction pattern was obtained from the stained pad using the micro-diffractometer with a 75 micrometer collimator in a reflection mode (chromium radiation). A segment of the pattern is shown in Figure 3 along with a segment taken from an adjacent pad showing no stain. The diffraction pattern indicates that the stain is due to partially undiffused gold (diffusion, partially blocked by an organic layer, formed a gold-nickel solid solution, but it had not gone to completion).

The second example is a transmission-reflection pattern obtained from a single sheet of sintered alumina. The pattern was collected using a 100 micrometer collimator, copper

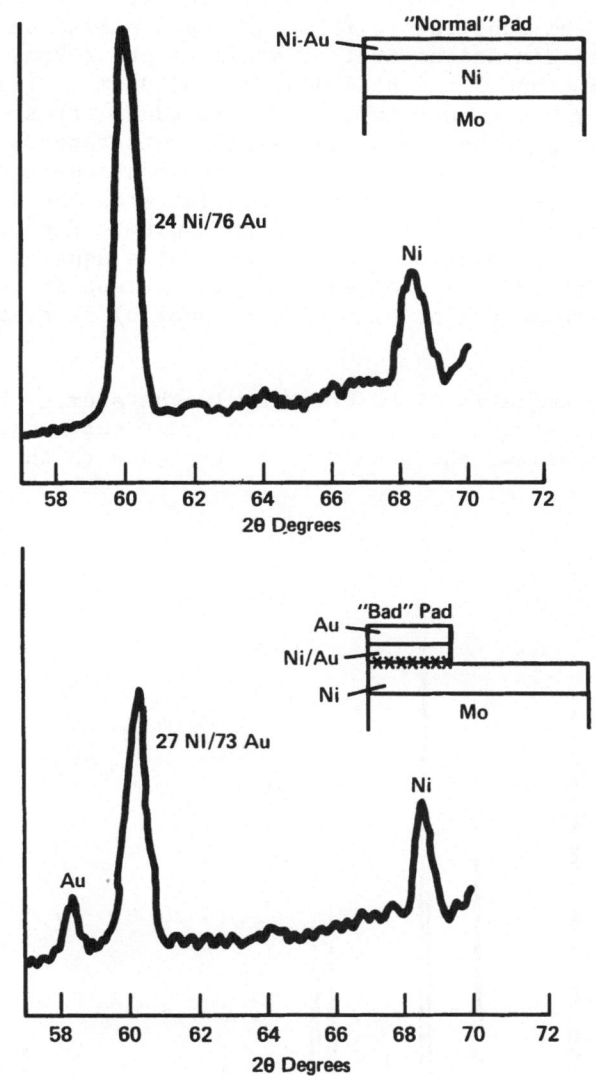

Fig. 3.   X-Ray Pattern Obtained from Stained Pad and Adjacent Pad

radiation, and the gas filled detector. Figure 4 shows that the
sheet is composed of a single crystalline phase (alpha alumina).

The third example involves the I/O pins for the microelectronic package. These pins are designed to be brazed to the
bottom surface of the package to supply electrical power/signal
connections to the substrate. The pins are a palladium plated
iron alloy whose dimensions are: overall length 2.2mm, a shank
diameter of 0.3mm and a pin head diameter of 0.7mm. Tests on the
pins supplied by one vendor indicated a residual stress problem
in the palladium plating. Residual stress measurements were made
in the plating on the pin head. The measurements were made using
the gas filled detector and copper radiation with the 100 micrometer collimator on the (331) line of palladium. Figure 5 shows
the results of peak position plated versus sine squared psi for
the plating obtained from two vendors. The stress in the
palladium plating on the problem pins (Vendor B) is clearly much
higher.

The last example involves cracking in the alumina between
engineering-change pads (a set of contact pads surrounding each
chip site which allows the module to be tested with the chips in

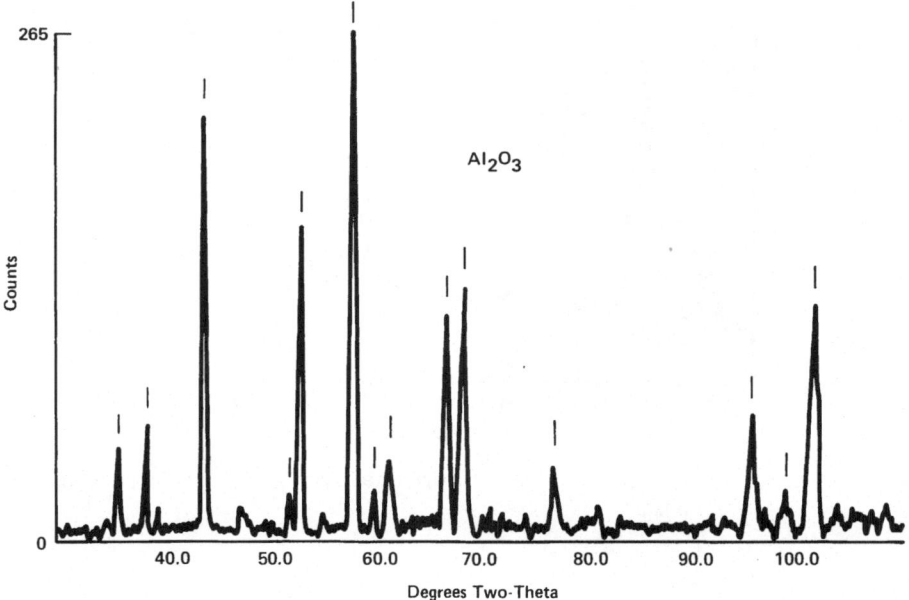

Fig. 4.   Scan Obtained from Alumina Sheet

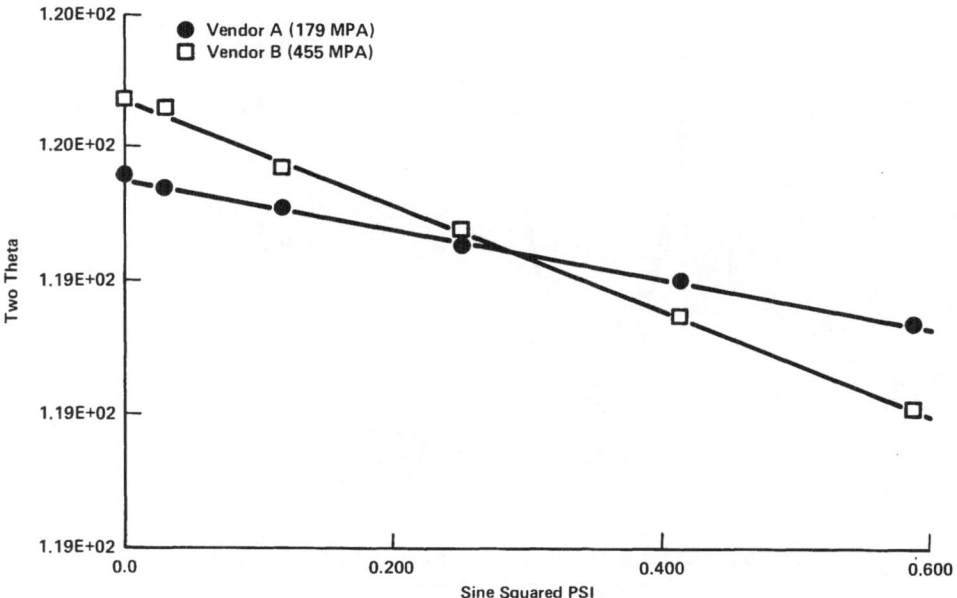

Fig. 5.   Stress Results from Pd Plating

place). Each substrate contains several thousand pads, with a
pad-pad spacing at either 250 micrometers or 100 micrometers.
The alumina between pads spaced at 100 micrometers exhibited a
slight tendency to crack from pad to pad. Residual stress
measurements in the alumina were attempted using a conventional
goniometer with a collimated X-ray beam. With this technique,
the residual stresses in the alumina measured near zero. The
residual stress was then mapped between pads using the micro-
diffractometer with a 30 micrometer collimator and copper
radiation. The mapping was done starting with a residual stress
measurement at a pad edge and taking several measurements across
the interval, eventually reaching the adjacent pad (there is
overlap in the measurement since the step size is smaller than
the beam diameter). The results of the stress mapping for both
pad-pad spacings are shown in Figure 6.

    The residual stress at the pad-alumina interface is approxi-
mately identical regardless of the pad-pad spacing. The residual
stress at the mid-point between the pads changes considerably,
becoming near zero with a 250 micrometer spacing, but remaining
high on the pads spaced at 100 micrometers. Apparently, the
mid-point stress is high enough on closely spaced pads to allow a
crack to propagate through this region extending from pad to

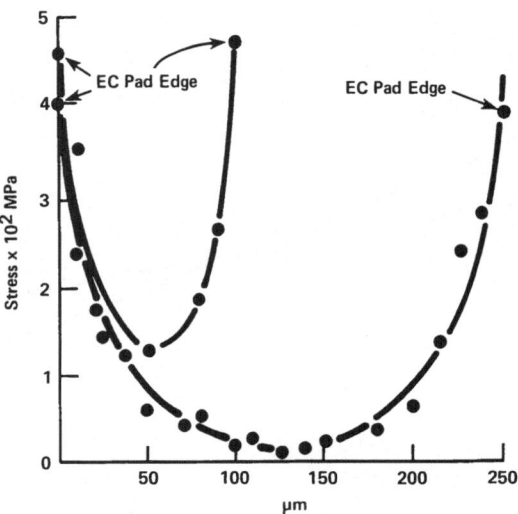

Fig. 6.   Stress Distribution in Alumina Between
EC Pads with Different Spacing

adjacent pad. Material and process changes were made to reduce
this stress. There has not been a field failure of this package
in the four years since its introduction.

DISCUSSION

      Some limitations associated with the use of this microbeam
technique should be discussed. The irradiated area on the
specimen is a function of collimator size, focal spot size, and
source to collimator distance. With the 100 micrometer
collimator and the standard 0.5mm X 10mm focal spot size, the
size of the irradiated area is approximately
105 X 105 micrometers. The irradiated area on the specimen is
not reduced with a smaller focal spot size, e.g. 0.1mm X 1mm, but
the intensity is drastically reduced. However, the use of the
30 micrometer collimator in conjunction with the 0.5mm X 10mm
focal spot does reduce the irradiated area on the specimen to
approximately 32 X 50 micrometers. A reduction in the focal spot
size to 0.1mm X 1mm results in a decrease in specimen irradiated
area to approximately 32 X 32 micrometers, but decreases the
X-ray intensity roughly an order of magnitude. For example, the
peak intensity from the palladium plating on the pins using the
100μm collimator and the 0.5 X 10mm focal spot size loaded to
40 KV-180ma is 400 counts per second.

X-ray intensity is not the only limit to reducing collimator size. There is also a trade off between the collimator size and specimen grain size which may result in having too few grains in the diffracting condition to represent the specimen's average condition. This has shown up in some stress measurements, where a large random scatter of some points is evident in the plots of two-theta vs. sine squared psi. We feel that the scatter of these points off the fitted least squares line is due to a lack of grains in the irradiated area with average strain (we are seeing grains with non-uniform strain).

Another problem area occurs in phase analysis. The patterns collected for phase identification show a background with a parabolic shape reaching a maximum at 90 degrees two-theta. This is partly due to the scanning of the detector along the X-ray beam path, which results in an apparent change in slit width with change in two-theta position (reaching a maximum slit width at 90 degrees two-theta). This yields a powder pattern with distorted relative X-ray intensities making search match procedures more difficult.

It is important to remember, with residual stress measurements, that the optics are Debye-Scherrer. When the sample is set perpendicular to the X-ray beam, the diffracting planes are not parallel to the substrate. Also, the planes that form each side of the Debye ring move, with psi change, in opposite directions in relation to the surface normal. To solve this problem, we mask part of the detector ring, using only the planes that form one side of the cone for strain measurement.

CONCLUSIONS

The use of the micro-beam diffractometer has been demonstrated with examples of phase analysis and residual stress applications. It is not intended to replace the conventional Bragg-Brentano goniometer, but to be used to extend the X-ray powder diffraction technique down to a level of less than 100 micrometers.

ACKNOWLEDGEMENTS

The authors would like to thank R. Martin and T. Bowmaster for some of the data collection and G. Scheer and T. Nunes for adding stepping motors for automation. Also thanks to T. Nunes for computer software support in data collection and reduction.

REFERENCES

1.   Rigaku, U.S.A., Inc., Danvers, Ma.
2.   G.A. Walker and C.C. Goldsmith, 16th Annual Proceedings of
     Reliability Physics, 56, (1978).
3.   Albert J. Blodgett, Jr., Scientific American, 249, 86,
     (1983).

RESIDUAL STRESS RELAXATION IN CEMENTED CARBIDE COMPOSITES

STUDIED USING THE ARGONNE INTENSE PULSED NEUTRON SOURCE

A. D. Krawitz, R. Roberts and J. Faber*

University of Missouri

Columbia, MO   65211

ABSTRACT

    Cemented carbide composites with a WC hard phase and a Co-Ni alloy binder phase have been subjected to monotonic and cyclic deformation and studied using the high resolution General Purpose Powder Diffractometer at the Argonne Intense Pulsed Neutron Source. Upon deformation, relaxation of bulk differential thermal residual stresses, tensile in the binder and compressive in the carbide, is observed to occur as a function of loading mode and plastic strain via shifts in diffraction peak positions. In addition, peak breadth behavior indicates broadening. At low plastic strain this is due primarily to a range of residual stress and at high plastic strain it is attributable to the plastic deformation alone since relaxation is essentially complete. The appropriateness of neutrons is discussed.

INTRODUCTION

    Cemented carbide composites (cermets) are produced by powder metallurgy methods using typical sintering temperatures of 1300-1400°C.[1] Upon cooling, differential thermal stresses are established between the carbide and binder phases by virtue of a substantial difference in the coefficients of thermal expansion, α, of the two phases. The actual levels of stress also depend

---

*Argonne National Laboratory
 Argonne, IL   60439

upon the elastic constants and volume fractions of the
constituents. The stress in phase i may be estimated from:[2]
$\sigma_i = K_i(\bar{\alpha}-\alpha_i)\Delta T$, where $\bar{\alpha}$ is the composite value. This relation
assumes isotropy andthe absence of cracks, interfacial shear and
plastic flow. Since the microstructures are fine-scale (WC grain
size is 2-3 μm, the stress state is essentially hydrostatic.

The discovery of stress relaxation in cermets was made in the
course of a study concerned with binder deformation mechanisms in
WC-(Co,Ni) cermets containing binders with 0, 15 and 30 wt. pct.
Ni.[3] As Ni content increases, the mechanism of binder deformation
shifts from a strain-induced (metastable) FCC-to-HCP martensitic
transformation to slip, dislocation generation and twinning. The
occurrence of relaxation appears to be quite general, i.e., has
been observed for all compositions. Only the WC-(Co,30 Ni)
material will be referred to in this article, with emphasis
placed on the basic character of the effect and the technique
used to measure it.

Neutron diffraction is the method of choice for a variety of
reasons[4]. First, neutrons are highly penetrating in most
engineering materials. This facilitates study of bulk phenomena
such as the balanced stresses between two phases in the present
case. For the cermet material under study a 1.5Å neutron beam
has a 50% absorption thickness of about 5 mm while the value for
an x-ray beam of the same wavelength (copper) is about 3μm. This
not only eliminates the need for surface preparation and its
problems, which are considerable in cermets, but enables study of
the actual stress state as opposed to study of an altered surface
stress. Second, neutron scattering cross-sections (b) do not
vary systematically with atomic number as for x-rays; relative b
values for W,C,Co and Ni are 0.47,0.66,0.25 and 1.03,
respectively. This greatly ameliorates the dominance of W that
occurs with x-rays. Third, neutron scattering cross-sections do
not decrease with scattering angle enabling access to high-angle
peaks. These aspects combined with the resolution, wavelength
and scattering angle capabilities of the GPPD instrument[5,6] have
made this study possible.

EXPERIMENTAL

Material

Cermets containing 17 wt. pct. binder of composition 70%
Co-30% Ni were produced from WC, Co and Ni powders. The starting
powders were milled, pressed into cylindrical form, sintered at
1375°C and hot isostatic pressed at 1300°C under 100 MPa of argon.
Cylinder dimensions are 12.70 mm diameter by 18.877 mm high.
Relevant properties of the constituents are shown in Table I.

Table I. Properties of the Constituents

Property	WC	Co	Ni
E (MPa)	7×10^5	2×10^5	2×10^5
K (MPa)	3.89×10^5	1.85×10^5	1.85×10^5
ν	0.2	0.32	0.32
$\alpha(°C^{-1})$	6.2×10^{-6}	13.8×10^{-6}	13.3×10^{-6}

Table II. Mechanical Treatment and Plastic Strain
for WC-(Co,30 Ni)

Mechanical Treatment	Stress/Strain Level	Number of Cycles	Plastic Strain (%)
As-Sintered	0	0	0
Low Fatigue	1.0 GPa	5×10^5	-0.10
High Fatigue	2.1 GPa	5×10^5	-0.61
Low Monotonic	0.75 %	1	-0.30
High Monotonic	5.00 %	1	-4.41

Mechanical Treatment

Mechanical conditioning and the resulting plastic deformation
are shown in Table II. Low and high levels of monotonic loading
were applied under strain control and low and high levels of cyclic
loading were applied for 5×10^5 cycles under stress control. The
low monotonic level was chosen to correspond to one cycle at the
high fatigue level. All treatments were in compression.

Neutron Diffraction

Neutrons are produced at the IPNS Facility via spallation by
bombardment of a uranium target with 500 MeV protons at 30 Hz.
The GPPD is a high-resolution powder spectrometer that operates
in a time-of-flight mode. The advantage of the time-of-flight
method is that all available d-spacings between 0.5 and 5.0 A may
in principle be utilized while preserving the high angle
sensitivity to d-spacing inherent in Bragg's law, i.e., since θ
is fixed and λ is varying, all peaks occur at, e.g., 151.8 2θ.
In addition, changes in peak breadth as well as position can be
readily monitored with high resolution.

Fixed detector banks are positioned at $2\theta = 30°$, 60°, 90° and
151.8° and are 1.5 m from the sample position. The instrument
resolution $\Delta d/d$, where Δd is full width at half maximum, is 0.25%

at 2θ = 151.8° and 0.40% at 2θ = 90°, the two detector banks
employed in this study. Peak position and breadth are obtained
from a multi-parameter non-linear regression analysis which fits
the entire peak profile using all available data points. Further
details may be found in refs. 5 and 6.

In this study determinations of the positions and breadths of
several peaks from each phase were made for all mechanical
treatments as well as for the undeformed (as-sintered) material.
Measurements were made at room temperature and the samples were
fixed, i.e., did not rotate. Considerations of resolution and
data collection time led to the use of the 151.8°2θ data bank
for determination of carbide positions, carbide breadths and
binder breadths; binder peak positions were determined using the
90°2θ bank.

RESULTS AND DISCUSSION

Neutron Diffraction

The time-of-flight spectrum obtained for the as-sintered
material is shown in Fig. 1. The maximum in the background at
about 0.6 Å is due to the Maxwellian distribution of energy in
the incident neutron beam. A much narrower region is detailed in
Fig. 2. This is an overlay plot of the as-sintered and high
monotonic samples. It includes two binder and one carbide peak,
as indicated. It is clear that the binder peaks shift to lower
d-spacing after deformation, indicating a relaxation of the
tensile residual stress, while the carbide peak shifts a smaller
amount in the opposite direction. It is also evident that the
peaks from both phases are broadened- the binder a great deal and
the carbide only slightly. A matched set of fits for the binder
200 peak in the as-sintered and high monotonic states is shown in
Fig. 3 to illustrate the number of data points utilized and the
quality of fit.

Stress Relaxation

Values for relaxation in strain were obtained directly from
the peak position data and are presented in Table III for a
variety of binder and carbide peaks. Corresponding stress values
were obtained assuming isotropic, hydrostatic and elastic
behavior, i.e., using $\sigma_m = K\Delta$, where σ_m is hydrostatic or mean
stress and Δ is volume strain and are also shown in Table III.
Strain relaxation (ε_r) vs. plastic strain (ε_p) is plotted for the
binder 311 peak in Fig. 4. The pattern seen for this peak is
representative: The relaxation is greater for a given plastic
strain in cycled material. The data also show that the rate of

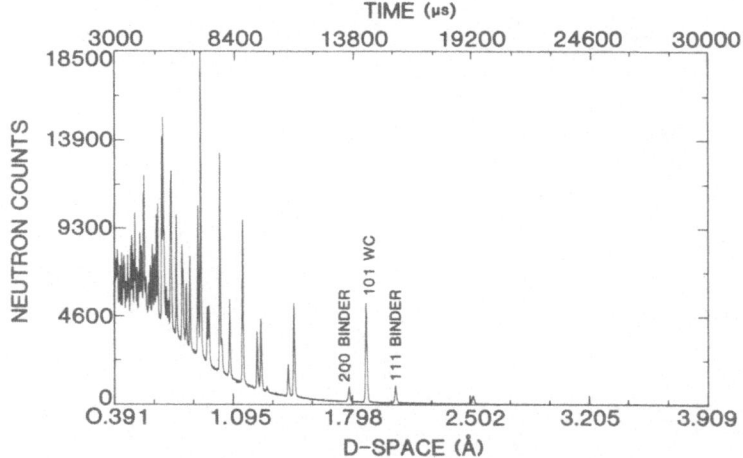

Fig. 1.   Experimental time-of-flight spectrum for as-sintered
(undeformed) sample.   Lower abscissa gives d-spacings of
peaks; upper abscissa gives time of travel for neutrons
over total flight path.

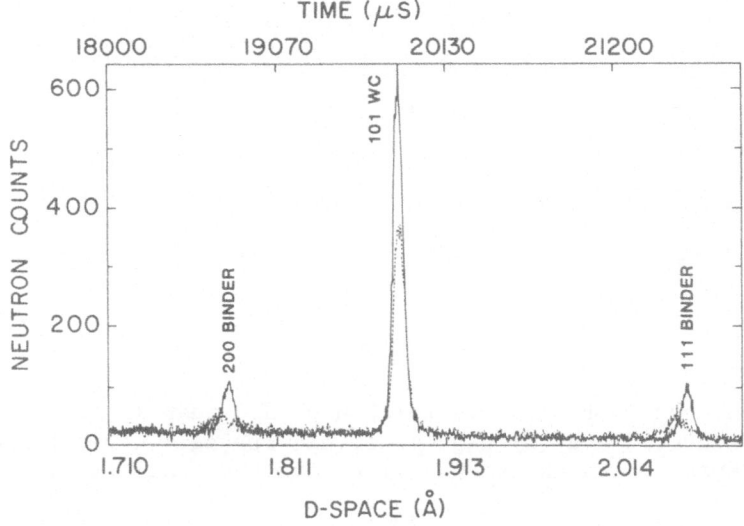

Fig. 2.   Overlay plot of narrow region of spectrum for as-sintered
(solid line) and high monotonic (dashed line) samples.

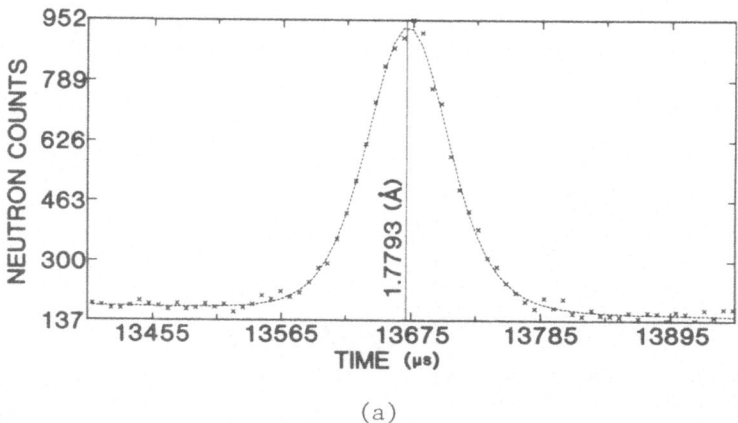

(a)

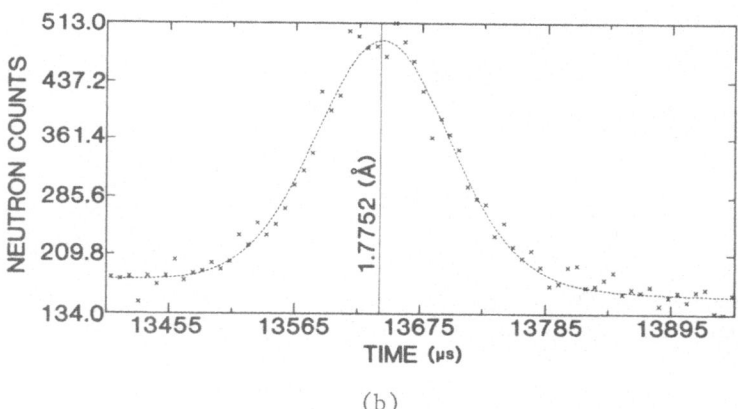

(b)

Fig. 3.   Matched sets of fits for binder 200 peak in (a) as-
          sintered and (b) high monotonic states.   Crosses are
          data points and dashed lines are the final fits.

relaxation is much greater in the early stages of plastic deformation, e.g., for monotonic loading, approximately 50% of the relaxation observed at 4.4% plastic strain has already occurred at 0.3% plastic strain. It is also noted that no real indication of anisotropy is shown in the binder data.

The carbide data show considerably reduced levels of stress with attendant increases in error; the low fatigue carbide peak shifts are too small to be reliably analyzed. There is also much greater peak to peak variation suggesting anisotropic effects. This is particularly true for the {002} and {110} planes. A possible explanation is that the coefficient of thermal expansion in [001] is greater than the average value for WC while the {110} are the slip planes.

Elasticity theory requires that at equilibrium the volume stresses must balance in the material, that is, for the present case: $v_c \sigma_c + v_B \sigma_B = 0$, where v_c and v_B are the carbide and binder volume fractions, respectively. Results are shown in Table IV using the values of stress from Table III with the exception of the fatigue data. The values for stress used were the simple averages for all peaks. The results show that, within error, the stresses are essentially balanced within the bulk volume of the phases, i.e., grain boundary effects appear to be negligible.

Peak Breadth Behavior

Many of the peaks exhibit the behavior plotted in Fig. 5 for the binder 311 peak breadths vs. plastic strain. Values for breadth ΔB assume the peak is ideally Gaussian and the FWHM values are approximately corrected for instrumental broadening via subtraction of the corresponding widths of annealed silicon powder peaks. The breadths for the monotonic treatments steadily increase while those for the cyclic treatments pass through a minimum, i.e., the peaks actually sharpen as a result of the low fatigue treatment then broaden again after the high fatigue

Table IV. Volume-Averaged Stress for Low Monotonic, High Fatigue and High Monotonic Cases (in MPa)

	Low Monotonic	High Fatigue	High Monotonic
Average Binder Stress	−258	−387	−501
Average Carbide Stress	+68	+116	+185
$v_B \sigma_r^B + v_c \sigma_r^C$	−20	−20	0

Table III. Values of Strain and Stress Relaxation for Various
Carbide and Binder Peaks after Mechanical Treatment

$$\epsilon_r \times 10^5$$

$$(\sigma_r, \text{ MPa})$$

hkl	Low Fatigue	Low Monotonic	High Fatigue	High Monotonic
		BINDER		
111	−85±8 (−157±15)	−141±8 (−261±15)	−205±8 (−379±15)	−295±10 (−546±18)
200	−94±10 (−175±18)	−138±10 (−255±18)	−205±8 (−401±17)	−236±14 (−437±26)
311	−88±8 (−163±15)	−141±8 (−261±15)	−217±7 (−401±13)	−279±9 (−516±17)
331	−88±16 (−163±30)	−139±15 (−257±28)	−198±15 (−366±28)	−273±18 (−505±33)
		CARBIDE		
101	3±5 (12±19)	13±4 (51±16)	19±4 (74±16)	33±5 (128±19)
110	5±7 (19±27)	21±7 (82±27)	36±7 (140±27)	56±7 (218±27)
002	10±13 (39±51)	28±12 (109±47)	46±12 (179±47)	80±13 (311±51)
201	5±6 (19±23)	17±5 (66±19)	30±5 (117±19)	40±6 (156±23)
112	−2±4 (−8±16)	10±4 (39±16)	17±4 (66±16)	33±4 (128±16)
211	4±8 (16±31)	15±7 (58±27)	30±7 (117±27)	43±8 (167±31)

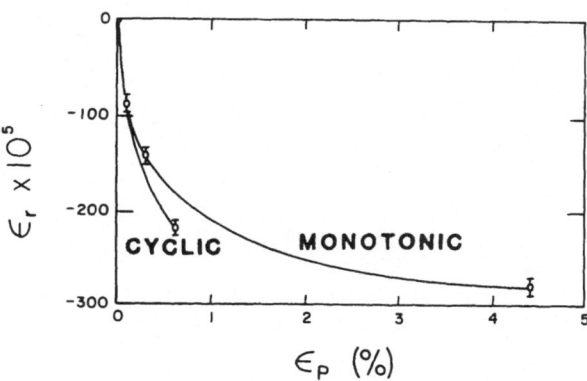

Fig. 4. Strain relaxation vs. plastic strain for binder 311 peak.

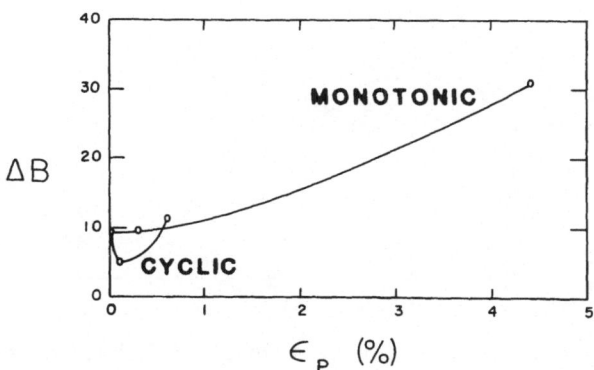

Fig. 5. Peak breadth (FWHM) vs. plastic strain for the binder peak. $\Delta B = (B^2_{binder} - B^2_{standard})^{\frac{1}{2}}$.

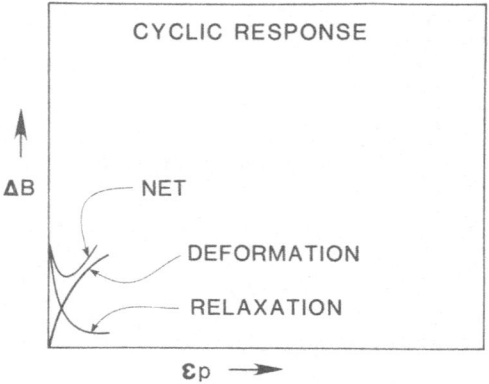

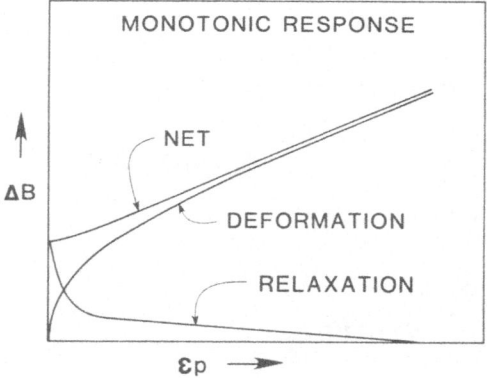

Fig. 6.   Schematic models for peak breadth behavior vs. plastic
strain showing contributions of stress relaxation and
plastic deformation for cyclic and monotonic loading.

treatment. Furthermore, peaks for the as-sintered material also
show broadening. A model offering a possible explanation for
this behavior is schematically illustrated in Fig. 6 and is
described as follows. Broadening is initially due to a residual
strain distribution caused by local variation in microstructure;
there is essentially no deformation in as-sintered material. As
plastic strain increases, residual stress, and thus the
distribution in stress, decreases and deformation becomes the
primary source of broadening. The minimum arises from the more
rapidly decreasing residual stress in the cyclic case.
Variability in the observed behavior, e.g., no minimum is observed
for the 200 peak, may be due to differences in the rate of
accumulation of plastic damage normal to particular planes.

ACKNOWLEDGEMENTS

    The support of Reed Rock Bit Co., which produced the samples
and performed the mechanical conditioning, is gratefully
acknowledged, especially the contributions of Dr. E. F. Drake.
The technical assistance of R. L. Hitterman of Argonne National
Laboratory was invaluable. Partial travel support was provided
by Argonne Universities Association and IPNS.

REFERENCES

    1.   H. E. Exner, Physical and Chemical Nature of Cemented
          Carbides, Intl. Met. Revs., 27(4):149 (1979).
    2.   W. D. Kingery, H. K. Bowen and D. R. Uhlmann,
          "Introduction to Ceramics, 2nd Ed.," John Wiley & Sons,
          New York (1976).
    3.   C. H. Vasel, A. D. Krawitz, E. F. Drake and E. A. Kenik,
          The Deformation Microstructure of WC-(Co,Ni) Cemented
          Carbide Composites, to be published.
    4.   G. E. Bacon, "Neutron Diffraction, 3rd ed.," Oxford
          University Press, London (1975).
    5.   G. H. Lander and B. S. Brown, "IPNS-1 Users' Handbook,"
          Argonne National Laboratory, Argonne, IL (1982).
    6.   S. R. MacEwen, J. Faber and A. P. L. Turner, The Use of
          Time-of-Flight Neutron Diffraction to Study Grain
          Interaction Stresses, Acta Metall., 31(5):657 (1983).
    7.   G. E. Dieter, "Mechanical Metallurgy, 2nd ed.,"
          McGraw-Hill Book Co., New York (1976).
    8.   M. K. Hibbs and R. Sinclair, Room Temperature Deforma-
          tion Mechanisms and the Defect Structure of Tungsten
          Carbide, Acta Metall., 29(9):1645 (1981).

STRESS DETERMINATION IN AN ADHESIVE BONDED

JOINT BY X-RAY DIFFRACTION*

Paul Predecki and Charles S. Barrett

University of Denver Research Institute

Denver, Colorado 80208

INTRODUCTION

At the present time there is considerable interest in the use of adhesive bonded joints for structural applications.[1] The design of such joints is based mostly on finite element calculations and failure tests rather than on strain or stress measurements. To our knowledge no measurements have been made of stresses at or near the adhesive/adherend interface (where many bond failures occur) because of difficulties with accessibility. The purpose of this work was therefore to try to make such measurements using X-ray diffraction.

The approach taken was to gain access to the bond by making one of the adherends in a single lap joint to be relatively transparent to X-rays and the other relatively opaque. Incident X-rays then penetrated the first adherend and the adhesive and were diffracted from gains in the second adherend adjacent to the adhesive/adherend interface. Stresses resulting from curing of the adhesive were determined first, after which a load was applied and the stresses redetermined.

MATERIALS AND EXPERIMENTAL METHODS

Since most polymeric adhesives do not possess high strengths, it was anticipated that the stresses to be measured would not be large and would require careful and accurate X-ray measurements. To facilitate this, Al was chosen as the opaque (diffracting) adherend

*Work supported by ARO on Grant No. DAAG29-81-0150.

and Be as the transparent adherend with $CuK\alpha_1$ the X-radiation to be used.   The adhesive chosen was FM-73M* (rubber modified epoxy with polyester matte), a common structural sheet adhesive.   Earlier work had shown that the 511 + 333 peak from Al had the highest sensitivity to stress (defined as the peak shift, $\Delta 2\theta_{\phi=\psi=o}$ per unit applied stress) of several cubic materials investigated using $CuK\alpha_1$ radiation.[2]

Some preliminary tests showed that the 6061 alloy in the T-4 condition (solution treated and quenched) was most suitable; the 511 + 333 peak with $CuK\alpha_1$ radiation occurred at a high angle ($\sim$ 162.4°2θ), was sharp (FWHM $\simeq$ .5°2θ) and sufficiently above background intensity ($I_p/I_b \simeq$ 1.75) when measured through a 0.79 mm (1/32") thick Be sheet and .091 mm (.0036") thick FM-73M to allow accurate peak position determinations to be made.   Solution treatment reduced the yield strength of the alloy to $\sim$ 193 MPa (28,000 psi) but substantially improved the peak sharpness and the α_1, α_2 separation as seen in Fig. 1.

Using these materials single lap joints were fabricated using a brass jig which maintained alignment and a uniform adherend separation during curing.   The dimensions of one such bonded specimen (ARO-23) used for this study are shown in Fig. 2.   The Be was made 1/4 the thickness of the Al since the Young's modulus of Be is 4 times that of Al.   The joint was therefore approximately of balanced stiffness in tension.   Procedures for adherend preparation and

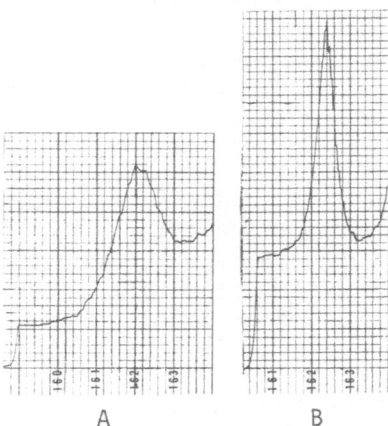

A   B

Fig. 1.   The $CuK\alpha_1$ diffraction peak from 511 + 333 planes of 6061 Al; (A) in the T-6 condition and (B) in the T-4 condition. Data were taken at ψ=45 using a single lap joint with 0.79 mm thick Be and 0.091 mm thick FM-73M adhesive.   Part of the $K\alpha_2$ peak is visible at the right of each trace.

*American Cyanamid, Havre de Grace, MD

bonding were developed which gave sound pore-free bonds as verified by dissolving away the adherends and examining the adhesive layer in an optical microscope. These are described elsewhere.[3] After cool-down from curing 1 hr at 125°C, end tabs of 6061-T6 Al were attached with Hysol EA9309 epoxy. The spew at the lap joint was machined off and the specimen stored in a desiccator.

To make X-ray measurements, the specimen was placed in the Clevis type grips of a small, manually-operated tensile frame with load cell. The frame was mounted on the goniometer table of a Siemens Kristalloflex 4 diffractometer. The frame held the specimen so that the incident and diffracted beams were in the 1,3 plane and the load was applied in the 1 direction (Fig. 2). Alignment adjustments on the frame allowed the specimen to be positioned so that the Al/adhesive interface contained the rotation axis of the diffractometer. Positioning was aided by viewing at low magnification the approach of the specimen to a centrally located removable reference point. The incident beam was unfiltered Cu radiation; diffracted beams were passed through a curved graphite monochromator inserted in front of a scintillation counter. The $K\alpha_1$ peak positions were determined by step scanning and least squares fitting of a parabola to 5 steps, each of 0.1°2θ, over the top ∿ 50% of the

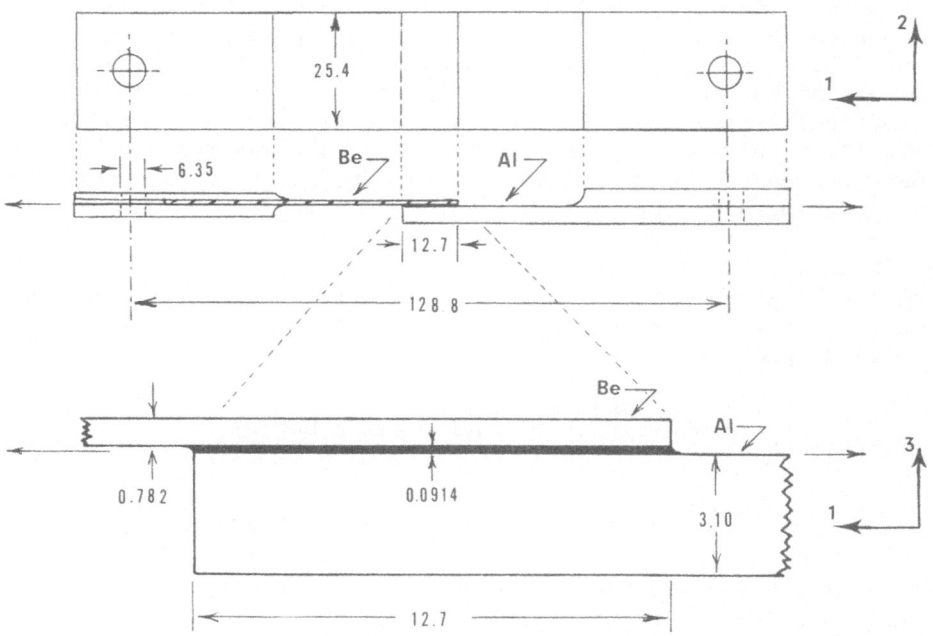

Fig. 2. Dimensions in mm of the single lap joint specimen (ARO-23) used for stress determinations. Tab thicknesses were such that the center-plane of the section at each clevis pin was coplanar with the FM-73M bond.

peak.[4] The scanning range was chosen where there would be approximately equal intensities at the first and fifth points.

Measurements were made in the 1',2',3' coordinate system which
was related to the 1,2,3 specimen axes by the angles ϕ and ψ in the
usual manner;[5] ψ tilts about the 2θ axis were employed. Measurements
were made at various places on blank adherend samples solution
treated identically to the bonded adherend to obtain "stress free"
values of the peak positions at all the ϕ and ψ settings used. The
standard deviation, s, of these peak positions was $\sim \pm .035°2\theta$ and
was associated mainly with point-to-point variations in the Al, since
standard deviations for repeated determinations at the same point
were much smaller; $\sim \pm .001°$.

To reduce s; specimens were rocked $\pm 1°$ about the 2θ axis during
exposure (this reduced s to $\sim .010°2\theta$), the air temperature in the
vicinity of the specimen was controlled at $27.5 \pm .1°C$ (a ΔT of $1°C$
produces a $\Delta 2\theta$ of $-.018°$ for Al) and the slit system was optimized
empirically.

Measurements were made on the bonded specimen along the center-
line in the 1 direction at the following angles: $\phi = 0$, $\psi = 0$, ± 45
and $\phi = 90$, $\psi = \pm 45$. Near the ends of the bond the ψ angles at
$\phi = 0$ were changed to $\psi = 0,-30,-45$ (near the end of the Al) and
$\psi = 0,30,45$ (near the end of the Be) to avoid passing the beam
through the ends of the bond. Irradiated areas defined by a move-
able mask at the specimen were as follows: 1.4 mm wide (in the
1 direction) x 7 mm high (in the 2 direction) for $\phi = 0°$ and 2.8 mm
wide x 1.6 mm high for $\phi = 90°$. The beam width was reduced to .56
mm near the ends of the bond for $\phi = 0$. The Al grain size determined metallographically was ASTM #6.

Small corrections were made to ψ for rocking and for rotation
of the joint when load was applied. LPA corrections were made by
dividing measured intensities by the following factor derived for
the bonded specimen:

$$\frac{\text{LPA}}{\text{factor}} = \left(\frac{1 + \text{Cos}^2 \, 2\theta \, \text{Cos}^2 \, 2\alpha}{\text{Sin}^2\theta}\right)[1 - \text{Tan} \, \psi \, \text{Cot} \, \theta]$$

$$\cdot \exp\left[- \frac{2}{1 - \text{Tan}^2 \, \psi \, \text{Cot}^2 \, \theta} \, (\mu_{Be} t_{Be} + \mu_g t_g)\right]$$

where α is the monochromator 2θ angle (beam bent towards X-ray tube
target), μ_{Be} and μ_g are the linear absorption coefficients of Be
and adhesive measured as 2.57 and 8.34 cm^{-1} respectively and t_{Be} and
t_g are the Be and adhesive thicknesses. It turned out that because
of the small step interval and small t_g the LPA correction was only
significant for $\psi = \pm 45$, shifting the peak a constant amount of
$-.0014°2\theta$ for $\psi = 45$ and $.0014°$ for $\psi = -45$.

RESULTS AND DISCUSSION

The strains $\varepsilon_{\phi\psi}$ whether residual (due to curing) or residual + applied were determined from

$$\varepsilon_{\phi\psi} = \frac{\text{Sin } \theta^\circ_{\phi\psi}}{\text{Sin } \theta_{\phi\psi}} - 1 \tag{1}$$

where $\theta_{\phi\psi}$ is the diffraction angle in the as-cured or the as-cured + load applied condition and $\theta^\circ_{\phi\psi}$ is the corresponding angle in the "stress free" condition. Using the notation of Cohen,[5] these are the strains $\varepsilon_{i'j'}$ in the 1',2',3' system (the measurement or laboratory system). To convert these to strains, $\varepsilon_{k\ell}$ in the 1,2,3 (specimen) system, the equation for the transformation of strain components is used:[6]

$$\varepsilon_{i'j'} = \sum_{k=1}^{3} \sum_{\ell=1}^{3} \varepsilon_{k\ell} \, \ell_{i'k} \, \ell_{j'\ell} \tag{2}$$

where $\ell_{i'k}$, $\ell_{j'\ell}$ are direction cosines. For the cases where $\phi = 0$ and $\phi = 90$, eq. (2) yields:

$$\varepsilon_{0,\psi} = \varepsilon_{11} \, \text{Sin}^2\psi + \varepsilon_{13} \, \text{Sin } 2\psi + \varepsilon_{33} \, \text{Cos}^2\psi$$

$$\tag{3}$$

$$\varepsilon_{90,\psi} = \varepsilon_{22} \, \text{Sin}^2\psi + \varepsilon_{23} \, \text{Sin } 2\psi + \varepsilon_{33} \, \text{Cos}^2\psi$$

Eqns. (3) were solved for ε_{11}, ε_{22}, ε_{33}, ε_{13} and ε_{23} using the ψ values employed in the measurements. It was assumed that ε_{12} (as cured) was zero and that ε_{22} (cured + load applied) = ε_{22} (as cured), i.e. that ε_{22} remained constant with applied load, an assumption made by most analytical and finite element treatments of single lap joints[7-9] (the adherends behave like plates in bending). The latter assumption made it unnecessary to make measurements under applied load in the $\phi = 90$ orientation.

From these strains, the normal stresses σ_{11}, σ_{22}, σ_{33} and the shear stresses σ_{13} and σ_{23} were determined from isotropic elasticity using the equations:[5]

$$\sigma_{ij} = \frac{\varepsilon_{ij}}{\frac{1}{2} S_2} \qquad \text{for shear stresses}$$

$$\tag{4}$$

$$\sigma_{ii} = \frac{\varepsilon_{ii}}{\frac{1}{2} S_2} - KE \qquad \text{for normal stresses}$$

where

$$K = S_1 [\tfrac{1}{2} S_2 (\tfrac{1}{2} S_2 + 3S_1)]^{-1}$$

and

$$E = \varepsilon_{11} + \varepsilon_{22} + \varepsilon_{33}$$

The elastic constants $\frac{1}{2}$ S_2 and S_1 were determined in a separate
experiment from the slope and intercept of a $Sin^2\psi$ vs ε_{33} plot in
the usual manner. The values obtained from a tensile specimen of
the Al adherend material were: $\frac{1}{2}$ S_2 = 1.911 x 10^{-5} MPa^{-1} (131.81 x
10^{-9} psi^{-1}) and S_1 = -5.148 x 10^{-6} MPa^{-1} (-35.5 x 10^{-9} psi^{-1}).

The normal and shear stresses calculated from eqns. (4) are
shown in Figs. 3 and 4 for the as-cured condition and in Figs. 5 and
6 for the cured plus 2678N (600 lbs) applied load case. The normal
curing stresses, $\sigma_{11}{}^c$ and $\sigma_{22}{}^c$ are quite substantial, as would be
expected from the dissimilar expansion coefficients of Al and Be
(α_{Al} = 23.6 x 10^{-6}/°C, α_{Be} = 11.6 x 10^{-6}/°C). Both show decreases
towards the ends A and B of the bonds due to edge effects ($\sigma_{11}{}^c$ and
$\sigma_{22}{}^c$ must go to zero at the edges of the bond area). $\sigma_{11}{}^c$ is smaller
than $\sigma_{22}{}^c$ because the bond area dimension is smaller in the 1-direc-
tion than in the 2-direction (12.7 mm vs 25.4 mm) so the edge effects
are more pronounced and probably extend across the bond in the 1-
direction. Bending about the 2 axis as in a bimetallic strip would
not explain why $\sigma_{22}{}^c$ is larger than $\sigma_{11}{}^c$. Bending would increase
$\sigma_{11}{}^c$ more than $\sigma_{22}{}^c$ since the neutral surface for such bending is on
the Al side of the bond. The curing shear stress $\sigma_{23}{}^c$ is about zero

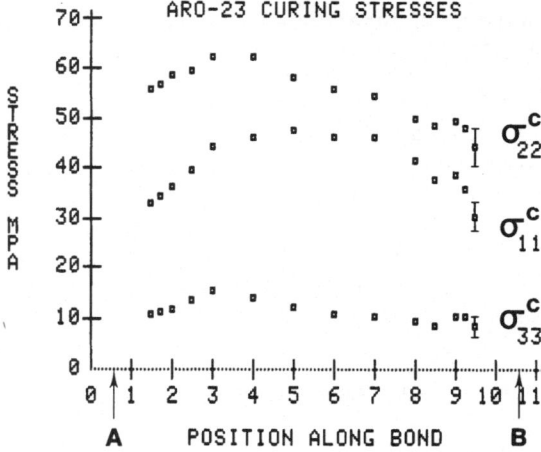

Fig. 3.   Curing normal stresses in the Al adherend at the adherend
          adhesive interface of the joint shown in Fig. 2. In this
          and subsequent figures, points A and B mark the ends of the
          Al and Be adherends respectively, stress measurements were
          made at 27.5°C (curing temperature was 125°C).

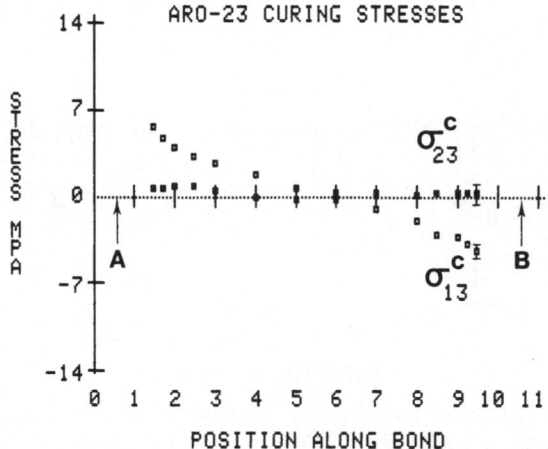

Fig. 4. Curing shear stresses in the Al adherend at the adherend/
adhesive interface of the joint shown in Fig. 2.

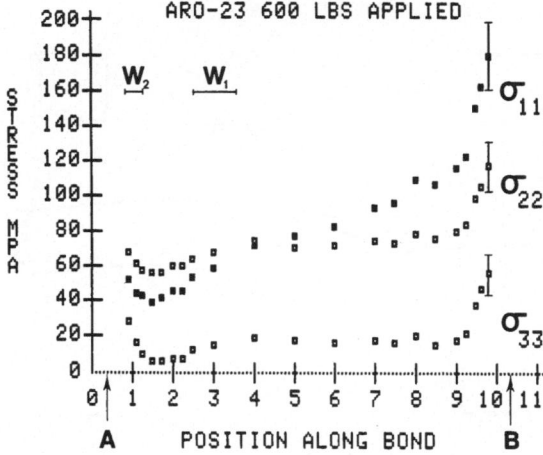

Fig. 5. Normal stresses in the Al adherend at the adherend/adhesive
interface of ARO-23 with a tensile load of 2678N (600 lbs)
applied in the 1 direction. Beam widths W_1 and W_2 used in
the middle and at the edges of the bond are also shown.

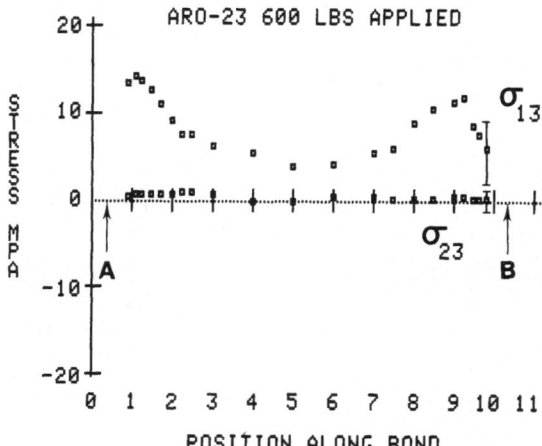

Fig. 6.   Shear stresses in the Al adherend at the adherend/adhesive
interface of ARO-23 with a tensile load of 2678N (600 lbs)
applied in the 1 direction.

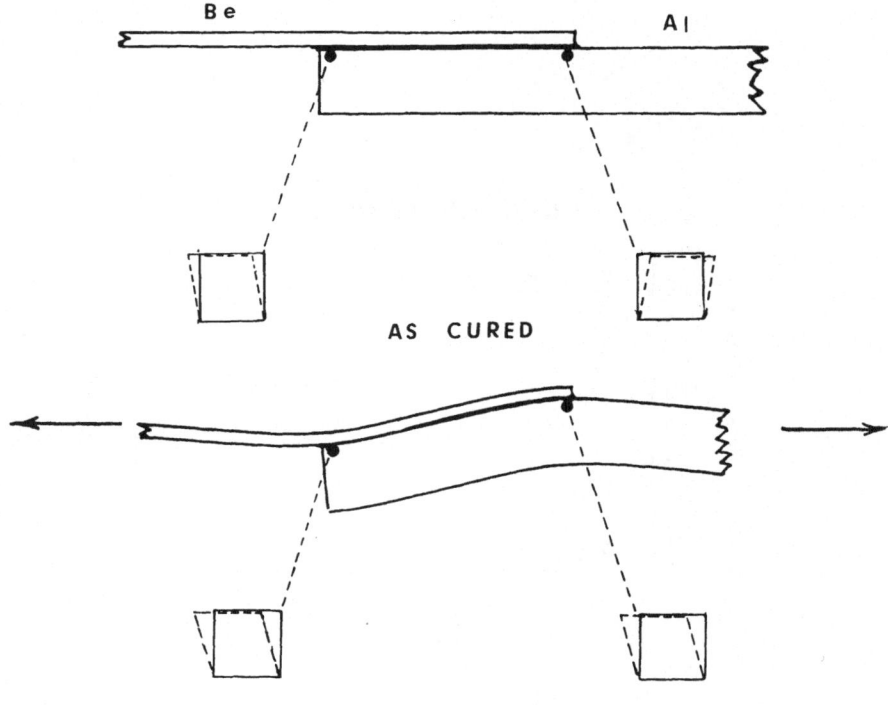

Fig. 7.   Schematic representation of the signs of shear strains in
the Al adherend near the edges of the bond. Effects of
Al thermal contraction are shown above and thermal contraction plus applied load below.

as expected, but $\sigma_{13}{}^c$ is negative at one end of the bond and positive at the other (Fig. 4). This is again due to edge effects in the thermal contraction of the Al as illustrated in Fig. 7.

When a load of 2678N (600 lbs) is applied, the normal stresses σ_{11} and σ_{22} increase towards the B end of the bond (Fig. 5) where the bending in the Al is greatest (Fig. 7). The peel stress σ_{33} increases at each end of the bond as expected from theory.[7] The trends in σ_{11} and σ_{33} shown in Fig. 5 agree qualitatively with finite element calculations of Francis et al.[10] for an Al/FM-73M/Al single lap joint of balanced stiffness and slightly different geometry, but the magnitudes of the finite element stresses are somewhat smaller than the measured ones. The shear stress σ_{23} remains at zero as expected and σ_{13} exhibits two peaks, one near each end of the bond, again in qualitative agreement with the results of Francis et al.[10] Subtracting $\sigma_{13}{}^c$ from σ_{13}, and averaging the net σ_{13} across the whole bond gives a value close to 8.3 MPa (1200 psi), which was the nominal shear stress applied.

With regard to errors, estimates of these were obtained by randomly introducing the standard deviation of the peak position ($\Delta 2\theta = \pm .01°$) into the 2θ values used for the stress calculation at one bond position. The resulting stresses were obtained and averaged for 7 such random introductions. The standard deviations of these averages are shown as error bars in Figs. 3-6. They varied from 0.7 to 19 MPa (0.1 to 2.7 ksi) depending on the stress. Since all the stresses except σ_{22} necessarily go to zero either at the back surface of the Al adherend or at the neutral axis, gradients in these stresses probably exist in the vicinity of the Al/adhesive interface. An error arises because the penetration depth (95%) for $CuK\alpha_1$ in Al is 0.11 mm for $\psi = 0$ and 0.075 mm for $\psi = 45$; however, this discrepancy is small compared with the adherend thickness (3.1 mm) so the error is probably small.

CONCLUSIONS

The X-ray method presented here permitted measurement, we believe for the first time, of the triaxial stresses at or near the adhesive/adheren interface of a single lap joint, and confirmed the theoretically predicted peaking of the normal stresses σ_{11}, σ_{22} and σ_{33} and shear stress σ_{13} at or near the ends of the joint when load is applied. Stresses after curing were smaller, largest around the middle of the joint and substantially influenced by edge effects. The required precision was reached by optimizing slits, temperature control, specimen heat treatment, and specimen rocking, by monochromatizing the CuKα radiation, by using a high 2θ reflection, and by making one adherend of Be.

ACKNOWLEDGEMENT

We are grateful to W. B. Jones of Texas Tech University and
L. J. Hart-Smith of Douglas Aircraft Co. for several valuable discussions and to Richard Miller for help with specimen preparation.

REFERENCES

1. See for example G. Graff, The Glued-Together Airplane,
    High Technology 3:67 (1983), September issue.
2. Charles S. Barrett and Paul Predecki, Stress Measurement
    in Graphite/Epoxy Uniaxial Composites by X-Rays,
    Polymer Composites 1:2 (1980).
3. Paul Predecki and Charles S. Barrett, Progress Reports
    #2 and 3 on ARO Grant No. DAAG29-81-0150. Jan. 1982-
    Jan. 1983. Manuscript in preparation.
4. Residual Stress Measurement by X-Ray Diffraction, SAE
    Information Report #SAE J784a. Aug. 1971.
5. See for example, J. B. Cohen, H. Dolle and M. R. James,
    Stress Analysis from Powder Diffraction Patterns, in
    "NBS Special Pub. 567: Accuracy in Powder Diffraction,"
    S. Block and C. R. Hubbard, eds. pp. 453-477 (Feb.
    1980).
6. F. A. McClintock and A. S. Aragon, "Mechanical Behavior
    of Materials," Addison-Wesley, p. 61 (1966).
7. R. D. Adams and N. A. Peppiatt, Stress Analysis of Adhes-
    ive-Bonded Lap Joints, Jnl. of Strain Analysis 9:185
    (1974).
8. Viscoelastic Stress Analysis Including Moisture Diffusion
    for Adhesively Bonded Joints, J. Romanko, P.I.,
    MRL Report FZM-6838, General Dynamics, Fort Worth,
    (1980).
9. W. J. Stronge and O. E. R. Heindahl, Stress in Single Lap
    Joints of Dissimilar Materials, Naval Weapons Center,
    China Lake, Report NWC TP 6242 (1981).
10. E. C. Francis, W. L. Huffered, D. G. Lewin, R. E. Thompson,
    W. E. Briggs, Time Dependent Fracture in Adhesive
    Bonded Joints, Interim Report #AFWAL-TR, CSD 2769-IR-01,
    (May 1982).

EVALUATION OF STRAIGHT AND CURVED BRAUN* POSITION-SENSITIVE PROPORTIONAL COUNTERS ON A HUBER-GUINIER X-RAY DIFFRACTION SYSTEM

R.A. Newman, P. Moore Kirchhoff, and T.G. Fawcett

The Dow Chemical Company

Midland, Michigan

ABSTRACT

The interfacing of both straight and curved Braun Position-Sensitive Proportional Counters (PSPC's) to a high resolution Huber-Guinier camera system has been accomplished, resulting in a 10 to 100-fold decrease in data collection times when compared to conventional Guinier (film or scintillation counter) detector techniques.

Various factors causing line broadening were evaluated for both PSPC Guinier systems. The depth of the PSPC gas chamber was found to have the greatest influence on line profiles. An 80% increase in peak half-widths was observed for PSPC-Guinier data compared to our highest resolution Guinier film data, but still yielded significantly better resolution than conventional powder diffractometer data obtained in our laboratory.

INTRODUCTION

The utility of high resolution Huber-Guinier cameras has been well documented (1-4). While the use of film as a detector provided a simple means of data retention, it was unfavorable in terms of data collection time (1-8 hours) and subsequent handling (developing and optical densitometering) to produce a computer data file for analysis. In order to eliminate these problems,

---

*Distributed in the U.S. by Innovative Technology, Inc.

the decision was made to evaluate a PSPC detection system. Since
the Braun PSPC's were readily available, and had already been utilized on a modified Huber-Guinier system by Goebel (5), as well as on
a diffractometer (6,7), they seemed the logical choice for our purposes. Two Braun PSPC's were recently installed and evaluated on
our Huber-Guinier camera equipped with an Fe-LFF X-ray source. One
of the detectors was straight, while the second was curved to a 100
mm radius. Also, the detector chamber depths were different: 3 mm
for the straight detector; 10 mm for the curved one. Due to the
diffraction geometry and focusing radius (57.3 mm) of the Guinier
system, two effects were observed when the PSPC's were installed:
(A) defocusing, and (B) parallax broadening. An alignment optimization was performed using 5 μ α-quartz to evaluate and minimize
these line broadening effects for both the straight and curved detectors. All data obtained during the optimization process were
transferred to the computer and analyzed by profile-fitting methods
(8) to obtain accurate peak FWHM values as well as peak symmetry
information.

EXPERIMENTAL

     All diffraction data were measured on a Huber-Guinier system.
The standard Huber powder camera was removed and replaced by the
Huber monocrystal chamber attachment (#631) which utilizes a goniometer head for sample mounting. The standard film cassette was removed and replaced by a Braun PSPC mounted on an adjustable brass
support designed and built at Dow. The system was aligned in the
transmission-subtraction geometry (Figure 1) to minimize asymmetric
line-broadening effects often observed with other systems (9).
Vertical divergence was reduced through the use of the focusing
Guinier geometry. Absorption and flat sample errors were reduced
by using small, thin samples (ca. 2 x 2 x 0.2 mm) in the transmission mode.

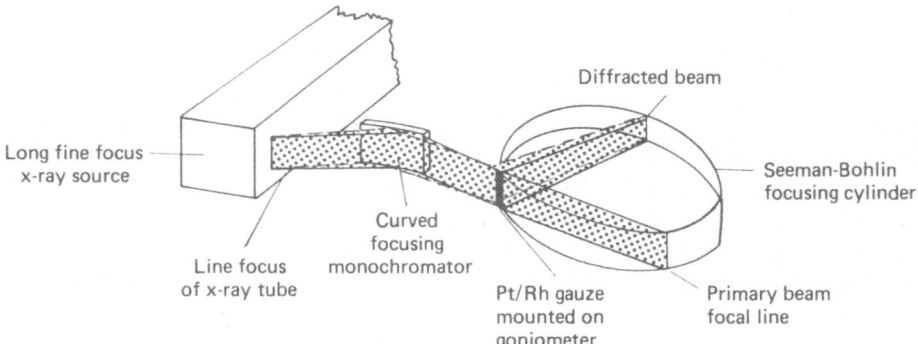

Figure 1.   Huber-Guinier High Resolution X-Ray Diffraction System,
            Transmission - Subtraction Geometry

A sample of 5 μ α-quartz was lightly pressed into a 2 x 2 mm
area of 0.2 mm thick 80 mesh Pt/10% Rh gauze mounted on a goniometer
head. Samples were centered using an alignment telescope, and a
1-3 mm diameter collimated X-ray beam was used for analysis.

The PSPC evaluations and alignment were performed in the range
from ∿15°-35° 2θ. In order to minimize line broadening effects, a
brass mounting bracket was designed which allowed both radial and
tangential adjustment of the PSPC's with respect to the center of
the focusing circle. For the tangential evaluations, data were
collected in 5° increments of tilt in the range from -15° to 15°,
where positive tilt direction is clockwise with respect to the
incoming X-ray beam. For the radial evaluation, measurements were
made in 3 mm increments from 47 mm to 65 mm radius. Data were
collected for 10 minutes/scan. All X-ray diffraction systems discussed here utilize a long-fine-focus X-ray source operating at
40 kV/20mA. For the Guinier systems, monochromatic $FeK\alpha_1$ (PSPC-
Guinier) and $CuK\alpha_1$ (film-Guinier) radiation were obtained from
curved focusing monochromators. The diffractometer is a typical
Philips θ-2θ system operating with $CuK\alpha_1\alpha_2$ radiation. Calibration
factors of 0.0165° 2θ/channel and 0.0190° 2θ/channel were obtained
for the straight and curved PSPC's, respectively. These values differ
from the .01° 2θ/channel quoted by Braun because we have reduced the
radius from 100 mm to 57.3 mm.

RESULTS AND DISCUSSION

The Braun PSPC's operate on the same principle as a conventional proportional counter, but the counter tube has been replaced
by a counting chamber containing a 5 cm long carbon-coated 25μ diameter quartz fiber with high resistance (8000 Ohm/mm) and a high degree
of positional resolution (50μ). Two line broadening effects not
observed with film detection were predicted for the PSPC system. The
first effect is defocusing, which is due to the fact that the detector "wire" is centered on the Guinier focusing circle at only one
point. With a detector curved to the Huber-Guinier radius, this defocusing would not be observed. Due to design and construction problems, however, a 100 mm radius detector was substituted for the 57.3
mm radius curved detector which was ordered. Optimum values from
the radial evaluation were 56 mm radius and 58 mm radius for the
straight and curved PSPC's, respectively. These values are, within
our positional error, equal to the 57.3 mm radius of the Guinier
system. The second effect, parallax broadening or "smearing", arises
because the entire depth of the PSPC detection chambers detect X-rays,
and diffracted X-rays enter the chamber at various angles. In the
Guinier transmission-subtraction geometry only one position on the
focusing circle (2θ = 45°) is perpendicular to the diffracted X-rays.
We optimized the parallax for the entire 20° range of the detector,
so that minimum broadening occurs at the midpoint of the detector
wire. However, it is possible to approach the resolution of film

data for a small ($\sim 5^{\circ}$) region of 2θ, by optimizing the detector only
for that region. Results of the tangential evaluation are shown in
Figure 2.

A summary of data for 5μ α-quartz is given in Table I. The σ_1
and σ_r values in the table refer to the fraction of peak FWHM which
falls to the left and right of a vertical line passing through the
peak maximum. R is an agreement factor between calculated and observed peak profiles. The R, σ_1 and σ_r values are obtained from the
internally developed profile-fitting routine. In comparing the
Guinier and diffractometer FWHM values, it is important to remember
that the diffractometer values are 2θ-dependent, because the α_1-α_2
separation is a function of the 2θ angle.

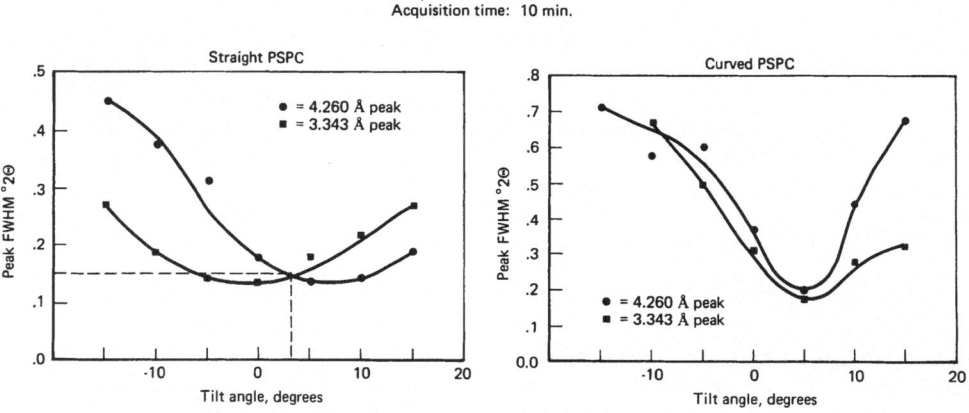

Figure 2.   Data from Angular Optimization of Straight
and Curved Braun PSPC's

A comparison of the film and PSPC data shows that the line broadening is minor; however the degree of asymmetry, particularly tailing,
is substantial. Diffraction from a powdered, randomly oriented sample
produces a series of diffraction cones with various radii at the focusing distance from the sample. The entire 10 mm height of the PSPC
detects diffracted X-rays from the segment of a diffraction cone which
impinges upon it. This results in a 2θ-dependent broadening and asymmetry due to the changing degree of curvature of the detected cone
segments. There could also be other instrumental errors which contribute to peak asymmetry and tailing, but they should not be 2θ-de-
pendent. Experiments utilizing a 2 mm slit between the sample and the
PSPC indicate that this tailing can be eliminated.

TABLE I – α–Quartz Data Comparison

Detector	Acquisition Time	Peak	$\sigma_1{}^a$	$\sigma_r{}^a$	Full 20° Optimization (PSPC) FWHM^a	R	Point Optimization (PSPC) FWHM^a	R
CEA film	6 hr	3.343 Å	0.037	0.038	0.075	.013	–	–
	6 hr	4.260 Å	0.028	0.047	0.075	.017	–	–
Straight PSPC	10 min	3.343 Å	0.086	0.050	0.136	.035	0.121^c	.049
	10 min	4.260 Å	0.100	0.043	0.143	.052	0.108^c	.089
Curved PSPC	10 min	3.343 Å	0.136	0.060	0.196	.040	0.174^c	.039
	10 min	4.260 Å	0.135	0.059	0.204	.073	0.185^c	.062
Diffractometer	45 min	3.343 Å	N.A.	N.A.	0.171^b	N.A.	N.A.	N.A.
(¼° rec. slit)	45 min	4.260 Å	N.A.	N.A.	0.162^b	N.A.	N.A.	N.A.

^{a} Units of °2θ

^{b} Measured manually on strip chart

^{c} With 2mm receiving slit

  N.A. = Not Available

TABLE II – System Comparisons

	Film-Guinier	PSPC-Guinier	Philips Diffractometer
Acquisition (range of 60° 2θ)	4 – 8 hr	3 sec – 15 min	30 min*
Data Handling	Developing, drying, etc.	Transferred to computer file	Strip chart or computer file**
Total Analysis Time	8 hr	30 min	1 – 4 hr

*Using 1° receiving slit.
**Usually requires longer data acquisition time (2-8 hours).

The data in Table I show that the degree of line broadening observed on changing from film to PSPC detection is significant. However, when comparing the two detection systems, the primary application
of the instrument should also be considered. A comparison of a typical
sample analysis using the two detection systems as well as a Philips
diffractometer, is shown in Table II. While the X-ray film clearly
has better resolution, it is much slower in obtaining good counting
statistics in data collection when compared to the PSPC. Another
important advantage of the PSPC system is the ability to observe and
control data collection, enabling immediate correction of mechanical
and alignment errors. With film, all experiments are "blind" because
they must be completed before determining whether or not they have
failed. Obviously, the strength of the PSPC system is its speed –
enabling the analysis of materials undergoing solid state transformations on a time scale of minutes.

CONCLUSIONS

We have shown that a straightforward combination of two commercially available instruments resulted in a new system with excellent
capabilities in time-dependent diffraction studies. The new system
is simple to operate and eliminates all of the manual handling required with film detection. Also, the PSPC system is:

1. interactive
2. fast in data collection
3. good in resolution
4. economical to operate
5. durable

The results of this work show that state-of-the-art capabilities can be attained, in certain cases, with a minimum amount of
development time.

REFERENCES

1. J. W. Edmonds and W. W. Henslee, Adv. in X-ray Analysis,
    22:143 (1979).
2. R. L. Snyder and J. W. Edmonds, Symposium on Powder Data Col-
    lection and Analysis, XII IUCr Congress, Ottawa, Canada (1981).
3. A. Brown, J. W. Edmonds and C. M. Foris, Adv. in X-ray Analysis,
    24:111 (1981).
4. A. Brown, Symposium on Powder Data Collection and Analysis XII
    IUCr Congress, Ottawa, Canada (1981), also see articles by A.
    Brown in Adv. in X-ray Analysis, 26:11 and 26:53 (1983).
5. H. E. Goebel, Adv. in X-ray Analysis, 25:315 (1982).
6. H. E. Goebel, Adv. in X-ray Analysis, 22:255 (1979).
7. H. E. Goebel, Adv. in X-ray Analysis, 24:123 (1981).
8. A. Brown and J. W. Edmonds, Adv. in X-ray Analysis, 23:361 (1980).
9. H. P. Klug and L. E. Alexander, "X-ray Diffraction Procedures",
    John Wiley and Sons, N.Y., N.Y. (1974).

A NEW IN SITU, AUTOMATIC, STRAIN-MEASURING X-RAY

DIFFRACTION APPARATUS WITH PSD

Louis Castex, Jean-Michel Sprauel, and Marc Barral

Ecole Nationale Supérieure d'Arts et Métiers
151, Boulevard de L'Hôpital
75640 Paris, France

INTRODUCTION

In order to increase the versatility of in-situ X-ray equipment for stress measurements, we decided to design and construct our own easily transportable apparatus. Similar equipments are already commercialized: (i) in Japan (RIGAKU STRAINFLEX, SHIMADZU); (ii) in West Germany (SIEMENS); (iii) in the USA (TEC). After the commercialization of PSD or PSSC, several prototypes were developed among which, the apparatus of Ruud[1], James[2], CETIM[3]... Our own apparatus has been developed for two main reasons: (i) For 1971, we have been working to ameliorate our own equipments effectively and frequently used for in situ measurements.
(ii) We need a very precise and rapid apparatus amainable for the greatest number possible of applications.

The best features of this instruments can also be found in this one, namely:
(i) An automated Ψ-goniometer, which allows the choice of the values of the incidence angle Ψ to be adapted to the material studied. This is particularly useful in the case of textured or triaxially stressed specimens. This automated goniometer may also be made to oscillate around each Ψ position, again a particularly useful feature for dealing with a sample with coarse grains.
(ii) A position sensitive detector (PSD) with a metallic anode wire which records the complete diffraction peak very rapidly (in some 5 sec.) and facilitates energy discrimination. This is very helpful in the case of poor diffraction conditions.
(iii) The use of small or ordinary X-ray tubes and an easily transportable X-ray generator.

(iv) A microprocessor with a sophisticated software able to process even very poor diffraction peaks (having a peak background ratio of less than 0.5)

The prototype we have developed meets all these demands very admirably and has been in trouble-free operation in our laboratory since october 1982.*

GENERAL DESCRIPTION OF THE APPARATUS

The Figure 1 gives a general view of the prototype. The main components of the system will now be briefly described:
Goniometer: to avoid the twin problems of X-ray beam defocusing and variation of X-ray absorption with the incidence angle Ψ, we chose a Ψ-geometry. The angle Ψ may vary between 0° and 60° and may be made to oscillate by anything up to ±5°. The system can handle up to 21 values of Ψ for each value of Φ. The goniometer is currently undergoing modification to enable the angle Ψ to be varied between -45° and +45°.
X-Ray Generation: we employ a compact X-ray generator (50 kV, 10 mA) which requires no heat exchanger, and standard tubes. This will shortly be replaced by a yet more compact generator using high frequency converters, and miniature X-ray tubes.
Detection: we use a position sensitive detector with a metallic anode wire (W-PSD)[5]. The PSD allows us to work with very small X-ray spots (down to 0.2 mm²). The PSD is connected to a multichannel analyser which controls the acquisition of the diffraction peak.
Microprocessor: we began by using an MC 6800 (8 bits) microprocessor with EPROM. Currently, we are in the process of substituting the MC 6800 by an MC 68000 (32 bits) and a more sophisticated software. The microprocessor controls the goniometer drive, the data acquisition and treatment, and the printout of the results on a miniprinter.

SOFTWARE

To control the different functions of the apparatus, the mi-

---

*The apparatus has been undergoing development in our laboratory – since 1971. The first prototype[5] utilised the Ω-geometry and a classical counter with step-scanning. Since 1974, a position sensitive detector has been used. Then, a non-automated apparatus with an Ω-geometry was designed and constructed between 1974 and 1977[6]. In 1977, the Ψ-geometry was chosen and the goniometer fully automated by liaison with a mini-computer which was later replaced by a MC 6800 microprocessor. It is this last prototype which is presented here.

croprocessor uses the algorithm shown in figure 2. The main parts of this algorithm are classical dealing with stress measurements. Nevertheless, some routines are original and deserve a more detailed description.

Calibration

Prior calibration is necessary as no spatial reference with adequate precision is available. The function of the calibration is to link the channel number of the multichannel analyser to the 2θ angle. This relation is non-linear because the PSD is not curved; we, thus, have to use a second degree polynomial expression whose coefficients define the origin of the cell, the mean step per channel and the curvature effect. For the determination of the three calibration parameters, we use three perfectly known standard powders whose diffraction angles are located, respectively, at the beginning, at the middle and at the end of the PSD cell.

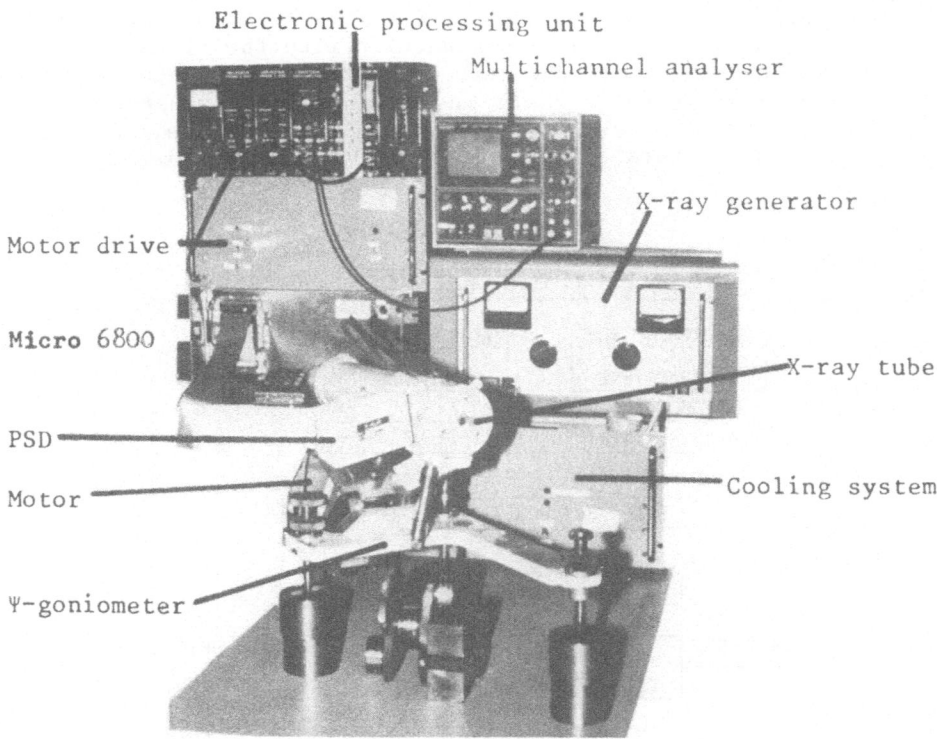

Figure 1   General view of the prototype

Peak treatment

    After a linear subtraction of the background we perform the
Lorentz-polarisation (LP) correction by an approximated formula.
For that purpose, we consider the LP function as a third degree
polynomial which increases considerably the efficiency of the calculations while introducing an error of less than 0.1 %.

    The analysis of the peak then starts by approximating succes-
sive portions of the profile by third degree polynomial whose coefficients are determined by least squares.   For this purpose, a
portion of the profile is defined by a zone around a given channel
number whose size depends on the width of the diffraction peak.
This process defines precisely each point of the profile as well as
its first and second order derivatives, thus allowing us to determine all the parameters of the diffraction peak: the position of
the maximum of the peak, the peak height, the full width at $^2/_5$
height and its middle point, the integrated intensity and the integral breadth.   The middle point of the full width at $^2/_5$ height is
used to define the position 2θ of the diffraction peak.   Treatment
of one peak needs about 40 seconds with the MC 6800 microprocessor
and should be reduced to about 1 second with the MC 68000.

Stress Calculation

    The stresses are calculated by a linear least squares fitting
to the value of 2θ vs $\sin^2\psi$ in conjunction with the classical $\sin^2\psi$
law.   It is envisaged that the next model of the apparatus will use
other types of treatment such as an elliptic least squares for the
determination of shear stresses and a more sophisticated method for
the analysis of triaxial stress states and stress gradients.

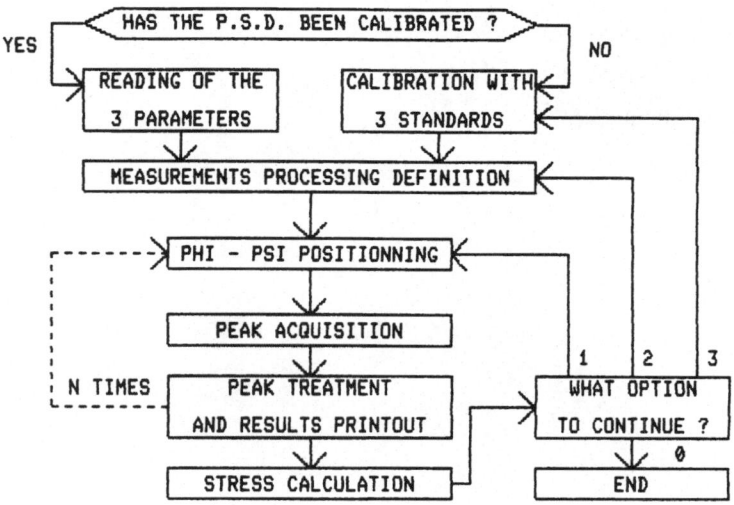

Figure 2   The general algorithm.

The error bars on the stress value are calculated from the deviations between the experimental 2θ values and its fitted points with a confidence level of 80 %. These errors are cumulative and are attributed to three different factors: the statistical error on the determination of 2θ, systematic instrumental error and deviation due to the heterogeneity of the material. Different experimental studies of this aspect of the measurements have demonstrated to us that if the experimental conditions are well chosen, the effect of heterogeneity predominates over the other two factors (Figure 3).

		Annealed SAE 1010	Shot peened 10N SAE 1010		Shot peened 10A SAE 4135	
		(2)	(1)	(2)	(1)	(2)
Stress in MPa		14	−547	−550	−663	−660
Standard deviation in MPa	(3)	4	21	23	39	31
	(4)	12	53	46	70	56
	(5)	7	8	14	32	12

Figure 3 Mean value of 10 consecutive measurements carried out on 3 different steels: (1) Ψ fixed, (2) Ψ oscillations ±3°
(3) standard deviation for 10 identical measurements,
(4) maximum of the error bars for 1 measurement,
(5) minimum of the error bars for 1 measurement.

Given good diffraction conditions, the total time of measurements for one value of Φ and six of Ψ is about 8 min; a figure which should be reduced to about 3 min with the MC 68000 microprocessor.

CONCLUSION

The current version of our X-ray apparatus for in-situ strain measurements has given complete satisfaction since October 1982. The precision of the results obtained with this apparatus is generally better than that from the other three Ω-geometry installations with PSD also used in our laboratory (Figure 4).

With the apparatus described in this paper, we have been able to conduct measurements on a diversity of difficult components such as a crankshaft, large components in titanium alloy, combustion

Type	detection	geometry	nber of angles Ψ	stress in MPa	error bars in MPa
RIGAKU STRAINFLEX	Scanning	Ω	5(> 0)	-390	54 [x]
RIGAKU/ENSAM	W-PSD	Ω	11(≥ 0)	-390	62
CGR/ENSAM	F-PSD	Ω	7(≥ 0)	-388	35
CETIM	W-PSD	Ω	7(> 0)	-389	7 [+]
This apparatus	W-PSD	Ψ	5(> 0)	-388	20

Figure 4 Comparison of the results obtained with 6 different apparatus for the measurement of a 4135 steel shot-peened 6N.
special formulas of RIGAKU (X) and of CETIM (+)

chambers of gaz turbines in Hastelloy X with a mechanically worked
surface and thread root on components in Inco 750 X.
    The prototype described in this paper should shortly be com-
mercialized.

REFERENCES

1- C. O. RUUD, P. S. DIMASCIO and D.M. MELCHER, Application of a
    PSD for non-destructive residual stress measurements inside
    stainless steel piping, Adv. in X-ray Anal., 26, 1983, pp.233
    -243
2- M. JAMES and J. B. COHEN, PARS: a portable X-ray analyser for
    residual stresses, J. T. E. V. A., 6, n°2, 1978, pp.91-97
3- P. BARBARIN, F. CONVERT and B. MIEGE, Mesure des contraintes
    residuelles par rayons X sur les structures métalliques (X-
    ray residual stress measurements on metallic structure),
    Rech. et Develop. Contr. Ind., may 1982, pp.71-78
4- L. CASTEX, J. L. LEBRUN and S. BRAS, A new position sensitive
    detector developed in France, Adv. in X-ray Anal., 24, 1981,
    pp.139-141
5- M. BARBIER, J. BENS, Etude et réalisation d'un appareillage de
    radiocristallographie destiné à la mesure non-destructive des
    contraintes superficielles 'in-situ' en métallurgie (Study
    and realization of an X-ray apparatus for 'in-situ' non-des-
    tructive measurements of superficial stresses in metallurgy),
    contract DGRST n° 71-7-2671, 1974
6- L. CASTEX, La détermination des contraintes par diffractométrie
    X (Stress determination by X-ray Diffractometry), Rev. Prat.
    de Contr. Ind., 83, 1977, pp. 25-28

A MINIATURE INSTRUMENT FOR RESIDUAL STRESS MEASUREMENT

C. O. Ruud, P. S. DiMascio, and D. J. Snoha

Materials Research Laboratory
The Pennsylvania State University
University Park, PA 16802

ABSTRACT

The Pennsylvania State University has developed and tested a miniature x-ray instrument for the measurement of residual stress (strains). The stress head including x-ray source and detection surface is approximately four and one-half inches long and can be inserted into an orifice less than two and one-half inches in diameter. This head is on the end of a several-foot long cable and is extremely manipulatable.

The instrumentation design is based upon a unique position sensitive scintillation detector which is capable of applying the single-exposure x-ray stress measurement technique. Total lapse time data collection periods for aluminum alloys are less than ten seconds with x-ray tube powers of 120 watts. Results from a four-point bend test on an aluminum alloy specimen showed excellent precision. The configuration of the instrument and the procedures for its application are described.

INTRODUCTION

The x-ray diffraction method is the only time-proven, truly nondestructive method for the measurement of residual stresses in crystalline materials [1,2]. Proof of its reliability lies in documentation of its use by thousands of engineers and scientists over the past several decades [3]. These applications have spanned from stress analysis of uranium/zirconium fuel rods to aluminum-alloy landing-gear components, and have included measurement of welding stresses in stainless steel piping as well as tempering

273

evaluations of carburized steels. The Society of Automotive Engineers considers the method of sufficient practical importance to have printed a Handbook Supplement on the subject three times [4], and the Society of Experimental Stress Analysis has formed a subcommittee on XRD residual stress measurement.

Even so, this nondestructive technology has been largely restricted to the laboratory, at least up to the last decade, because the specimens to be studied had to be of such size and shape that they could be placed upon a diffractometer. As a consequence, a number of innovative concepts of demountable diffractometers have been designed and marketed. These have included the Fastress and Strainflex, introduced in the sixties or early seventies; which are marketed by American Analytical Co. of Cleveland, Ohio, and Rigaku/USA of Danvers, Massachusetts, respectively. However, in spite of the advantages that these aforementioned devices offer, none can provide stress measurement in confined situations, for example, inside of piping smaller than approximately fifteen inches inside diameter.

A unique concept for a position sensitive scintillation detector (PSSD) was introduced a few years ago [5] and this concept has evolved into a dependable, highly accurate x-ray detection instrument at the Materials Research Laboratory of The Pennsylvania State University [6]. Recently, an x-ray residual stress measurement device incorporating this Ruud-Barrett PSSD was applied to nondestructive residual stress measurement on the inside surface of pipe less than ten inches inside diameter [7]. This paper describes the interfacing of this same unique PSSD* [6] with a miniature low wattage, x-ray tube** to provide unprecedented compactness and portability for XRD stress measurement.

INSTRUMENTATION

Figure 1 shows a photograph of the miniature XRD stress measurement head. The x-ray tube body is 4.1 inches (104 mm) long and 1.25 inches (32 mm) in diameter, and is water cooled. The anode end of the x-ray tube is to the left of Fig. 1 and that same figure shows a wrapping containing the water hoses and high-voltage cables. Attached to the anode end of the tube is the incident beam collimator and attached circular segment holding the x-ray sensitive ends of the fiber optics [6]. Figure 2 shows a mechanical drawing of the miniature stress head with its dimensions. An important and unique feature of the PSSD used in

---

*The PSSD is the Denver X-Ray Instruments Model D-1000-A detector.
**The tube is the Kevex model K3050S-W x-ray tube.

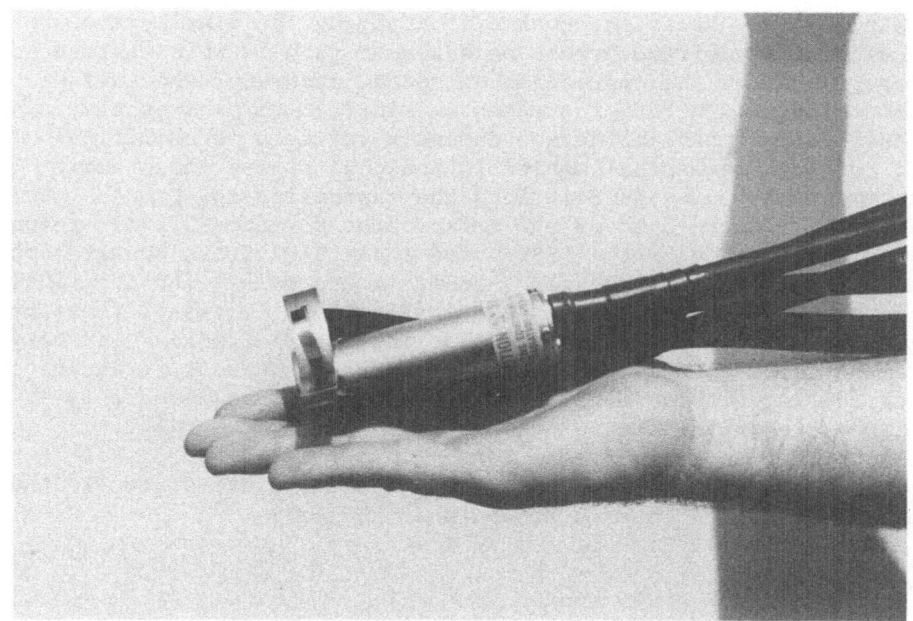

Fig. 1   Photograph of the miniature XRD stress measurement head
         showing the x-ray sensitive components of the PSSD mounted
         upon the low-wattage x-ray tube.

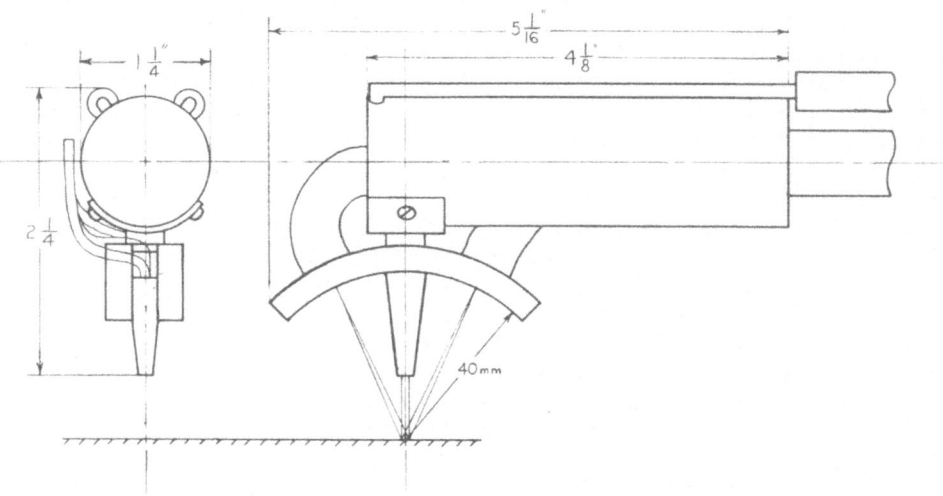

Fig. 2   Engineering drawing of the miniature XRD stress measurement
         head showing major dimensions.

this study is that it is capable of applying the single-exposure technique (SET) of XRD stress measurement [3,4]. This feature is a major reason for the unprecedented speed, accuracy, and ease of application of R-B PSSD instruments, since it eliminates the necessity of mechanical motion during x-ray data accumulation. The need for such mechanical motion in residual stress measurement instrumentation greatly restricts the capability for remote operation. Figure 3 shows the x-ray paths for the SET with respect to its application by the PSSD based miniature stress measurement head. The two diffracted x-ray beams intersecting with the fiber optic x-ray sensors represent two different psi angles. These psi angles are defined by the angle that the incident x-ray beam makes with the surface, i.e., the beta (β) angle, and the diffraction of Bragg (θ) angle. The equations for the two psi angles are $\psi_1 = \beta - 90 + \theta$ and $\psi_2 = \beta + 90 - \theta$.

The equation used for calculating the absolute stress in the specimens examined in this study is as follows:

$$\sigma = K \frac{S_{o_2} + S_2 - S_{o_1} - S_1}{R_o \sin^2\theta_o \sin 2\beta}$$

where S_{o_2}, S_2, S_{o_1}, and S_1 are defined in Fig. 3; R_o is the distance between the irradiated spot in the specimen and the fiber optic sensing surface; θ_o is the Bragg angle for unstressed aluminum; and beta is the angle between the incident beam and the specimen surface.

The PSSD is interfaced to a PDP 11/03 computer with 32 K memory. Peripherals include a teleprinter terminal, dual floppy discs, and CRT display. The PSSD control and residual stress data reduction program is on floppy disc, and programing runs under RT11 control. The software employed in the miniature stress measurement instrument uses a number of algorithms to ameliorate electronic noise and artifacts as well as to correct for mechanical misalignments and x-ray focusing errors. This paper will not describe in detail the derivation and/or basis for the algorithms, but the following is a brief description of the sequence of corrections as displayed on the CRT during data refinement.

Upon command, the instrument will initiate x-ray data collection; this normally requires about 3 seconds at 40 KV and 30 Ma for aluminum alloys, using a full wave rectified x-ray power supply, and about 6 seconds for the miniature version. At the conclusion of the collection period, the x-ray raw data is displayed upon the CRT, Fig. 4a. At this point the operator gives the command to calculate stress if the displayed data is satisfactory. The raw data is then corrected for electronic noise and gain inconsistencies in the detector. Next, a group of data points on both sides of each peak is used to provide a linear least square

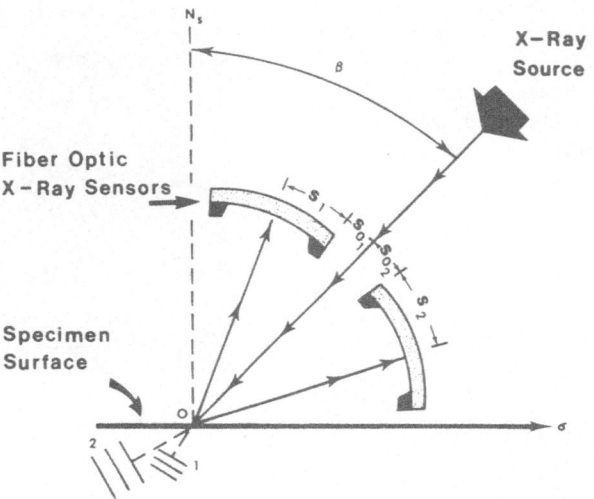

Fig. 3   Single-exposure technique (SET) of x-ray diffraction
         residual stress measurement. The numbers 1 and 2 denote
         certain crystallographic planes immediately below the
         specimen surface. These planes are diffracting the
         incident x-ray beam (note the arrows pointing toward the
         specimen) backward to the fiber optic x-ray sensors (note
         the arrows pointing toward S_1 and S_2). The angles between
         the incident and diffracted beams are $S_1 + S_{01}$ and $S_1 +$
         S_{02} and the stress at the surface of the specimen at point
         O is directly related to $(S_2 + S_{02}) - (S_1 + S_{01})$. The
         angle between the specimen surface normal (N_s) and the
         incident x-ray beam is beta (β).

Shown are the (511,333) peaks of
a 2014 alloy powder at 162.5° 2θ
with CuKα; data accumulation time
30 seconds at 30 KV and 4 Ma.

(a) Photo 1: Raw data
(b) Photo 2: Noise and gain
     corrected with background
     subtracted
(c) Photo 3: Same as Photo 2;
     LPA corrected and peak fit
     with parabola

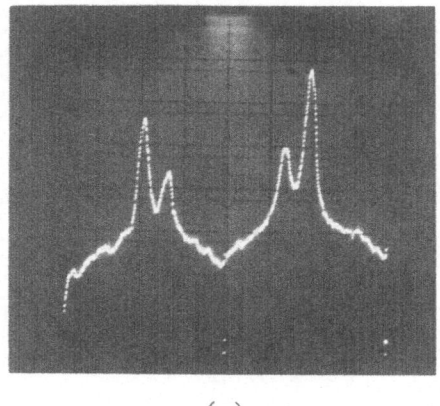

(a)

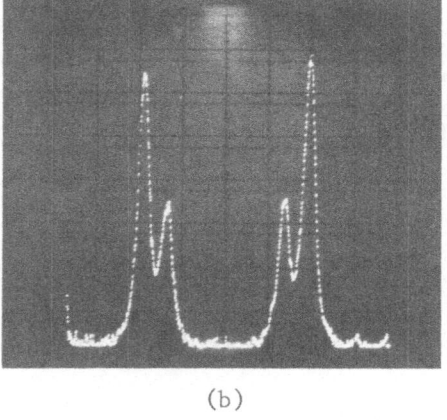

(b)

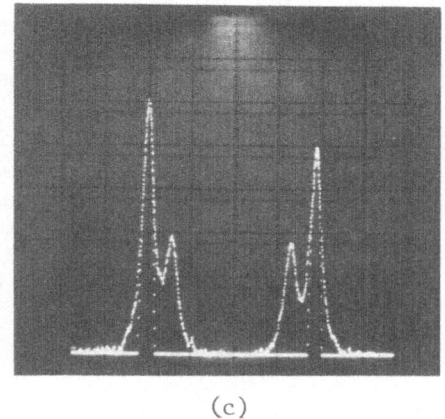

(c)

Fig. 4   Aluminum XRD peaks using and R-B PSSD detector and a
         Kevex K3050 x-ray source.

fit for x-ray background correction. Finally, the calculated
background is subtracted point by point across all of the collected
data and a parabola is fit by a least square routine to the upper
half of the x-ray peak, Fig. 4b. A parabola is then fit to the
major peak detected by each of the two sensing surfaces, Fig. 4c,
and the position of the apex of these parabolas is defined as the
diffracted peak position. Figure 5 shows a typical printout of
data from the PSSD's computer.

EXPERIMENTAL PROCEDURE AND RESULTS

    Evaluation of the miniature stress measurement head was
performed by determining the mean and standard deviation of stress
measurements on a loose aluminum alloy powder, i.e., zero stress
specimen, and through determining the standard deviation of a
linear regression fit to data from a four-point bend test.

    The beta angle and therefore the psi angles, were established
through setting the attitude of the x-ray tube so that the line of
the incident beam provided an angle of 60° with the horizontal
plane. The line of the x-ray beam was established through use of a
pointer which slipped over the incident beam collimator and which
was parallel to the line of the x-ray beam. That the pointer was
parallel to the line of the incident x-ray beam had been previously
established through use of a fluorescent screen. The 60° angle to
the horizontal was confirmed using a template, i.e., triangle.
Then to insure that the x-ray beam would intercept the specimen
surface at the desired beta angle, that surface was arranged
horizontally and this was confirmed with a bubble level. Thus, any
new specimens selected for stress measurements need only to have
the surface of interest place horizontal; then the stress head
could be moved near the surface and measurements commenced with no
need of contact between the specimen and stress measurement head.

    The distance from the irradiated point on the sample and the
x-ray sensitive fiber optic surfaces, i.e., R_0 in equation 1, was
not critical because the SET is inherently insensitive to small
errors in R_0 [4]. Therefore, the R_0 distance can be roughly
measured using a template of the proper length or through
adjusting the specimen to stress head distance so as to center the
diffracted x-ray peaks on the R-B PSSD instrument's CRT display.
The latter technique is preferred but is dependent upon adequate
pre-calibration of the instrument for a particular R_0 distance.

    The collimator selected for this study provided an irradiated
area of 2 x 10 mm. The x-ray tube target was Cu and the (511,333)
line of aluminum, at approximately 162.5° two-theta, was used.

Zero Stress Tests

A 2014 aluminum alloy, -400 mesh, powder was selected as a
specimen to evaluate the miniature instrument's accuracy and
precision for zero stress measurement. Such a loose powder has
been recognized as a reliable zero stress reference in ASTM
E915-83. The powder was placed in a powder specimen holder and
its smoothed surface made horizontal. Figure 4 shows a CRT display
of a typical set of XRD peaks from this powder at the two psi
angles, nominally, 12.2 and 47.8 degrees. The previously calibrated instrument was placed at approximately 40 mm from the powder
and upon initiation of data collection, and subsequent data
reduction, and it was determined that the R_O was 39.4 mm. The
measurement was repeated three times at this distance. Then the
instrument was raised 0.5 mm, as measured by a dial gage, where
three more stress measurements were made. Three measurements were
also made at 40.2 mm. Table 1 shows the mean and standard
deviation of the data at the three distances and shows the relative
insensitivity of the measured stress to variations in R_O. Each
value given in the table was the result of three stress readings
and each reading required 6 seconds total lapse time to acquire the
x-ray data.

Table 1. Zero Stress Powder Data

R_O (mm)	σ (KSI)
39.4	0.6 ± 0.4
39.9	-0.8 ± 0.5
40.2	1.5 ± 0.2

Mean and S.D. of nine readings = 0.4 ± 1.1

Four-Point Bend Tests

A 1-inch (25.4 mm) x 4-inch (101.6 mm) specimen was cut from a
0.062-inch (1.6 mm) thick sheet of 6061-T6 aluminum alloy. An
electrical strain gage was then adhered to the specimen, as
described by Prevey [8], and placed in four-point bending in a
device also described in the same paper. Figure 6 shows the
results of stressing this specimen from between zero and 20 KSI
(141 MPa). The standard deviation of a linear regression fit to
this data is 1.1 KSI (7.7 MPa). Each data point required six
seconds total lapse time to acquire the x-ray data.

```
ENTER COMMAND -- 6
PROFILE COLLECTION -- are you sure ?    Y
DATA COLLECTED
DATA COLLECTED
DATA COLLECTED
Do You Want to WRITE to DISK?   Y
DISK WRITE
Enter Filename (max. of 6 chars.) -- G4FAZ1

ENTER COMMAND -- 30
DISPLAYING PROFILES WITH CALCULATED BACKGROUNDS
FIT used 103 points
LEFT APEX = 256.42
FIT used 98 points
RIGHT APEX = 252.11
RO = 39.9   STRESS = -20.4 KSI
DISPLAYING PARABOLA FIT
FIT used 103 points
FIT used  98 points
FIT used 103 points
FIT used  98 points

ITERATIONS =  2    FINAL RO = 40.1    FINAL STRESS = -18.8 KSI

                  LEFT          RIGHT
PEAKS           256.75         252.78
THETA           78.2172        78.2784
SIN SQ PSI       0.0977         0.4429
DISPLAYING LPA CORRECTED DATA AND PARABOLA
```

Fig. 5 Typical data collection printout from R-B PSSD computer.

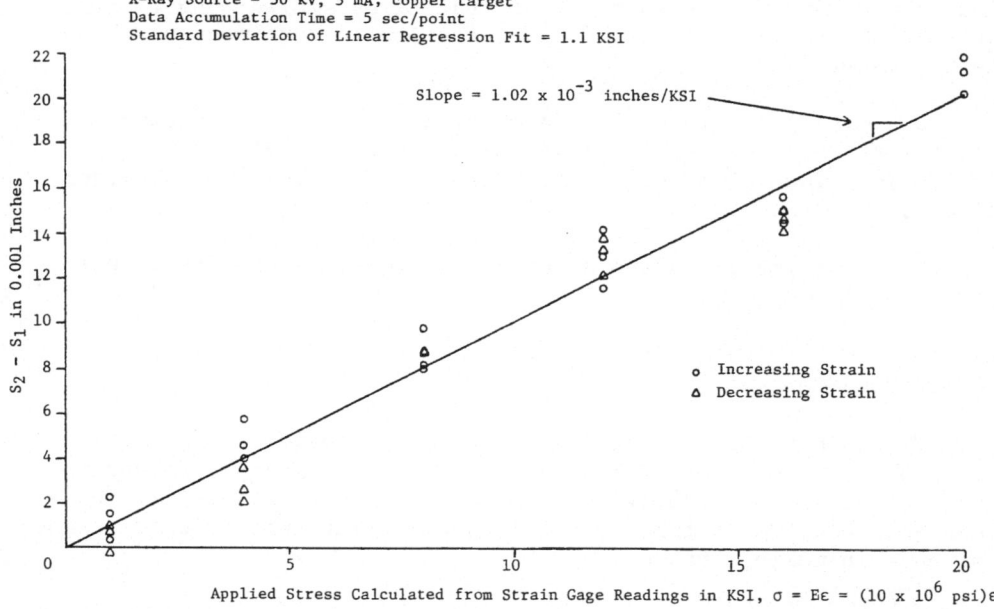

Fig. 6 Four-point bend test results using the miniature x-ray
stress head and a 6061-T6 aluminum alloy sheet specimen
under first increasing strain, circles, then decreasing
strain, triangles. Cu K-alpha radiation was diffracted
from the (511,333) planes of aluminum.

Remarks

It should be noted that no attempt was made to optimize the alignment of the collimator and tube target or the x-ray sensitive areas of the fiber optics to the miniature tube. The collimator and fiber optics were simply removed from a standard Dunlee Chromium target x-ray tube and affixed to the miniature tube. Optimization of alignment would no doubt improve precision and accuracy of stress measurement and reduce the time for data accumulation.

The miniature instrument described herein is small enough to fit through a 2.5-inch (64 mm) orifice and can measure stresses inside of diameters of less than 4 inches (102 mm). This capability for application in confined areas as well as its speed of data collection, accuracy of stress measurement, and remote application features provide an instrument with unprecedented capabilities.

CONCLUSIONS

1.   The operation of a miniature x-ray diffraction stress measurement instrument which is several times smaller and therefore considerably more versatile for application in confined areas has been demonstrated.

2.   The proficiency for data collection of the miniature instrument described is excellent as demonstrated by data collection times of 6 seconds for stress measurements in aluminum alloys.

3.   The precision and accuracy of the miniature instrument described is excellent.

ACKNOWLEDGEMENTS

The authors are indebted to Brian Skillicorn of Kevex tube division for the generous loan of the miniature x-ray tube and for his enthusiasm in encouraging this study. Also, the authors thank Lois Annechini-Moore for her skill in preparing this manuscript and patience in making changes.

REFERENCES

1. Ruud, C.O., "X-Ray Analysis and Advances in Portable Field Instrumentation," J. of Met. 31(6), 10 (1979).

2. Ruud, C.O., "A Review of Selected Nondestructive Methods for Residual Stress Measurement," NDT Int., 15 (1982).

3. Barrett, C.S. and Massalski, T.B., Structure of Metals, 3rd Ed., McGraw-Hill, New York, pp. 466-467, 1966.

4. SAE, "Residual Stress Measurement of X-Ray Diffraction-SAE J784a," Soc. of Auto. Eng. Inc., 1971.

5. Steffen, D.A. and Ruud, C.O., "A Versatile Position Sensitive X-Ray Detector," Adv. in X-Ray Anal. 21, 309-315.

6. Ruud, C.O., "Position-Sensitive Detector Improves X-Ray Powder Diffraction," IR&D, January 1983.

7. Ruud, C.O., DiMascio, P.S., and Melcher, D.M., "Application of a Position-Sensitive Scintillation Detector For Nondestructive Residual Stress Measurements Inside Stainless Steel Piping," Adv. in X-Ray Anal. 26, pp. 233-243.

8. Prevey, P.S., "A Method of Determining the Elastic Properties of Alloys in Selected Crystallographic Directions for X-Ray Diffraction Residual Stress Measurement," Adv. in X-Ray Anal., 20, pp. 345-354, 1977.

A LOW-HIGH TEMPERATURE CAMERA FOR IN-SITU X-RAY DIFFRACTION STUDIES OF CATALYSTS

J.Pielaszek* and J.B.Cohen

Department of Materials Science and Engineering
The Technological Institute
Northwestern University, Evanston Illinois 60201

INTRODUCTION

X-Ray diffraction studies of substances under controlled atmospheres and at different temperatures are of great importance in many research areas. This is especially true in the area of catalysis, where the correlation of structural and catalytic properties is needed. The camera described here was made for this purpose although any sample in the powdered form can be studied as well. Many catalysts are in the form of highly dispersed metal deposited on a granulated support. The content of metal may vary from a few tenths to several percent. In a camera used by Janko and Borodzinski[1] a small amount of catalyst was spread out on a porous silica glass sample holder which then was placed in a high temperature XRD camera with flowing gas of controlled composition. In our laboratory a sample holder was designed[2] with a section where catalyst could be heated under controlled atmosphere and afterwards transferred to an XRD cell without exposure to air. The cell itself has a mica window for the X-rays. A modification of this cell[3] allows in situ catalytic measurements at room temperature from a small amount of catalyst.

The camera discussed here was constructed to allow both low and high temperature measurements on the same sample under controlled atmosphere. Before any catalytic reaction a catalyst is usually pretreated under some standard conditions. In most cases this involves oxidation and/or reduction at a temperature higher than that of the catalytic reaction. In some cases however, dur-

---

*) On leave from the Institute of Physical Chemistry Polish Academy of Sciences, Warsaw, Poland

ing a catalytic reaction, there is a possibility of formation of
phases which are unstable at room or higher temperatures. In
other cases it may be desirable to lower the reaction rate by
lowering the temperature[3]. For these reasons the camera was
provided with a liquid nitrogen attachment allowing the sample to
be cooled far below room temperature.

When doing XRD studies of catalysts some facts should be remembered. First the concentration of catalyst metal is very low,
in most cases not exceeding a few percent. Next the intensity of
X-rays scattered from the support (background noise) is relatively strong and in most cases the ratio of the signal to background is of the order of 1:10, or even less. Because the
dispersion of a catalyst is also usually high (i.e the particle
size is small) the diffraction peaks are very broad. Therefore
proper subtraction of the background is of crucial importance.
Sometimes the signal is so weak and peaks are so broad that
conventional X-ray sources are not adequate, and more powerful
sources should be used (rotating anode X-ray generators or
synchrotron radiation)[2]. In the particular case of silica and
alumina supports and Mo K_α radiation, some geometrical factors
are also important. The great penetration depth for this
radiation causes the irradiated volume to be large and the
intensity of the signal to be greater than with Cu K_α . But the
greater penetration depth also causes the peaks to be
broadened and shifted. Because of these factors the adjustment of
diffractometer receiving slits should be done carefully[4]. The
catalysts are normally in the form of loose powder or fine
granules, and usually one wishes to study some definite
amount of catalyst. This means that the sample holder should
have one of its walls transparent to X-rays. In the sample
holders mentioned earlier[2,3], this was a thin mica window.
For the construction of the present camera a thin Be foil has been
chosen.

DESCRIPTION OF THE CAMERA

The camera was designed to mount on a standard horizontal
diffractometer. It consists of a cylindrical body about 76.2 mm
in in diameter and 152.4 mm in high to which is attached a liquid
nitrogen dewar (Fig.1e) from the outside and the sample holder
from the inside (Fig.2). The X-ray window in the camera body
(Fig.1a), is covered with a thin sheet of mylar supported by a
thin nylon grid. A clamp with a soft rubber seal fastens the window to the camera (Fig.1b). A viton O-ring is placed between the
cover and the cylinder. The camera body is water cooled (Fig.1c).
During the measurements a rotary pump is used , providing a vacuum
of 5×10^{-3} Torr but tests with a diffusion pump showed that a dyna-
mical vacuum of 10^{-5} Torr can be maintained. The sample holder
(Fig.2a), a copper block, is suspended from the cover on three

thin-wall stainless steel tubes which are attached to the cover by
nylon screws. The sample cell (Fig.2b) is appproximately 40 x 24
x 3 mm in size. Its front wall is a 0.5 mm thick Be foil which
separates the sample cell from the evacuable part of the camera
and is fitted to the copper block with a stainless steel frame. A
soft copper gasket is sandwiched between the Be foil and sample
holder body. Using Be foil causes some problems which will be
discussed below but this is the most appropriate material for
these specific purposes. The sample holder is provided with an
inlet (Fig.2c) which allows the sample to be changed without removing the Be foil. The gas inlet and outlet tubes (Fig.1d and
2d) allow a constant flow of gas. To heat the gas to the temperature of the sample holder before bathing the sample, part of the
inlet tube is wrapped around the rear part of the sample holder.
The temperature of the sample holder can be altered either by a
heater placed in a niche at the rear of the sample holder, or by
lowering the dewar via a flexible vacuum- tight screw mechanism
(Fig.2e) until its tip (Fig.2f) reaches the sample holder . The
temperature is controlled through three iron-constantan thermocou-
ples placed at the gas inlet, inside the sample cell and in the
sample holder body (Fig.2g).

EXPERIMENTAL DETAILS

     A rotating anode high intensity Rigaku-Denki X-ray source
with a Mo anode was used at 50 KV and 180 mA. Step scanning was
employed with a counting time of 100 to 200 seconds per point,
spaced 0.1 to 0.2 deg (2θ).

EXAMPLES

     As an example some studies of Co/SiO_2 and Co-Rh(50wt%)
catalysts are presented. The catalysts were prepared on 80-100
mesh silica gel by the impregnation of a penthane solution
of $Rh_2Co_2 (CO)_{12}$ and have total metal loading of 2.15wt% and
2.1wt% respectively. Their dispersion (defined as the percent of
metal atoms on the surface and measured by hydrogen chemisorption)
is 4% and 14% respectively. Catalysts were studied at room
temperature before and after reduction in situ at 823 K for
about 20 hrs. Fig.3 shows the spectrum of the Co/SiO_2
catalyst and the SiO_2 support as well as the resulting Co
spectrum obtained by subtraction of profiles. The signal to
background ratio is less then 10%. The particular Be foil used
in the experiments has a small amount of berylium oxide which
(due to the very pronounced 001 texture of the foil) gives
a relatively strong 101 reflection. Due to the presence of this
peak,normalization of the spectra was done with respect to it,
rather than to the background at the ends of the recording range.

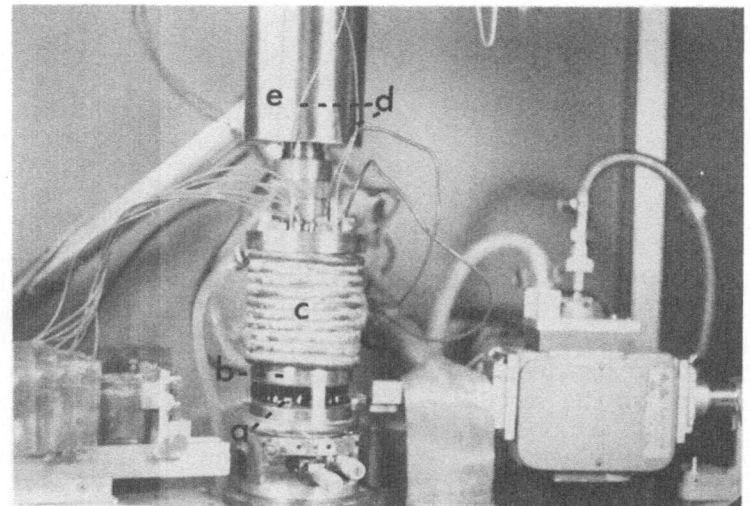

Fig.1 The camera mounted on horizontal diffractometer.

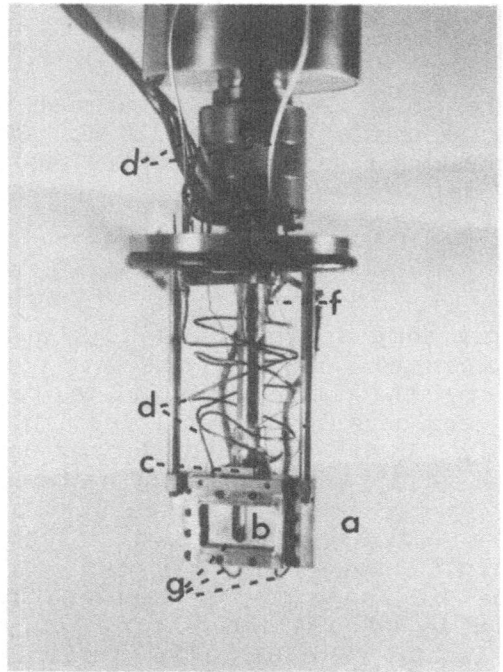

Fig.2 Sample holder attached to the camera cover.

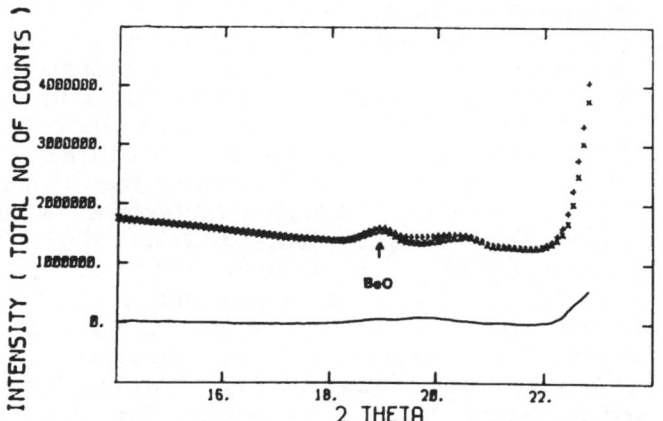

Fig.3 The X-ray diffraction spectrum from Co/SiO$_2$ catalyst:
crosses – after reduction, x – SiO$_2$ support. Solid line is
difference.

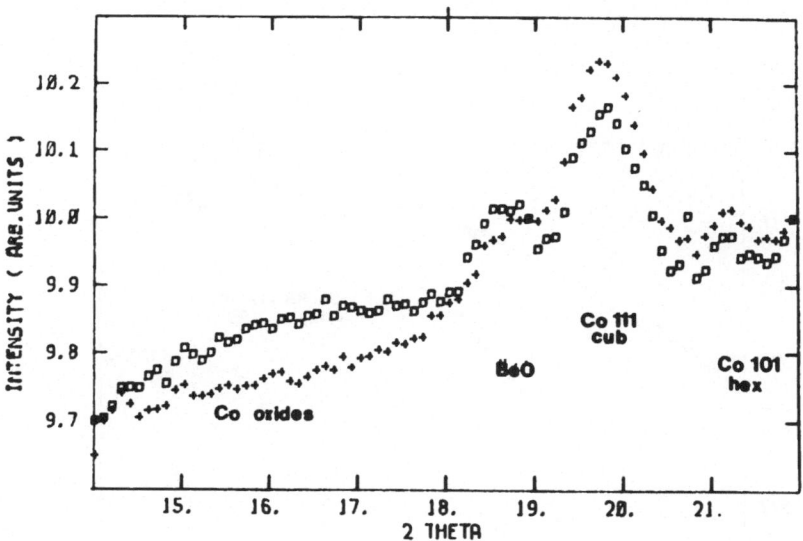

Fig.4 The X-ray diffraction spectra from cobalt catalyst after
subtraction of background: before reduction (squares) and
after reduction (crosses). This corresponds to an expanded
view of a section of solid curve on Fig.3.

Nevertheless, a shoulder from BeO is still seen in Fig.4, which represents the resulting cobalt spectrum before and after reduction.
This shoulder is less pronounced for the reduced cobalt catalyst
owing to the   stronger Co 111   peak.   The pronounced wide peak
from cobalt oxide at low 2θ angles disappeares after reduction
giving rise to   a   stronger Co 111   peak (from the cubic phase).
A 101 peak due to the hexagonal phase is also visible. For the
Co-Rh(50wt%) catalyst   the   reduction   causes   the   disappearence
of the broad peak from oxides. A rhodium-rich phase is identified
together with hexagonal cobalt (101 peak). The third peak
(Fig.5) can be interpreted as the superposition of Co and Co-Rh
alloy peaks , the latter from an alloy of about 20 to 40 at% of
rhodium (5).   The spectra in Figs 4 and 5 have decreasing
background towards low angles. This is due to the fact that
both   catalyst and substrate have different absorption
coefficients and different penetration depths for the radiation
used.   To have all the scattered radiation registered the receiving slits should be sufficiently wide[4]. In our particular case
the geometric arrangement of the frame supporting the Be foil put
a limit on the width of the receiving slit.   It   should   also   be
noted   that due to the high penetration depth of the X-rays, the
maxima in Figs 3, 4 and 5 are shifted towards low 2θ angles.

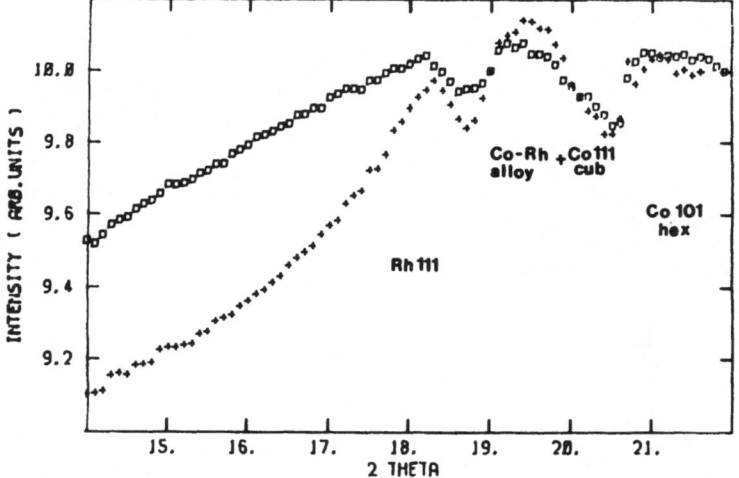

Fig.5 The X-ray diffraction spectra from cobalt-rhodium catalyst
      after subtraction of background: before reduction
      (squares) and after reduction (crosses).

CONCLUDING REMARKS

The camera has proven to be useful for in-situ studies at high temperatures. No actual low temperature measurements have yet been made , although the camera was tested and reached -140°C in about 2 hrs. The results indicate that special attention should be paid to the normalization procedure when subtracting background in such studies.

ACKNOWLEDGEMENTS

We thank Mr.J.Hahn and Mr.S.Jacobson for constructing this camera. Financial support was provided by DOE under grant No DE-ACOZ-77ER04254. The X-ray measurements were performed in the X-ray Diffraction Facility of Northwestern University's Materials Research Center, supported in part by NSF under grant No DMR-MRL-76-80897.

The catalysts were prepared in the Ipatieff Catalytic Labora-tory, Northwestern University, by Dr. Z.Karpiński.

REFERENCES

1. A.Borodzinski and A.Janko, Flow Reactor for Kinetic Studies with Simultaneous X-ray Phase Analysis of a Catalyst, Reac.Kinet.Catal.Lett. 7:163 (1977)
2. R.K.Nandi, F.Molinaro, C.Tang, J.B.Cohen, J.B.Butt and R.L.Burwell,Jr.,Pt/SiO - Pt.VI The Effect of Pretreatment on Structures, J. Catalysis 78:289 (1982)
3. R.K.Nandi,R.Pitchai, S.S.Wong, J.B.Cohen, R.L.Burwell,Jr. and J.B.Butt, Pd/SiO - Pt.1 In Situ X-ray Diffraction Studies during Treatment and Hydrogenolysis, J.Catalysis 70:298 (1981)
4. M.E.Milberg , Transparency Factor for Weakly Absorbing Samples in X-ray Diffractometry, J.Appl.Phys. 29:64 (1958)
5. W.Koster and E.Horn, Zustandbild und Gitterkonstanten der Legierungen des Kobalts mit Rhenium, Ruthenium, Osmium, Rhodium und Iridium, Zf.Metallk. 43:444 (1952)

A LOW COST REJUVENATION OF OLD G.E. AND PICKER X-RAY GENERATORS

G. Raykhtsaum and P. Georgopoulos

Dept. of Materials Science and Engineering, The
Technological Institute, Northwestern University,
Evanston, IL 60201

INTRODUCTION

General Electric (G.E.) and Picker generators are currently
installed in many research laboratories all over the United States.
These units are old and do not satisfy the modern requirements for
X-ray generators. Instabilities of high voltage and tube current
complicate the experiment. To maintain G.E. and Picker generators
is becoming more and more expensive due to the high cost of vacuum
tubes and replacement parts. To replace old generators with new
high performance commercial units can cost even more and requires
additional adaptations. In the present paper another alternative
is proposed, which assumes the replacement of vacuum tube mA and KV
controls with solid state circuits using the old G.E. or Picker
high voltage transformers. This alternative allows rejuvenation of
old generators without difficulties of reinstallation and does not
require adaptations to the existing experimental X-ray equipment.
The combination of old high voltage transformers with modern solid
state control devices provides high performance at low cost. For
this purpose the control device was designed and is described
below.

OPERATION

The old G.E. and Picker transformers operate with 60Hz AC
voltage. In order to use these transformers and at the same time
improve the parameters of the X-ray generator, the solid state AC
voltage phase control method was used in our device for mA and KV
controls. This method is very well known and is applied in many
modern commercial generators.

The simplified diagram of X-ray generator with Picker constant
potential high tension transformer is shown in Fig.1. To provide
for direct reading of high voltage, the divider R2 and R3 was
placed inside the transformer. A 3 digit KV meter measures the
value of high voltage reduced by 5 orders of magnitude. A 3 digit
mA meter measures the voltage drop across the resistor R1, which is
proportional to the X-ray tube current.

The control circuit consists of two independent channels of
voltage and current stabilization. Both channels utilize the
difference between the reference voltage and the voltage to be stabilized to provide phase control of the AC voltage on the primaries
of high voltage transformer HVTR and the filament transformer FTR.

Let us consider the current stabilizer, which consists of operational amplifier OA1, comparator COM1, pulse generator PG1,
solid state relay on the triac base SSR1, and the transformer TR1.
The time-signal diagrams are shown in Fig.2. The amplified difference between positive reference voltage REF1 and the positive voltage across R1 is compared with a saw signal which is synchronized
with AC voltage on the secondary of TR1 as shown in Fig.2. The duration of the pulses on the output of COM1 equals the distance
between crosspoints of OA1 output signal and saw signal. The pulse
generator produces two pulses, the first of which is synchronized
with the positive edge of the pulse from COM1 output, whereas the
second pulse is shifted by 8.33 ms. These pulses control the SSR1
which accomplishes the phase control of AC voltage on the primary
of FTR. As the reference voltage is connected to the noninverting
input of OA1, the more positive the difference of the signals on
the OA1 inputs, the the more power is applied to the tube filament.
This kind of negative feedback is due to the fact that only negative slope of the saw signal contributes to the generation of the
control pulses.

The voltage stabilizer is similar to the one described above
with the only differences that the voltage across R2 is negative,
the reference voltage is negative also, and the saw signal has been
inverted befor going to the input of COM2.

The filament current is measured as a mean-root-square value
of the voltage across the resistor R1. The value of this resistor
has been chosen to be numerically equal to the transformation ratio
of FTR in ohms, so the digital meter A reads the true filament current. The multiplier MT output equals the product of the signals
from R1 and R2, which is the power applied to the X-ray tube. The
logic circuit provides tube protection by means of cooling water
flow interlock, filament current limit threshold, and power limit
threshold. Power and filament current limits are settable in ac-
cordance to the specifications of a variety of X-ray tubes. When
the limits are exeeded, power to the X-ray tube is shut off.

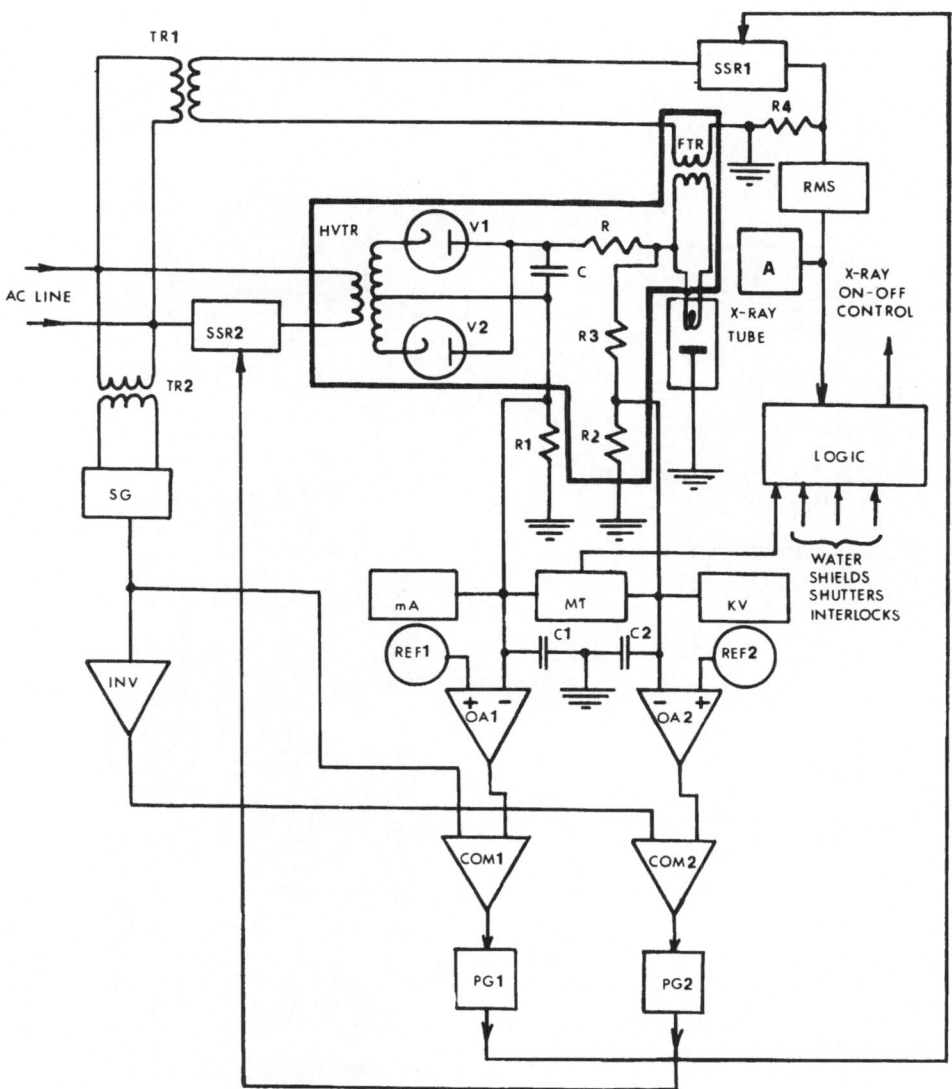

Fig. 1. Block-diagram of X-ray generator. HVTR-high voltage
transformer, FTR-filament transformer, SSR-solid state
relay, SG-sawtooth generator, INV-inverter,
MT-multiplier, REF-reference voltage, OA-operational amplifier, COM-comparator, PG-pulse generator,
RMS-root-mean-square converter.

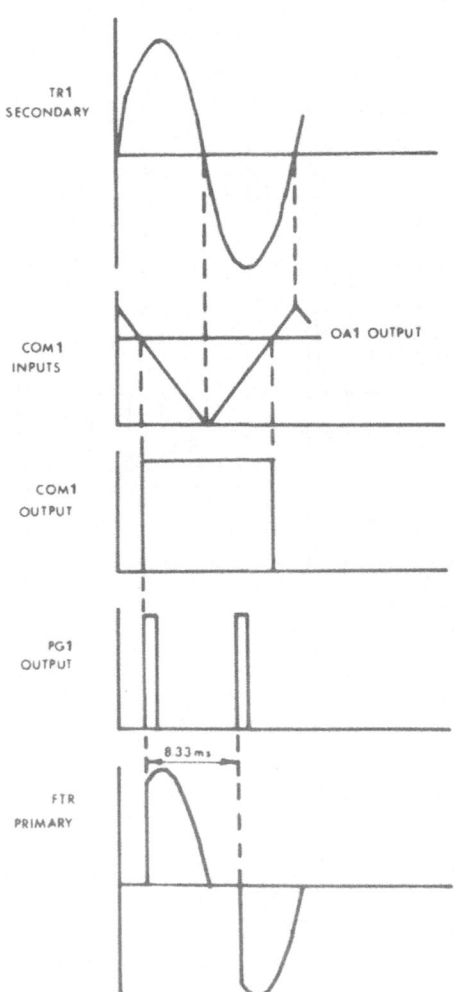

Fig. 2. Signal-time
diagram

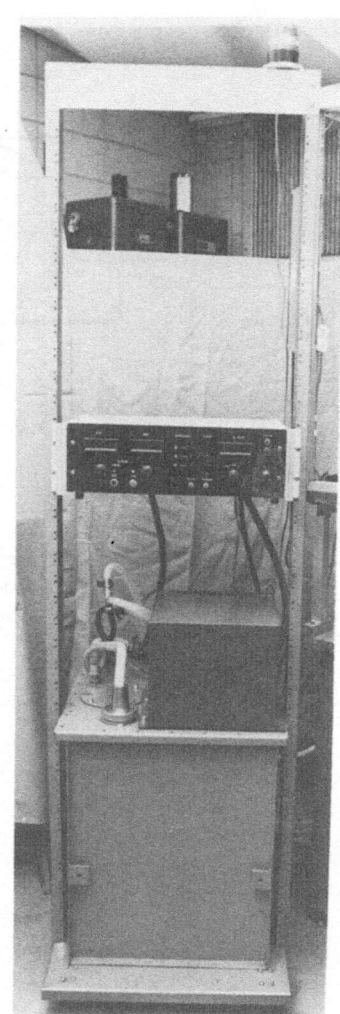

Fig. 3. X-ray generator with
control device and
Picker transformer

Safety interlocks for radiation shields and shutters are also connected to the logic circuit such that one can not turn X-rays on
until the safety conditions are satisfied.

The noninverting inputs of OA1 and OA2 can be switched to
external voltage sources (5 VDC for current channel and negative .5
VDC for voltage channel). This makes possible the external control
of tube current and high voltage. Switching the inverting input of
OA1 to the output of a ratemeter provides the stabilization of the
intensity recorded by a ratemeter rather than the tube current.

G.E. high voltage transformers do not contain condenser C.
High voltage and tube current in this case have a pulsed shape.
However capacitors C1 and C2 smooth the pulsation such that some
effective values of high voltage and tube current are stabilized.

SPECIFICATIONS

Fig.3 shows the Picker high tension transformer with the control device mounted on the rack. The generator requires single
phase 208 VAC, 60 Hz with power supply capacity of 25 A. Operation
of the generator is practically independent of line voltage variations and ambient temperature change within +/- 5° C.

Three modes of KV and mA control are available: internal,
external, and ratemeter control modes. The following stabilization
parameters are provided:

        high voltage range...........................15 - 50 KV
        long term voltage stability..................+/- .02%
        tube current range...........................1.5 - 50 mA
        long term current stability..................+/- .02%

SUMMARY

A modern X-ray generator based on the combination of existing
Picker or G.E. high voltage transformers with solid state control
devices has obvious adventages in comparison with the old units.
It has the same or even more features then the modern commercial
generators:

  high performance,
  internal, external, and ratemeter control modes of operation,
  direct reading of high voltage, tube and filament currents,
  does not need additional adjustments during the installation,
  has all necessary interlocks and tube protection circuits,
  can be used with any X-ray tube.

This work was sponsored in part by NSF through Northwestern
University's Materials Research Center, Grant No DMR-MRL-76-80847.

AN INEXPENSIVE IN-HOUSE EXAFS SPECTROMETER

P. Georgopoulos and C. H. Tang[*]

Dept. of Materials Science and Engineering
The Technological Institute
Northwestern University
Evanston, IL 60201

INTRODUCTION

Given the vast superiority of synchrotron radiation sources for EXAFS and near-edge measurements, one might assume that conducting such experiments in the laboratory is a waste of time and resources. Upon more careful consideration of the realities of everyday research, however, one can easily see that this is not true. In fact, many people have come to this realization, so much so that a whole conference[1] has been devoted to laboratory EXAFS facilities, their role in research and their relation to synchrotron facilities. The concensus after the conference was that laboratory instruments can and have been developed, with adequate performance for a variety of nontrivial experiments. They usually employ dedicated focusing spectrometers, rotating anode generators and fairly expensive monochromator crystals (see articles in ref. 1 and 2).

In this paper, we present an less expensive alternative, which makes use of equipment commonly present in an x-ray laboratory: The standard powder diffractometer. Furthermore, the spectrometer parts to be described below can be built as diffractometer attachments, which can be easily set up or removed.

----

* Present address: IBM Research Center, Yorktown Heights, NY 10598

EXAFS APPARATUS

In this section we will describe the various functional components of our EXAFS facility and briefly discuss their individual characteristics and performance. The performance of the system as a whole will be discussed in a later section.

X-ray source

The EXAFS apparatus is installed on a Rigaku RU-200PL rotating anode x-ray generator, equipped with a molybdenum anode. One major disadvantage of this generator is that it cannot be operated at voltages less than 20kV, because of space charge buildup in the filament-anode space. In fact, the maximum permissible anode current at low voltages is less than 200mA (120mA at 20kV, 160mA at 25kV, full current at 30kV and up, according to Rigaku). Hence, when working with light elements, one has to cope with the presence of higher harmonic radiation produced by the x-ray tube and passed by the monochromator. Still, one has recourse to a number of tricks which help eliminate the effects of higher energy components, such as selecting a monochromator crystal with a structure factor of zero for the second order reflection (such as Si(111), Si(311) etc.). In this case, one has to worry only about third order radiation and higher, which may have excitation voltages above 20kV. When the proper monochromator is not available, harmonic radiation can be discriminated against with careful design and operation of the x-ray detectors, as discussed later in this section.

Monochromator assembly

This is the most important part of the spectrometer, since it determines its major characteristics: Radiation intensity and energy resolution. Instruments described in the literature to date usually employ Johansson ground crystals. We have opted for the much simpler Johann geometry, in which the crystal is initially flat and simply bent elastically to a cylindrical shape. Although the Johansson geometry is free from focusing defects, it has been shown[3] that neither the brightness nor the resolution are seriously degraded in the Johann geometry. Most importantly in the latter case, one can freely change the bending radius to obtain near-optimal focusing, whatever the Rowland circle radius may be, whereas a Johansson ground crystal requires a fixed specific radius.

Once a known fixed Rowland circle radius is not an absolute requirement, the design of the spectrometer can be simplified tremendously, provided a way of continuously varying the bending radius can be devised. It is presicely the development of such a bending device that has allowed us to have a fully focusing spectrometer built on a conventional powder diffractometer.

Our crystal bending devices are based on a simple mechanical principle: When a diamond-shaped thin slab is supported along one of its diagonals and pushed at the other two corners, it bends into a cylinder. Figure 1 illustrates the structural details of our crystal bender. The crystal is pushed against the two front coaxial pins by means of a yoke assembly, equipped with knife edges in contact with the tips of the crystal. The position of the yoke and hence the bending radius of the crystal can be changed by sliding a wedge between the rear of the yoke and a stationary back plate. The up and down motion of the wedge is motorized, thus allowing for computerized continuous adjustment of the crystal radius to always conform to the Rowland circle radius, a function of diffractometer radius and scattering angle.

Detectors

The x-ray detectors in use on our spectrometer are gas proportional counters equipped with entrance and exit x-ray windows. The detectors are filled with an argon-methane mixture at a pressure that can be adjusted to achieve the required efficiency for the energy range of interest (see ref. 4 for more details). Ion chambers have improved and can be used profitably even down to low count rates (their high count rate capability is, of course excellent), but they totally lack energy discrimination. Due to the

Fig. 1   Motorized crystal bender.

limitations in low voltage capability of the Rigaku generator discussed earlier, our detectors must be capable of discriminating
against possible harmonic contamination. To this end, the detector
downstream from the sample was constructed with four times larger
diameter than the reference detector and both detectors are run at
the same gas pressure. When the pressure is adjusted such that the
reference detector stops 10% of the incident beam, the larger detector is only 35% efficient, but at the same time its efficiency
for higher harmonics is very low (the efficiency drops with approximately the cube of the energy). Whatever little contamination is
still detected can now be effectively eliminated electronically by
means of a pulse height discriminator.

The response of our detectors-counting electronics is adequate
up to several hundred thousand counts per second, although at rates
above 50,000-100,000 cps linearity suffers. This would not be a
serious problem if the measured intensities varied in a narrow
range, but the presence of characteristic and impurity (mainly
tungsten) lines in the tube spectrum makes such a situation very
unlikely. To compensate for these spectral variations, we have
made the current controller in the generator part of a feedback
loop, which constantly attempts to keep the reading of the reference detector constant. This technique was first described in ref.
4 and it is a very effective way of eliminating glitches due to
sudden flux variations.

Spectrometer

The EXAFS spectrometer is assembled on a standard Picker
powder diffractometer, shown in figure 2. The two major components
which convert the diffractometer into a part (or full) time EXAFS
spectrometer are the monochromator assembly which is mounted at the
center of the diffractometer and an experimental stage, carrying
the x-ray detectors and the sample chamber, which slides on the
standard receiving slit track of the diffractometer. Both can be
readily removed to convert the diffractometer back to its "native"
configuration. In order to reduce air absorption and provide radiation safety, the path of the beam from the x-ray tube to the monochromator housing and from the latter to the receiving slit is en-
closed and the beam tubes, along with the monochromator housing can
be filled with helium gas.

PERFORMANCE

The performance of this spectrometer has proved to be quite
respectable, adequate for a number of fairly sophisticated experiments. To date, we have collected EXAFS and near-edge spectra of

Fig. 2   EXAFS spectrometer.   The x-ray source is to the left.

3d   transition metal compounds, as well as vanadium oxides containing tungsten[5] (at the tungsten L(III) edge).   With our brightest
crystal (Si(111)),   we have obtained fluxes well in excess of 10^6
counts per second at 8keV, which allowed "easy" spectra, such as Fe
or Ni, to be collected with good statistics in less than one half
hour.

Figure 3 shows edge spectra of some vanadium compounds   (from
ref.   5), collected under the following conditions: 20kV, 120mA,
Si(220) crystal. Note that, under these conditions, both second
and third harmonics were present in the incident beam.   They were
eliminated by careful setting of the pulse height discriminator.
Each spectrum took about two hours (a Si(311) crystal would have
allowed much shorter times but was not available).   The estimated
resulution for these experiments was approximately 2eV, just a little worse than what is available at an unfocused synchrotron line.

Figure 4 shows the edge spectrum of the tungsten L(III) edge,
collected at CHESS (solid line) and our in-house spectrometer
(dashed line), along with the difference between the two spectra
(dotted line).   It is evident that the resolution at CHESS is
better than that of our spectrometer (estimated at 2eV versus 4eV),
but for comparisons of various spectra collected at the same reso-

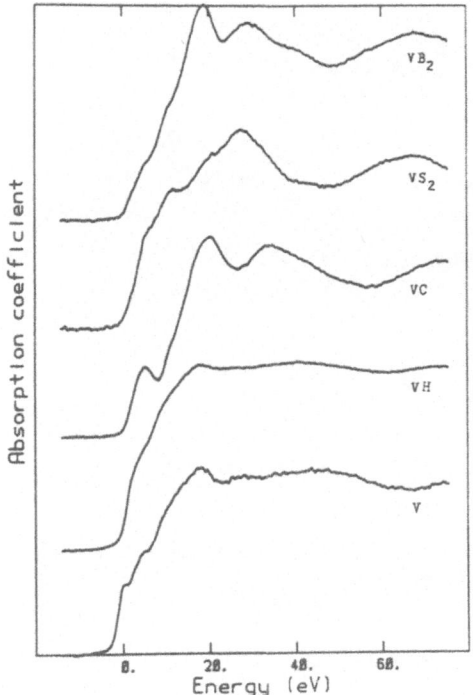

Fig. 3.   Edge spectra of various vanadium compounds.

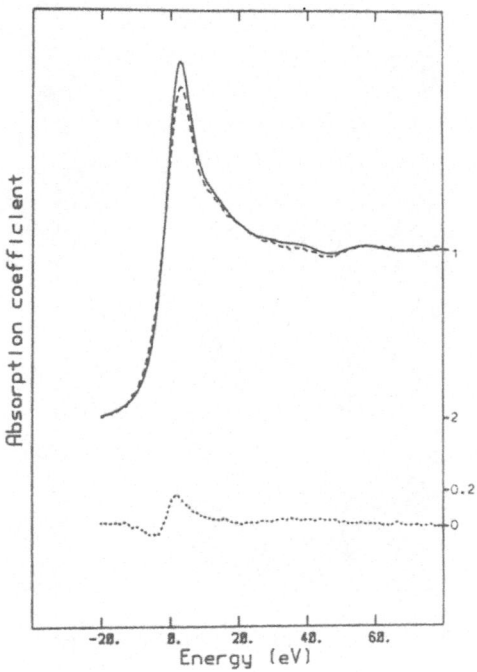

Fig. 4.   Comparison of the CHESS (solid line) and in-house (dashed line) L(III) edge spectra of tungsten metal. The lower curve is the difference between the two spectra.

lution, the performance is adequate. By deconvoluting the expected linewidth out of the in-lab spectrum, we have managed to essentially eliminate the difference in the two spectra, but the noise was blown up somewhat in the process. Higher signal to noise spectra are definitely possible in the laboratory, hence, even at the expense of increased data collection times, near-synchrotron quality can be had in many cases of interest. This should surprise certain scientists[6], who hold a dim view of what are the capabilities of a laboratory EXAFS spectrometer of modern design.

ACKNOWLEDGEMENTS

This research was supported by the Army Research Office under grant no. DAAG-29-80-C-0035. The x-ray measurements were performed in the X-ray Diffraction Facility of Northwestern University's Materials Research Center, supported in part by NSF under grant no. DMR-MRL-76-80897. The authors would like to thank Prof. J. B. Cohen and Dr. G. S. Knapp for many valuable suggestions throughout the course of this work. Mr. R. Hsieh has modified our software to control the motorized crystal bender.

REFERENCES

1. "Laboratory EXAFS Facilities – 1980" (Univ. of Washington Workshop, AIP Conf. Proc. no. 64, E. A. Stern, ed., New York (1980).
2. W. Thulke, R. Haensel and P. Rabe, Versatile Curved Crystal Spectrometer for Laboratory Extended X-ray Absorption Fine Structure Measurements, Rev. Sci. Instrum. 54:277 (1983).
3. K. -Q. Lu and E. A. Stern, Johann and Johansson Focusing Arrangements: Analytical Analysis, AIP Conf. Proc. no. 64, E. A. Stern, ed., New York (1980).
4. P. Georgopoulos and G. S. Knapp, Design Criteria for a Laboratory EXAFS Facility, J. Appl. Cryst. 14:3 (1981).
5. C. H. Tang, Local Atomic and Electronic Arrangements in $W_xV_{1-x}O_2$, Ph.D Thesis, Northwestern Univ., Evanston, IL (1983).
6. T. I. Morrison, G. K. Shenoy and D. Niarchos, Noise Level and Resolution Effects in EXAFS Spectra, J. Appl. Cryst. 15:388 (1982).

SUPPRESSION OF X-RAY FLUORESCENCE BACKGROUND IN X-RAY POWDER DIFFRACTION BY A MERCURIC IODIDE SPECTROMETER

J. Nissenbaum, A. Levi, A. Burger, and M. Schieber

School of Applied Science and Technology
The Hebrew University of Jerusalem
Jerusalem 91 000, Israel

Z. Burshtein

Nuclear Research Centre-Negev
P.O.Box 9001, Beer-Sheva 84 190, Israel

ABSTRACT

We have explored the merits of using a HgI_2 spectrometer as a detector in x-ray diffraction systems instead of a proportional gas counter, or a scintillation counter. The full width at half maximum energy resolution of the HgI_2 spectrometer used was about 1.1 keV for the CuK_α line (8.1 keV), and about 1.5 keV for the MoK_α line (17.4 keV). The energy resolution was utilized to eliminate x-ray fluorescence background from powder diffraction spectra. We demonstrate the suppression of Fe x-ray fluorescence in diffraction patterns of $ErFeO_3$ obtained with a Cu x-ray tube, and of Y x-ray fluorescence in diffraction patterns of Y_2O_3 obtained with a Mo x-ray tube. The peak height to background ratios were improved by about an order of magnitude in both cases.

INTRODUCTION

In a recent publication[1] we have introduced some advantages of using mercuric iodide x-ray spectrometers instead of proportional gas counters in x-ray diffractometry. The study of powder x-ray diffraction requires the use of monochromatric x-rays. This may be achieved by filtering the diffracted K_α lines from the white radiation, and particularly from the K_β lines, by using a crystal monochromator. Another option to eliminate the K_β radiation (but not

the entire white radiation) is to place a proper filter in the tube
exit window. The elimination of the remaining white radiation is
then accomplished by pulse-height analysis using a single-channel
analyzer. This step utilizes the energy resolution capabilities of
the x-ray detector used. Generally, either proportional gas counters or scintillation counters are used in x-ray diffractometers.

Mercuric iodide crystals are now being commercially* produced
as electronic gamma and x-ray spectrometers operating at room temperature.[2] So far, the applicability of this detector has been demonstrated in a variety of subjects, such as in x-ray fluorescence
measurements of high Z,[3-5] and low Z materials,[6-9] in thermal neutron detection,[10] for medical purposes such as x-ray computed tomography,[11] cerebral blood-flow measurements[12] and heart monitoring.[13]
HgI_2 has also been used as a photo-detector for scintillation spectrometry.[14] The operation of a HgI_2 detector requires a low d.c.
bias, of only 200-500 V, compared to the ∿1,500 V bias needed for
proportional gas counters. A photomultiplier operating in conjunction with a scintillator also requires a higher voltage, typically
1,000-1,500 V. The high stopping power of HgI_2[15] results in a maximum quantum efficiency. The energy resolution of the HgI_2 detector
allows the discrimination of unwanted radiation by pulse-height a-
nalysis. In contrast to a proportional gas counter, the pulse heights
in a HgI_2 detector are independent of the bias voltage over a large
voltage range.[16] In our previous publication[1] we demonstrated how
the CuK_β peaks in the powder diffraction spectra of Si or HgI_2 with
a Cu x-ray tube could be eliminated without the need for a monochromator or a filter.

In our present work we further explored the capabilities of
HgI_2 detectors in x-ray diffraction systems, particularly in supp-
ressing x-ray fluorescence background. Considerable x-ray fluorescence background is involved in x-ray diffraction when the incident
radiation photon energy is appropriate to excite fluorescence in
any of the atomic components of the studied sample. For example,
with a Cu x-ray tube (λ_{K_α} = 1.542 Å) considerable x-ray fluorescence will be excited in matrices containing Co, Fe or Mn; with a
Mo x-ray tube (λ_{K_α} = 0.710 Å), x-ray fluorescence will be excited
in Y,Sr and Rb.[17] The occurance of x-ray fluorescence not only introduces a high background to the diffraction patterns. It also reduces the radiation intensity available for diffraction. Diffraction
spectra then show a very poor peak to noise ratio. The x-ray fluorescence energy is shifted to lower energies compared to the exciting radiation. Therefore, the energy resolution capabilities of mercuric iodide spectrometers may be very useful to improve signal-to-
noise ratios in such cases.

---

*Yissum Research and Development Co., Jerusalem, Israel

EXPERIMENTAL

The measurements were made with a standard (Philips model PW 1140/00/60) x-ray diffractometer using Cu or Mo x-ray tubes. The tubes were normally operated at 40 kV and 30 mA. Detection of the diffracted radiation was carried out by a commercial* HgI_2 x-ray spectrometer. Our detector contact area was $4x10$ mm^2, and its thickness was 400 μm. The detector was operated under 550 V dc bias, and was coupled to a charge-sensitive preamplifier. The output signals were amplified with a linear amplifier (Tennelec model TC-205A), and fed into the counting system through a single channel analyzer (Elscint model SCA-N-3). The full width at half maximum signal height resolution for ∿8 keV photon energy (CuK$_\alpha$ emission line[17]) was about 1.1 keV. Powder diffraction characteristics were measured for orthorhombic ErFeO$_3$ [18] with a Cu x-ray tube, and for cubic Y$_2$O$_3$ [19] with a Mo x-ray tube.

RESULTS

(a) Cu x-ray tube

As mentioned, diffraction from materials containing, for example, Co, Fe or Mn with the CuK$_\alpha$ emission radiation involve a considerable background originating from x-ray fluorescence. As an example, we show in Fig. 1(a) a diffraction spectrum of orthorhombic ErFeO$_3$.[18] The counting rate of the HgI$_2$ detector is plotted vs scattering angle 2θ. In this spectrum the single-channel-analyzer base-line discriminator was set just above the electronic noise level. The upper-window discriminator was set fully open. The spectrum thus consists of both K$_\alpha$ and K$_\beta$ reflections, as marked in the figure, along with a large background noise from the x-ray fluorescence. The unmarked peaks in the figure correspond to a cubic (Garnet structure[20]) Er$_3$Fe$_5$O$_{12}$ impurity which was present in our sample. The ErFeO$_3$ (112) peak height to background ratio is about 3.3. To further analyse the source of background noise we show in Fig. 1(b) the energy spectrum of the countings in the ErFeO$_3$ (110) reflection. This reflection peak height is smaller than the main (112) peak. Therefore, for this peak the x-ray fluorescence background is relatively more accentuated. In Fig. 1(b) the counting rate is plotted vs. photon energy. The latter was selected by dialing the base-line discriminator of the single-channel analyzer. The upper-window discriminator was set constant, permitting a resolution of about 0.15-keV, as marked in the figure (1(b)). This spectrum reveals two peaks. The higher energy peak corresponds to the CuK$_\alpha$ energy (8.09 keV[17]). The calibration of the energy scale in the figure follows from this peak identification. The full width at half maximum of this peak is about 1.1 keV. The lower energy peak corresponds to the K$_\alpha$ x-ray fluorescence of Fe[17] (∿6.4 keV). The Cu K$_\beta$

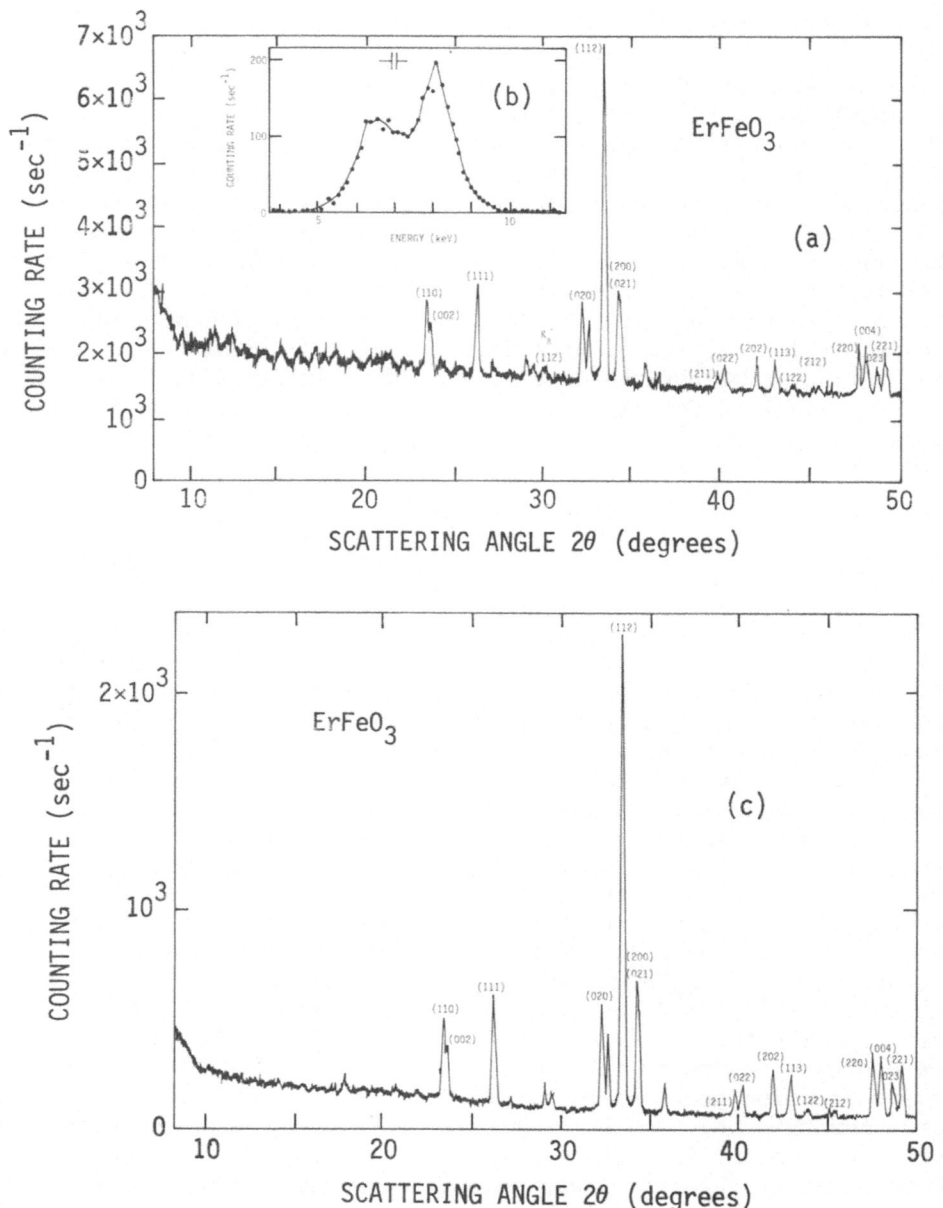

Fig. 1.   (a) Counting rate vs. scattering angle 2θ of $ErFeO_3$ powder;
(b) Counting rate vs. photon energy diffracted or scattered
from $ErFeO_3$ powder at $2\theta = 23.2^{\circ}$ ((110) reflection direction for CuK_{α} line);   (c) Counting rate vs. scattering
angle 2θ of $ErFeO_3$ powder. Base-line discriminator at
7.3 keV, upper-window fully open. All measurements made with
HgI_2 detector; Cu x-ray tube operated at 40 kV and 30 mA,
angular resolution 0.1°.

peak energy is not apparent here, as the reflection angle (23.4°) corresponds to a CuK_α reflection (Fig. 1(a)).

To eliminate CuK_β reflections from the diffraction spectrum (Fig. 1(a)) we have used a Ni filter between the tube exit window and the sample. Then, to reduce the x-ray fluorescence background the base-line discriminator of the single-channel analyser was set at 7.3 keV (see Fig. 1(b)). The upper-window dial was set fully open. The diffraction spectrum thus obtained is shown in Fig. 1(c). The counting rate of the HgI_2 detector is plotted vs. scattering angle 2θ. As seen, the background has been reduced drastically. Now, the (112) peak height to background ratio is about 30, compared to a ratio of 3.3 in Fig. 1(a). The K_β peakes are absent here, due to the Ni filter.

The experimental arrangement described above, which enables the reduction of unwanted radiation, is not the only one possible. For example, the upper-window discriminator may be used to reduce CuK_β reflections.[1] Else, a crystal monochromator may be inserted in the diffracted x-ray path. In the latter case, the advantage of a HgI_2 detector follows from its higher quantum efficiency (100%, compared to the measured 65% with our ordinary proportional gas detector).

(b) Mo x-ray tube

Diffraction measurements with the MoK_α emission radiation ($\lambda = 0.710$ Å[17]) of matrices containing e.g. Y, Sr or Rb[17] are associated with considerable x-ray fluorescence giving rise to a large background. As an example, we show in Fig. 2(a) a diffraction spectrum of cubic Y_2O_3 powder.[19] The counting rate of the HgI_2 detector is plotted vs. scattering angle 2θ. In this spectrum the single-channel analyzer base-line discriminator was set just above the electronic noise level. The upper-window discriminator was set fully open. The spectrum thus consists of both K_α and K_β reflections, as marked in the figure, along with a very large background noise from the x-ray fluorescence. Some of the cubic Y_2O_3 reflections[19] are marked in the figure. The prominent unkarked peaks correspond to the other hexagonal phase of Y_2O_3.[21] The cubic Y_2O_3 (400) peak height to background ratio is about 0.35. To analyze in detail the background source we show in Fig. 2(b) an energy spectrum analysis of the countings in the (400) reflection ($2\theta = 15.4°$). The counting rate is plotted vs. photon energy. The photon energy was selected by dialing the base-line discriminator of the single-channel analyzer. The upper-window discriminator was set constant, permitting a resolution of about 0.15 keV, as marked in the figure. The energy spectrum reveals three main peaks. The higher energy peak corresponds to the MoK_α emission (17.5 keV[17]). The calibration of the energy scale in the figure follows from this peak energy identification. The full width at half maximum of this peak is about

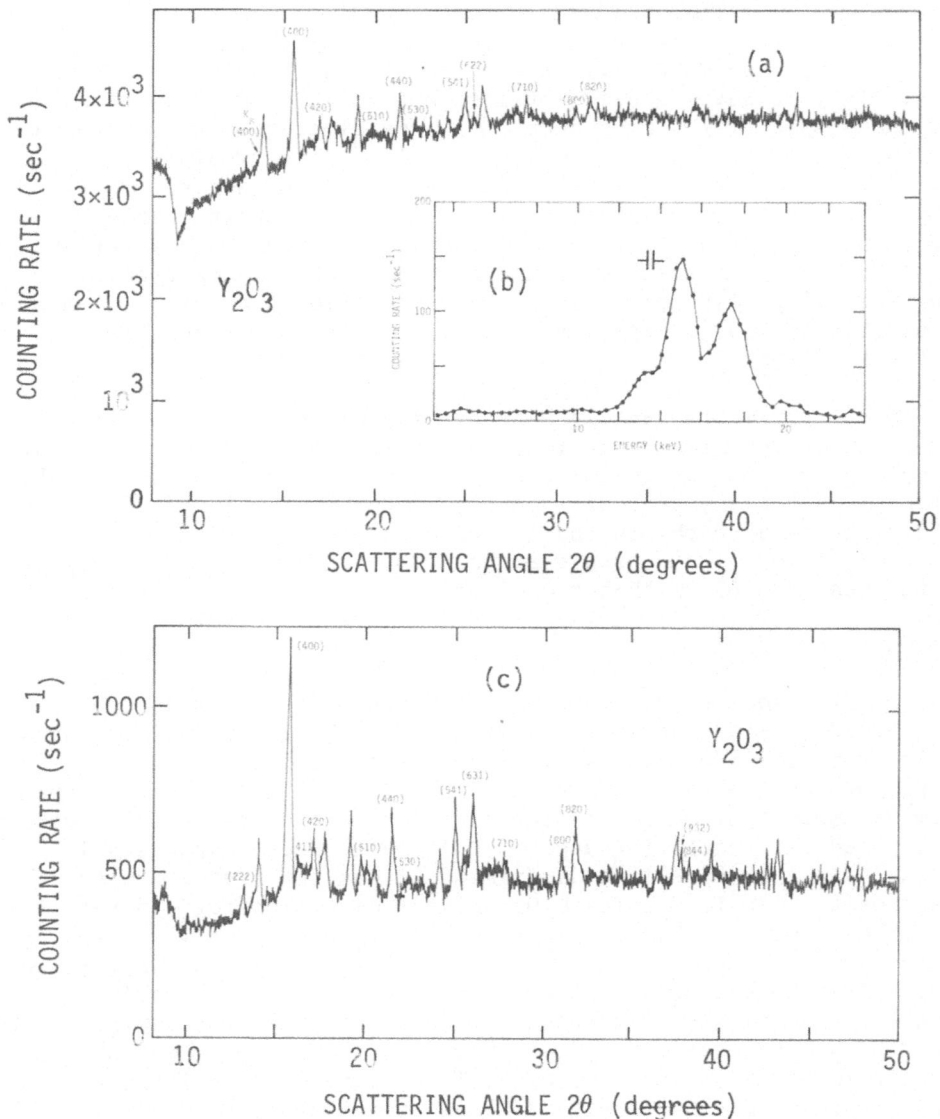

Fig. 2.   (a) Counting rate vs. scattering angle 2θ of Y_2O_3 powder;
(b) Counting rate vs. photon energy diffracted or scattered
from Y_2O_3 powder at $2\theta = 15.4°$ ((400) reflection direction
for MoK_α line);   (c) Counting rate vs. scattering angle
2θ of Y_2O_3 powder. Base-line and upper-window discriminators allow energies between 16.5 and 18.5 keV. All measurements made with HgI_2 detector; Mo x-ray tube operated at
40 kV and 30 mA, angular resolution $0.1°$.

1.5 keV. The lower energy peak centered at 15.0 keV corresponds to the K_α x-ray fluorescence of Y.[17] The small peak, centered at about 13.0 keV, is most probably a contribution of two effects taking place in the HgI_2 detector itself: an escape peak of the MoK_α radiation due to iodine L-shell transitions,[17] and an escape peak of the Y K_α radiation due to mercuric M-shell transitions. Some traces of the MoK_β line (19.6 keV[17]) are also apparent on the high energy side.

To reduce the x-ray fluorescence background, as well as the contribution of K_β reflections, the base-line and upper-window discriminators were set to allow the counting of photon energies between 16.5 keV and 18.5 keV only. The diffraction spectrum thus obtained is shown in Fig. 2(c). As seen, the background is drastically reduced. The (400) peak height to background ratio has increased from 0.35 in Fig. 2(a) to about 1.8 here. The K_β reflection interference in the peaks is also reduced. This is evident by the change in the peak height ratios (see particularly the cubic Y_2O_3 peak, and the unmarked (101) peak at 14.2° of the hexagonal Y_2O_3 phase in Figs. 2(a) and 2(c)).

DISCUSSION

Thus, we have demonstrated the method by which the x-ray fluorescence existing for certain matrices in x-ray diffraction spectra may be reduced by using the energy resolution capabilities of a HgI_2 detector. The HgI_2 detector used in our present work was of a relatively low energy resolution ($\sim$1.1 keV at 8 keV, see Fig. 1(b)), and should therefore be of low cost and commercially available. The best energy resolution (full width at half maximum) reported[22] for a HgI_2 detector with room temperature operating components is about 0.30 keV at 5.9 keV radiation. Such resolution, however, has been obtained with a detector whose contact area was only about 4 mm^2 (the thickness was about 400 μm). The typical active detector area dimensions needed for x-ray diffractometry is about 10x2 mm^2. The input capacitance observed by the preamplifier, as well as the leakage currents should thus increase. Both effects act as sources of noise. Using the compilation of noise figures of Dabrowski and Huth,[23] we estimate a value of 0.4-0.45 keV as an achievable energy resolution limit for a HgI_2 detector for the 5.9 keV radiation. In making this estimate we counted on our experience that leakage currents in detectors of the mentioned dimensions could be as small as 6-8 pA. With such high energy resolution one should be able to obtain complete discrimination of K_β reflections as well as scattered x-ray fluorescence from diffraction spectra.

The energy resolution of a scintillation counter in the 0-20 keV energy range is about 50% at best[24]. It could not therefore stand the competition of HgI_2. On the other hand, the energy resolution displayed by a proportional gas counter is similar to what has been shown above for a HgI_2 detector. Still, in a proportional gas counter, any drift in the bias voltage will change the peak position. A typical variation of peak position with bias voltage is about 5%/V.[25] The peak position in a HgI_2 detector is independent of voltage[16] over a large voltage range (>30 V). HgI_2 is also better from the point of view of counting efficiency. In a proportional counter the quantum efficiency is typically $\sim$70%, and varies with the photon energy.[25] When the x-ray source tube is changed one should consider the replacement of the proportional counter for a different type, of a different gas mixture and pressure, or for a scintillator, to obtain best counting efficiency. The HgI_2 detector counting efficiency is 100%, independent of photon energy up to about 50 keV (for typical 500 μm detector thickness[15]). Therefore, it should be suitable for any x-ray tube used in diffractometry.

An energy resolution which is much better than expected for HgI_2 is possible with a Si x-ray detector[26]. A typical figure for the resolution of a Si detector operating at liquid nitrogen temperature is about 0.15 keV.[26] This high resolution should make the Si detector as best suited for elimination of unwanted radiation from x-ray diffraction spectra. However, the complications involved by the necessary cryogenic arrangements[26] should be considered against the simplicity of room-temperature operation of HgI_2.

In summary, the present work demonstrates the utilization of a HgI_2 spectrometer to suppress Fe x-ray fluorescence in diffraction patterns of $ErFeO_3$ when using a Cu x-ray tube, and to suppress Y x-ray fluorescence in diffraction patterns of Y_2O_3 when using a Mo x-ray tube. We are indebted to Dr. M. Melamud of the Nuclear Research Centre - Negev, for calling our attention to the problem.

REFERENCES

1.   J. Nissenbaum, A. Levi, A. Burger and M. Schieber, Replacement of monochromator and proportional gas counter by HgI_2 detector in x-ray diffraction, J. Appl. Cryst. 16:136 (1983).
2.   M. Schieber, Fabrication of HgI_2 nuclear detectors, Nucl. Instr. Methods 144:469 (1977).
3.   M. Singh, B. C. Clark, A. J. Dabrowski, J. S. Iwanczyk, D. E. Leyden and A. K. Baird, Background and sensitivity considerations of x-ray fluorescence analysis with a room temperature HgI_2 spectrometer, Adv. x-ray Anal. 24:337 (1981).
4.   J. Nissenbaum, A. Holzer, M. Roth and M. Schieber, Fluorescence analysis of high z materials with mercuric iodide room temperature detectors Adv. x-ray Anal. 24:303 (1981).

5. M. Singh, Computer analysis of x-ray fluorescence spectra obtained with a room-temperature HgI_2 detector and two application studies, Nucl. Instr. Methods, Phys. Res., 193:135 (1982).

6. M. Singh, A. J. Dabrowski, G. C. Huth, J. C. Iwanczyk, B. C. Clark and A. K. Baird, X-ray fluorescence analysis at room temperature with an energy dispersive HgI_2 Spectrometer, Adv. x-ray Anal. 23:249 (1980).

7. A. J. Dabrowski, G. C. Huth, M. Singh, T. E. Economou and A. L. Turkevich, characteristic x-ray spectra of Na and Mg measured at room temperature using HgI_2 detectors, Appl. Phys. Letters 33:211 (1978).

8. J. C. Huth, A. J. Dabrowski, M. Singh, T. E. Economou and A. L. Turkevich, A new x-ray spectroscopy concept-room temperature HgI_2 with Peltier-cooled pre-amplification, Adv. x-ray Anal. 22:461 (1979).

9. J. S. Iwanczyk, J. S. Kusmiss, A. J. Dabrowski, J. B. Barton and G. C. Huth, Room temperature HgI_2 spectrometry for low-energy x-rays, Nucl. Instr. Methods, Phys. Res. 193:73 (1982).

10. M. Melamud, Z. Burshtein, A. Levi and M. M. Schieber, New thermal neutron solid-state electronic detector based on HgI_2 x-ray detector, Appl. Phys. Letters 43:275 (1983).

11. I. Beinglass, L. Kaufman, K. Hoisier and J. Hoenninger, Evaluation of HgI_2 detector for x-ray computed tomography, Med. Phys. 7:370 (1980).

12. A. Levi, M. Roth, M. Schieber, S. Lavy and G. Cooper, Development of HgI_2 gamma radiation detector for application in nuclear medicine, IEEE Trans. Nucl. Sci. NS-29:457 (1982).

13. A. Lahiri, J.C.W. Crawley and E. B. Raftery, A mercuric iodide detector for continuous monitoring of left venticular function, Talk presented at the Society of Nuclear Medicine, 30th Annual Meeting, St. Louis, Missouri, June 7-10, 1983.

14. I. S. Iwanczyk, J. B. Barton, A. J. Dabrowski, J. H. Kusmiss and W. M. Szymczyk, A. novel radiation detector consisting of an HgI_2 photodetector coupled to a scintillator, IEEE Trans. Nucl. Sci. NS-30:363 (1983).

15. H. L. Malm, T. W. Raudorf, M. Martini and K. R. Zanio, Gamma-ray efficiency comparisons for Si(Li), germanium, cadmium telluride and mercury(II)iodide detectors, IEEE Trans. Nucl. Sci. NS-15:500 (1973).

16. A. Levi, M.M. Schieber and Z. Burshtein, Carrier surface recombination in HgI_2 photon detectors, J. Appl. Phys. 54:2472 (1983).

17. J. A. Bearden, x-ray wavelengths p. E131 in "Handbook of Chemistry and Physics" 3rd edition, CRC press, Cleveland, Ohio, U.S.A. (1972-73).

18. File # 20-389, in "Powder diffraction file", Joint Committee of Powder Diffraction Standards, Philadelphia, U.S.A.(1967).

19.   File # 5-0574 in Ref. 18.
20.   File # 23-240 in Ref. 18.
21.   File # 20-1412 in Ref. 18.
22.   J. S. Iwanczyk, A. J. Dabrowski, D. Del Duca and W. Schnepple,
      A study of low noise pre-amplifier system for use with
      room-temperature HgI_2 x-ray detectors, IEEE Trans. Nucl.
      Sci. NS-28:579 (1981).
23.   A. J. Dabrowski and G. C. Huth, Towards the energy resolution
      limit of room-temperature low energy x-ray spectrometry,
      IEEE Trans. Nucl. Sci. NS-25:205 (1978).
24.   "Harshow" catalog of scintillation counters.
25.   "Philips" proportional detector probe instruction manual.
26.   "Ortec"/"Tractor"/"Kevex" catalogs of x-ray detectors.

COMPUTER-ASSISTED ALIGNMENT OF A GUINIER X-RAY

POWDER DIFFRACTION SYSTEM

J.J. Fitzpatrick and J.S. Pressnall

Chemical and Materials Sciences Division
Denver Research Institute
University of Denver
Denver, CO   80208

ABSTRACT

Digitized data acquisition and subsequent peak-profile fitting
are used in aligning and maintaining the Guinier diffraction system
in this laboratory. Utilization of a stepping motor driven goniometer along with profile analysis allows rapid and reproducible
alignment while providing a means to investigate the types and magnitudes of errors introduced by various misalignments of the optical system.

INTRODUCTION

Until recently, one of the most pronounced drawbacks to the
use of Guinier optics for routine powder diffraction work has been
difficulty in aligning and maintaining alignment of the X-ray optical system. The reliability and excellent spectral characteristics
of the newest generation of commercially available monochromator
crystals has, however, removed the major stumbling block to simple
alignment. Furthermore, with the advent of computer controlled
goniometers and digitized data acquisition, a straightforward and
expedient method of alignment free from the vagaries of subjective
judgement can be outlined which will result in reproducible intensity and positional data.

The procedure outlined in this paper is one which has proven
to be useful in both initial alignment and routine maintenance of
the Guinier system in this and several other laboratories. While
the procedure is utilized on the system manufacted by Huber
Diffractionstechnik, the principles are generally applicable to

317

any horizontal X-ray diffractometer system which utilizes Guinier
optics.

   Beyond the initial roughing stages of alignment which are
generally covered in the manufacturer's installation manual, we
have found that a Guinier powder diffraction system requires fine
tuning to achieve optimal results. The procedures described in the
second section of this paper are aimed at achieving this. Comments
in the first section are aimed at elucidation of points which may
be unclear in the installation manual and pitfalls which the first-
time operator may encounter which are not mentioned in the manual.

DESCRIPTION OF THE GUINIER SYSTEM

   A schematic of the entire data collection system is shown in
Fig. 1. The major features of this system include the Huber-Guinier
high-temperature powder diffraction system, the detector and detector
electronics, a microprocessor controlled sample furnace and the
Nicolet L-11 data processing and acquisition system. The original
slewing motor on the diffractometer has been replaced with a stepping motor which is under the direct control of the computer system
via a Slo-syn translator. The L-11 system features an LSI-11/23
processor, which is currently operating under TSX-Plus, with a 20
megabyte Winchester as the system device. Numerous other peripherals
are also available on the Q-bus for manipulation of data.

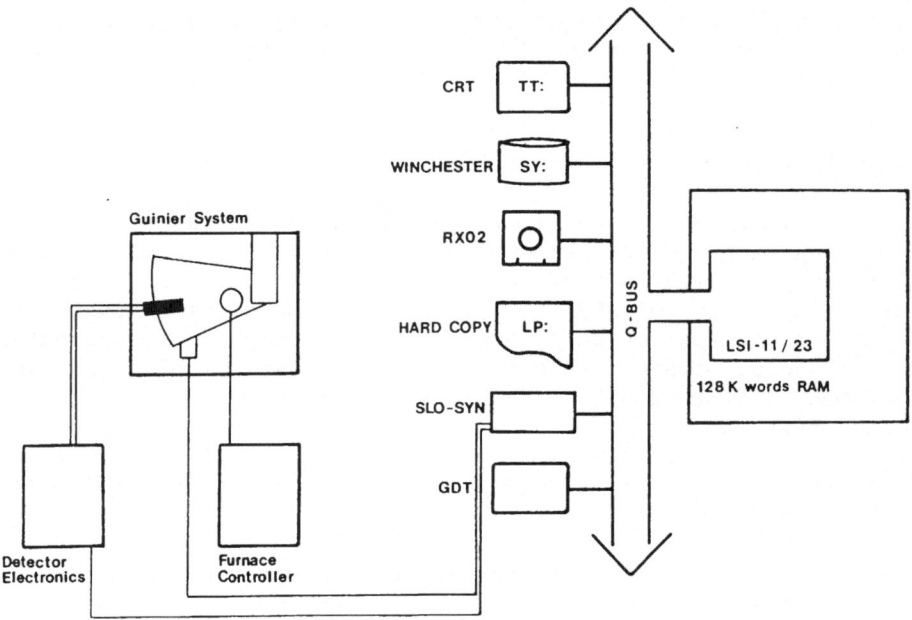

Fig. 1. Computerized X-ray system schematic.

A closer look at the diffraction system is given in Fig. 2. This figure details the degrees of freedom associated with the diffraction table itself. These include table tilt, height and position, the table zero, and the 'b' or monochromator to crossover distance adjustment.

The monochromator also has several degrees of freedom. On the monochromator manufactured by Huber, these include the 'a' or target to crystal distance, ϕ_1 and ϕ_2 positional axes and the θ or primary crystal rocking axis (Fig. 3).

Alignment of the optical system proceeds in two stages. The first stage of alignment, covered in the manufacturer-supplied operating manual, consists of roughing in the parameters mentioned

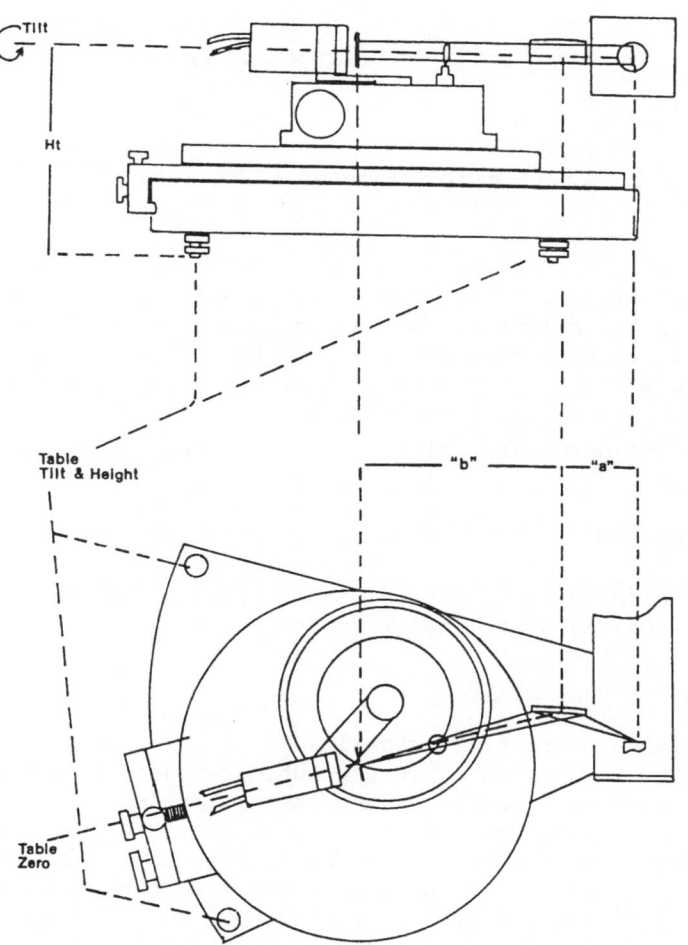

Fig. 2. Diffraction table alignment parameters.

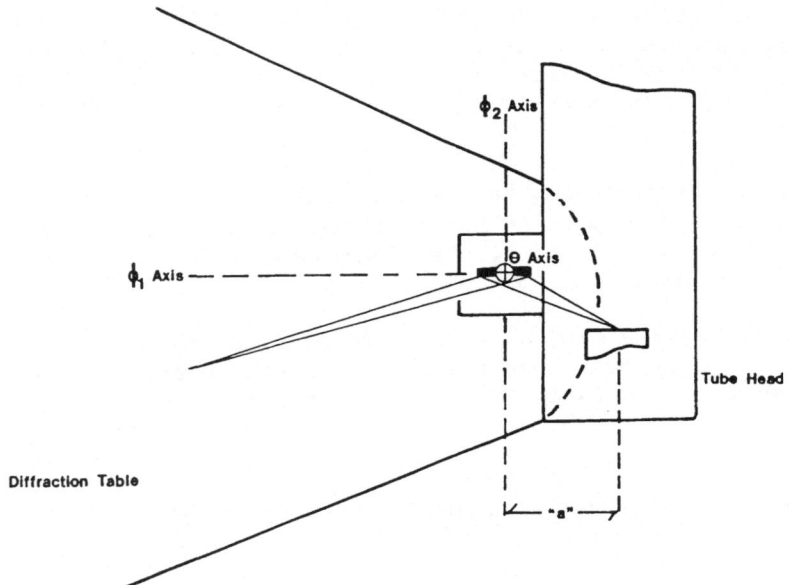

Fig. 3. Monochromator alignment parameters.

above using the alignment tools and fluorescent screens provided.
The second stage of alignment can be regarded as fine tuning the
system for maximum performance utilizing the detector system and
data reduction software. In both stages, it is the undeviated
$0°2\theta$ K_α or K_{α_1} from the monochromator which is examined. No
sample is present during the alignment procedure.

STAGE ONE ALIGNMENT

     Comments offered in this section are meant to augment informa-
tion already supplied by the manufacturer.

Roughing-in the Monochromator

     The manufacturer-supplied manuals discuss the general procedure
for mounting the monochromator on the tube head and roughly setting
the various adjustments on the monochromator housing. Before
mounting the monochromator, the operator should assure himself that
a long fine-focus X-ray tube is mounted in the tube head and that
the monochromator will reside at one of the line focus windows. It
is not possible to achieve crossover if a spot focus X-ray source
is used. The size of any spacing blocks supplied by the manufacturer which are designed to reside between the tube window and the

back of the monochromator should be checked to insure that the
total distance from the focal spot to the monochromator crystal
("a" distance), as designated by the manufacturer, is achievable
within the range of adjustment on the monochromator housing when
the spacing block is used. The size of these blocks is dependent
upon the monochromator crystal type and some adjustment in the
size of the spacing block may be required, depending on the type
of tube head used or the presence of any safety devices on the
windows.

Having performed these preliminary tasks, assemble the spacing
block and mounting plates on the tube head. Before mounting the
monochromator itself, it is advisable to open the exterior shutters
and run ϕ_1 and ϕ_2 through their full range of travel to insure that
both adjustments operate freely. Leave ϕ_1 and ϕ_2 at the mid-range
of their travel. Inspect the crystal for any damage and adjust the
crystal θ-axis so that the crystal will be roughly parallel to the
incoming X-rays when mounted in place. Push the monochromator onto
the mounting assembly on the tube head and run it in on the lateral
adjustment screw. Follow the recommended procedure for lateral
centering of the monochromator housing.

Set the distance between the center of the top tube window and
the θ-axis marker to the manufacturer's specified value of "a" with
a ruler,using the adjustment provided for this purpose on the
housing. Rock the crystal on its θ axis until the $K\alpha_1$, $K\alpha_2$ doublet
is seen on the fluorescent screen.

Roughing-in the Diffraction Table

The diffraction table base can now be set in position under the
tube head. The Huber base has an index pointer incorporated into
the base which is designed to allow accurate positioning of the
front end of the table under the monochromator. The table is physically positioned so that the top of the pointer, when raised, fits
perfectly into the bottom θ-axis marker on the monochromator housing. The positioning of the two rear feet of the table base is not
critical since the table provides ample secondary adjustment, but
it is convenient to locate the true zero two theta (θ_0) direction
in the approximate center of the table zero adjustment. Set the
"b" distance, as indicated on the scale in the table, to the manufacturer's recommended value.

An alignment bridge and several fluorescent screens are supplied
by Huber to assist in the next few steps. It should be remembered
at this point that the monochromator is aligned first and then the
diffraction table base is aligned to the monochromator. It is
therefore pointless to expend effort adjusting the table before the
crystal is shown to be in good alignment.

"a" distance check. Install the alignment bridge on the diffraction base. Place a small fluorescent screen at the end of the
bridge closest to the monochromator. The screen should be turned
away from the incoming X-ray beam so as to intercept it at a shallow
angle (Fig. 4). With the generator at minimum power and the shutter
open, rock the monochromator crystal through the $K\alpha_1$, $K\alpha_2$ maxima
several times and observe the image on the screen. Check to see
that the intensity distribution behaves as described in the manual,
that is, the intensity should appear in the center of the screen
and sweep symmetrically off the top and bottom of the screen. It
should not move from side to side, nor corner to corner as the
crystal is rocked. If the "a" distance has been set as previously
recommended, the crystal should be quite close to the proper setting
and no further adjustment until final alignment should be necessary.
If the "a" distance appears to be badly mis-set, recheck the crystal
to target distance with a ruler.

It is instructive to deliberately mis-set the "a" distance by
several turns of the adjustment screw on either side of the correct
setting to observe the effect on the image on the fluorescent screen.
This also helps establish that crossover can be reached from either
side, which should be possible if the monochromator, spacing block
and mounting plates have been properly assembled. Leave the monochromator at the correct "a" distance.

ϕ_1, ϕ_2 checks. Close the shutters and remove the small screen
from the bridge. Place the 6° screen in its receptacle on the
alignment bridge. Turn it so as to intercept the θ_0 beam at a shallow angle. Open the shutters and observe the image on the screen as

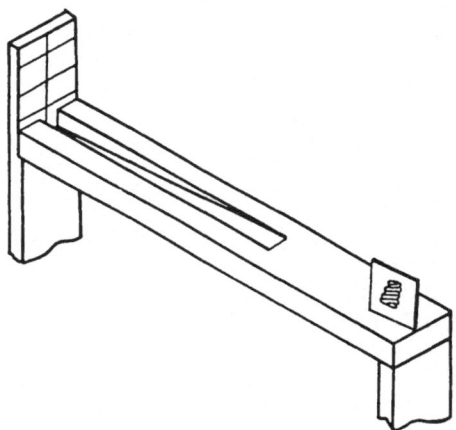

Fig. 4. Placement of front auxiliary screen
during "a" distance check.

the crystal is again rocked on its θ axis. With the θ axis adjusted
to provide maximum intensity, make whatever table adjustments
(table-zero and "b" distance) necessary to get the image centered
on the screen. With the intensity maximized, $K\alpha_1$ and $K\alpha_2$ should be
clearly resolved on the 6° screen. It is generally not possible to
resolve $K\alpha_1$ and $K\alpha_2$ over the entire diagonal length of the fluorescent screen, but resolution over half the screen should be
achievable. Check to see that the two lines are resolved and do
not appear to be twisted about each other or fan asymmetrically
either to one or the other corner of the screen. With ϕ_1 and ϕ_2
adjusted to the center of their travel it is generally not necessary to make further adjustments on these axes. This step is provided merely as a check to assure that no gross misadjustment of
the crystal press has occurred. Leave the crystal at a position
of maximum intensity.

Final Table Adjustments

Following the manufacturer's recommended procedure, the table
height and tilt should now be adjusted. The proper height is easily
achieved by synchronizing the image of the defocused θ_0 primary
beam from the crystal at two screens on the front and back of the
alignment bridge. To adjust the table tilt, we have found it convenient to utilize the 6° screen, in a flat or face-on position.
Utilizing the table zero adjustment, swing the table back and
forth so that the focal line repeatedly crosses the vertical center
scribe mark on the 6° screen. Because no fluorescent material is
present in the scribe mark, the beam will be seen to disappear as
it passes across the mark. Observe carefully how the beam disappears. When the table tilt is correct, it should extinguish completely over its entire length as it encounters the scribe line.
If the table tilt is not correct, the image of the focal line will
disappear either from the top down or the bottom up, depending on
the sense of misadjustment. Adjust the two back feet of the table
accordingly until the scribe line is perfectly parallel to the
focal line. Recheck the table height after this adjustment is
completed.

When this point is reached it is advisable to mark the position of all three base feet on the table top in order to allow easy
recovery of the position of the diffraction base should it become
necessary to move it.

STAGE TWO ALIGNMENT

With the optical system now roughed in, the procedure for fine
tuning can be carried out. It is advisable to allow the X-ray
generator to stabilize at full operating power for several hours

before beginning this part of the alignment as the monochromator is extremely sensitive to small shifts in the position of the focal line on the anode.

The diffractometer assembly should be mounted onto the base plate and any axial limiting slits (Soller slits, etc.) should be removed. Open the receiving slit to $\sim$ .1 mm and introduce an attenuator between the monochromator and the detector. (Twenty mils (.02") of aluminum sheet is usually sufficient). The detector should be driven to 0° and the table zero adjusted so that the θ_0 primary maximum intensity is located at 0°2θ. This is accomplished by observing the θ_0 signal on the detector system ratemeter while adjusting the table zero. Perform a wide scan (-1° to +1°) using a .005° step size and a 1 second acquisition time. The acquired profile should show Kα_1 and Kα_2 clearly resolved.

"a" Distance Final Adjustment

Close down the receiving slit to .025 mm (halfway between completely closed and the first unmarked division on the receiving slit adjustment dial). Fine tune the "a" distance by repeatedly step scanning between -.2° and +.2° 2θ, readjusting "a" by 1/4 turn each time. After each scan, recover the absolute integrated intensities, full-width at half maximum and peak separations for Kα_1 and Kα_2, utilizing the peak-profile fitting software supplied by the vendor. If this software is unavailable, these values can be recovered from the displayed peak traces on either a graphics terminal or a strip chart recorder; however, this greatly increases the amount of time necessary to perform the fine-tuning aspects of the alignment. We have found that a 1/4 turn displacement in the "a" adjustment can lead to a 5-6% loss of intensity in Kα_1. The final "a" setting should be chosen to maximize intensity and minimize the FWHM. The separation between Kα_1 and Kα_2 ($\Delta 2\theta$) is relatively insensitive to small errors in "a" and will deteriorate only if errors as large as 1 full turn or more occur.

"b" Distance Final Adjustment

The "b" or crystal-to-crossover distance is dictated by the manufacturing specifications of the monochromator crystal. It is the "focal length" of the monochromator. As such, it is necessary that the receiving slit be exactly "b" millimeters away from the θ axis of the crystal for maximum resolution and intensity.

This test is based on the resolution of the Kα_1-Kα_2 doublet, therefore the θ-axis of the monochromator should remain, as in the previous steps, at a position which provides maximum intensity. Using the adjustment provided on the diffraction base, set the detector 6mm outside the focussing circle (manufacturer's recommended value plus 6mm). Note that the apparent position of the θ_0

primary will shift every time "b" is changed, so that it will be
necessary to continually recenter θ_o at the table zero. This can
be done by eye, as before, using the ratemeter. Scan from -.2° to
+.2° 2θ as before, stepping the "b" distance in one millimeter at a
time until the receiving slit is an indicated 6mm inside the focussing circle. The results should resemble those in Fig. 5. It will
be apparent from these scans that misadjustments in "b" cause rapid
deterioration of resolving power and loss of intensity. Note that
the separation, $\Delta 2\theta$ between $K\alpha_1$ and $K\alpha_2$ is not affected. We have
chosen to utilize the ratio between the peak count rate of $K\alpha_1$ and
the minimum residual count rate at the trough between $K\alpha_1$ and $K\alpha_2$
as the parameter to maximize during this step. This parameter is
sensitive to 2% for a misset of 1mm close to the correct "b"
setting.

Table Tilt Final Adjustment

Colinearity of the receiving slit and focal line must be established to insure the best possible resolution. A final check on the
table tilt can be carried out utilizing the FWHM of the $K\alpha_1$ line
from the monochromator. Insert a 3° axial divergence limiting slit
immediately ahead of the receiving slit in the carrier provided. If
Soller slits are to be part of the desired final configuration, introduce them at this time. Scan the $K\alpha_1$, α_2 doublet as before and calculate the FWHM of the $K\alpha_1$ component. Adjust the table tilt in steps
of the 1/2 turn making sure to compensate the table height by drop-
ping one back foot of the table the same number of turns as the
other is raised. The effect on the half-width of the $K\alpha_1$ line should
be noticeable within 1 full turn of the adjustment feet on the base.
Calculate the $K\alpha_1$ FWHM after each adjustment. Position the table so
as to minimize this value. (See Fig. 6).

Monochromator Rocking Curve - Final Adjustment

The final $K\alpha_1$ profile can now be selected. Utilizing a starting point of maximum intensity, begin rocking the θ axis of the crystal away from $K\alpha_2$ (counterclockwise when viewed from the front of
the monochromator housing) in very small increments. Reset the table
zero as necessary and scan the θ_o primary. Repeat this process until
all remnants of $K\alpha_2$ have been tuned out of the θ_o signal but as much
of the $K\alpha_1$ intensity as possible is retained. If the crystal is
rocked too far away from $K\alpha_2$, a small $K\alpha_{3-4}$ peak will appear on the
high 2θ side of $K\alpha_1$. This is also undesirable, and should be
avoided. This process is shown graphically in Fig. 7, which details
the rocking curve behavior of the Ge monochromator used in this
laboratory during this phase of the alignment procedure. The final
θ position is correct when no trace of $K\alpha_2$ remains and the intensity
of $K\alpha_1$ is maximized.

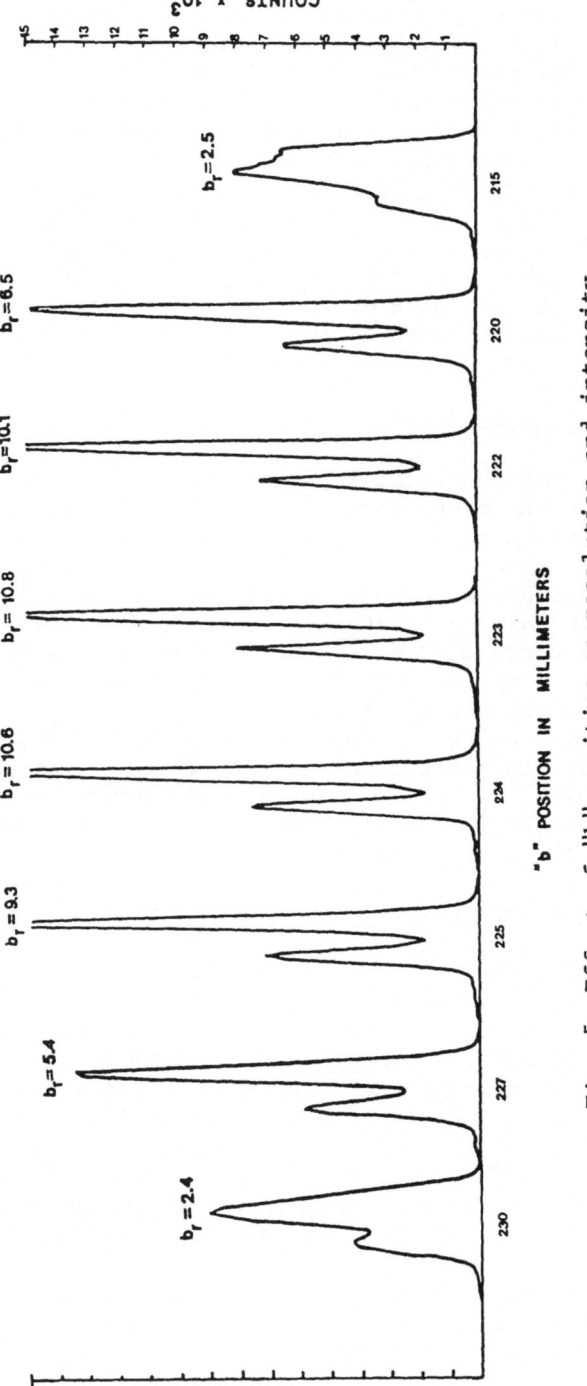

Fig. 5. Effect of "b" position on resolution and intensity.

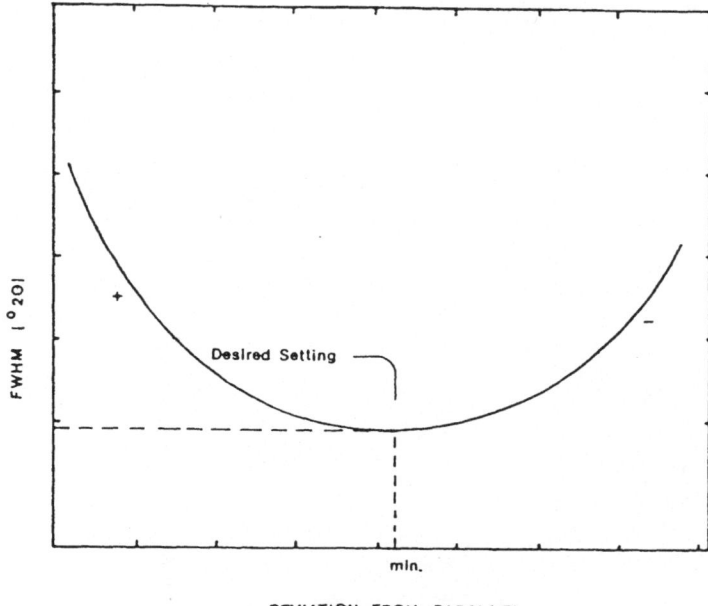

Fig. 6. Effect of table tilt on $K\alpha_1$ half-width.

Receiving and Divergence Slits

The setting of the slit configuration will depend upon the
immediate application of the X-ray system. For high resolution
work with the large, oscillating sample holder, the slit system can
be shut down when intensity is not at a premium. Utilization of
the goniometer head and platinum loop, however, requires that the
slit system be configured for intensity maximization, not minimization of peak half-widths. It is recommended that the user collect
data for a calibration chart such as that shown in Figure 8. A plot
of maximum count rate and half-width vs. receiving slit width such
as the one shown will enable the intelligent operator to choose the
appropriate balance of receiving slit width, axial pre-slit window
and Soller slits for the immediate application at hand.

Table Zero

The final adjustment to be made is the positioning of the table
zero. This must be set to coincide with the θ_0 primary $K\alpha_1$ peak
position. Drive the detector to $0°2\theta$, scan the θ_0 primary by hand,
using the table zero adjustment. Maximize the θ_0 intensity on the
ratemeter. Now scan over the θ_0 primary from $-.2°$ to $+.2°$ 2θ,
utilizing the smallest step possible (usually $.0025°2\theta$). Note the
position of the peak maximum. Adjust the table zero until the position of maximum intensity falls in the $0.0000°$ step. (See Figure 9).

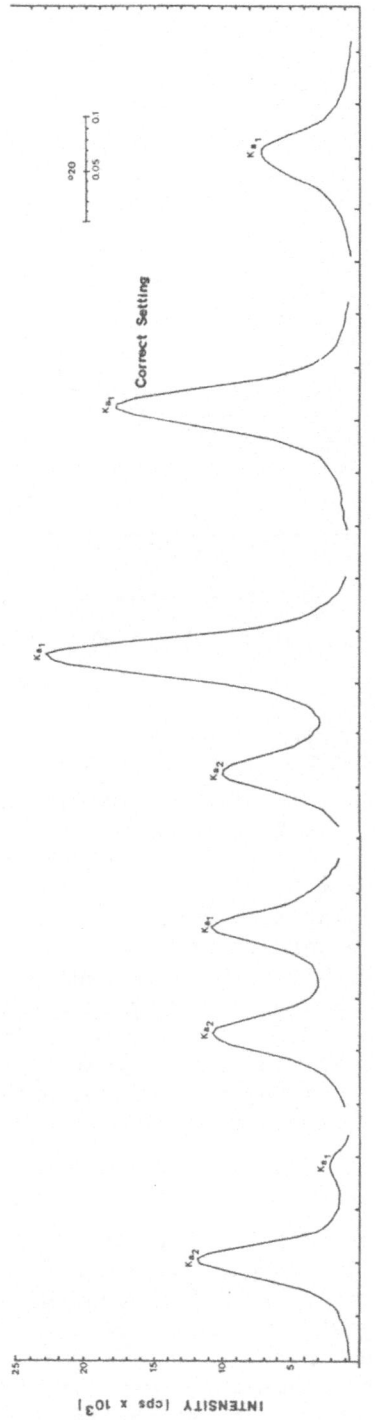

Fig. 7. Monochromator crystal rocking curve.

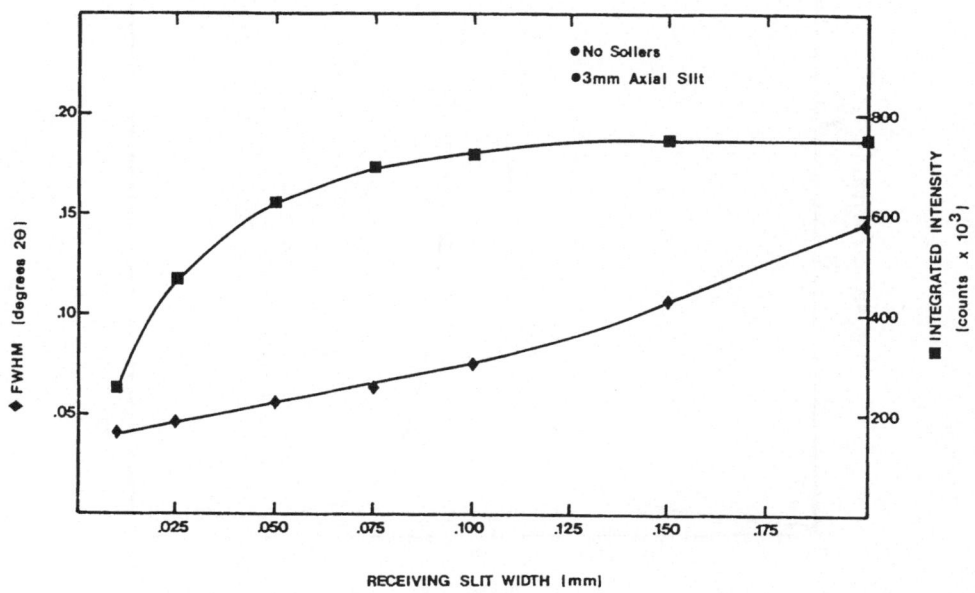

Fig. 8. Effect of receiving slit width on intensity and half-width.

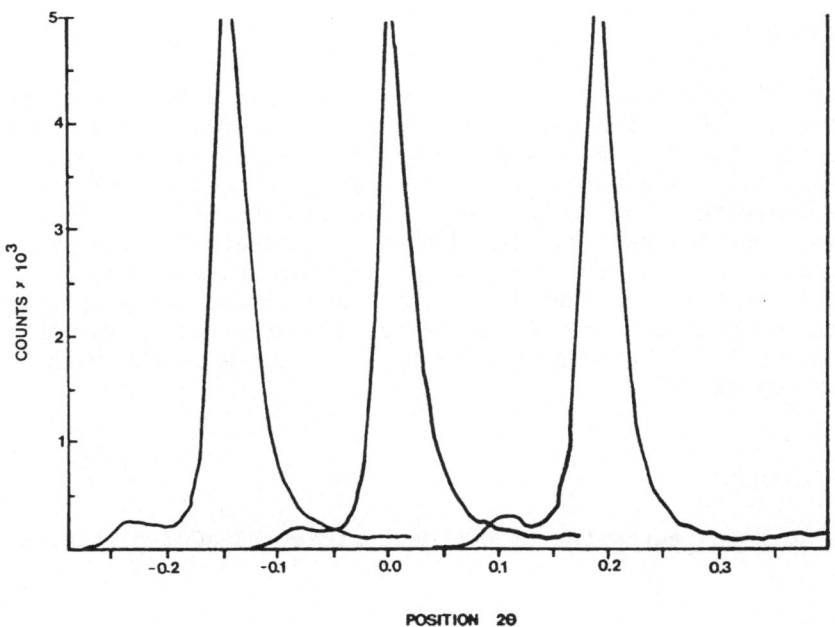

Fig. 9. Final table zero adjustment.

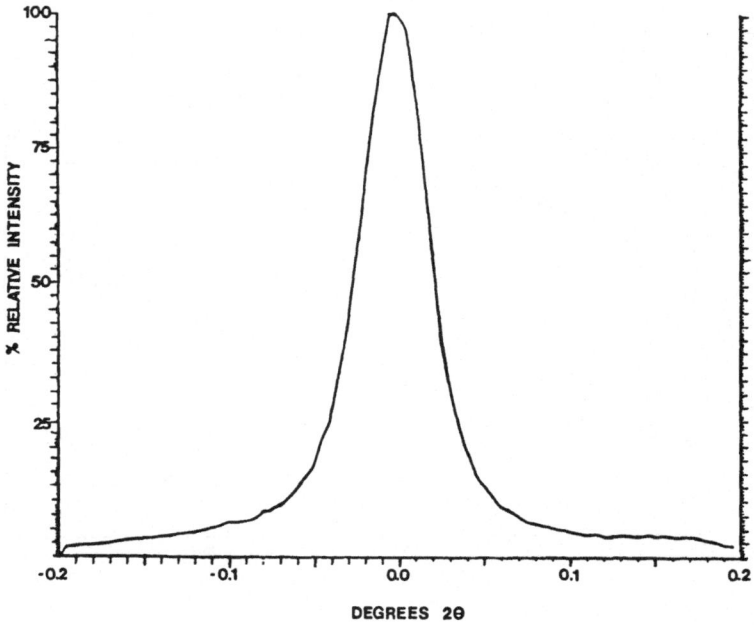

Fig. 10. Final $K\alpha_1$ profile.

MAINTENANCE

After the procedure outlined above has been carried out, the alignment should periodically be checked. A record of the final $K\alpha_1$ profile should be kept for comparison during routine maintenance (Fig. 10). The parameters most likely to need a weekly touchup are the table zero and the rocking curve adjustment. If the generator has been powered down and restarted, a check of these two parameters is essential. The X-ray generator should always be allowed to stabilize for at least two hours before changing any of the table or monochromator adjustments. If high accuracy work is underway, cycling the generator during the course of the work is not recommended.

ACKNOWLEDGEMENT

This work was supported by DOE on contract DE-ACO2-81ER10896.

ENERGY DISPERSIVE DIFFRACTION IN A DIAMOND ANVIL HIGH PRESSURE CELL USING SYNCHROTRON AND CONVENTIONAL X-RADIATION

David R. Black, Carmen S. Menoni, and Ian L. Spain

Department of Physics, Colorado State University
Fort Collins, Colorado 80523

INTRODUCTION

A wide range of structural studies have been carried out in high pressure diamond anvil cells using x-rays. The most common experimental geometry is shown in Fig. 1a. The incident x-ray beam passes axially through the first diamond and enters the sample, typically 100-300 µm in diameter and 20-100 µm thick; the diffracted x-rays exit via the second diamond. Energy-dispersive detection techniques (EDXRD) have been used.[1] However the intensity of diffracted radiation from the sample is weak, so that typical exposure times with a conventional, fixed anode, x-ray source are typically one to several days.[1] Accordingly, higher intensity radiation from synchrotron sources has been used for these experiments.[2-5]

In this paper we compare the Energy Dispersive X-ray Diffraction experiment performed with an optimized conventional source to that performed with a synchrotron source. The effect of the synchrotron beam on sample heating is also considered.

DIFFRACTION OPTIMIZATION USING CONVENTIONAL FIXED ANODE SOURCE

The laboratory diffraction geometry is shown schematically in Fig. 2. The source slit of width w_s and height $2h_s$ defines the area A' on the anode from which x-rays can reach the sample. This slit also defines the horizontal and vertical divergence angles $\Delta\theta_i$ and $2\Delta\phi_i$, respectively. The diffracted information is collected by a solid state detector, fixed at an angle $2\theta_0$. The detector slit defines the horizontal and vertical divergence

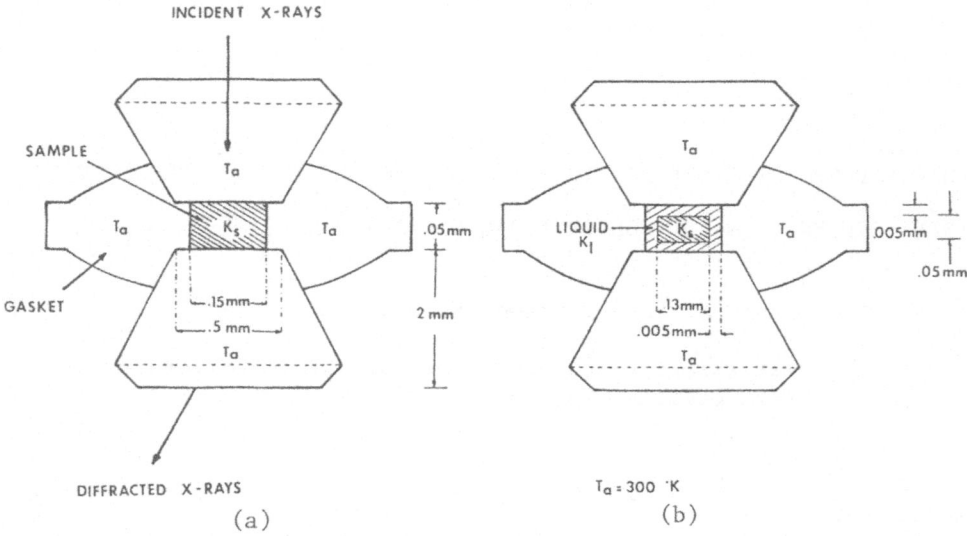

Fig. 1.   Standard geometry for x-ray diffraction in a diamond
          anvil cell. Different configurations considered in
          the calculations of the temperature rise (a) only
          sample between the diamonds, (b) sample immersed in a
          fluid.

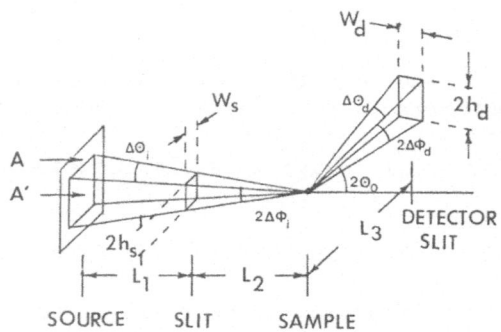

Fig. 2.   Schematic of the laboratory arrangement,
          with nomenclature used in the text.

angles $\Delta\theta_d$ and $2\Delta\phi_d$. For larger values of $\Delta\theta_i$, A' is increased and more photons can reach the sample. Similarly, if $\Delta\theta_d$ is increased, the count rate also increases. However, $\Delta\theta_i$ and $\Delta\theta_d$ also cause broadening of the diffraction lines. Therefore a compromise between count rate and resolution must be made. We have chosen a value of $\Delta\theta_i = \Delta\theta_d = 2 \times 10^{-3}$ radians. This value is chosen such that the main contribution to the broadening of the diffraction peak is due to the detector electronics.[6] The broadening due to the vertical divergence $2\Delta\phi_i$ and $2\Delta\phi_d$ is asymmetric and shifted to the low energy side. If the source slit is divided into M intervals in the vertical direction indexed by m, and the detector slit is divided into N intervals indexed by n, then the correction to the diffraction angle $2\theta_o$ from the element m to n is given by:

$$\cos 2\theta'_{n,m} = \cos 2\theta_o \left[1 - \Delta\phi^2 (\frac{n^2}{N^2} + \frac{m^2}{M^2} - \frac{2nm}{NM\cos 2\theta_o}) \right] . \qquad (1)$$

The diffraction profile can then be obtained by summing over the indices. We find that a value of $\Delta\phi = 0.025$ radians causes little additional broadening, increasing the FWHM from 300 to 350 eV, for example, and shifts the peak downward in energy a fractional amount.

It is interesting to compare this optimized system to that utilizing synchrotron radiation. The comparison is given by the ratio of the rates at which diffraction information can be collected. Some workers claim that ratios as high as 10^6 should result, whereas values $\sim$200-5000 are obtained experimentally.

In the conventional case, the number of photons/sec incident on the sample in a 1% bandpass is given by:

$$\dot{N}_{LAB} = D(\varepsilon) \frac{I}{q} \frac{S}{A'} \frac{\varepsilon}{100} T(\varepsilon) 2\Delta\phi_i \Delta\theta_i \qquad (2)$$

where $D(\varepsilon)$ is the number of photons/incident electron/steradian/eV,[7] I is the x-ray tube current, q is the electronic charge, S is the sample area perpendicular to the beam, A' is the anode area, ε is the energy, $T(\varepsilon)$ is the transmission of the first diamond and $2\Delta\phi_i$, $\Delta\theta_i$ are the vertical and horizontal divergence angles. The rate at which diffraction information is collected is then given by:

$$R_{LAB} = \dot{N}_{LAB} f(\theta,\varepsilon) T_d(2\theta,\varepsilon) 2\Delta\phi_d \Delta\theta_d \qquad (3)$$

where $f(\theta,\varepsilon)$ is a function that describes the diffraction process,

$T_d(2\theta,\varepsilon)$ is the transmission of the second diamond and $2\Delta\phi_d$, $\Delta\theta_d$ are the vertical and horizontal divergence angles defined by the detector slit.

In the case of a synchrotron source the x-rays form a slightly diverging beam which impinges upon the diamond cell. A collimator is used so that only part of the sample is illuminated. The number of photons/sec reaching the sample can be determined from the spectral distribution function $G(\varepsilon)$ in photons/sec/mrad/ma/10% bandpass[8] and the experimental geometry. The angular dispersion in the vertical direction is given by γ^{-1} where $\gamma = E/mc^2$ and m is the electron rest mass, c is the speed of light, and E is the electron energy. The sample subtends a horizontal angle given by $2r/R$ in radians, where r is the radius of the sample and R is the distance to the source. Therefore, the number of photons/sec incident on a sample of area S in a 1% bandpass is:

$$\dot{N}_{sync} = 100\ G(\varepsilon)\ \frac{I\gamma S}{R^2}\ T_d(\varepsilon) \tag{4}$$

where I is the synchrotron current and $T_d(\varepsilon)$ is the transmission of the first diamond. Again the rate at which diffraction information is collected is given by:

$$R_{sync} = \dot{N}_{sync} f(\theta,\varepsilon) T_d(2\theta,\varepsilon) 2\Delta\phi_d \Delta\theta_d\ . \tag{5}$$

The comparison between synchrotron and conventional sources is given by:

$$\alpha = \frac{R_{sync}}{R_{LAB}}\ . \tag{6}$$

A plot of α for both CHESS and SPEAR is given in Fig. 3. Table 1 gives typical values of parameters used in the calculations. It should be noted that no detector limitations have been considered in Eq. (6). If the sample under consideration is an efficient scatterer it is likely that the synchrotron count rate will exceed that of the detector and, therefore, the synchrotron experiment will be detector-limited.

In the preceding discussion we have shown that a compromise is needed between increasing count rate and decreasing resolution. An analysis of the line broadening due to both horizontal and vertical angular divergence was given and acceptable values stated. A comparison between the conventional and synchrotron experiments was made and this showed that the synchrotron should be $\lesssim 5000$ times faster than the conventional case. However, detector limitations can substantially reduce this value.

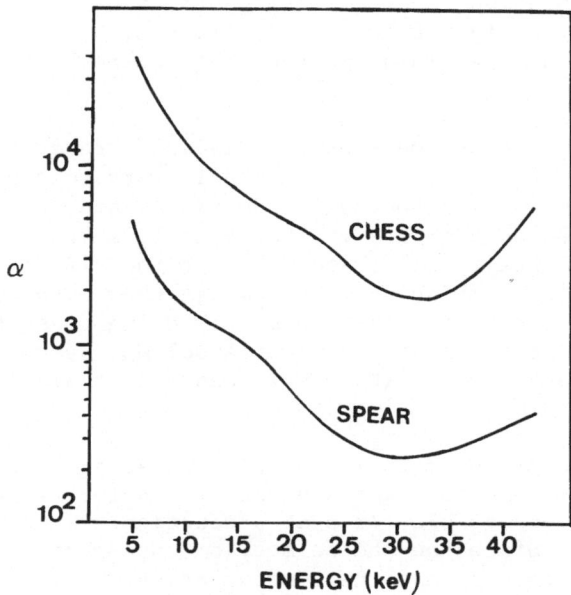

Fig. 3.   The variation of the effective ratio α,
          as a function of x-ray energy for CHESS
          and SPEAR.

Table 1.   Typical Parameters used in Calculations

	Conventional, Fixed-Anode X-Ray Source*	Synchrotron Source CHESS	SPEAR
Beam energy	45 kV	5 GeV	3.5 GeV
Beam current	40 mA	10 mA	30 mA
Incident beam $\Delta\theta_i$	2.0×10^{-3}	3.2×10^{-6}	5.8×10^{-6}
Divergence (rdns) $\Delta\phi_i$	2.5×10^{-2}	9.8×10^{-5}	15×10^{-5}
Diffracted beam $\Delta\theta_d$	2.0×10^{-3}	2×10^{-3}	0.25×10^{-3}
Divergence (rdns) $\Delta\phi_d$	2.5×10^{-2}	3×10^{-2}	4×10^{-3}
Source to sample distance R	–	11.5 m	17.3 m
Sample to detector	100 mm	500 mm	600 mm

*Tungsten source FAAQ 60 tube.

HEATING EFFECTS OF THE SYNCHROTRON BEAM

In the calculations of the heating effect of a high photon flux, two sample configurations are considered; one in which the sample is immersed in a fluid and the other where no fluid is present (Fig. 1).

In our model, we assume that the diamonds are at a constant temperature $T_a = 300°K$. This assumption is very reasonable since diamonds have a very high thermal conductivity and are relatively transparent to photon energies greater than 10 keV. Since the incident beam is usually collimated down to the size of the sample, the gasket is not absorbing photons and can therefore be considered to be at constant temperature T_a. Using these boundary conditions the heat transfer equation is solved, assuming that the total power absorbed by the sample can flow radially and axially.

The total power absorbed is calculated taking into account that the inelastic scattering processes (photoelectric and Compton scattering) are responsible for heat production. We consider the sample to be Ge. Since the photoelectric cross-section for this material is much larger than the Compton cross-section in the energy range 0-70 keV, we have not considered this latter process.

The number of photons reaching the sample was calculated using expression 4. The calculations were performed numerically, and the photoelectric cross-sections were computed using the fitting parameters given in the McMaster tables.[9] Standard operating conditions were considered at SPEAR, (3.5 GeV, 30 mA) and projected at CHESS (8 GeV, 100 mA). The sample was located 17.3 m and 12.0 m from the electron beam at SPEAR and CHESS, respectively. For the beam conditions chosen, the power absorbed by a sample of area 1 mm^2 and thickness 0.05 mm is P = 0.10 W/mm^2 for SPEAR and P = 5.4 W/mm^2 for CHESS. The Ge sample absorbs approximately 75% of the SPEAR beam energy reaching it after passing through a 2 mm thick diamond, while the absorption is approximately 84% for CHESS. A temperature rise of .5°K is expected for CHESS, and approximately .01°K for SPEAR when only the solid sample is considered. However, many experiments are carried out using ethanol:methanol mixtures, for which the thermal conductivity in the liquid state, is $K_\ell \sim 0.15$ W/m°K.[10] The idealized case depicted in Fig. 1(b) is considered. In this case the predominant heat loss mechanism is found to be by conduction rather than convection, since the Rayleigh number is approximately 80 if a temperature difference of 100°K between the sample and the diamonds is assumed. A critical value of 1700 is considered the threshold value for predominantly convection loss.[11] With these considerations, a temperature rise of 1°K is obtained for SPEAR, and 52°K for CHESS, with only a small temperature gradient across the sample.

The foregoing analysis shows that heating in the diamond anvil cell may be more serious than has been allowed for previously when using radiation intensity levels projected at CHESS. These effects are more important when the sample is immersed in a fluid. At pressures in excess of 100 kbar, where the usual ethanol-methanol mixture used in these experiments freezes, the thermal conductivity of the mixture approximates to the value of the thermal conductivity of the sample. In this situation heating effects are small.

REFERENCES

1. E. F. Skelton, C. Y. Liu, and I. L. Spain, "Simple Improvements to a D.A.H.P. Cell for X-Ray Diffraction Studies," High Temp.-High Press. 9:19-26 (1977).
2. I. L. Spain, S. B. Qadri, C. S. Menoni, A. W. Webb, E. F. Skelton, "Structural Studies at High Pressure and Temperature Using Synchrotron Radiation," pp. 73-80, in: "Physics of Solids Under Pressure," J. S. Schilling and R. N. Shelton, eds., North Holland Publ. Co., Amsterdam (1981).
3. A. L. Ruoff and M. A. Baublitz, "Physics of Solids Under Pressure," pp. 81-90, J. S. Schilling and R. N. Shelton, eds., North Holland Publ. Co., Amsterdam (1981).
4. M. A. Baublitz, V. Arnold, and A. L. Ruoff, "Energy Dispersive X-Ray Diffraction from High Pressure Polycrystalline Specimens Using Synchrotron Radiation," Rev. Sci. Inst. 52:1616-1624 (1981).
5. E. F. Skelton, S. B. Qadri, A. W. Webb, C. W. Lee, and J. P. Kirkland, "Improved System for Energy-Dispersive X-Ray Diffraction with Synchrotron Radiation," Rev. Sci. Instr. 54:403-409 (1983).
6. B. Buras, J. Staun Ölsen, L. Gerward, G. Will, and E. Hinze, "X-Ray Energy-Dispersive Diffractometry Using Synchrotron Radiation," J. Appl. Cryst. 10:431-438 (1977).
7. D. B. Brown, J. V. Gilfrich, and M. C. Peckerar, "Measurement and Calculation of Absolute Intensities of X-Ray Spectra," J. Appl. Phys. 46:4537-4540 (1975).
8. H. Winick and S. Doniach, "Properties of Synchrotron Radiation," Plenum Press, New York (1979).
9. W. H. McMaster, N. Kerr Del Grande, J. H. Mallett, and J. H. Hubbell, "Compilation of X-Ray Cross Sections," National Bureau of Standards, UCRL-50.74, Sec. II, Rev. 1.
10. CRC Handbook of Chemistry and Physics, 63rd edition, (1982).
11. B. Gebhart, "Heat Transfer," McGraw-Hill, Hightstown, NJ (1961).

REFERENCE INTENSITY QUANTITATIVE ANALYSIS USING

THIN-LAYER AEROSOL SAMPLES

Briant L. Davis

Institute of Atmospheric Sciences
South Dakota School of Mines and Technology
Rapid City, South Dakota 57701-3995

INTRODUCTION

Interest in the reference intensity ratio (RIR) method continues as a result of the potential capabilities of the technique for rapid multi-component quantitative analysis. The theoretical basis for the RIR technique is now well established (Chung, 1974; Hubbard et al., 1976; Davis, 1980, 1981; Davis and Johnson, 1982). Major areas for which the method can still be greatly improved include the methods used for sample preparation, the measurement of accurate intensities, and the use of an internally consistent set of reference intensity constants (designated RIR, or k_i). In the methodology developed at the Institute of Atmospheric Sciences (IAS), South Dakota School of Mines and Technology, sample preparation centers about the suspension of the pulverized sample into an aerosol and collection onto filter media. Because of this step, intensities must be corrected from their raw values to intensities representative of "infinite thickness" and volumetrically constant conditions of normal sample diffraction. The measurement and correction of intensities and sample preparation methodology is the subject matter of the present paper.

AIR SUSPENSION METHOD FOR SAMPLE PREPARATION

The attainment of random orientation of crystallites throughout the diffracting volume of the sample under study is perhaps the most important key to successful application of the RIR technique. A comparison of the number of methods used for reducing preferred orientation has been presented by Calvert et al. (1983), with the most successful technique being the creation of spray-dried solid aggregates, even though this technique requires considerable preparatory effort.

The air suspension technique used by IAS involves simply reduction of the sample to particulate matter less than 10 μm diameter, followed by suspension of the particles into an aerosol in a 4- or 8-liter jar, followed by rapid collection of the particulate matter onto, preferably, a glass fiber substrate. When the technique was first proposed by Davis and Cho (1977), the mineral substance used as a test for randomness of particle orientation was calcite. The results of Davis and Cho are reproduced here in Table 1. In spite of the seemingly good results, it was concluded that calcite, in fact, is not as severe a test of the technique as is a platelike material such as a mica, or MoO_3. We here present in Tables 2 and 3 intensities obtained from aerosol suspensions of a Black Hills muscovite, and the National Bureau of Standards (NBS), MoO_3 (Calvert et al., 1983) for samples collected on glass fiber filters, in comparison with powder diffraction file (PDF) data and calculated data. In Table 2, no PDF data are given because of the highly variable nature of muscovite

TABLE 1:   Relative Intensities of Several $CaCO_3$ Peaks Obtained by the Air Suspension Chamber (ASC). [From Davis and Cho, 1977]

hkℓ	$100\ (I_{hk\ell}/I_{014})$		
	Calculated[*]	ASC	PDF[†]
012	8.1	8.0	12
014	100.0	100.0	100
006	2.3	2.0	3
110	14.2	12.2	14
113	20.9	17.4	18
022	15.6	15.2	18
116	22.7	22.5	17

[*]From Graf (1961).

[†]From Card 5-0586 in X-ray Powder Diffraction File (1970).

TABLE 2: Relative Integrated Intensities -- Black Hills Muscovite

hkl	2θ-OBS[‡]	I/I$_0$ IAS-OBS[*]		I/I$_0$ Calculated[†]	
		HV	GF/C	2M$_1$ (1)	2M$_1$ (2)
002	8.83	82.1	79.7	38.7	43.9
004	17.75	18.5	13.3	10.7	12.2
020,110,$\bar{1}$11	19.95	66.7	64.0	55.3	45.3
021	20.3		6.6	3.3	6.3
111	20.74	9.4	9.2	11.3	10.0
022	21.70	6.2	5.5	8.7	10.9
112	22.51	3.1	4.2	4.7	3.2
$\bar{1}$13	23.00	18.9	22.5	27.3	23.5
023	23.90	21.5	23.8	28.0	23.1
113	24.96	2.0	3.6	1.3	0.9
$\bar{1}$14	25.58	29.9	39.6	36.7	33.0
024,006	26.78	68.3	74.9	57.3	64.3
114	27.95	31.9	36.6	35.3	33.5
$\bar{1}$15	28.71	2.4	2.0	3.3	2.7
025	29.96	37.4	32.2	38.0	35.8
115	31.35	16.6	25.8	26.7	24.4
$\bar{1}$16	32.10	14.9	18.6	20.7	19.9
$\bar{1}$31,200,116, $\bar{2}$02,131	34.9±	100	100	100	100
008,$\bar{1}$17	35.89	5.1	7.9	6.7	4.5
$\bar{1}$33,202	36.62	13.8	19.3	22.7	29.4
$\bar{2}$04,113,027	37.81	18.5	16.5	22.0	24.9

[*]Suspension on Hi-Vol (HV) glass fiber substrate and on Whatman (GF/C) glass fiber substrate.

[†]Borg and Smith, 1969.   [‡]CuKα radiation.

TABLE 3:   Relative Intensities -- MoO_3

| | | | $100\ I/I_o$ | | | | |
| | | | IAS Filter | | Calculated* | | |
hkl	2θ-deg (CuKα)	PDF[†]	PH	INT	PH	INT	SD[‡]
020	12.77	34	30	28	36	31	--
110	23.35	82	67	65	76	73	72
040	25.72 ⎫	61	42	60	38	34	39
210	25.91 ⎭				31	25	--
021	27.36	100	100	100	100	100	100
130	29.72	13	6	5	7	7	--
101	33.15	19	13	13	15	15	--
111	33.76	35	23	23	26	27	26
140	34.40	6	1	1	3	3	--
041	35.52 ⎫	12	7	7	7	7	--
121	35.66 ⎭				<0.7	1	--
131	38.61	12	7	7	8	8	--
060	39.01	31	21	23	21	22	19
150	39.69	18	9	11	13	14	--
141	42.42	9	2	4	7	8	--
160	45.44	4	1	2	2	2	--
200	45.78	13	7	8	8	8	9
210	46.30 ⎫	17	11	15	11	8	--
061	46.32 ⎭					8	--
151	46.96	--	0.4	0.5	1	1	--
220	47.74	--	0.3	0.4	1	1	--
002	49.28	21	15	20	15	17	15
230	50.09	11	4	8	9	11	--

*Data from C. R. Hubbard, National Bureau of Standards.

[†]JCPDS (Powder Diffraction File) No. 5-508.

[‡]Spray-dried aggregated (Calvert et al., 1983).

varieties; however, two calculated sets are provided from Borg and Smith (1969), representing the $2M_1(1)$ and $2M_1(2)$ structures. For this sample, the muscovite particle size was reduced by comminution with a file and the vertical airflow in the chamber adjusted to collect only particles under 10 μm diameter. The suspensions were completed on a precision scientific hi-vol glass fiber substrate and a Whatman GF/C glass fiber substrate, the latter representing a much finer matte of fibers than the former. Comparison of the various sets of data in Table 2 indicates that intensities from the GF/C preparation are, with few exceptions, in very good agreement with the $2M_1(2)$ structure intensities.

Perhaps the most severe test of preferred orientation comes with the preparation of MoO_3 (molybdite), which is an orthorhombic substance having a tabular habit with excellent pinacoidal cleavage. The sample used here was provided by Dr. Camden R. Hubbard of the National Bureau of Standards. After grinding under alcohol, the particulate matter was suspended and the large particles allowed to settle for approximately 30-sec duration before collection. This resulted in very nearly monodisperse particulate collection with optical examination revealing 6 μm average length and 2 μm average width in the two dimensions of optical view. Table 3 presents the results of the intensity measurements from the IAS Precision Scientific "hi-vol" glass fiber filter, compared to the powder diffraction file values and calculated intensities provided by Dr. Hubbard. Both peak height (PH) and integrated intensities (INT) are presented in the table. Also presented in the table are the peak height data for seven lines reported by Calvert et al. (1983) for the spray-dried (SD) aggregate preparation. The χ^2 test relative to the calculated set for the NBS data results in the value of 1.8, and for the IAS data, 3.9, indicating a somewhat better fit of the peak-height intensities for the spray-dried method, although this may not be necessarily so for the integrated intensity data.

The comparison of both peak height and integrated intensity data from the aerosol suspension method to the calculated data indicates that the air suspension technique is an appropriate one to use for sample preparation in quantitative analysis, and this is especially important in view of the simplicity of the air suspension procedure. The success of the method appears to be dependent upon the relative dimensions of the particles being suspended and the "pores" of the fibrous substrate. A schematic illustration of this interaction and the hypothesized buildup of particles beyond the fibrous substrate is illustrated in Fig. 1. In this two-dimensional illustration, tabular particles similar to those of MoO_3 are trapped in the fiber matrix of the filter in such a manner as to statistically attain random orientation. As the layer is built up to, and even beyond, the substrate surface, subsequent particles will maintain a certain degree of randomness because of the antecedent orientation

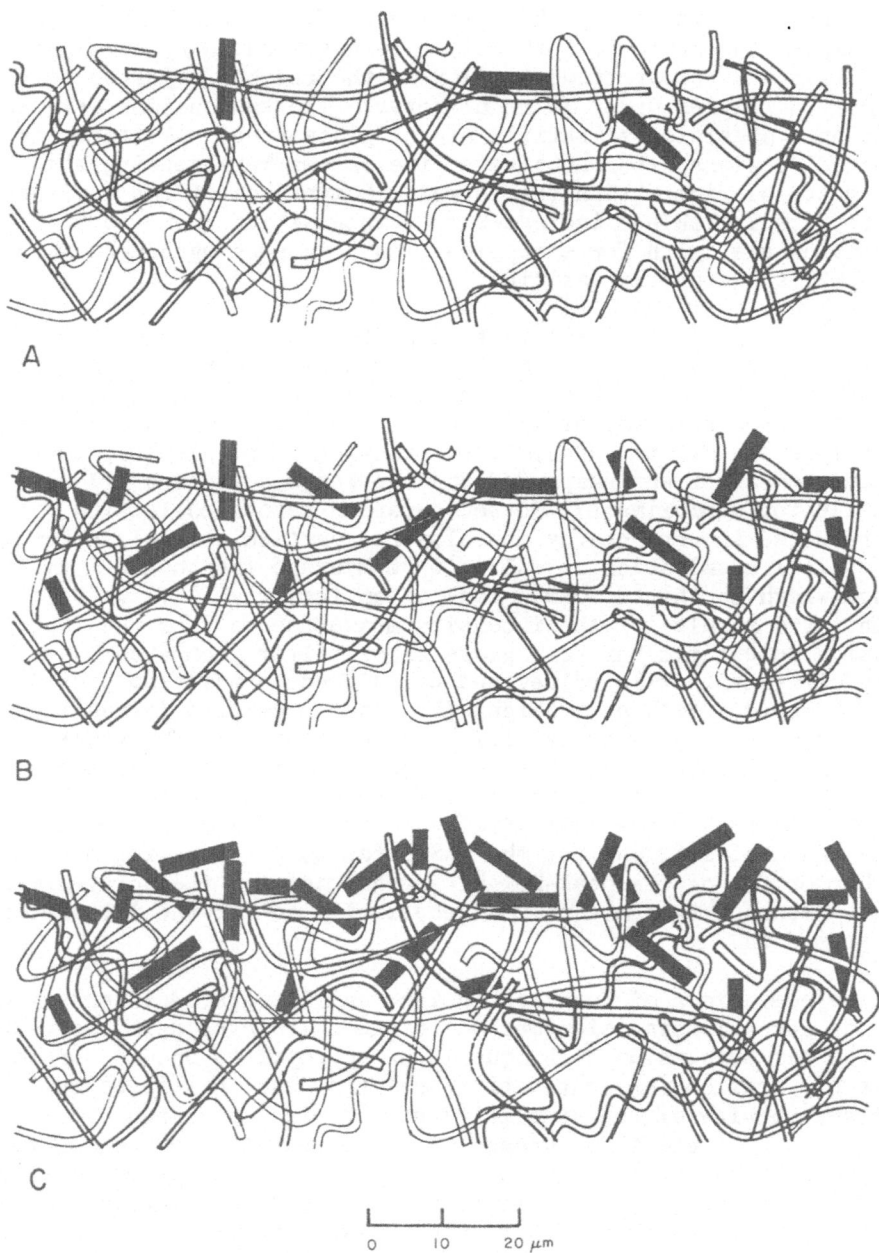

Fig. 1.   Schematic representation of particle entrapment within
         the glass fiber substrate. At stage C, the influence
         of earlier loaded particles on the orientation of
         later ones is illustrated.

set by particles trapped earlier in the fiber matte. The extent
to which this condition is maintained with increasing layer thickness beyond the substrate surface is conjecture, however, and
additional study is needed.

COMPONENT INTENSITY CORRECTIONS

The intensities obtained from the filter-collected aggregates
described in the previous section were corrected for transparency
and matrix effects before the relative intensities were calculated.
From the raw integrated intensities I_i^a, the corrected intensities
representing "infinite thickness" conditions are given by (Davis
and Johnson, 1982):

$$I_i^\infty = \frac{I_i^a \mu_H^* \xi}{W_B^H \mu_B^* \left[1 - e^{-2\mu_H^* \breve{M}_B / W_B^H \sin\theta}\right]} \tag{1}$$

where μ_B^* is the mass absorption coefficient of the sample material,
μ_H^* is the mass absorption coefficient of the sample and substrate
material down to the limit of penetration, W_B^H is the weight fraction
of the sample material in the sample-matrix composite, $\breve{M}_B$ is the mass
mass per unit area of sample on and within the substrate, θ' is half
the Bragg angle, and ξ is the volumetric factor to convert area-
constant intensities now obtained with use of a theta compensator to
intensities representing constant volume condition as a function of
Bragg angle θ. If we define the composite layer as one having all
pore space removed (yielding "effective" thickness), then the param-
eter W_B^H is given by $\breve{M}_B / (\breve{M}_B + t_f \rho_f)$, where t_f and ρ_f are effective
thickness and density of the substrate (filter) material,
respectively.

This very important equation has been derived (Davis, 1981) and
used for many years without adequate experimental testing. Since
this equation, in effect, converts the intensities for a very thin
layer to those represented by the "infinite thickness" case, the
best test of the equation would be to compare weight fractions in a
multi-component analysis obtained from several sample preparations
of the same bulk sample but with filter load weights $\breve{M}_B$ varying
from near the detection limit for minor components (approximately
150 μg cm^{-2}) to levels approaching that for infinite thickness
conditions.

For this test, we have taken a granitic rock from the southern
California batholith which has been extensively studied previously
(Davis and Walawender, 1982) and completed seven analyses of varying
weights, the results of which are illustrated in Fig. 2. In this

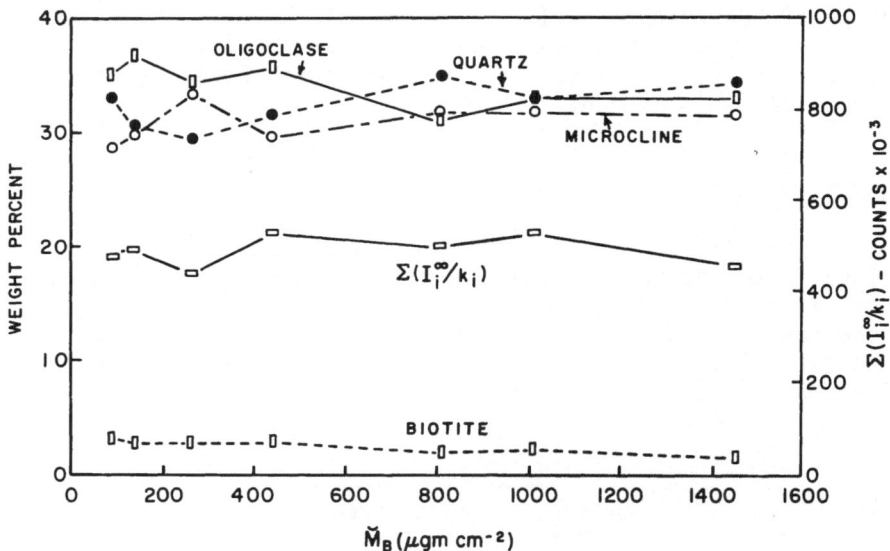

Fig. 2.   Variation of weight fractions of four components and
specific diffracting power with mass loading for a
granite rock sample loaded on glass fiber filters
and analyzed by the RIR procedure.

figure, we see that the various levels of W_i remain essentially the
same throughout the span of $\check{M}_B$ although, as expected, the scatter in
the values for individual W_i components become increasingly large
with smaller $\check{M}_B$. In addition to the four mineral components, the
quantity $\Sigma(I_i^\infty/k_i)$ is plotted as well since it represents the specific
diffracting power or, according to Chung (1974), the sample "unified
sensitivity." Therefore, if Eq. (1) is properly correcting the raw
intensities, all lines on Fig. 2 should be horizontal, including
the specific diffracting power.

The linear regressions for the five data sets of Fig. 2 were
completed to determine the component slopes. The slopes for the
four mineral components are -0.0011, 0.0027, 0.0011, and -0.0026
percent-cm^2 μgm^{-1} for biotite, quartz, microcline, and oligoclase,
respectively. The slope for the specific diffraction power is
7.6 count-cm^2 μgm^{-1}. Although the maximum value of $\check{M}_B$ in the series
falls short of the 0.019 gm cm^{-2} required for t^∞, it was not
experimentally feasible to work with heavier loads.

Although possible minor perturbations are apparent in these
data, we can conclude that the equation provides the proper corrections to the raw intensities for thin aerosol layers embedded in
fibrous substrates.

CONCLUSIONS

We have demonstrated in several studies with mineral particulate samples that preferred orientation of inequant particles can be greatly suppressed by loading the particles in the form of an aerosol onto an irregular fibrous substrate such as a glass fiber filter. Successful application of this technique to the formation of random aggregates has been shown for calcite, muscovite, and molybdite. Best results for this method will be achieved if mean particle size is no greater than 10 μ (diameter) in order that the individual particles be of the same order of magnitude in size as the voids in the substrate surface.

In all quantitative x-ray diffraction work, including the preparation of x-ray spectra where accurate intensities are required, the raw measured intensities obtained from filter preparations must be corrected for transparency and matrix effects. Experimental verification of the validity of the correction equation used in our laboratory in the past has been presented in the form of a series of analyses plotted as a function of specific sample loads collected on the filter substrate. The linear regression of the individual component variations with load resulted in slopes which were nearly zero in magnitude, as would be expected in such a verification.

ACKNOWLEDGMENTS

I am grateful to Dr. Camden R. Hubbard of the National Bureau of Standards for providing the samples and calculated intensities for molybdite. This research was supported by the State of South Dakota.

REFERENCES

Borg, I. Y., and Smith, D. K., 1969, Calculated x-ray powder patterns for silicate minerals, Geol. Soc. Amer. Mem. No. 122.
Calvert, L. D., Sirianni, A. F., Gainsford, G. J., and Hubbard, C. R., 1983, A comparison of methods for reducing preferred orientation, Adv. in X-ray Analysis, 26 (in press).
Chung, F., 1974, Quantitative interpretation of x-ray diffraction patterns of mixtures, J. Appl. Cryst., 7:519-531.
Davis, B. L., 1980, "Standardless" x-ray diffraction quantitative analysis, Atmos. Environ., 14:217-220.
Davis, B., L., 1981, A study of the errors in x-ray quantitative analysis procedures for aerosols collected on filter media, Atmos. Environ., 15:291-296.

Davis, B. L., and Cho, N-K., 1977, Theory and application of x-ray diffraction compound analysis to high-volume filter samples, Atmos. Environ., 11:73-85.

Davis, B. L., and Johnson, L. R., 1982, Sample preparation and methodology for x-ray quantitative analysis of thin aerosol layers deposited on glass fiber and membrane filters, Adv. in X-ray Analysis, 25:295-300.

Davis, B. L., and Walawender, M. J., 1982, Quantitative mineralogical analysis of granitoid rocks: a comparison of x-ray and optical techniques, Amer. Mineralogist, 67:1135-1143.

Graf, D. C., 1961, Crystallographic tables for the rhombohedral carbonates, Amer. Mineralogist, 46:1283-1316.

Hubbard, C. R., Evans, E. H., and Smith, D. K., 1976, The reference intenstiy ratio, I/I_c, for computer simulated powder patterns, J. Appl. Cryst., 9:169-174.

QUANTITATIVE ANALYSIS: A COMPARATIVE STUDY USING

Mo AND Cu RADIATION ON COAL RELATED MINERALS AND FLY ASH

J. Amenson, J. Benson, T. Demirel and R. Jacobson

Ames Laboratory-USDOE and
Iowa State University
Ames, Iowa 50011

ABSTRACT

Our experience using Mo Kα radiation in coal related research indicates that minerals can be detected and quantitatively measured at the 0.2% level in an amorphous substance such as coal. Since the linear absorption coefficient for most elements is about a factor of eight lower for Mo radiation than it is for Cu radiation, the matrix affect is greatly reduced and the intensity is a more linear function of concentration.

Although there is considerable overlapping of peaks when complex mixtures are analyzed using Mo radiation, the use of second derivative techniques to locate peaks, and least squares techniques to fit peaks can resolve many of these problems. The reduced time needed to collect a diffraction pattern, and the increased intensity obtained more than compensate for the extra computer time needed to analyze the data.

INTRODUCTION

Five years ago we shared an automated Picker theta-theta diffractometer with a group which was doing extended liquid diffraction studies. Because of the nature of the latter study, we did not wish to modify the instrument, especially the x-ray source everytime we switched applications. Our specific goal was to attempt to quantify the pyrite in coal without going through a low temperature ashing step. To our pleasant surprise we found that using Molybdenum radiation we obtained a linear relationship between the peak height above background and the pyrite concentration for up to 25% pyrite.

Furthermore we could detect little or no matrix effect arising from
the other common minerals in coal. As is shown below, in going from
Cu Kα to Mo Kα radiation, intensities do not significantly decrease
as might be expected, but can even increase. Because of our favor-
able experience using Mo Kα radiation on this instrument, its usage
in this configuration has increased markedly over the past few years;
a recent survey indicates that we now have 20 different users running
on the machine, with samples ranging from fly ash, low temperature
ash, ceramic and metallurgic samples, and catalytic materials, to
those of our own coal research.

GENERAL THEORY

    The diffraction intensity, as well known, varies with λ^3, i.e.
if one doubles the wavelength of the radiation, there is an eight-fold
increase in the diffracted intensity. The general intensity expression

$$I(hkl) = I(o)Cm\lambda^3F(hkI)^2VLp/u$$

where $I(hkl)$ = the resulting diffracted intensity
          $I(o)$ = direct beam intensity
          C   = constant for all wavelengths
          m   = the multiplicity of the reflecting plane
          λ = wavelength of the radiation used
          $F(hkI)$ = structure factor for planes diffracting
          V   = total volume of the diffracting crystals
          Lp   = Lorentz-polarization factor
          u   = linear absorption coefficient

indicates that if one were to change from Cu to Mo radiation, given
the same $I(o)$, there would be a reduction in $I(hkl)$ of 10 to 24-fold
due to wavelength. A computation of the Lp factor for a given d
spacing using both wavelengths (Table 1), shows a five to six-fold
increase in $I(hkl)$ obtained for Mo radiation. An examination of the
linear absorption coefficients for different common minerals indicates
that the diffracted intensity is about eight to nine times larger for
Mo radiation (Table 2). There is also an increase in the volume of
sample that is diffracting due to the greater penetration of the short
wavelength radiation. This all adds up to an increase in $I(hkl)$ of
four to five, instead of the expected eight-fold decrease due to
lambda alone.

EXPERIMENTAL

    We decided that we would test the quantitative analysis ability
of our two different instruments by collecting data using "standard
conditions" for each instrument. The Cu radiation data was collected
on a Siemens D500 at 50 KV and 25 MA. A one degree slit system was
used with a 0.05°slit before the graphite monochromator and a 0.15°

Table 1. The Lp factor for se-
          lected d spacings for
          Mo and Cu radiation.

d spacing	Mo	Cu	Ratio
14.21	3210.2	679.7	4.7
7.70	941.6	198.8	4.7
3.32	174.7	36.2	4.8
2.70	114.8	23.5	4.9
2.0	62.6	12.4	5.0
1.29	25.6	4.5	5.7
1.04	16.0	2.4	6.6

Table 2. Linear absorption
          coefficients for
          selected minerals.

Mineral	Density	Mo	Cu
Kaolinite	2.59	8.3	77.2
Quartz	2.65	9.8	91.2
Al_2O_3	3.99	13.4	124.2
Calcite	2.71	21.8	192.3
Pyrite	5.00	115.5	956.9

slit in front of the detector. The step-size was 0.03 degrees/step,
with three seconds per step counting time. The software was Siemens
DIFFRAC V. The Mo radiation data was collected using an automated
Picker theta-theta diffractometer at 50 KV and 16 MA. A one degree
slit system was used, with a 0.005° slit in front of the graphite monochromator. The step-size was 0.04 degrees per step, with one second
counting time per step. Software was of local origin.

The methods used for quantitative analysis were the methods of
matrix flushing and the adiabatic principle as described by Frank
Chung[1,2]. Pure samples for five of the more common minerals in coal
were chosen along with Al_2O_3 as the internal standard; a series of
50-50 mixtures of gypsum, kaolinite, quartz, calcite and pyrite with
Al_2O_3 were prepared from -200 mesh samples. An "unknown" was also
prepared from four of the above samples. All of the x-ray data were
collected for d spacings in the range 16Å to 1.5Å. The total time
required for each scan was approximately 90 minutes for Cu radiation
and 20 minutes for Mo. The sample holders for the two instruments
are sufficiently different that a single sample mounting could not be
used in this study.

Intensity ratios were calculated for several different lines in
each diffraction pattern (Table 3). If one compares the intensity
ratio for calcite using Cu radiation that we obtained with the ratio
observed by Chung in his paper[1] it is apparent that there was some
preferred orientation in that sample. This will affect the Cu results
for the "unknown" for all phases using the adiabatic principle but
only the calcite results in the matrix flushing technique.

RESULTS

Data were obtained via one quick pass through the diffraction
pattern to better approximate the procedures routinely used in

Table 3.   Intensity ratios I/Ic for
the minerals in this study.

Mineral	d spacing	Mo	Cu
Kaolinite	2.085	1.13	0.71
	1.601	1.14	0.74
Quartz	2.552	3.84	3.45
	2.085	3.73	3.59
Calcite	2.552	2.56	4.49
	1.601	2.20	4.74
Gypsum	2.552	3.56	4.63
	1.601	3.48	4.76
Pyrite	2.552	2.21	1.21
	2.085	2.17	1.23

acquiring diffraction data. The results obtained (Table 4) indicate
that Mo radiation yields as good, if not better data than, Cu in this
instance. Both results could certainly be improved if more care was
taken in sample preparation and data collection. There are still
some overlap problems not accurately enough accounted for in the Mo
data, but these data suffer much less from preferred orientation than
do the Cu data. Work is still in progress on the development of
better least squares techniques to fit overlapping peaks, especially
in the choice of starting parameters for the various functions that
are being used to describe the peak shape. A bifurcated Pearson VII
function was used to obtain the area of the peaks in the Mo data,
while the Siemens software was used for the peak area in the Cu data.
We could not decide whether alpha 2 stripping was superior to alpha 1-
alpha 2 curve fitting, but we did use alpha 2 stripped data in our
calculations.

CONCLUSIONS

We feel that for routine analysis (either qualitative or quantitative) the use of Mo radiation in x-ray diffraction can greatly
reduce the time necessary to collect a diffraction pattern, and
greatly increase the through-put on any diffractometer. The advent
of digital filters and second derivative techniques to smooth the
data and to locate peaks means that a greater degree of overlap can be
tolerated now than in the past. For large cells, or for work with
completely unknown compounds, Cu radiation still is the first choice.

Both instruments detected 0.5% quartz in the "unknown", but
neither instrument could readily detect Al_2O_3 in a 95% gypsum-5% Al_2O_3
mixture. The Al_2O_3 pattern was below the experimental cut-off level
in the Siemens software, and was just barely detectable in the Mo data.

Table 4.   Results of quantitative analysis
           on "unknown" sample.

| Mineral | Adiabatic Principle | | Cu |
	Mo % found	% known	% found
Kaolinite	5.0	7.8	7.5
Quartz	0.7	0.5	1.7
Calcite	26.1	25.6	35.9
Pyrite	30.0	33.0	21.6
Al_2O_3	38.0	32.9	33.2
Matrix Flushing			
Kaolinite	4.6	7.8	7.8
Quartz	0.6	0.5	1.7
Calcite	23.8	25.6	37.6
Pyrite	25.9	33.0	21.4
Al_2O_3	--	32.9	--

There appears to be several types of materials that exhibit
different diffraction patterns when one uses Mo radiation instead of
Cu.   The most obvious difference is in material with large absorption
coefficients.   Figure 1 is from a 50-50 mixture of Al_2O_3 and pyrite,
with the pattern obtained with Mo $K\alpha$ displayed at the top of the
figure, and the Cu pattern at the bottom.   There are two features
about this pattern that deserve mention.   The first is that the complete Mo set of data is finished before the first pyrite peak is
reached using Cu.   The second feature to notice is the reversal of the
intensity of the first two peaks in the pattern.   In the Mo data,
pyrite is the stronger diffracting mineral, while in the Cu data Al_2O_3
is almost equal in intensity to pyrite.   This is evident in the intensity ratio also, 2.2 for Mo, and only 1.2 for Cu.

The second type of material that shows vastly different patterns
is amorphous material.   This is apparent in the coal diffraction
pattern in Figure 2.   The upper pattern was made using Mo radiation
with one second per step counting time and the counting rate approaches 1000 counts per second; with the Cu radiation at three
seconds per step (lower pattern), the rate is only 100 counts per
second.

The third type of material that shows differences is kaolinite
and perhaps other clay minerals.   In the few clay samples that we have
compared using the two radiations, we seem to observe much greater
intensity for the hk0 and hk1 types of reflections using Mo radiation
than with Cu.   We plan to pursue this observation with some further
research in the near future.

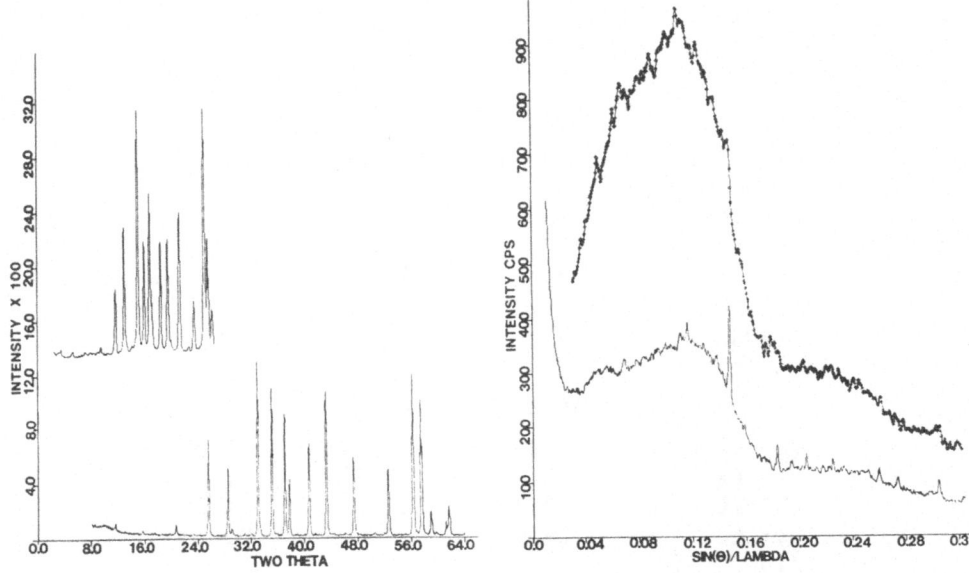

Figure 1. Mo and Cu pattern for   Figure 2. Mo and Cu pattern for
           $FeS_2-Al_2O_3$ mixture.   an Illinois #6 coal.

ACKNOWLEDGEMENTS

    Ames Laboratory is operated for the U.S. Department of Energy by
Iowa State University under Contract No. W-7405-Eng-82. This work was
supported by the Office of Fossil Energy and by the Engineering
Research Institute, Iowa State University. The authors gratefully
acknowledge the support given which made this work possible.

REFERENCES

1.   F. H. Chung, "Quantitative Interpretation of X-ray Diffraction
      Patterns of Mixtures. I. Matrix-Flushing Method for Quantita-
      tive Multicomponent Analysis," J. Appl. Cryst. (7),519 (1974).
2.   F. H. Chung, "Quantitative Interpretation of X-ray Diffraction
      Patterns of Mixtures. II. Adiabatic Principle of X-ray
      Diffraction Analysis of Mixtures," J. Appl. Cryst. (7),526
      (1974).

MULTIPHASE QUANTITATIVE ANALYSIS OF COLORADO OIL SHALES INVOLVING OVERLAP OF THE DIFFRACTION PEAKS

D. K. Smith, S. M. Sterner, and D. M. Kerrick

Department of Geosciences
The Pennsylvania State University
University Park, PA 16802

W. S. Meddaugh

Gulf Research and Development
Pittsburgh, PA 15230

ABSTRACT

The method of intensity ratios can be used for quantitative x-ray diffraction analysis even when none of the diffraction peaks of any phase can be individually resolved. The method uses intensity values from a finite number of discrete integration intervals and appropriate reference-intensity-ratios measured on a set of reference samples to yield a best fit composite pattern that is matched to the unknown by least-squares refinement. Solid-solution phases are analyzed by using multiple reference patterns and determining the best choice during the pattern matching procedure. The method may be extended easily to employ whole diffraction traces by using a large number of small intervals.

INTRODUCTION

Quantitative phase analysis by x-ray powder diffraction, QXRPD, is one of the few instrumental analytical methods which can provide phase composition information. The extensive use of computer-controlled powder diffractometers has rejuvinated its use. Unfortunately, its successful employment has been limited by many factors, one of which is overlap of the diffraction maxima of the different phases in the mixture diffractogram. Most QXRPD procedures, all of which are based on the pioneering work of Klug and

and Alexander,[1] attempt to use fully resolved maxima for each of the
phases of interest and usually incorporate an internal reference
standard to eliminate the absorption problems. The first effort to
utilize diffraction peaks which involved overlap was by Copeland and
Bragg[2] in their analysis of cement. Although their method allows
overlap of the analyte maxima, it does require resolution of the
internal standard maxima. Phases yielding complex diffraction spectra usually preclude this requirement. In our study of oil shales
we must consider several phases with complex patterns and contend
with some whose compositions are also variable. These complications
imply that all experimental intensity maxima have the potential that
every phase may be a significant contributor.

The solution of this condition, using a method now known as
the intensity-ratio method, was first reported by Karlak and
Burnett.[3] A diffractogram of a multiphase mixture may be considered
as a series of segments, i, each of which can be integrated between
defined 2θ limits to yield a set of intensity values, I_i. These
segments do not have to be equal in width nor do they have to cover
the whole pattern. The intensity in each segment is the sum of the
contributions of all the phases present

$$I_i = \sum_{j=1}^{n} a_{ij} \, C_{oj} \, w_j \, .$$

The quantities $a_{ij} = I_{ij}/I_{oj}$ are the ratios of the intensity of the
ith segment to the most intense segment in a pattern of phase-pure
j. The relative-absolute scale factors $C_{oj} = K_{oj}/\mu^* \rho_j$ are the
ratios of the Klug and Alexander[1] intensity factor of the strongest
region to the product of the effective mass absorption coefficient
and the phase density. The quantity w_j is the weight fraction of
phase j. Karlak and Burnett show that this system of equations can
be solved by the addition of an internal standard, s, of known
weight fraction, w_s. With substitutions, the intensity equation
takes the form

$$I_i = C_{os} \, w_s \sum_{j=1}^{p} a_{ij} \, C_{js} \, w_{js} \, ,$$

where $C_{js} = C_{oj}/C_{os}$ the well known reference-intensity-ratio and w_{js}
$= w_j/w_s$. The equation is not useful in this form, but by forming
the ratios $R_{ik} \equiv I_i/I_k$, where k is a selected segment of the experimental pattern which has a non-zero intensity, a set of equations of
the form

$$0 = \sum_{j=1}^{p} C_{js} \, (a_{ik} - a_{kj} R_{ik}) \, w_{js} \qquad i=1,2,\ldots k-1,k+1,\ldots n$$

is obtained. The equation for i = k is identically zero and of no use. This set of equations may be solved simultaneously if the number of phases p = n-1, or by least-squares refinement if p $\leqslant$ n. Ideally, for least-squares refinement, p $\ll$ n is desirable.

Because $w_{ss} = 1$, the equations are most usefully written

$$R_{ik}\, a_{ks} - a_{is} = \sum_{j=1}^{p} C_{js}\,(a_{ij} - R_{ik}\, a_{kj})\, w_{js} \,. \qquad i \neq k$$

The quantities a_{ij} and C_{js} must be determined initially from phase-pure samples and from known mixtures of the pure phases and the internal standard. For solid-solution phases, several samples along the compositional series are necessary to cover the range of variation in the samples under study. The weight fraction, w_i, in the mixture without the internal standard can be determined from the known weight fraction of the added standard.

There are several ways to employ this technique. The first way is to define segments which span well-defined fully-resolved peaks including the standard and whose limits all correspond to background. This approach corresponds to the Copeland and Bragg[2] method when some segments include multiple lines and to the internal standard method when there is no overlap. The second way is to use segments which may include several maxima and possibly the internal standard with limits selected at a minimum between strong peaks but not necessarily at "true" background. In this method care must be taken to define true background from the whole diffraction pattern. This second approach is the subject of the present paper. A third way to use this technique is to divide the whole diffraction trace into small equal segments and essentially use the entire pattern for quantitative analysis. This latter approach is probably the most powerful and is currently under study in a related project.

SAMPLE PREPARATION AND DATA COLLECTION

In theory QXRPD is sound, but in practice it is limited by sample behavior and counting errors. In this study the reference standards were crushed to -400 mesh and either mixed with silica gel to obtain the constants a_{ij} or with 33 wt. % α α-Al_2O_3 to obtain the reference-intensity-ratios C_{js}. The samples were backloaded or side loaded with great care and multiple samples were made on several phases to examine reproducibility. Mechanical mixtures of the reference phases used to test the method and the oil shale samples under study were mixed with 33 wt. % α Al_2O_3 and treated the same way. Preferred orientation did not appear to be a problem especially in the oil shales because of their small natural crystallite size, < 20 μm.

Using the complete set of phase-pure patterns, including the internal standard, segments of integration were selected which included all the major peaks of all the possible phases. Twenty segments were used in this study, a limit imposed by the diffracto-meter system. Each segment contained significant contributions from one or more phases but not necessarily all of them. Once the segments had been defined, the a_{ij} values were determined by correcting the measured intensity for true background including the silica gel. The background was obtained by fitting a smooth profile to the full diffractometer trace using only points that were free of peak inter-ference. Appropriate background values were then estimated for each scan segment endpoint. This method was used because the full trace was not available in computer readable form. The a_{ij} values were scaled to $a_{oj} = 100$.

The C_{js} values were determined from the I_i values measured from a sample of the phase mixed with 33 wt. % α-Al_2O_3 and the a_{ij} values obtained for the phases and α-Al_2O_3. Initial estimates of C_{jn} were made from two peaks in the pattern, then the equations

$$I_i = k_1 a_{ij} w_j + k_2 a_{is} w_s \qquad i = 1, 20$$

were solved by least-squares regression for k_1 and k_2. The refer-ence intensity ratio becomes $C_{js} = k_1/k_2$. Several comments are in order concerning the a_{ij} matrix. Only a single interval, i=1, has only one contributing phase which is illite. Most have several con-tributors. Some phases like the feldspars contribute to most of the intervals. The α Al_2O_3 standard peaks are overlapped in all cases. This pattern of data justifies the need for this new approach. The C_{js} values are similar to literature values where available, but they cannot be directly compared because they are based on inte-grated intensities over segments which are different than the re-ported data.

COMPUTER PROGRAM

The program KBQUAN solves the quantitative analysis equations using the method of Karlak and Burnett.[3] It uses the a_{ij} and C_{js} matrices generated through KBBKG and KBCOR, respectively. The experimental I_i matrix is processed by KBQUAN which calculates the ratios R_{ik}. The only other input data required is the weight frac-tion of the internal standard, the estimated sum of the weight frac-tion of the crystalline components (usually 1.0 but it may be smaller), the ratio line k, and selection parameters which identify the phases desired in the analysis.

The core of the program is a subroutine LSEI, written by Haskell and Hanson.[4] This routine allows three types of solutions to be tried. The first pass uses only the n-1 intensity equations. The second pass uses the equality constraint $\Sigma w_j = N$ where N is the estimated sum of the weight fractions of the crystalline portion of the sample. Equalities such as the sum of carbonate or other chemical information could also be used. The third pass uses the equality constraints along with additional inequality constraints such that $0 < w_j < N$ for all j. The results of each pass may then be examined for consistency. If the results do not agree, experience has indicated that the phase selection was incorrect. Figure 1 shows the flow of the data through the program.

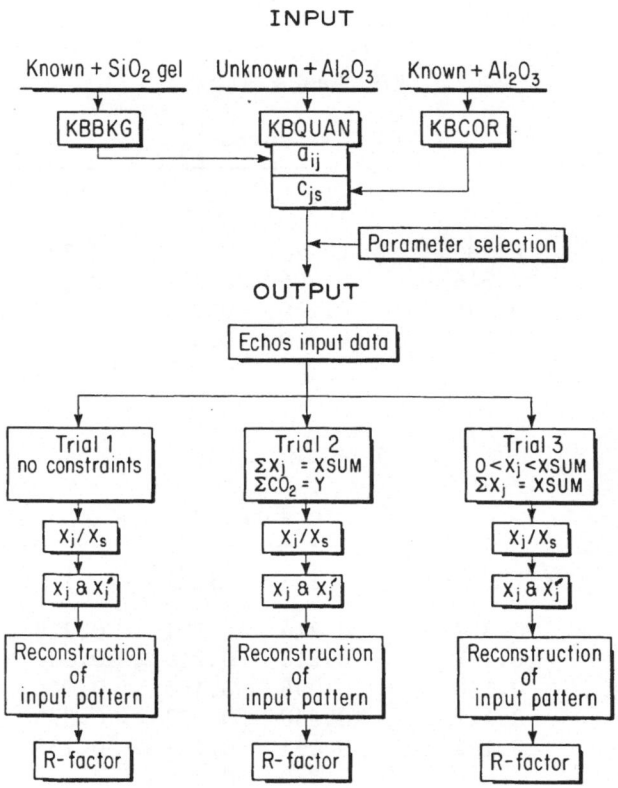

Figure 1. Flow of data through computer program.

RESULTS

Because it is difficult to verify the true phase compositions of oil shale samples, the program was first tested using prepared mixtures of the same materials used for the reference data.   Three and five phase mixtures were prepared with 33 wt. % α Al_2O_3 as the internal standard.   The results of these tests are listed in Table I.   In Test 1, three well crystallized phases abundant in oil shale were mixed in equal proportions.   The results of the passes with equality and inequality constraints were identical and only slightly different from the unconstrained pass.   The constrained results were insignificantly closer to the mixture proportions, and the R-factor of the reconstructed pattern was also slightly lower.   Test 2 contained five phases including the feldspar albite.   The results show some differences between the constrained and unconstrained passes. In this test, the unconstrained results are closer to the mixture values.   The R-factor indicates a better intensity fit for the unconstrained results, and it may be the best indicator of the correctness of the weight fraction estimates.

Table I.   Test Examples

Test 1
3 phases

Phases	Actual Xj	Measured X'_j		X_j	
		1	2,3	1	2,3
Corundum	0.330	0.225	0.226	0.335	0.337
Calcite	0.346	0.206	0.214	0.307	0.319
Dolomite	0.334	0.240	0.231	0.358	0.344
Corundum		0.330	0.330		

R = 0.072, 0.071

Test 2
5 phases

Phases	Actual Xj	Measured X'_j		X_j	
		1	2,3	1	2,3
Quartz	0.327	0.214	0.201	0.319	0.300
Calcite	0.308	0.218	0.218	0.326	0.326
Albite	0.130	0.096	0.105	0.144	0.157
Dolomite	0.113	0.050	0.029	0.075	0.043
Analcime	0.122	0.091	0.117	0.136	0.174
Corundum		0.330	0.330		

R = 0.088, 0.111

DISCUSSION

     The purpose for developing this program was to establish a
reliable method for measuring phase proportions in very complex oil
shales. Previous efforts[5] to analyze oil shales have met with
limited success because of the solid-solution effects of several of
the minerals, especially dolomite, siderite ("ferroan") and the
feldspars. These minerals show both shifting of the diffraction
peaks and variation of intensity values with changing composition.
By defining scan ranges which include the peak shifts and by having
several representative reference samples along each solid solution
series, better phase estimates will be possible than have been
achieved to date. Intensity variations with composition of the
strongest group of peaks among the feldspars are not significant (B.
L. Davis, private communication), but the Mg/Fe ratio does have a
strong effect on intensities in both dolomite and siderite. These
phases will require several patterns in the reference set for
several compositions covering the range of compositions, and the
best fit must be determined during the analysis most probably based
on the R-factor which appeared to be the most sensitive indicator in
the test examples. More experience with oil shale samples will be
needed to define a better fitting parameter.

     The problem of solid solutions is not to be underestimated. In
mineral systems, especially those involving the rock-forming minerals, solid solutions are more the rule rather than the exception.
Attempts to correct pure phase data for compositional deviations has
not proved effective, thus a reference pattern must be obtained
which agrees very closely with the phase in the unknown samples. If
the phase varies in composition from sample to sample, then reference patterns must be obtained for a number of samples of the solid
solution which span the compositional range. The problem then
becomes selecting the proper reference for each unknown sample. If
there is a significant change in cell dimensions with composition,
d-spacing shifts of key peaks may be used. However, the shifts
require additional measurements to locate the peaks and usually are
too small to be useful. Thus the method of selective fitting of the
best reference data set based on intensity matching has proved more
effective. Although the use of limited discrete intervals can
choose among several reference patterns, undoubtedly the use of the
whole diffraction trace should prove even more sensitive.

CONCLUSIONS

1.   The method of intensity ratios allows quantitative analysis with
     unresolved standard patterns.

2.   The method does not rely on peak resolution or on knowledge of
     any peak profile.

3. The method may be used with solid solutions by employing multiple reference patterns from samples with different compositions and determining the best reference through the final fit of the reconstructed mixture pattern with the unknown pattern.

4. The method may be extended to using the whole diffraction pattern.

ACKNOWLEDGMENTS

The authors would like to thank Dr. Camden R. Hubbard for making the NBS QUANT82[6] package available from which the LSEI subroutine was extracted. He also provided many helpful suggestions based on his experience with x-ray quantitative analysis.

REFERENCES

1. H. P. Klug and L. E. Alexander, Anal. Chem. 20:886 (1948).

2. L. E. Copeland and R. H. Bragg, Anal. Chem. 30:196 (1958).

3. R. F. Karlak and D. S. Burnett, Anal. Chem. 38:1741 (1966).

4. R. J. Hanson and K. H. Haskel, Sandia Laboratories Tech. Reports SAND77-0552 (1978) and SAND78-1290 (1979).

5. N. B. Young, J. W. Smith, and W. A. Robb, U.S. Bureau of Mines Report of Investigations 8008 (1975).

6. C. R. Hubbard, C. R. Robbins, and R. L. Snyder, Adv. X-ray Anal. 26 (1983).

EVIDENCE FOR STACKING FAULTS IN MULTIAXIAL STRAINED ALPHA-BRASS*

R. B. Roof

Los Alamos National Laboratory

Los Alamos, New Mexico 87545

INTRODUCTION

As part of a program studying the effects of large strain deformations resulting from multiaxial loading to a variety of materials, a thin walled tube (0.46" O.D. x 0.02" wall thickness) of 70-30 Brass was subjected to strain deformation in the following directions 1) along the tube axis, ε_z = 0.3393; 2) circumferential around the tube surface, ε_θ = -0.0121; 3) perpendicular to the wall thickness, ε_R = 0.3514. This report describes the results of an x-ray examination of the external surface of the tube by the line broadening technique.

EXPERIMENTAL TECHNIQUES

Examination of the shape of an x-ray diffraction line profile is a technique whereby information can be obtained concerning the condition of the material on an atomic scale. Crystallite size, lattice strain, and lattice stacking faults are items that can be obtained. Since descriptions of the general procedures usually employed are available in standard literature references[1-4] the detailed procedures will not be described further in this report. A Fourier coefficient computer program was employed to aid in the analysis of the peak shape. A unique feature of the program is the inclusion of equations due to Wilson[5-6] to calculate standard deviations of the Fourier coefficients directly in terms of the experimental intensity expressed as counts per second. The standard deviations were further propagated (with covariance included) through

---

*Work performed under the auspices of the Department of Energy.

the complex division necessary to arrive at Fourier coefficients free from instrumental abberations.

The experimental sample used in the x-ray diffraction examination was formed as follows. Three pieces were cut from the strained tube approximately 0.50" long in the axial direction and 0.125" wide in the peripheral direction, mounted in plastic, and metallographically polished to provide a flat surface of 0.5" x 0.5" of the external face of the tube. Similar sections obtained from an unstrained tube were used as reference material.

α-Brass is a face-centered-cubic material and the lattice constant, obtained from least-squares extrapolation techniques,[7] for the strained sample is a_o = 3.6851±3 $\overset{o}{A}$. For the reference material the value is a_o = 3.6840±1 $\overset{o}{A}$.

For the material under examination the reflections 111 and 220 are relatively sharp while 200 and 311 display broadening. The reflections 222 and 400 have extremely weak intensities, barely above background, and reliable measurements on these reflections cannot be obtained. The reflections 331 and 420 are available but with weak intensities and they are not included in the present analysis.

A full detailed analysis of all the diffraction data was made utilizing the general procedures presented previously.[8] From this analysis it was determined that the following approximations are valid for the present investigation: the double stacking fault probability, α'', is essentially zero; the residual stress, σ, is virtually nil; and the crystallite size is in excess of approximately 2000 $\overset{o}{A}$, the practical upper limit for determination by x-ray diffraction techniques. These assumptions/approximations allow considerable simplifications of the equations for determining the separation of the compound stacking fault parameters from one another.

DATA ANALYSIS

The normalized unfolded Fourier coefficients $(C=(A^2+B^2)^{\frac{1}{2}})$ for the four reflections examined were plotted as a function of L. Two general sets were observed; those coefficients belonging to the relatively sharp lines 111 of 200 and those in the second set belonging to broadened lines 200 and 311.

The log of the Fourier coefficients were next plotted as a function of $h^2+k^2+l^2$ for various values of constant L. It was noted that the Fourier coefficients did not fall on a common curve. The departure from a common curve is indicative of stacking faults in the material. It is desired to fit the data to the general exponential equation y = a.exp(bx). This is usually done for the different orders of reflections from the same set of parallel lattice planes. Since no second or higher orders are available the fitting is done with the selected pairs (111, 220) and (200, 311).

The root mean squared strain, ε_F, can be calculated from the average slope of the log of the Fourier coefficient plotted as a function of $h^2+k^2+l^2$ utilizing the following equation

$$\varepsilon_F^2 = \frac{a_o^2}{L^2 . 2\pi^2}(-slope) \tag{1}$$

where a_o = 3.685 Å for α-brass.

The strains determined in this manner are plotted as a function of L. Extrapolation to L=0 results in an estimated local strain of 0.00225.

The Fourier crystallite size coefficients were plotted as a function of L. The intersection on the L axis of a line drawn through the linear portion of the curve yields an effective crystallite size, D_e. This results in D_e = 450 Å for (111,220) and 200 Å for (200, 311).

The separation of compound stacking faults and true crystallite size from the effective crystallite size is accomplished by plotting $1/D_e$ as a function of V(hkl). V(hkl) is a constant for a given value of hkl and reflects the contribution to the total observed line profile of the various signed permutations of hkl that are affected by stacking faults. Tables of V(hkl) are available.[2-3]

In Figure 1 $1/D_e$ is plotted versus V(hkl) for 111 and 200, the first hkl's of the selected pairs of reflections. The intercept in Figure 1 is the reciprocal of the true crystallite size, D; by fixing it at zero we satisfy the assumption that the crystallite size exceeds 2000 Å. The slope of Figure 1 is equal to $(1.5 \alpha' + 1.5 \alpha'' + \beta)/a_o$ where α' = the probability of finding a single stacking fault between neighboring 111 planes; α'' = the double stacking fault probability; and β = the twin fault probability. With α'' assumed to be negligible the slope reduces to $(1.5 \alpha' + \beta)/a_o$. From Figure 1 the value of $(1.5 \alpha' + \beta)$ = 0.01850±12.

A value for a stacking fault combination has been obtained and a second value from a different equation is needed to separate the stacking faults. The lattice parameter, a_{hkl},

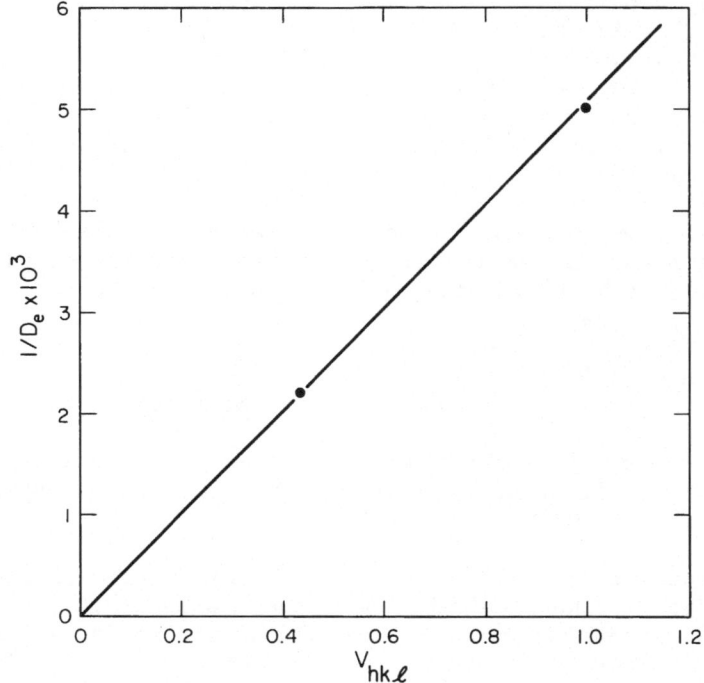

Fig. 1.   Separation of true crystallite size and compound stacking fault parameter from effective crystallite size.   See text.

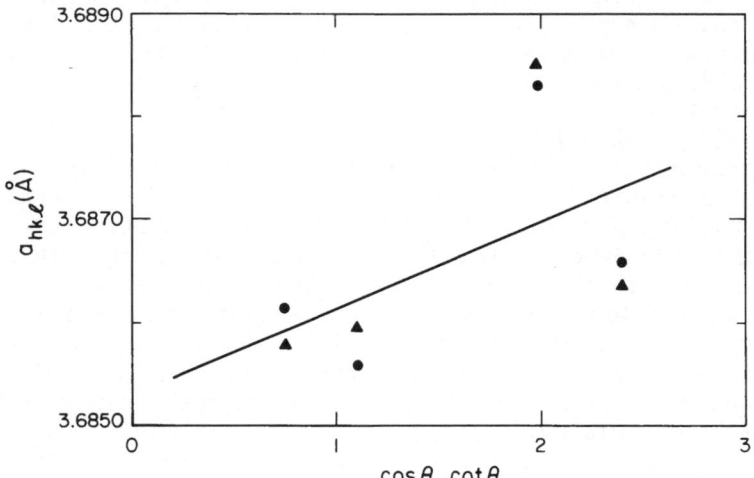

Figure 2.   Calculated latice constant, a_{hkl}, versus
$\cos\theta\cot\theta$ for plane strained alpha-brass. ▲ =
experimental; ● = effect of $\alpha' = 0.005$.

calculated from a peak position of the individual hkl reflections of fcc materials depends on the true lattice parameter, a_o, the difference $(\alpha' - \alpha'')$, the residual stress, σ, and the geometrical abberations of the diffractometer. It has been shown[2] that these quantities are related according to

$$a_{hkl} = a_o + (S_1)_{hkl}^A \cdot a_o \cdot \sigma + G_{hkl} \cdot a_o \cdot (\alpha' - \alpha'') + m.f(\theta) \quad (2)$$

where $(S_1)_{hkl}^A$ and G_{hkl} are constants which depend on the planes hkl. Tables of G_{hkl} are available.[3-4] With σ and α'' assumed to be negligible equation (2) reduces to

$$a_{hkl} = a_o + \alpha' \cdot a_o \cdot G_{hkl} + m.f(\theta) \quad (3)$$

or

$$y = P_1 x_1 + P_2 x_2 + P_3 x_3$$

where $x_1 = 1.0$, $x_2 = G_{hkl}$, and $x_3 = \cos\theta\cot\theta$ for diffractometer focusing geometry. Values for a_{hkl} (i.e., y) were calculated from 2θ values representing the centroids of the observed experimental intensity data. Values for the remaining terms (i.e., x_j, $j = 1,3$) were calculated or taken from published tables.[3-4]

The data processed by equation (3) is usually plotted as y versus x_3 with deviations from a straight line representing the contributions of stacking faults of the lattice parameter (see Figure 2). Using the data and solving equation (3) by least-squares techniques results in $P_1 = 3.6853\pm8$, $P_2 = 0.020\pm8$, and $P_3 = 0.00083\pm50$.

The stacking fault parameter of interest, α', is obtained from the ratio, P_2/P_1, and equals 0.0054 ± 22. From $(1.5 \alpha' + \beta) = 0.0185\pm1$, β is calculated to be 0.0104 ± 22.

CONCLUSIONS

Two different techniques of X-ray diffraction have been applied to the examination of multiaxial strained alpha-brass. From an examination of the unfolded Fourier coefficients describing the shape of the diffraction profile it has been determined that the true crystallite size probably exceeds 2000 Å, a practical upper limit for determining crystallite size by x-ray methods. The localized strain is approximately 0.225% and a combined stacking fault probability $(1.5 \alpha' + \beta) = 0.0185\pm1$.

From lattice constant variations the single stacking fault probability, α', has been determined to be 0.0054 ± 22. The twinning stacking fault, β, is thus 0.0104 ± 22. The reciprocal of the probability is the number of planes of atoms between the indicated stacking fault. The magnitude of the numbers indicates that twinning is twice as common as single stacking faults.

REFERENCES

1.   H. P. Klug and L. E. Alexander, "X-ray Diffraction Procedures", (Wiley, New York, 1954), p. 491.
2.   C. N. J. Wagner, in "Local Atomic Arrangements Studied by X-ray Diffraction", edited by J. B. Cohen and J. E. Hilliard, (Gordon and Breach, New York, 1966), p. 219.
3.   B. E. Warren, "X-ray Diffraction", (Addison-Wesley, Reading, Massachusetts, 1969), p. 251.
4.   N. C. Halder and C. N. J. Wagner in "Advances in X-ray Analysis", Vol. 9, (Plenum Press, 1966), p.91.
5.   A. J. C. Wilson, Acta Cryst. $\underline{23}$ (1967), p. 888.
6.   Idem, ibid $\underline{A25}$ (1969), p. 584.
7.   R. E. Vogel and C. P. Kempter, Acta Cryst. $\underline{14}$ (1961), p. 1130.
8.   R. B. Roof and R. O. Elliott, J. Mat. Sci. $\underline{10}$ (1975), p. 101.

LINE BROADENING STUDIES ON HIGHLY DEFECTIVE Al$_2$O$_3$

PRODUCED BY HIGH PRESSURE SHOCK LOADING†

B. Morosin, E. J. Graeber and R. A. Graham

Sandia National Laboratories
Albuquerque, New Mexico 87185

INTRODUCTION

Enhanced solid state reactivity of materials both during and after shock compression has been attributed to the introduction of large numbers of defects into the crystalline lattices and to reduction in the particle and crystallite size of powders [1,2]. Line broadening of x-ray diffraction profiles provides a means to determine the residual lattice strain resulting from such defect concentrations as well as a means to determine the coherent crystallite size. Various diffraction studies on shock-loaded powder materials have previously been reported and much of this work primarily by Soviet and Japanese scientists has recently been reviewed [2]. Cohen has reported results on shock-loaded copper [3]. In prior work, however, shock pressures have not typically been quantified and there are few detailed line broadening investigations of refractory inorganic powders [1,4,5]. The present study on shock-loaded alumina powders is a detailed investigation of the influence of shock loading on residual lattice strain and coherent crystallite size.

BACKGROUND

The basic principles for the analysis of peak profiles are well known and have recently been examined and discussed [6]. For such analysis, the Fourier coefficients for a peak profile $\underset{\sim}{f}$ free from instrumental effects may be obtained by complex division

†Work supported by U.S. DOE contract # DE-AC04-76DP00789.

("deconvolution" or "unfolding") of the Fourier coefficients for
the broadened h and standard g profiles.

Previous studies have considered the microstructural effects
of shock-wave loading on Al_2O_3 powders [1,4,5,7,8]. Peak profile analysis showed a reduction in larger crystallite sizes (by
1/3 - 1/2 to ~ 200-250Å) and an increase in microstrain (by an
order of magnitude to ~ $3x10^{-3}$). These magnitudes are similar to
those found in heavily cold-worked metals with similar dislocation
densities and elastic strain energies [4,5]. Transmission electron
microscopy on shock-loaded Al_2O_3 showed both dislocation generation
and twin formation as active deformation modes, with dislocations
largely confined to the basal plane, with Burgers vectors of the
type (1120) and (1010), though some on prism, (1100), and rhombohedral, (1101), planes was also noted [7]. Twinning was confined
to the basal plane. The large dislocation densities probably
form polygonized structures from the localized increases in temperature during or immediately following shock-loading. At higher
temperatures, ductility of the Al_2O_3 allows microstructural
refinement similar to the work-annealing processes previously
studied [8].

METHODOLOGY

A generally used method to interpret size-strain broadening
of x-ray diffraction lines is that given by Warren and Averbach
[6,9]. This model yields information on the effective size with
contributions from both domain length and faulting as well as
information on the non-uniform strains in the form of mean-squared
strains as a function of correlation length. Interpretation of
microstructural defects responsible for these mean-square strains,
is limited to an estimation of dislocation density. When combined
with TEM or electron spin resonance measurements, however, somewhat
more detailed interpretations are possible.

In the Warren-Averbach model, regions in the sample which
scatter x-rays coherently are designated as domains. Each domain
is thought to consist of a column of unit cells, a distance L, perpendicular to the reflecting lattice planes, (hkl), whose spacing
is d. It has been shown that the Fourier transform F(L) of a line
profile can be written for a particular L as:

$$F(L) = A^S(L)F^D(L)$$

where $A^S(L)$ expresses the so-called size broadening, which is
independent of the order of (hkl), and where $F^D(L)$ expresses as an
exponential the strain broadening term, which is dependent on the
order of (hkl). Series expansion of $F^D(L)$ and restriction to the
leading term with respect to strain gives

$$|F(L)| \simeq A^S(L)(1-2\pi^2(L/d)^2 \ (<e^2(L)>-<e(L)>^2)$$

where e(L) is the strain resulting from a displacement of two unit cells at a distance L. If two orders of a (hkl) plane are available, $A^S(L)$ and $<e^2(L)>-<e(L)>^2$ are obtained from a plot of $|F(L)|$ versus $1/d^2$. Further, if the centroids of the line profiles of the broadened and standard reflection are properly determined and the origin of the unfolded profile coincides with its centroid, $<e(L)>=0$. Thus values of $A^S(L)$ and $<e^2(L)>$ as a function of L can be determined.

Experimental curves of $|F(L)|$ and $A^S(L)$ versus L often show negative curvature near L values of zero, a behavior which is theoretically impossible. This has been termed a "hook" effect. Experimental errors leading to such hook effects result from truncation of the line profile and estimation of the background of the line profile at too high a value. Approximations mentioned above may also be significant.

The mean average area-weighted size, $<D_a>$, normal to the (hkl) plane is given by the initial slope of the Fourier coefficient curves, i.e.,

$$<D_a> = -\left\{\left.\frac{dA^S(L)}{dL}\right|_{L\to 0}\right\}^{-1} .$$

In the presence of a small "hook" effect, the usual method consists of extrapolating the more or less straight portion of the Fourier coefficient curve at small values of L.

The corresponding mean volume-weighted size $<D_v>$, can be determined from the integral breadth B^S of the size-broadened profile. B^S can also be obtained from the sum of the size Fourier coefficients. The relationship between these two sizes has been shown to be

$$<D_v> = <D_a^2>/<D_a>$$

with $<D_v>/<D_a> \geqslant 1$ and with differences between both measures of size easily of the order of 100%. Further, $<D_a>$ can be used to estimate $<D>$, the average size of the coherently diffracting domains normal to the (hkl) plane as well as α and β, the deformation and twin fault probabilities, respectively, provided the proper coefficients V_{hkl} are known.

$$(1/<D_a>=(1/<D>)+(1.5\alpha+\beta)V_{hkl}/d,$$

the V_{hkl} coefficients depending on hkl and the number of reflection components affecting and not affecting stacking faults. These coefficients are available for cubic materials but will not be considered further in this study. Usually such $\langle D \rangle$ values are approximately equal to $\langle D_v \rangle$ values.

A root mean square strain e_L can be obtained by averaging the values of $\langle e^2(L) \rangle$ over D_a obtained by the standard Warren-Averbach analysis.

Strain values e_B can also be obtained from the integral breadth B^D of the strain-broadened profile. With the assumption of a Gaussian strain profile e_B can be related to e_L values

$$e_B = \sqrt{\pi/2} \; e_L \; .$$

Integral breadths B^F can be obtained by summing over the Fourier coefficients $F(L)$ of the unfolded profile. The values obtained by the Warren-Averbach model and those from the classical methods, the Hall-Williamson and the Gauss squared, are of interest. The first assumes that both size and strain broadening produce a Cauchy peak profile while the second assume a Gaussian peak profile. The Hall-Williamson method employs a linear (Cauchy) relation

$$B^F \cos\theta/\lambda = 1/D_c + 4e_c \sin\theta/\lambda$$

where θ and λ are the usual diffraction angle and wavelength and D_c and e_c are the Cauchy size and strain values. Similarily under the square (Gaussian) relationship, i.e., Gauss squared method,

$$(B^F \cos\theta/\lambda)^2 = (1/D_G)^2 + 16e_G^2 (\sin\theta/\lambda)^2$$

and D_G and e_G are the corresponding size and strain values. Plotting of these functions readily yields the corresponding size and strain. The Cauchy relationship results in larger size values and smaller strain values than those obtained from the Gauss-square method. Values of $\langle D \rangle$, $\langle D_v \rangle$ and D_G are approximately equal.

EXPERIMENTAL

The Al_2O_3 powder was Reynolds RC-HP-DBM, which has been characterized and studied using various methods [10].

The powders were pressed into compacts to densities of about 60% of solid density, subjected to various controlled explosive loadings at peak pressures from 4 to 27 GPa and preserved for post-shock analysis as previously described [11].

The peak profiles of the x-ray diffraction lines for Al_2O_3 were obtained with a Siemens D-500 automated diffractometer equipped

with a monochromator on the detector set so that both α_1 and α_2
were obtained. The aperture and diffraction slits are 1° and those
adjacent to the diffracted beam monochromator are both 0.15°. The
intensity data were taken by stepping in increments of 2θ equal to
0.01° and counting for a fixed time, generally 10 sec except for
particular weak peaks (to 100 sec). Instrumental broadening was
determined using a 1500°C annealed sample of our starting material
as the standard.

Fourier analysis of the peak profiles was aided by a computer
program UNFOLD which unfolds the broadened and standard profiles
to give the Fourier coefficients free from instrumental effects [12].
For this particular study the integral breadth B^F of the peak profile was obtained from the sum of these unfolded Fourier coefficients. Also a computer code to perform the Warren-Averbach
analysis of the Fourier coefficients as outlined in METHODOLOGY
above was employed to separate the size and strain contributions.
The multiple-line Fourier methods employed the sets (012) and
(024), (110) and (220), and (113) and (226) for Al_2O_3.

RESULTS

The anisotropy of the hexagonal crystal structure for Al_2O_3
appears to influence the corresponding plots of the Fourier coefficients F(L) vs. the square reciprocal interplanar spacing
or alternately the integral breadth B^F vs. scattering angle plot
(Fig. 1). The strain value determined from the (012) and (024)
set appears to correspond to the maximum strain value for the 3
sets in the lattice (Fig. 1). These data show a systematic scatter
or pattern when compared with similar displays using different
shock-loading pressures in the other Al_2O_3 samples. For example,
in Fig. 1 the (106), (124) and (030) integral breadths B^F result
in points lying below the straight lines determined by the (012)-
(024) set or even the (113)-(226) set. Correspondingly, the values
for the Fourier coefficient lie above the lines yielding the size
coefficient $A^S(L)$ and average mean square strain value $\langle e^2(L) \rangle$.
The indices for the profiles examined are shown at the top of Fig. 1.

Both the (110) and (030) planes are normal to the basal plane
on which dislocations were shown by TEM to concentrate [7].
Figure 2 shows values for the size coefficients $A^S(L)$ for the 17
GPa sample. Values of $\langle D_a \rangle$, the mean average-area-weighted size,
which is obtained from the extrapolation at small L values of the
linear portion of the curve with positive curvature, for the (012)-
(024) reflection set are slightly larger than those for the (110)-
(220) set. As can be seen in our data, small errors are introduced
since a small hook effect is evident. It is our belief that the
major portion the a hook effect results from the background of the
broadened profiles not being extended sufficiently from the centroid.

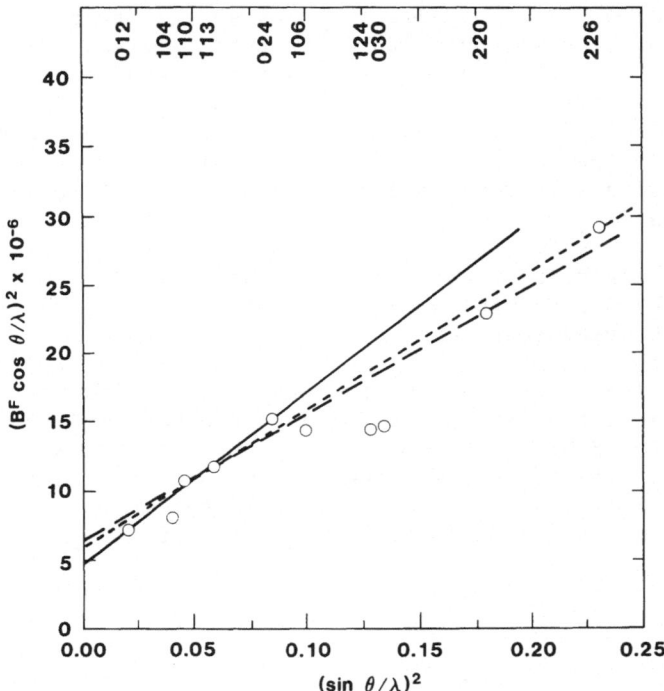

Fig. 1   Typical plot of integral breadth data on shock-loaded
          Al_2O_3 (17 GPa sample shown). Solid, dashed or dotted
          straight lines are drawn through the (012)-(024), (110)-
          (220) or (113)-(226) pairs, respectively. For comparison,
          the Hall-Williamson method yields a crystallite size D_C
          = 690, 580 and 740 Å and a strain e_c = 2.12, 2.10 and
          1.82 (all x 10^{-3}) for the (012), (110) and (113) planes,
          respectively.

     Examination of $A^S(L)$ size coefficients for the Reynolds
as-received and/or various shock-loading pressures indicate a
marked decrease occurs with increasing shock-loading pressure. As
indicated above, the slope of such curves at small L value yield
the mean average-area-weighted size $\langle D_a \rangle$. The values for the
Reynolds as-received material result from ball milling in the
preparation of the material. Table I summarizes such $\langle D_a \rangle$ size
coefficients and compares them with crystallite size values, D_G,
determined from the Gauss squared method employing integral breadth
B^F values.

     Figure 3 shows the typical values for the root average mean
squared strain values $\langle e^2(L) \rangle^{1/2}$ obtained from the Warren-Averbach
analysis of the Fourier coefficients as a function of L values.
Again as in Fig. 2, values are shown for two sets on our 17 GPa
sample. The values are not shown for L values below 40 Å which

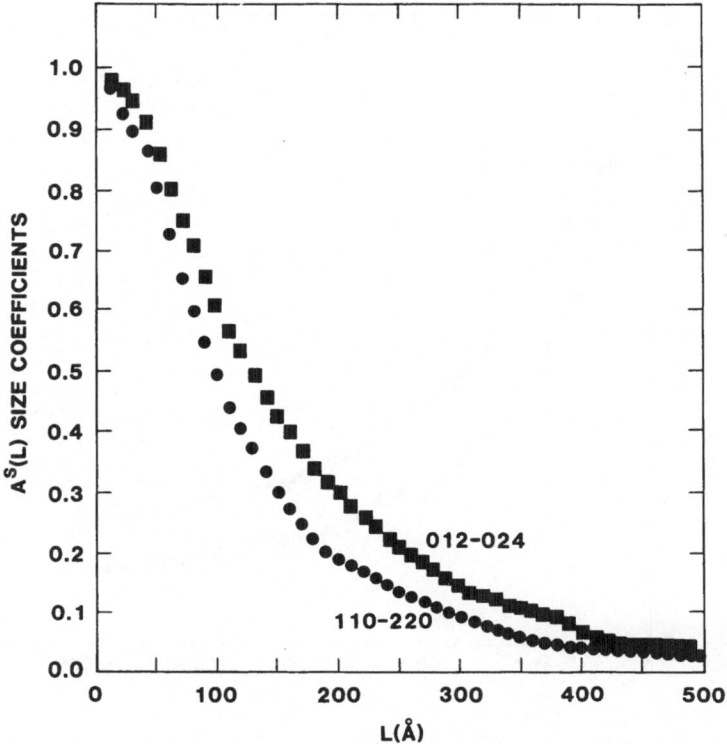

Fig. 2   Values for the Warren-Averbach size coefficients as a
function of L for the 17 GPa sample. Corresponding values
for the (113)-(226) set essentially overlap those for
(110)-(220) and are not shown.

Table 1. Warren-Averbach average area-weighted size $\langle D_a \rangle$ and
Gauss squared crystallite size D_G values

Shock Pressure GPa	012-024		110-220		113-226	
	$\langle D_a \rangle$	D_G	$\langle D_a \rangle$	D_G	$\langle D_a \rangle$	D_G
4	341	790	242	538	238	577
17	228	480	162	392	158	408
20	184	450	165	378	143	375
22	182	420	154	343	135	319
27	148	320	143	297	130	339

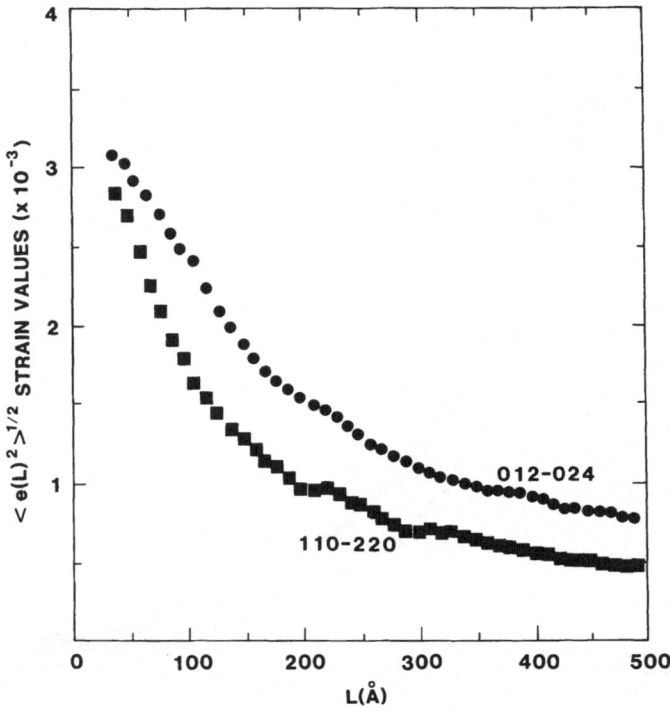

Fig. 3   Warren–Averbach values for the root average mean squared
          strain values obtained in our 17 GPa sample. Our values
          for the (113)–(226) set essentially overlap with those
          for the (110)–(220) set and are not shown.

were found to be erratic in some cases, probably because of the
"hook" effect as could be seen in Fig. 2 in various data sets. Data
in Fig. 3 suggest that crystalline anisotropy probably influences
the strain fields in Al_2O_3, preventing a more detailed analysis
of the $A^S(L)$ coefficients to obtain information on deformation
and twin fault probabilities as can be done for isotropic materials.
The fact that the strain distribution is smaller along ⟨110⟩ is
consistent with TEM results mentioned above [7]. Higher shock-
loading pressures result in larger strain values. A root mean
square strain e_L can be obtained by averaging the values of
$⟨e^2(L)⟩$ over the interval of 40 Å to the appropriate mean-area-
weighted size $⟨D_a⟩$ (Fig. 2 and Table I). Table 2 tabulates
such root mean square strain values together with strain values
(e_B) obtained by the Gauss squared method.

In the METHODOLOGY section, the Gauss squared strain e_B was
shown to be equal to about 1.25 times the Warren–Averbach root mean
square strain values e_L. Examination of Table 2 suggests this holds
for our higher shock-loading pressures. At lower shock-loading
conditions, departure of these two strain values occurs. It appears

Table 2. Warren-Averbach Root-mean square strain e_L and Gauss squared e_B strain values

(all x 10^{-3})

Shock Pressure GPa	012-024		110-220		113-226	
	e_L	e_B	e_L	e_B	e_L	e_B
4	1.70	1.80	1.65	1.32	1.50	1.36
17	2.20	2.80	1.73	2.38	1.92	2.48
20	2.47	3.85	2.25	3.12	2.25	3.25
22	2.77	4.15	2.30	3.42	2.36	3.51
27	2.90	4.50	2.47	3.61	2.57	4.29

that at small values of the size D_a, the averaging procedure employed may not be valid.

The as-received Reynolds material was ball-milled subsequent to its preparation and showed a significant residual strain value. Shock treatment increased the residual strain by about an order of magnitude. The 3 to 4×10^{-3} strain values are similar in value to those reported by Heckel and Youngblood [4] for Al_2O_3 employing undefined shock-loading pressures.

DISCUSSION

The present data appear to confirm the crystallite size reduction by 1/3 to 1/2 noted previously by Heckel and Youngblood [4]. That study employed starting material of several sizes. Assuming peak pressure loading similar to our 20 to 25 GPa values, the present results are quite consistent with these values. Pruemmer and Ziegler also found finer material decreasing in size with shock-loading pressure under less well characterized shock loading [5]. Their coarse material did not show a large reduction in size with pressure. In fact, these authors were able to ballmill material to crystallite size values smaller than all but their highest shock-loading pressure. For our alumina, size appears to be reduced essentially linearly with pressure. Strain values are essentially increased by an order of magnitude, consistent with previous results [4,5]. The mean-squared strain plots as a function of L are quite similar to those reported on cold-worked aluminum [13]. These data demonstrate that the short duration but intense shock-compression may significantly alter strain in refractories beyond that observed in lower intensity but repeated deformation by extensive ball-milling or grinding. The degree of shock modification is strongly dependent on specific shock conditions.

REFERENCES

[1] Bergmann, O.R. and Barrington, J., Effect of Explosive
     Shock Waves on Ceramic Powders, J. Am. Ceram. Soc. 49 (1966)
     502-507; Also see Barrington and Bergmann, Preparation of
     Brittle Inorganic Polycrystalline Powders by Shock-Wave
     Techniques, U.S. Patent #3,367,766 (1968).
[2] Morosin, B. and Graham, R. A., Shock-Induced Inorganic
     Chemistry, in Nellis, W. J., Seaman, L., Graham, R. A. (eds.),
     Shock Waves in Condensed Media - 1981, Menlo Park, AIP
     Conference Proceedings, No. 78, AIP, New York (1982) pp
     4-13 and references therein.
[3] Cohen, J.B., X-ray Line Broadening from Explosively Loaded
     Copper, Tran. AIME 218 (1960) 1135-1136.
[4] Heckel, R. W. and Youngblood, J. L., X-ray Line Broadening
     Study on Explosively Shocked MgO and α-Al$_2$O$_3$ Powders,
     J. Am. Ceram. Soc. 51 (1968) 398-401.
[5] Pruemmer, R. A. and Ziegler, G., Structure and Annealing
     Behavior of Explosively Compacted Alumina Powders, Proc.
     of 5th International Conf. on High Energy Rate Fabrication,
     June 24-26, 1975, Denver.
[6] Delhez, R., De Keijser, Th. H., and Mittemeijer, E. J.
     Determination of Crystallite Size and Lattice Distortions
     Through X-ray Diffraction Line Profile Analysis, Fresenius,
     Z., Anal. Chem. 312 (1982) 1-16; references therein.
[7] Yust, C. S. and Harris, L. A., Observation of Dislocation
     and Twins in Explosively Compacted Alumina, in Shock Waves
     and High-Strain Rate Phenomena in Metals: Concepts and
     Applications, (eds) Meyers, M. A. and Murr, L. E., Plenum
     Press, NY (1981) pp 881-893 and references therein.
[8] Heuer, A. H., Sellers, D. J., Rhodes, N. H., Hot-Working of
     Aluminum Oxide: I, Primary Recrystallization and Texture.
     J. Amer. Ceramic Soc. 52 (1969) 468-474; references therein.
[9] Warren, B. E., "Chap. 13-Imperfect Crystals" in X-Ray Diffrac-
     tion, Addison-Wesley Pub. Co. (1969) Menlo Park, Calif.
[10] Graham, R. A.; Morosin, B.; Venturini, E. L.; Beauchamp, E.
     K.; Hammetter, W. F., Shock-Induced Modifications of Inorganic
     Powders in Conf. Proc. Emergent Process Methods for High
     Technology Ceramics, Nov. 8-10, 1982 North Carolina State Univ.
[11] Graham, R. A., and Webb, D. M., Fixtures for Controlled
     Explosive Loading and Preservation of Powder Samples, in
     Proceedings, American Physical Society Topical Conference on
     Shock Waves in Condensed Matter, Santa Fe, NM, July 18-21, 1983.
[12] Roof, R. B. and Elliott, R. O., Evidence for the Existence
     of Faulting in a splat-cooled δ-Pu(Ti) alloy. J. Mater.
     Sci. 10 (1975) 101-108.
[13] Turunen, M. J., de Keijser, Th. H., Delhez, R., van-der Pers,
     N. M., "A Method for the Interpretation of the Warren-Averbach
     Mean-Squared Strains and Its Application to Recovery in Alumi-
     num," J. Appl. Cryst. 16 (1983) 176-182.

LINE BROADENING STUDIES ON HIGHLY DEFECTIVE TiO$_2$ PRODUCED BY HIGH PRESSURE SHOCK LOADING

B. Morosin, E. J. Graeber and R. A. Graham

Sandia National Laboratories

Albuquerque, New Mexico 87185

INTRODUCTION

Enhanced solid state reactivity of materials both during and after shock compression has been attributed to the introduction of large numbers of defects into the crystalline lattices and to reduction in the particle and crystallite size of powders [1]. In particular, orders of magnitude increases in the catalytic activity has been observed in shock-modified TiO$_2$ [2]. Line broadening of x-ray diffraction profiles provides a means to determine the coherent crystallite size and the residual lattice strain resulting from defect concentrations. The present study on shock-loaded rutile is a detailed investigation of the influence of shock loading on residual lattice strain and coherent crystallite size. Annealing of shock-modified rutile powders is also studied.

BACKGROUND

The basic principles for the analysis of peak profiles are well known and have recently been examined and discussed [3]. In a companion paper, the generally used method to interpret size-strain broadening is outlined and the identical description of variables is employed for this paper [4]. For such analysis, the Fourier coefficients, $F(L)$, for a peak profile $\tilde{f}$ free from instrumental effects may be obtained by complex division ("deconvolution" or "unfolding") of the Fourier coefficients for the broadened $\underset{\sim}{h}$ and standard g profiles.

†This work performed at Sandia National Laboratories supported by the U.S. Dept. of Energy under contract # DE-AC04-DP7600789.

Microstructural information is obtained in the direction
normal to the particular (hkl). The more thorough methods for
profile analysis need at least two orders of the particular Bragg
peak, or the (hkl) plane, considered. Furthermore, various (hkl)
planes should be examined, their integral breadths determined and
plotted according to one of several suggested methods [3]. The departure of specific data points (integral breadths times an appropriate scattering length function) from a linear behavior, based on
either higher order (hkl) sets or the entire collection of points
suggests strain and/or crystallite size along those directions to
be influenced by anisotropy in the crystal structure or strain distribution. Structurally anisotropic materials require information
from several different (hkl) planes if detailed information is
needed. In the absence of such higher order (hkl) planes, one may
consider an average crystallite size and strain, determined on the
basis of the available (hkl) planes. This procedure may obscure
the additional information on dislocation distribution (or concentration with respect to particular crystallographic planes) which
is contained in the individual broadened (hkl) peak profiles.
This is particularly the case for the unusually large dislocation
densities found in shock-modified materials.

Previous shock-loading studies on rutile have concerned
Hugoniot measurements and a high pressure transformation, [5,6]
effects of porosity (and temperature) [7], shock-modification manifested in catalytic behavior [2], electron paramagnetic resonance
[8] or thermal measurements [9]. Linde and DeCarli [6] suggested
that the presence of stacking faults in shock-synthesized orthorhombic phase TiO_2 was responsible for the fact that (111) peak
profiles were twice as broad as (110) and (020). It is believed
that our results are the only detailed studies on crystallite
size and strain in shock-modified rutile.

METHODOLOGY

Integral breadths B^F can be obtained by summing over the
Fourier coefficients F(L) of the unfolded profile. The values
obtained by the Warren-Averbach model [10] and those from the
classical methods, the Hall-Williamson and the Gauss squared, are
compared in the companion paper [4]. The Hall-Williamson method
assumes that both size and strain broadening produce a Cauchy
peak profile while the other assumes a Gaussian peak profile.
The Hall-Williamson method employs a linear (Cauchy) relation

$$B^F \cos\theta/\lambda = 1/D_c + 4e_c \sin\theta/\lambda$$

where θ and λ are the usual diffraction angle and wavelength and
D_c and e_c are the Cauchy size and strain values. Similarily under
the square (Gaussian) relationship, i.e., Gauss squared method,

$$(B^F \cos\theta/\lambda)^2 = (1/D_G)^2 + 16e_G^2 (\sin \theta/\lambda)^2$$

and D_G and e_G are the corresponding size and strain values. Plotting of these functions readily yields the corresponding size and strain. The Cauchy relationship results in larger size and smaller strain values than those obtained from the Gauss-square method.

EXPERIMENTAL

The TiO_2 employed in this study is Johnson-Matthey "Spec-pure" grade with a total metallic impurity content below 20 ppm by weight. X-ray analysis of this powder showed only the rutile phase present.

The shock modification experiments were carried out by cold pressing the powders in copper sample recovery capsules. When these capsules are placed in appropriate fixtures they preserve the sample intact for post-shock study and permit a controlled, repeatable quantifiable loading [11]. In the present work, peak pressures of 20 and 27 GPa were induced in powder compacts of either 44 or 54% solid density. The effect of changing powder compact density is to change the estimated bulk temperature rise from 150 to 450°C at 20 GPa and from 450°C to 650°C at 27 GPa. Temperature is expected to influence the degree of plastic deformation and may also act to anneal defects in both the shocked state and in the immediate post shock state.

Initial x-ray diffraction studies were performed using Ni filtered Cu Kα radiation and a standard 114.5 mm Norelco powder camera. This served to survey the shock-loaded materials for structural phase or compositional changes. The peak profiles of the x-ray diffraction lines for TiO_2 were obtained with a Siemens D-500 automated diffractometer equipped with a monochromator on the detector set so that both α_1 and α_2 were obtained. The aperture and diffraction slits are 1° and those adjacent to the diffracted beam monochromator are both 0.15°. The intensity data were taken by stepping in increments of 2θ equal to 0.01° and counting for a fixed time, generally 10 sec except for particular weak peaks (to 100 sec). The various (hkl) peaks examined are indicated below on the appropriate initial figure.

Our annealing study employed a Paar-Hi temperature furnace on the Seimens diffractometer with the sample mounted on a platinum filament. Instrumental broadening for the TiO_2 peaks was assessed from peaks obtained on the as-received material which had been prepared at high temperature and which gave exceedingly sharp lines at high 2θ as observed on films and diffractometer traces. A separate standarization run on material loaded in the high temperature furnace was obtained for assessing annealing studies. The

annealing studies were carried out by collecting data at temperature and equilibrating for an hour before profiles were collected.

Instrumental broadening was determined on the as-received
material. This material exhibited sharp lines at the highest
observed scattering angles; these are as sharp as those for Al_2O_3
annealed at 1500°C. For this standard, peak profile broadening
results from instrumental effects and contributions due to crystallite size and strain effects may be assumed negligible.

Fourier analysis of the peak profiles was aided by a computer
program UNFOLD which unfolds the broadened and standard profiles
to give the Fourier coefficients free from instrumental effects
[12]. For this particular study the integral breadth B^F of the
peak profile was obtained from the sum of these unfolded Fourier
coefficients. The multiple-line method employed the (110) and
its second order (220) as well as the (101) and its higher order
(202) for TiO_2.

RESULTS

Shock-Loaded TiO_2

Several features of shock-loaded TiO_2 are evident upon
examining the appropriate integral breadth-scattering angle plots.
Figure 1 shows such Gauss-squared data for TiO_2 shock loaded
to 20 GPa. The corresponding strain value normal to (101) planes
is larger than that normal to (110) planes,. possibly due to the
tetragonal crystal system. Note that the point for (200) planes
is significantly above the lines towards a larger integral breadth
value. This difference could correspond to either an apparent
smaller crystallite size or a possibly much larger strain value.
Extrapolation along a line with the slope obtained for the (110)
data (equal strain) would yield an apparent crystallite size near
230 Å. This value appears unlikely based on the tetragonal
structure and the 540 Å value corresponding to the 110 peaks.
One would expect a √2 relationship because of the tetragonal symmetry should stacking faults (or other size terminating effects)
be present. Furthermore, in rutile cleavage is distinct on {110}
and much less on {100} which suggests the size length along <100>
be larger than that along <110>. Since this behavior is reproducible with respect to certain (hkl) planes in other shock-loaded
TiO_2 samples, it is not an artifact of the experimental procedure.
A similar structure-related difference is observed for the other
experiment performed at 20 GPa but with a different initial
powder packing density (Figure 2). Here the value of the strain
is slightly larger, possibly due to the more ductile behavior
at the more elevated temperatures.

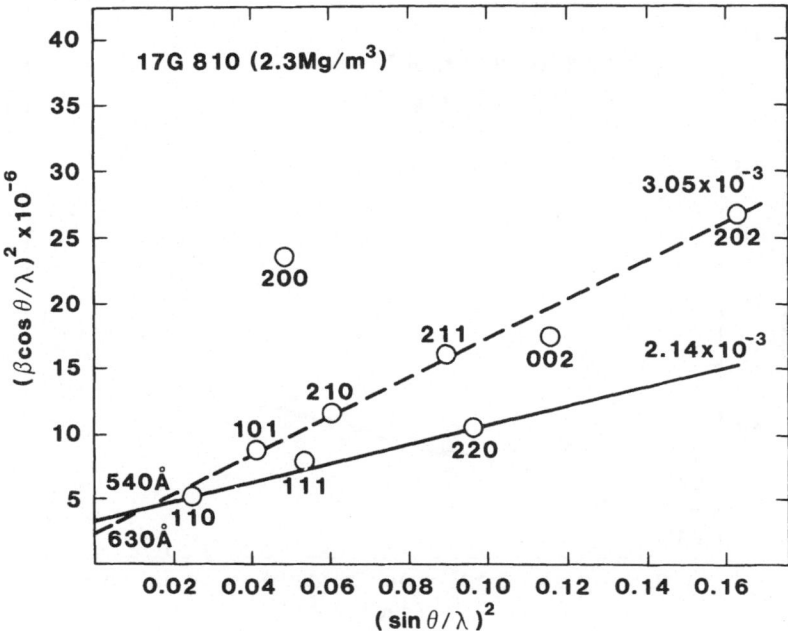

Fig. 1. Plot of the integral breadth determined from the sum of
the unfolded Fourier coefficients shows considerable
scatter due to anisotropy in the structure. Two sets
of peaks were observed with their second order (110),
(220) and (101), (202) to illustrate relative magnitude
of such strain differences. The indices of peak profiles
examined are indicated.

The results shown in Figure 3 for two experiments at 27 GPa
again show behavior influenced by annealing. Here the differences at various (hkl) is much smaller than for the 20 GPa experiments; again with the lower initial packing density, resulting
in a slightly larger retained lattice strain. In data not reported here, lower shock-loading pressures yield line broadening
data which lie nearly on the (110-220) line and anisotropies
arising from the crystal structure are not seen.

Annealing Study

In our annealing study on shock-loaded TiO_2, poorer quality
profiles were obtained due to the background scattering and to
the smaller sample volume resulting from the use of the platinum
foil. Again the (110)-(220) set is employed for determination of
apparent crystallite size and strain for this study. Reasonable

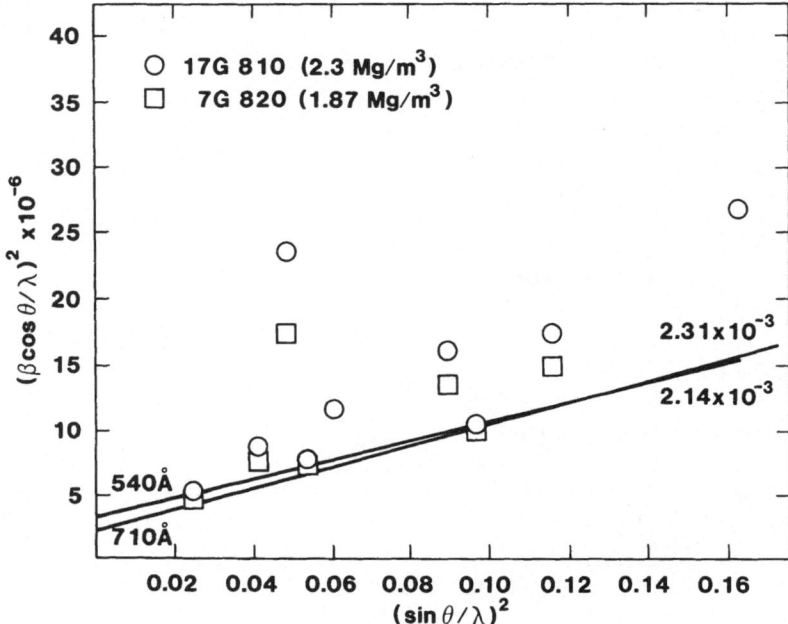

Fig. 2.   Comparison of rutile shock loaded at 20 GPa at two
          different initial sample packing densities (1.87 and
          2.3 Mg/m^3). A similar pattern for the scatter in the
          data points is found; the deviations from the (110)
          straight line is smaller for the lower density experi-
          ment. The shock-induced increase in temperature is
          about 300°C greater for the lower density experiment.

trends are observed for increasing temperature; strain decreases
as dislocations are annealed, crystallite size increases as dislocations which had formed sub-grain boundaries are annealed out.

     Residual strain and apparent crystallite size as a function
of temperature are shown in Figs. 4 and 5, respectively. Changes
are noted even at 325°C and this is in qualitative agreement with
the reduction of defects as determined by EPR and DTA measurements
on similar shock loaded samples [8,9]. Both of these functions
show rather smooth behavior.

DISCUSSION

     Our data on rutile show anisotropy resulting from the crystal
system as well as from dislocation and slip systems determined

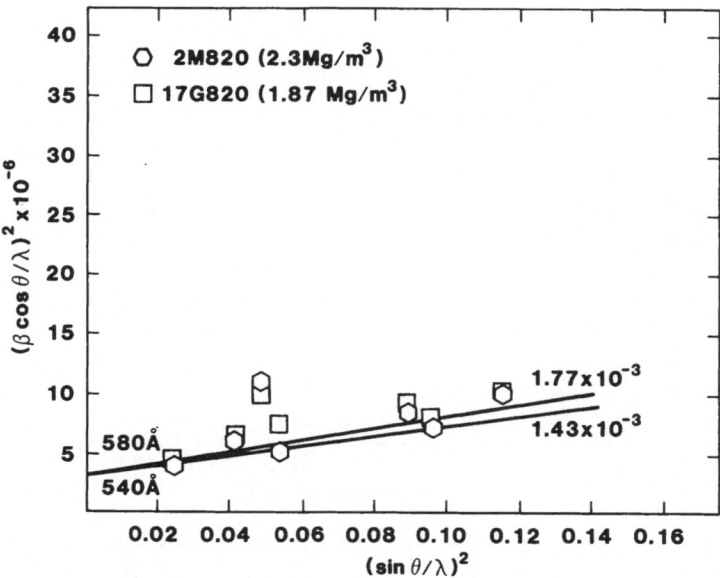

Fig. 3.   Comparison of rutile shock loaded at 27 GPa at two different initial packing densities (1.87 and 2.3 Mg/m^3).
Differences in data points is smaller and strain values
are smaller than for the 20 GPa experiments.

for the material. Extrapolation of crystallite size using the
(200) data point and using the (110)-(220) strain value appears
to yield a value smaller than expected for a material with
tetragonal symmetry. One might propose that the internal bonding
of the crystal structure might be responsible. In rutile,
chemical bonding is strongest in planes parallel with (110).
This suggests that the internal structure might be strained along
⟨100⟩, essentially affecting only angles, in a manner similar to
that observed in expansivity or compressibilities of materials.
That is, in many materials containing chemically bonded groups of
atoms, the bond distances are not altered very much, however,
angles change significantly when materials are subjected to temperature or hydrostatic pressure changes.

The shock modification pressure dependence of the strain and
crystallite size in rutile is more difficult to properly evaluate
than was the case for Al_2O_3 [4]. The annealing studies show
that higher temperature reduces strain and results in slightly
larger crystallite sizes, as dislocations which were formed are
annealed. We believe that immediately after shock compression

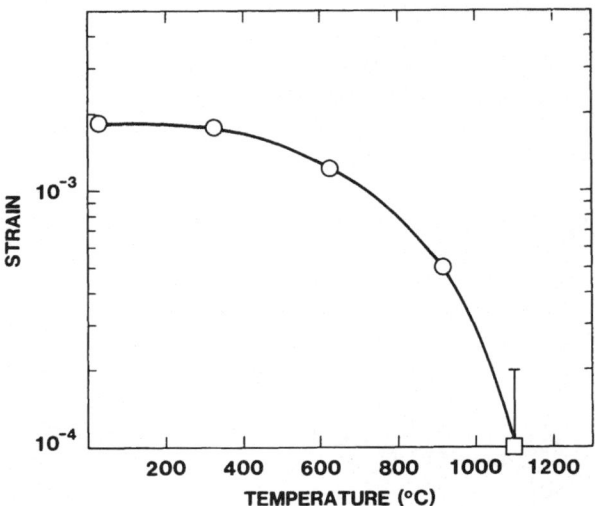

Fig. 4.   Gauss-squared strain values of shock-loaded rutile as
          a function of subsequent annealing temperature.   The
          point at 1100°C represents data from a separate experi-
          ment obtained using x-ray film and estimating the
          strain value on the basis of the highest Bragg lines.

the strain values for the lower initial packing density may have
been significantly higher than the values on the recovered material.
The shock-induced temperature is sufficient to begin annealing of
dislocations.   For the 27 GPa pressure loading the temperature is
larger than that for 20 GPa.   Thus at 20 GPa, one has nominally
more strain retained in the recovered material than at 27 GPa.
With increase in temperature excursion, strain is reduced with
respect to the lattice as well as with respect to the proposed
internal or crystal bonding structure.   These temperature effects
are manifested in a less periodic scattering in the plot seen in
Fig. 3 than in Fig. 2.

     Dislocation densities, ρ, may be calculated from either
crystallite size or strain values employing certain assumptions
and procedures [13].   For rutile, ρ_D is equal to $1 \times 10^{11}/cm^2$
for D = 550 Å and ρ_e is equal to $1.2 \times 10^{11}/cm^2$ for e = 2.3×10^{-3}.
Similarily, the stored energy can also be calculated from strain

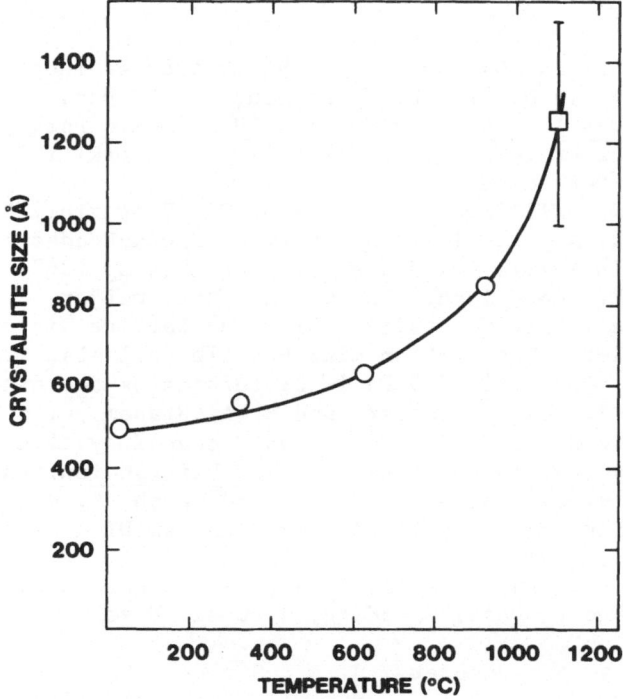

Fig. 5.   Gauss-squared crystallite size as a function of temperature on the annealing of shock-loaded rutile shows a
gradual increase with temperature.

values under some assumptions [14]. The stored energy for these
strain values is 0.32 cal/g.

These data demonstrate the high dislocation densities
generated by the short duration, intense shock compression.
Such dislocation densities and stored energy values are similar
to those found in heavily cold work metals and are higher than
can be obtained in ceramics by other means. Annealing behavior
of the strain is consistent with a reduction of defect levels as
determined by other techniques [8,9]. The defect and dislocation
levels are strongly dependent on specific shock conditions and
resulting temperature rises.

REFERENCES

[1]   Morosin, B. and Graham, R. A., Shock-Induced Inorganic
      Chemistry, in Nellis; W. J. Seaman, L., Graham, R. A. (eds.),
      Shock Waves in Condensed Media - 1981, Menlo Park, AIP
      Conference Proceedings, No. 78, AIP, New York (1982) pp
      4-13 and references therein.
[2]   Golden, J., Williams, E., Morosin, B., Venturini, E. L. and
      Graham, R. A., Catalytic Activity of Shock-Loaded TiO_2
      Powder, in Nellis, W. J., et al. loc cit. pp 72-76.
[3]   Delhez, R., De Keijser, Th. H., and Mittemeijer, E. J.
      Determination of Crystallite Size and Lattice Distortions
      Through X-ray Diffraction Line Profile Analysis, Fresenius,
      Z., Anal. Chem. 312 (1982) 1-16; references therein.
[4]   Morosin, B., E. J. Graeber, and R. A. Graham, Line Broadening
      Studies on Highly Defective Al_2O_3 Produced by High pressure
      Shock Loading, in Advances in X-Ray Diffraction, this volume.
[5]   McQueen, R. G., Jamieson, J. C., and Marsh, S. P., Shock-Wave
      Compression and X-ray Studies of Titanium Dioxide, Science
      155 (1967) 1401-1404.
[6]   Linde, R. K., and DeCarli, P. S., Polymorphic Behavior of
      Titania Under Dynamic Loading, J. Chem. Phys. 50 (1969)
      319-325.
[7]   Bugaeva, V. A.; Podurets, M. A.; Simakov, G. V.; Telegin,
      G. S.; Trunin, R. F., The Dynamic Compressibility and
      Equation of State of Rutile-Structure Minerals, Isvestiya;
      Earth Phys. 15 (1979) 19-25
[8]   Venturini, E. L., Morosin, B., Graham, R. A., "Paramagnetic
      Defects in Shock-Loaded TiO_2", in Nellis, W. J., et al.,
      loc cit. pp 77-81.
[9]   Hammetter, W. F., unpublished results.
[10] Warren, B. E., "Chap. 13-Imperfect Crystals" in X-Ray Dif-
      fraction, Addison-Wesley Pub. Co. (1969) Menlo Park, Calif.
[11] Graham, R. A., and Webb, D. M., Fixtures for Controlled
      Explosive Loading and Preservation of Powder Samples, in
      Proceedings, American Physical Society Topical Conference on
      Shock Waves in Condensed Matter, Santa Fe, NM, July 18-21,
      1983.
[12] Roof, R. B. and Elliott, R. O., Evidence for the Existence
      of Faulting in a splat-cooled δ-Pu(Ti) alloy. J. Mater.
      Sci. 10 (1975) 101-108.
[13] Williamson, G. K. and Smallman, R. E., Dislocation Densities
      in Some Annealed and Cold-Worked Metals from Measurements
      on the X-ray Debye-Scherrer Spectrum, Phil. Mag. 1 (1956)
      34-46.
[14] Faulkner, E. A., Calculation of Stored Energy from Broadening
      of X-ray Diffraction Lines, Phil. Mag. 5 (1960) 519-521.

EXAMINATION OF THE REACTION KINETICS AT SOLDER/METAL INTERFACES VIA HIGH TEMPERATURE X-RAY DIFFRACTION

P. W. DeHaven, G. A. Walker, and N. A. O'Neil

IBM General Technology Division, East Fishkill
Hopewell Junction, N.Y. 12533

INTRODUCTION

There is a great deal of interest by the electronics industry in understanding the reactions occurring at the interface between a solid metal and a liquid solder or braze. This is due to the complex nature of current microelectronics packaging, where soldering or brazing operations often involve complex metallurgies and tightly controlled furnace profiles. However, to date only a few studies have been carried out on the reaction kinetics at a liquid metal-solid metal interface.[1] This is due in part to the difficulty in carrying out such experiments. In the past, two main techniques have been used to obtain solid-state kinetic data. The first, quantitative metallography, is slow and tedious to perform, and generally of limited accuracy. The second technique involves measuring the change of some property, such as electrical resistivity, that is a function of the concentration of one of the phases. The main disadvantages of these techniques are that absolute values of concentration are not obtained, and that the relationships between a property and constitution are rarely available. In addition, most physical properties are sensitive to factors other than constitution, and interference from these factors can often result in erroneous data.

We have developed a technique which uses high temperature X-ray diffraction to obtain kinetic data from liquid metal-solid metal reactions. The method permits us to monitor the reaction _in situ_, and allows us not only to collect kinetic data, but also to obtain structural data on the formation of intermetallic phases. An important advantage of the X-ray technique results

389

from the fact that the physics of diffraction from
polycrystalline materials is well understood. This enables one
to correlate the volume of a particular phase to the diffracted
beam intensity of a reflection from that phase. High temperature
diffraction has already been successfully used to obtain kinetic
data from solid state reactions.[2,3] We are currently using the
technique to investigate the interaction between various metals
and a liquid lead/tin solder (40 weight % Pb - 60 weight % Sn).
To date two metals, nickel and gold, have been studied, the
results of which are presented here.

SAMPLE PREPARATION

     Since we were interested in observing the reactions
occurring at the solder/metal interface, it was necessary to
fabricate the samples so that the X-ray beam could penetrate to
the interface. This was accomplished by mounting a thin foil
(less than five microns) of the metal over the solder (in the
form of 1.3mm thick preformed strips). In order to maintain
rigidity in the foil, two different techniques were developed.
For the case of nickel, we used strips of a copper/beryllium
sheet, approximately 1.5mm thick, which had been plated on both
sides with two microns of nickel. Suitably sized samples
(5 X 15mm) were first cut from the strip. The top layer of
nickel was then removed, after which a selective etch was used to
remove most of the copper. The remaining copper formed a "frame"
which kept the bottom nickel layer rigid. For the case of gold,
we were able to directly place a one micron foil over both the
solder and the sample holder. Adhesion of the foil to the sample
holder during subsequent heat treatments was found to be
excellent.

DATA COLLECTION/ANALYSIS

     The diffraction studies employed a Rigaku[4] Theta-Theta Wide
Angle goniometer with an attached furnace. The theta-theta
geometry permits the sample to remain stationary in the
horizontal plane, which is essential if liquid phases are to be
studied. Sample heating is accomplished by the sample holder,
which doubles as a resistance heater. A second heater is mounted
above the sample to ensure uniform heating. Both the goniometer
and the furnace are interfaced to an IBM Series/1 computer. This
permits simultaneous control over the diffraction scans and
sample heating.

     All experiments to date have been carried out in a reducing
atmosphere (N_2 spiked with 10-15% H_2). Monochromated copper
radiation and a scintillation detector were used for data

collection. To study the reaction kinetics, each sample was
rapidly heated (2–5°C/second) to the desired temperature.
Reaction data was obtained by monitoring the intensity of a
single, well resolved, and reasonably intense reflection as a
function of time. While it was theoretically possible to monitor
either the growth of an intermetallic or the disappearance of the
base metal, we found the latter to be the most practical. Some
characteristic diffraction profiles are illustrated in Fig. 1 for
the reaction of nickel with 40/60 lead/tin at 400°C.

RESULTS

    It has been found empirically that a wide variety of
reactions can be described by a generalized equation (termed the
Johnson–Mehl equation) of the form:[5]

$$X_m(t) = 1 - \exp\,[-K(T)t^n] \tag{1}$$

where $X_m(t)$ represents the volume fraction of metal m
isothermally transformed at time t, while $K(T)$ and n are
constants, independent of both time and X. The value of n
represents the order of the reaction, and may take on any
positive value (0.5 to 2.5 is the most common range). Equation
one can be re-written as:

$$\log\,\log\,[1/(1-X_m(t))] = n\,\log\,t + \log\,K(T) \tag{2}$$

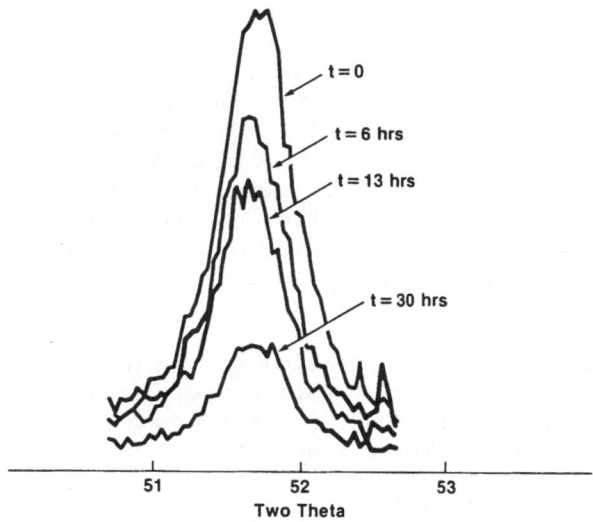

Fig. 1.   Reaction of nickel with 40/60 Pb/Sn at 186°C.
          The intensity profiles are of the Ni (200)
          reflection as a function of time (t).

If a straight line results from a plot of log log $[1/(1-X_m(t))]$
versus log t, then the reaction conforms to the Johnson-Mehl
equation with order n (obtained from the slope of the line). For
the case of a sufficiently thin film, one can, to a first
approximation, define $X_m(t)$ in terms of the intensity (in
arbitrary units and corrected for background) of a scanned
reflection, $I_{hkl}(t)$:[2]

$$X_m(t) = [I_{hkl}(t=0) - I_{hkl}(t)]/I_{hkl}(t=0). \qquad (3)$$

This permits one to obtain kinetic information from a
diffraction experiment, provided that the time required to scan
the reflection is small compared to the rate of the reaction. In
addition, the activation energy for the reaction can be obtained
using an Arrhenius plot based on the equation:

$$\ln t_{0.5} = E_a/k_b T + const.$$

where $t_{0.5}$ is the time required for 50% transformation at
temperature T, and k_b is the Boltzman constant.

Figures 2 and 3 summarize the reaction kinetics of pure
nickel with 40/60 Pb/Sn solder. The average value of n obtained
from Fig. 2 was found to be 0.69. Interpretation of this value
is difficult, as any model must take into account the effect of
two simultaneous interfacial reactions. One of these involves
the liquid solder/intermetallic interface, while the second
involves the solid nickel/intermetallic (Ni_3Sn_4 or Ni_3Sn_2)
interface. Most current theoretical models are concerned with
solid state transformations and involve only a single interface.
The activation energy, E_a, obtained from the Arrhenius plot
(Fig. 3) was found to be 6.5 Kcal/mole. This value is notably
higher than that reported for intermetallic growth between nickel
and liquid tin (4.1 Kcal/mole).[1] However, as our study involved
a lead/tin alloy as opposed to pure tin, the discrepancy may be
the result of an additional energy term, which represents the
energy required to transport the tin to the solid/liquid
interface.

We discovered that the gold reacted too rapidly with the
solder for us to obtain good quantitative data (dissolution rates
of as high as one micron per second at approximately 200°C have
been reported elsewhere[6]). We were able to obtain qualitative
data by fixing our detector at the most intense point in the
diffraction profile of the Au (111) reflection. The intensity at
this point was then monitored at one-second intervals until all
the gold had been consumed. The result of a typical experiment
is illustrated in Fig. 4, along with a typical run involving
nickel. Note that even though the temperature at which the

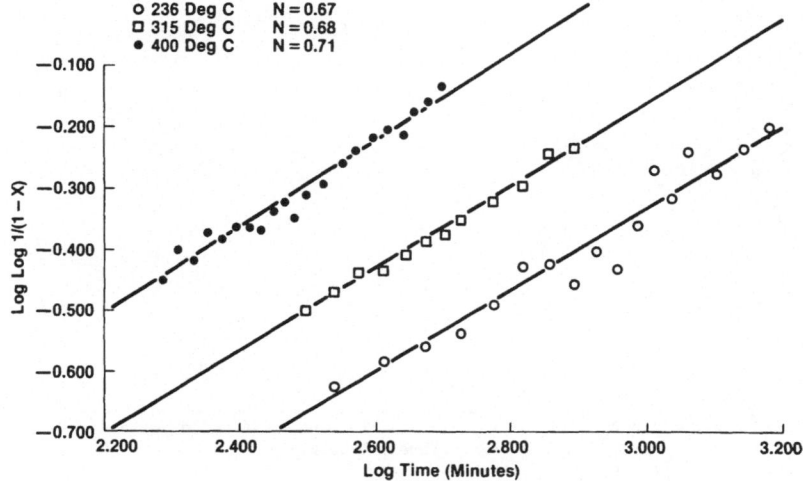

Fig. 2.   Reaction of nickel with 40/60 Pb/Sn.
           Determination of the reaction order, n.

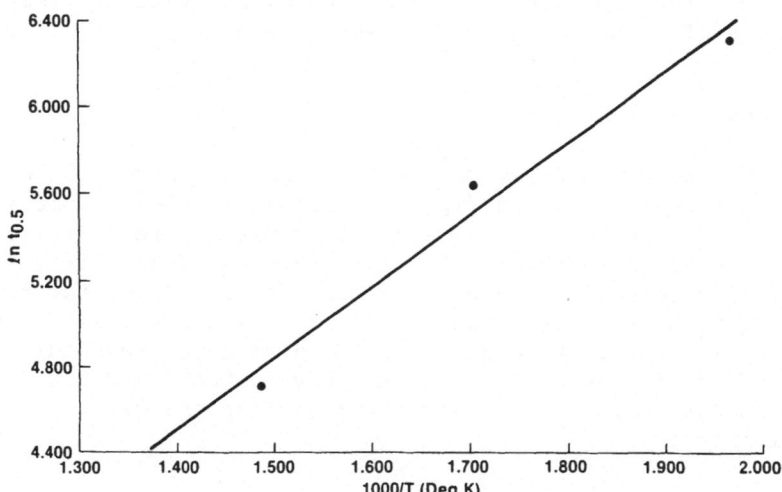

Fig. 3.   Reaction of nickel with 40/60 Pb/Sn.
           Arrhenius plot of time ($t_{0.5}$) for 50%
           transformation vs reciprocal of
           temperature (T).

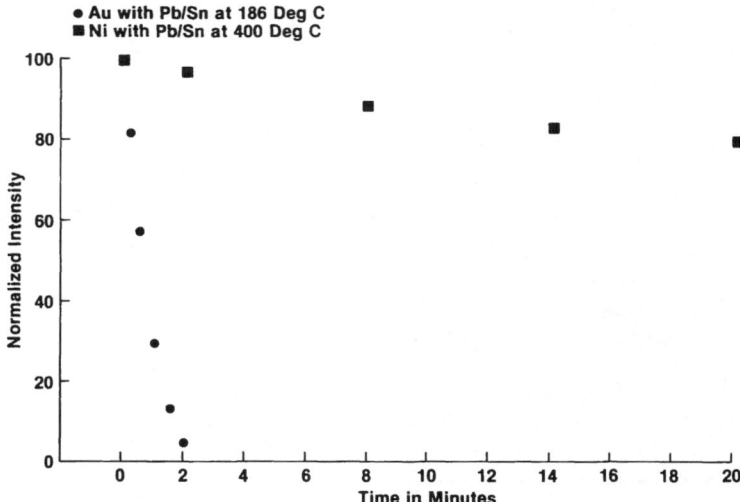

Fig. 4.   Comparison of reaction rates of nickel and
          gold with 40/60 Pb/Sn solder.

nickel/solder kinetics were studied was 400°C, while the
gold/solder kinetics were studied at 186°C, the rate of reaction
of the gold with the solder was notably faster.

CONCLUSIONS

    High temperature X-ray diffraction can be a powerful tool in
studying the reactions occurring at a liquid metal-solid metal
interface. The technique allows one to monitor a reaction in
situ, and to obtain both kinetic and structural data. If a
scintillation detector is used, the main limitation of the method
results from the time required to scan a reflection (limits one's
ability to monitor fast reactions). We plan to address this
problem by installing a position-sensitive detector on our
diffractometer, which should significantly improve data
collection rates.

ACKNOWLEDGEMENTS

    We are indebted to R. Anderson, M. Sullivan and E. Harris
for their assistance in sample procurement and preparation. In
addition, we are grateful for the valuable advice provided by
C. Goldsmith, T. Bowmaster and T. Nunes.

REFERENCES

1.   S.K. Kang and V. Ranachavabrav, Scripta Metallurgica,
      14, 421 (1980).
2.   A.K. Sinha and T.E. Smith, J. Appl. Phys., 44(8), 3465
      (1973).
3.   C.C. Goldsmith, G.A. Walker and M.J. Sullivan, 16th Annual
      Proceedings Reliability Physics IEEE, p. 64 (1978).
4.   Rigaku USA, Inc., 3 Electronics Avenue, Danvers, MA 01923.
5.   J. Burke, "The Kinetics of Phase Transformations in Metals",
      Pergamon Press, Oxford, 1965.
6.   E.W. Brothers, The Western Electric Engineer, 25(2), 48
      (1981).

HIGH TEMPERATURE GUINIER X-RAY DIFFRACTOMETRY FOR THERMAL EXPANSION MEASUREMENTS IN THE HEXAGONAL FORM OF CORDIERITE ($2MgO \cdot 2Al_2O_3 \cdot 5SiO_2$)*

J. S. Pressnall**, J. J. Fitzpatrick and Paul Predecki

University of Denver, Denver, CO   80208

ABSTRACT

A computer-controlled high temperature Guinier diffractometer system for accurate determination of lattice thermal expansion is described. A critical test of the system using α-Al_2O_3 (0.3μ polishing alumina) showed close agreement with the single crystal expansion data of Wachtman et al.[4] Lattice thermal expansion of cordierite doped with the following dopants: Ge^{+4}, P^{+5}, Zn^{+2}, Li^{+1} and Ca^{+2} was investigated. Of these the Li^{+1} at the 5% level (5% of Si^{+4} replaced by Li^{+1} + Al^{+3}) produced the largest decrease in mean lattice expansion.

INTRODUCTION

Cordierite ($2MgO \cdot 2Al_2O_3 \cdot 5SiO_2$) has generated considerable interest in the ceramics industry due to its low coefficient of thermal expansion. This property along with its good corrosion resistance and relatively good mechanical strength (MOR, 12-20 KSI) makes it a candidate for applications in heat exchanger systems and as a substrate material for automotive catalytic converters. The most common synthetic form of $2MgO \cdot 2Al_2O_3 \cdot 5SiO_2$ is hexagonal (P6/mcc) and is termed indialite whereas cordierite refers to the slightly distorted and ordered orthorhombic form (Cccm).[2] It is common practice, however, to refer to the hexagonal form as cordierite also.

---

*Work supported by DOE on contract DE-AC02-81ER10896.
**Present address: Systems Planning Corp., Arlington, VA

Values of the mean polycrystalline thermal expansion coefficient, $\overline{\alpha}_p$ for cordierite reported in the literature vary widely from 9 to > 30 x 10^{-7}/°C (25-800°C).   The lattice expansions α_a and α_c also vary, as shown in Fig. 1.   The reasons for these variations are thought to be in part due to differences in impurity content.   A study has therefore been undertaken to examine the thermal expansion of analytically pure cordierites doped with varying amounts of foreign ions which were initially selected by virtue of size and charge to reside on specific sites in the structure and thus reveal the sensitivity of the lattice expansion behavior to the occupancy of those sites.   The method chosen to investigate the thermal expansion behavior of these doped cordierites was X-ray diffraction utilizing a system unique to this laboratory, that of computer-controlled high-temperature Guinier diffractometry.

INSTRUMENTATION

A schematic of the overall data-collection system is shown in Fig. 2.   Salient features of the system include a conventional Huber-

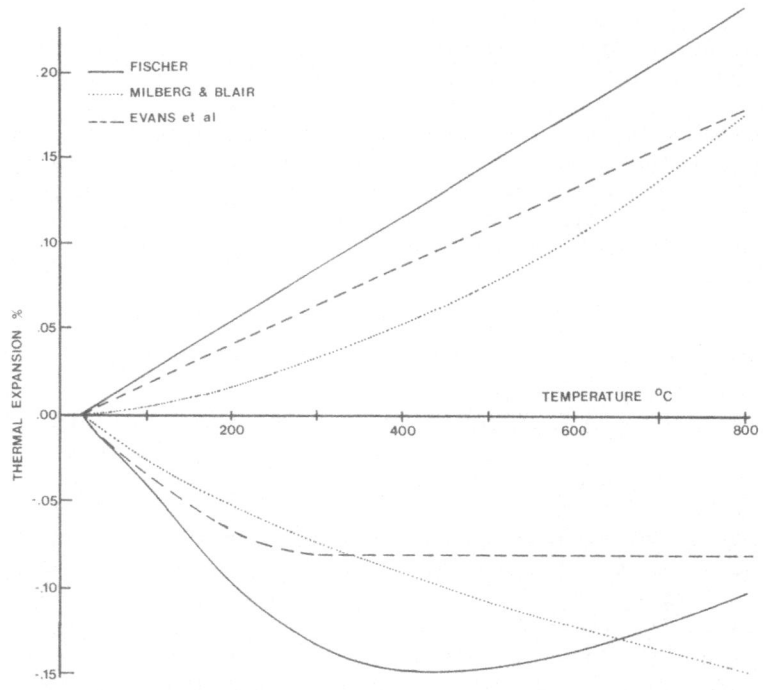

Fig. 1.   Thermal expansion of hexagonal cordierite along a and c (from D. L. Evans et al., A. Ceram. Soc. Bull., <u>63</u>, 629 (1980).   M. E. Milberg and H. D. Blair, J. Am. Ceram. Soc. <u>60</u>, 372 (1977)).

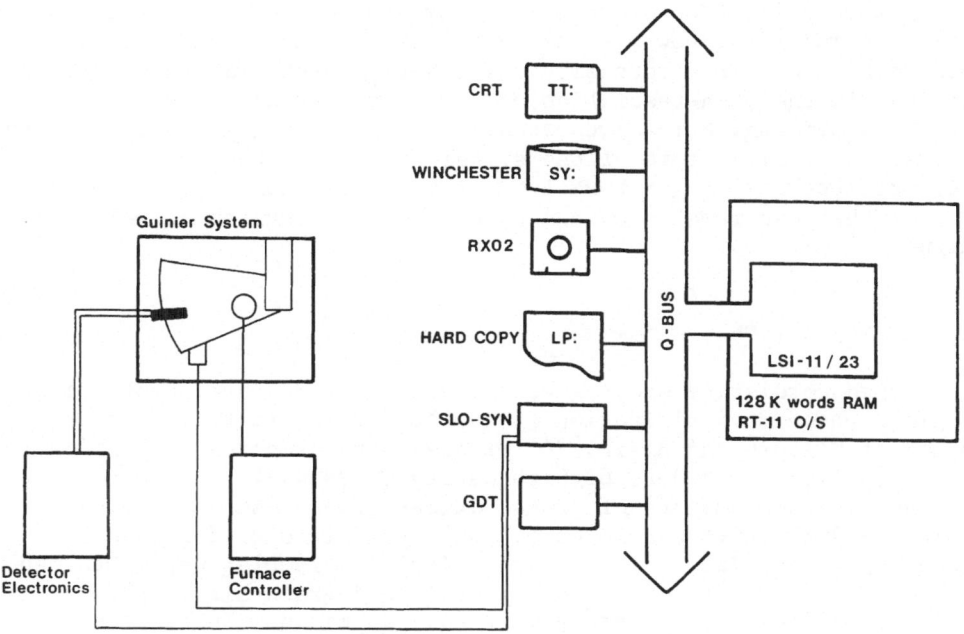

Fig. 2.   Computerized X-ray system schematic.

Guinier powder diffractometer in which the 2θ slewing motor has been
replaced with a stepping motor which is under direct computer control; the sample furnace which is currently under the control of an
off-line microprocessor; and the computer system, an LSI-11/23
operating under TSX-Plus, a multi-tasking RT-11 upgrade. The standard Huber furnace was reconstructed so as to double the horizontal
dimensions of the hot zone and allow use of a larger sample. The
overall system has many peripheral devices which serve to store and
display data. Communication between the 11/23 and the X-ray diffractometer is accomplished through an interface board and the slo-
syn translator utilizing a CRT as the console terminal. Data acquisition and reduction are carried out through utilization of vendor
supplied software. (Nicolet Dual L-11 System software).

   The Guinier X-ray diffraction system must be carefully aligned
in order to produce reliable data with good counting statistics.
The procedure used to align the system has been outlined in another
paper in these proceedings and will not be repeated here. One important point regarding final alignment of the system should, however, be made. Final tuning of the X-ray optics depends on the
application for which the system will be used. If the separation of
closely spaced peaks is desirable for the intended application, the
optics should be left in a high-resolution configuration at the end
of the alignment procedure. This is accomplished by setting all
limiting apertures and slits, especially the receiving slit, to the
minimum opening which still provides an acceptable peak-to-background.

During the utilization of this X-ray system for thermal expansion
work on cordierite, however, peak-to-peak resolution was not as
important as acquired intensity. The X-ray system was therefore not
utilized in the high-resolution mode, but rather in an intensity-
maximized mode which involved opening the receiving and axial divergence slits to allow more of the signal to pass to the counter. This
step results in a larger FWHM for any one peak; however, the position of the peak remains unchanged provided the system is well
aligned.

SAMPLE PREPARATION

Doped cordierite samples were prepared from constituent oxides
by glass phase devitrification at 1000°C for 24 hrs following the
method of Milberg and Blair.[3]   Glasses were prepared by melting
constituent oxide powders of high purity (> 99.999%) well mixed
with appropriate amounts of dopant oxides of highest available
purity (> 99%) in Pt crucibles at 1600°C for 12 hrs, followed by
quenching in deionized water. All devitrified samples were examined
in the SEM for phase content homogeneity and grain size. After
devitrification,samples were ground to < 5μ mixed with polyethylene
oxide binder and packed into a layer 7 x 2.5 x 0.3 mm thick within
a Pt loop. This loop was then mounted in the goniometer head on
the Guinier system and aligned tangent to the focussing circle. The
furnace was then slowly ramped up to 600°C to burn out the binder before starting a run.

SYSTEM SHAKEDOWN

The measurement of small shifts in peak position on a system
such as this can be subject to many possible inaccuracies. These
inaccuracies take the form of random and systematic errors involving
sample movement during the heating cycle, inaccuracies in temperature measurement, inaccuracy in sample position relative to the X-
ray optics and, in the case of cordierite, required use of peaks in
the inherently inaccurate low 2θ region. Much effort was expended
early in the project to minimize these effects wherever possible.

Temperature was measured with a fine Pt/Pt-13% Rh thermocouple
standardized and placed adjacent to the center of the sample. Sample
temperature fluctuated < ± 1°C at setpoints. The temperature distribution in the hot zone was mapped to determine thermal gradients
and the furnace was eventually packed with fiberfrax to minimize heat
loss through the bottom opening.

Inaccuracies in room temperature sample positioning relative to
the X-ray optics were corrected for by use of an internal Si standard (NBS SRM-640). A critical test of the whole system including

the possibility of sample motion during heating or cooling, consisted of determining the lattice expansion of α-Al_2O_3 (Buehler Ltd.
#40-6352, 0.3μ AB, α-polishing alumina). Data were acquired over
the same 2θ range as used subsequently for the cordierite. The
results of these runs (Fig. 3) agree remarkably well with the single
crystal quartz dilatometer data of Wachtman et al.[4] and indicate
that sample motion during heating or cooling was not an important
factor in overall accuracy. We feel that the small difference in
the lattice parameters between our data and those of Wachtman et al.
is probably due to small differences in composition.

DATA COLLECTION AND REDUCTION

    X-ray scans of the (004), (224), (512) and (433) peaks indexed
on the indialite structure were collected using a step-width of
0.03°2θ for times varying from 50 to 150 seconds per step so as to
obtain integrated intensities $\geqslant 10^4$ counts per peak. Data were
taken in $\sim$ 100°C increments from room temperature to 810°C using
$CuK\alpha_1$ radiation.

    The peaks were profile fitted using a learned peak shape algo-
rithm available in the Nicolet software to locate the exact peak
position. The d-spacings thus calculated were used as input to a
least squares cell refinement program: ELST*. The room temperature
d-spacings of each sample were corrected in separate experiments
using an internal Si standard. The standard deviations of the cell
constants ranged from 1 part in 10^5 to 6 parts in 10^4). The latter
error was confined to the 5% Zn^{+2} doped samples and is felt to be
due to the onset of a structural distortion.** Details of the data
reduction and sample preparation are given elsewhere.[5]

    Data in hand at the present time are summarized in Table 1.
The 1% and 5% values in Table 1 are as follows: for tetrahedral
dopants, 1% (or 5%) of the host tetrahedral cations are substituted
for by the dopant (Ge^{+4} for Si^{+4} and $P^{+5} + Al^{+3}$ for $2Si^{+4}$). For
octahedral dopants, 1% (or 5%) of the host Mg^{+2} are replaced by the
dopant (Zn^{+2}). For c-axis channel dopants, 1% (or 5%) of the Si^{+4}
are substituted for by the dopants ($Li^{+1} + Al^{+3}$ for Si^{+4} and $\frac{1}{2}Ca^{+2} +$
Al^{+3} for Si^{+4}).

    As an example of the expansion data, the results for Li doped
samples are shown in Fig. 4. The c-axis expansions for most of the
dopants are non-linear and show trend reversals with both temperature

---

*We are very grateful to Eric Gabe of NRC Ottawa, Canada, for pro-
  viding this program.
**G. V. Gibbs, personal communication.

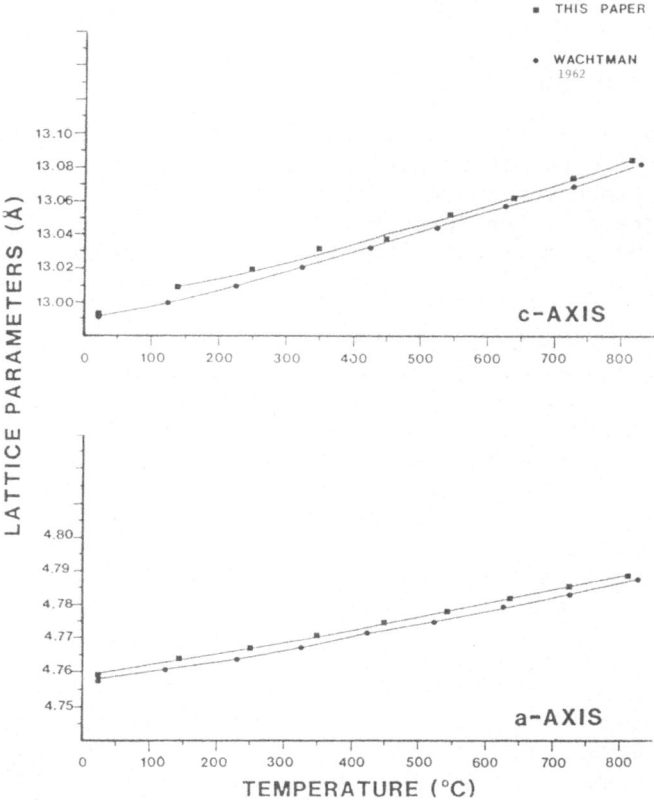

Fig. 3.   Linear thermal expansion of α-Al_2O_3. Lattice parameters
were obtained from a least squares cell refinement using
the 113, 024 and 030 reflections.

Table 1. Summary of Doped Hexagonal Cordierite
Expansion Data

Dopant	Expected Site	$\alpha_a \times 10^6 (C^{-1})$	$\alpha_c \times 10^6 (C^{-1})$	$\overline{\alpha} = (\frac{2}{3}\alpha_a + \frac{1}{3}\alpha_c)$ $\overline{\alpha} \times 10^6 (C^{-1})$	Range (°C)
Pure		3.27	+ .28	2.27	25–639
Ge–1%	T	2.94	− .40	1.83	25–810
Ge–5%	T	2.54	− .30	1.59	25–810
Li–1%	C	3.52	− 1.70	1.78	25–810
Li–5%	C	2.32	− 3.09	.52	25–725
Zn–1%	O	2.56	− .75	1.46	25–810
Zn–5%	O	2.21	− 1.23	1.06	25–810
Ca–1%	C	2.63	+ .05	1.77	25–810
P –5%	T	3.64	− 2.65	1.54	25–725

T = tetrahedral, O = octahedral, C = c=axis channel.

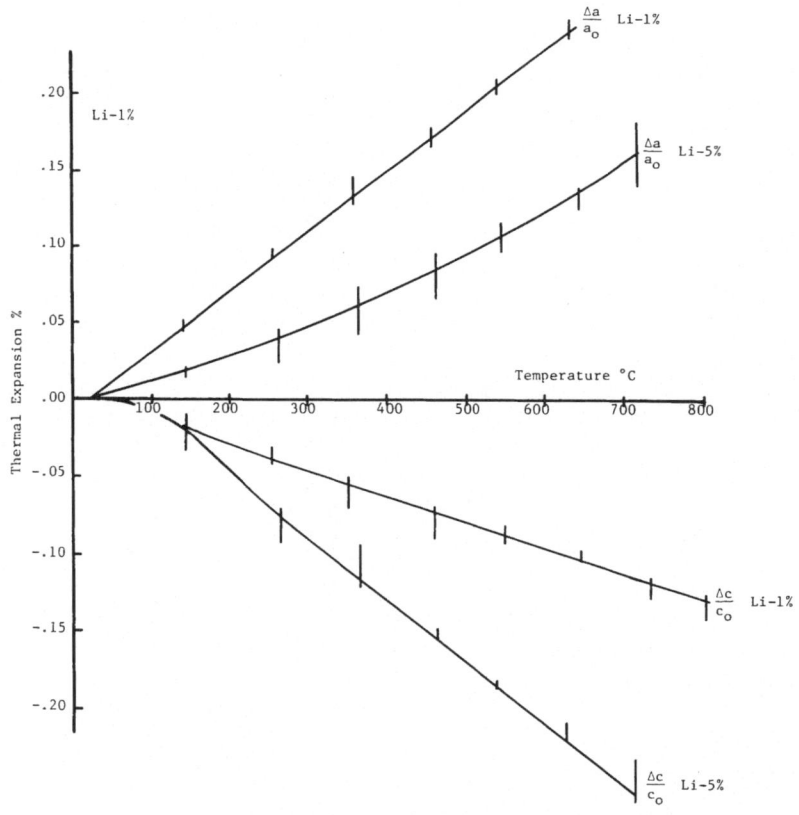

Fig. 4. Lattice expansion data for Li^+ doped hexagonal cordierite
with 1% and 5% of the Si^{+4} replaced by (Al^{+3} + Li^{+1}).

and dopant level. A discussion of the dopant effects in terms of possible expansion models will be given in a future paper.

REFERENCES

1.   E. P. Meagher and G. V. Gibbs, The Polymorphism of Cordierite I:
       The Crystal Structure of Low Cordierite, <u>Canad</u>. <u>Mineralogist</u>
       15:43 (1977).
2.   G. V. Gibbs, The Polymorphism of Cordierite I: The Crystal
       Structure of Low Cordierite, <u>Amer</u>. <u>Mineralogist</u> 51:1068
       (1966).
3.   M. E. Milberg and H. D. Blair, Thermal Expansion of Cordierite,
       <u>J</u>. <u>Am</u>. <u>Ceram</u>. <u>Soc</u>. 60:372 (1977).
4.   J. B. Wachtman, Jr., T. G. Scuderi, and G. W. Cleek, Linear
       Thermal Expansion of Al_2O_3 and ThO_2 from 100 to 1100°K,
       <u>J</u>. <u>Am</u>. <u>Ceram</u>. <u>Soc</u>. 45:319 (1962).
5.   J. S. Pressnall, M.S. Thesis, Univ. of Denver (Dept. of Physics),
       (1983).

CONTINUING DEVELOPMENT OF MERCURIC IODIDE

X-RAY SPECTROMETRY

J.S. Iwanczyk, A.J. Dabrowski, G.C. Huth, and W. Drummond*

University of Southern California
Institute for Physics and Imaging Science
4676 Admiralty Way, Suite 932
Marina del Rey, CA 90231

*Tracor Xray Inc.
Mountain View, CA 94043

INTRODUCTION

The continuing development in recent years of mercuric iodide room temperature x-ray spectrometers has led to steady improvment in the spectral energy resolution capability of these systems. This has been due largely to the continuing reduction in the electronic noise level of the associated preamplification electronics. It has been demonstrated that a system consisting of a mercuric iodide detector in combination with a pulsed-optical feedback preamplifier provides superior energy resolution performance in x-ray spectrometry in comparison to the other types of preamplification (1). Previously, results have been reported by us of the energy resolution of such systems with both the detector and input FET of the preamplifier at room temperature (1-3) and with the input FET cooled by liquid nitrogen and the mercuric iodide x-ray detector slightly cooled (4,5). This paper presents results recently obtained with the preamplifier input FET cooled by liquid nitrogen and the mercuric iodide detector at room temperature. Spectra of manganese K x-rays as well as spectra of different samples exited with an x-ray generator are presented. Additionally, x-ray spectra obtained with such a mercuric iodide spectrometer are compared to similar spectra obtained with a cryogenically cooled, state-of-art Si(Li) spectrometer.

THEORETICAL COSIDERATIONS

One of the most important parameters characterising a solid state radiation detector is its spectral energy resolution. The energy resolution is measured as a full width at half maximum (FWHM) as related to the variation in the amplitude of pulses in the response the monoenergetic ionization radiation. Total variance of the pulse amplitude distribution can be written as a sum of three components:

$$\sigma^2 = \sigma_{el}^2 + \sigma_{gen}^2 + \sigma_{tr}^2 \qquad (1)$$

where: σ_{el}^2 -the variance due to electronic noise of the system

σ_{gen}^2 -the variance due to the electron-hole generation process in the detector

σ_{tr}^2 -the variance due to trapping phenomena in the charge collection process

The electronic noise component is independent of energy and contributes equally at all photopeak energies in a symmetrical way. The statistical spread associated with the charge generation process is symmetrical and proportional to the square root of the energy of incident radiation. The function describing the effect of charge loss due to trapping is complex. In the case of low energies, when the incident radiation interacts near the entrance electrode resulting in one charge carrier (the electron) collection, and the detrapping effect is neglected, the spread in the drift lengths of carriers due to inhomogeneity of the detector material usually plays a major role in the trapping component. This contribution depends on the detailed nature of the spread in drift length λ (where $\lambda = \mu\tau E$; $\mu\tau$ is mobility-lifetime of the charge carrier and λ describe the mean distance which carrier will drift in an electric field E before it is trapped). The shape of the photo-peak is affected by spread in λ through the Hecht relation (6) and will be proportional to the energy.

The electronic noise contribution to the energy resolution is relatively easy to determine by measuring the width of an electronic pulser peak. On the other hand, the statistical variation in the number of generated charge carriers and the spread in number of

trapped charge carriers is difficult to separate, since
value of the Fano factor for mercuric iodide is not
known and neither is the form of the inhomogeneity
distribution.

At extremely low ("ultralow") x-ray energies if
there are no problems related to the window effect (2)
the photopeak resolution should approach the electronic
noise linewidth because effects which are proportional
to $E^{1/2}$ or to E become negligible and equation (1) will
take the form $\sigma^2 = \sigma^2_{el}$.

The best values obtained for the spectral energy
resolution of a mercuric iodide radiation detector at
room temperature, with all of the associated
electronics at room temperature are 295 eV (FWHM) for
the Mn K_α line (5.9 KeV) and 245 eV for the Mg or Al
K line (1,2). In both cases, the electronic noise
level ΔE_{el} of the system as measured from the
linewidth of a pulser peak, was 225 eV (FWHM). On the
other hand, the electronic noise level characteristic
of Si(Li) spectrometers with detector and input FET
cryogenically cooled is lower than 100eV. Leakage
current and capacitance of mercuric iodide detectors
operating at room temperature conceptually allow
to approach the same noise levels as cooled Si(Li)
systems. Using preamplifier with its FET cooled by
liquid nitrogen and a mercuric iodide detector at room
temperature offers the possibility of significant
improvement in energy resolution of mercuric iodide
x-ray spectrometry system. For ultra-low x-ray
energies the linewidth of x-ray peaks should approach
the linewidth of the pulser peak where there is large
room for improvement. For higher energies and if the
trapping effect is neglected equation (1) becomes:

$$\sigma^2 = \sigma^2_{el} + \sigma^2_{gen} = \sigma^2_{el} + FwE \qquad (2)$$

where: w=4.2 eV is mean energy required for creation
                 of an electron-hole pair in HgI_2
          F   is the Fano factor
          E   is an energy of the incident
                 radiation.
Since there is a normal distribution of both
electronic and statistical components, equation (2) can
be rewritten in terms of energy resolution FWHM where :

$$FWHM = 2.35\ \sigma = 2.35\ (\ \sigma^2_{el} + FwE\ )^{1/2}$$

For an energy of 5.9 keV and an electronic noise
level of 100 eV (FWHM) and assuming a Fano factor value
of 0.19 (7), the energy resolution of a mercuric
iodide spectrometer should approach a value of 190 eV
(FWHM).

An experiment was designed to justify the above
assertions. The   purpose of this experiment was to
attempt to come closer to the theoretical limits of
energy resolution for mercuric iodide spectrometers
and to verify other factors which would conceptually
impose restrictions on attainable energy resolution
level such as trapping-detrapping phenomena and excess
electronic noise due to factors and the methods of
detector construction and fabrication.

EXPERIMENTAL

The experimental arrangement consisted in part of
a standard Si(Li) liquid-nitrogen cryostat modified to
accommodate   the   mercuric   iodide detector. The first
stage FET was thus at a temperature close to its
optimum operating point. The mercuric iodide detector
used had 1 mm^2 . and a thickness of 0.4 mm.   It was
mounted on an alumina substrate together with a heater
and thermistor for temperature control and monitoring.
The temperature of the mercuric iodide detector was
held constant at 300 K.

Standard Tracor nuclear electronics were utilized.
These consisted of a Model 505 pulsed optical feedback
preamplifier, a TX 221232 amplifier/pulse processor,
and a TN 2010 ADC and computer system. A high quality
Si(Li) 10 mm^2 detector was used for comparison with
mercuric iodide arrangement. Fluorescent x-ray spectra
from different samples were obtained by irradiation
using an x-ray generator with a silver anode.

Figure 1 shows a spectrum obtained using Mn K
x-rays (5.9keV) and the mercuric iodide detector
coupled to a pulsed-light feedback preamplifier. The
solid line represents the spectrum obtained with the
preamplifier input FET at room temperature. The energy
resolution is 380eV (FWHM). The dotted line represents
the spectrum obtained with the input FET cooled with
liquid nitrogen.   The energy resolution is 175eV
(FWHM).   In both cases the Mn K$_\beta$ peak is clearly
visible. Figure 2 shows on a logarithmic scale Mn K
x-ray spectra taken with the two spectrometers.
Spectrometer "A" represented by the solid line, is the

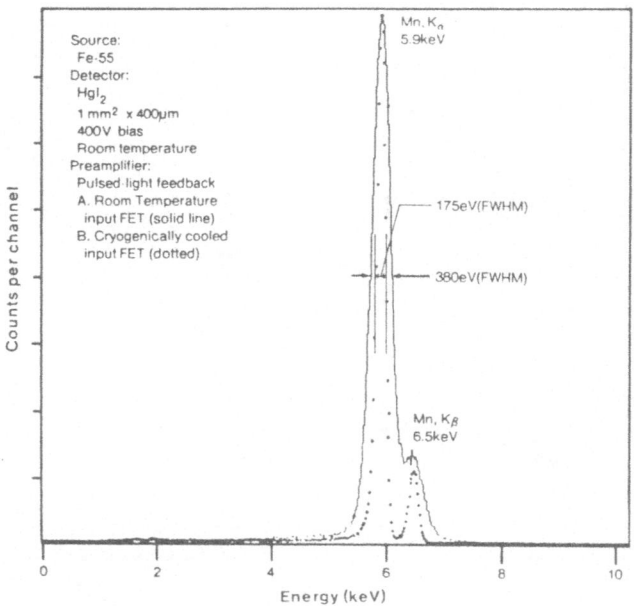

Figure 1.   A spectrum of Mn K x-rays (5.9keV) taken
with an mercuric iodide x-ray detector coupled to a
pulsed-light feedback preamplifier. The solid line
represents the spectrum obtained with the preamplifier
input FET at room temperature. The dotted line
represents the spectrum obtained with the input FET
cooled with liquid nitrogen.

mercuric iodide detector coupled to the   pulsed-light
feedback preamplifier with cryogenically cooled input
FET. Spectrometer "B" represented by the dotted line
is the cryogenically cooled Si(Li) system. The peak to
background ratio of about 400 and 2000 for the mercuric
iodide and Si(Li) spectra respectively is indicated.
The smaller active area of the mercuric iodide detector
results in a pronounced fringing electric field effect
and some trapping phenomena would seem to be the cause
of the higher background counts in the mercuric iodide
spectrum.   No attempts have yet been made to optimize
the mercuric iodide detector geometry to this end (as
has been done in the case of silicon). Figure 3 shows
a spectrum for the Al K x-ray taken with the mercuric
iodide detector (again with cryogenically cooled input
FET and the pulsed-light preamplifier). The energy
resolution is 145eV (FWHM). Figure 4 is a spectrum of a
sample of "old paint" taken with the mercuric iodide
spectrometer. Figure 5 shows a spectrum of 0.14%
sulfur in crude oil taken with the same spectrometer.
Figure 6 shows a spectrum of 100 ppm trace metals in

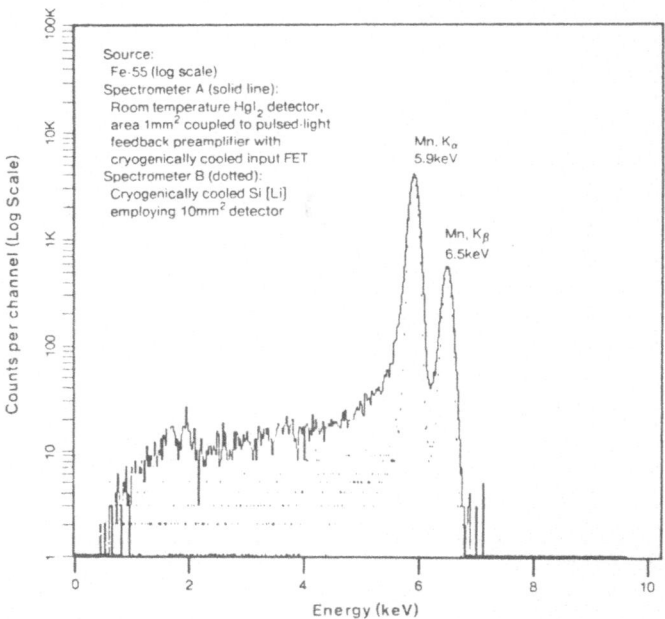

Figure 2.   A spectrum of Mn x-rays (Fe-55 source) taken
with   two   spectrometers   in   the   logrithmic   scale.
Spectrometer "A",   represented by the solid line, is an
HgI₂ detector coupled to a pulsed-light feedback
preamplifier   with   cryogenically   cooled   input   FET.
Spectrometer "B", represented by the dotted line, is a
cryogenically cooled Si(Li) system.

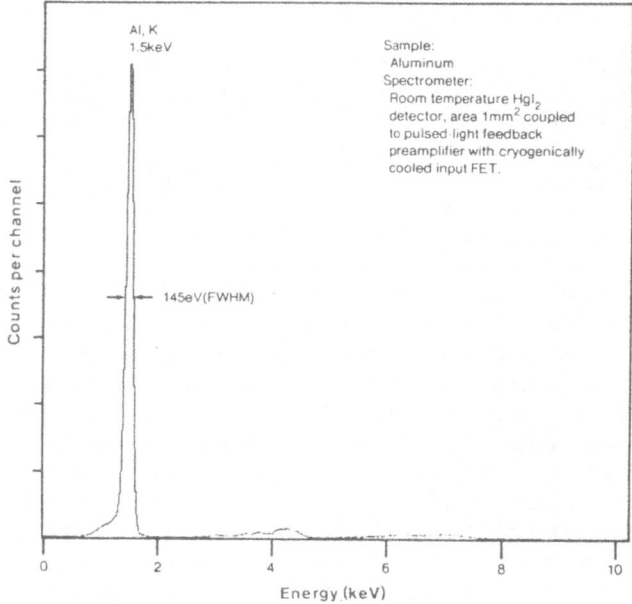

Figure 3.   A spectrum of the Al K x-ray taken with an
mercuric iodide spectrometer.

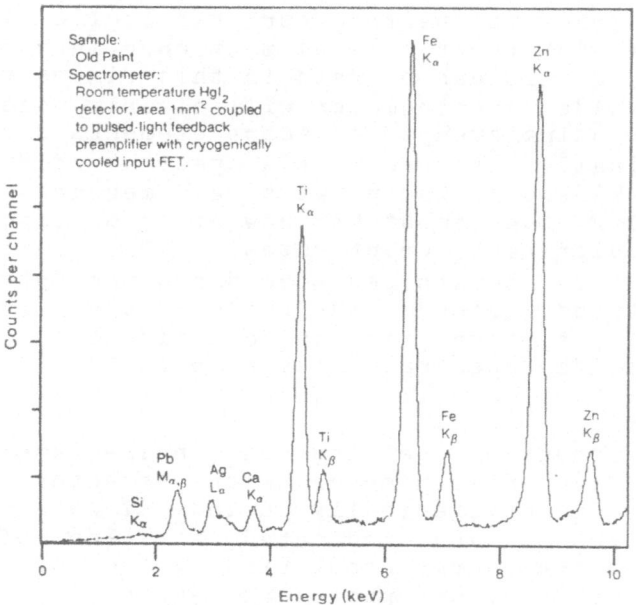

Figure 4.   A spectrum of a sample of an "old paint"
taken with the mercuric iodide spectrometer.

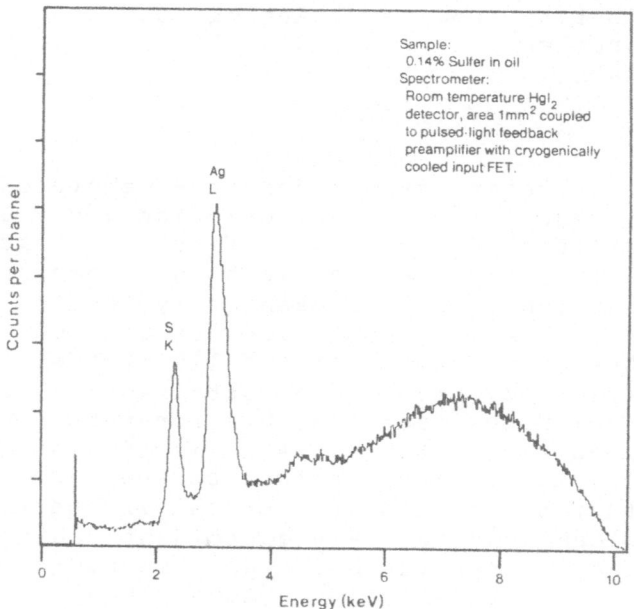

Figure 5.   A spectrum of 0.14% sulfur in oil taken with
the mercuric iodide spectrometer.

crude oil taken with spectrometer "A" (solid line) and
spectrometer "B" (dotted line) mentioned in connection
with Figure 2. It can be seen in this figure that the
mercuric iodide spectrometer will clearly resolve all
peaks coresponding even to adjacent elements such as Ni
and Cu. Finally, Figure 7 is a spectrum of Mn x-rays
from an Fe-55 source taken with a mercuric iodide
detector (with the input FET now at room temperature)
using two different count rates. The solid line
represents 1000 counts per second and the dotted line
10,000 counts per second. The two curves are identical
indicating that there is no deterioration in the
mercuric iodide spectra at least up to 10,000 counts
per second.

A measurement of the electronic noise level of the
mercuric iodide spectrometer with detector at room
temperature and cryogenically cooled FET was made by
the pulser method and indicated a value of 135eV
(FWHM). Using the energy resolution value of 175 eV
(FWHM) for the Fe-55 photopeak and value of 135 eV
(FWHM) for the electronic noise, a new value of the
Fano factor for mercuric iodide can be calculated from
equation (2). The value 0.1 is thus obtained which is
much smaller than the 0.19 value reported previously.
A lower measured value for the Fano factor reflects a
lower contribution from the trapping phenomena in the
x-ray spectra and a better quality of grown mercuric
iodide crystals.

CONCLUSIONS

The spectrometer system described above comprised
of a room temperature mercuric iodide x-ray detector
and cryogenically cooled first stage of
preamplification has been shown to be comparable in
energy resolution to cryogenically cooled Si(Li)
spectrometers. Energy resolution values of 175 eV
(FWHM) for the Mn K_α photopeak and 145 eV (FWHM) for Al
K x-ray spectrum have been obtained. A new lower
value for the Fano factor of 0.1 for mercuric iodide
has been measured. This value is similar to the
experimental value of Fano factor obtained for silicon.
This means that the statistical spread in the number of
charge carriers produced by the incident radiation can
limit energy resolution of mercuric iodide
spectrometers practically no more than in Si(Li)
systems. Various spectra presented show the usefulness
of mercuric iodide in extra-laboratory x-ray
fluoresence elemental analysis applications. Future

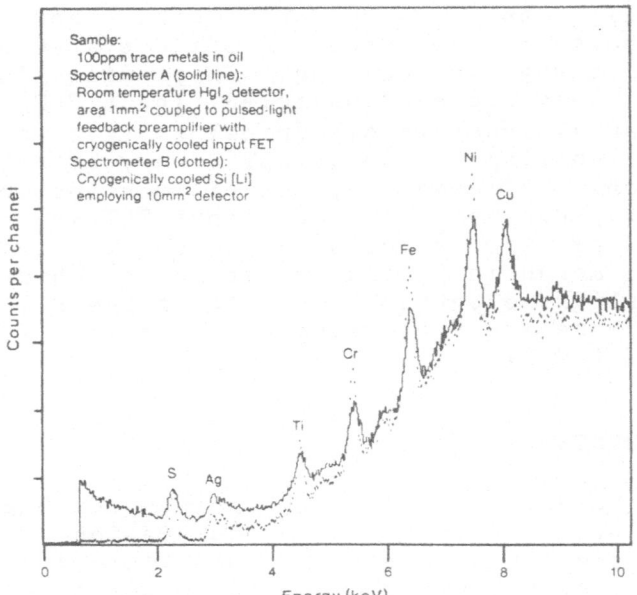

Figure 6.   A spectrum of 100 ppm trace metals in oil taken with spectrometer "A" (solid line) and spectrometer "B" (dotted line) mentioned in connection with Figure 2.

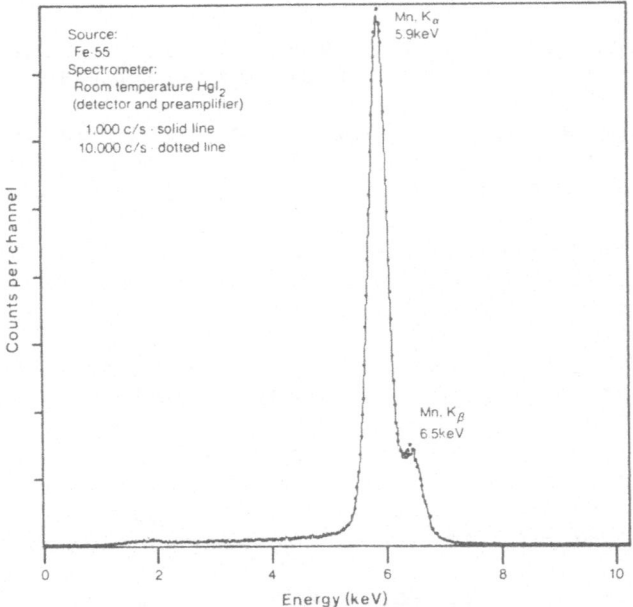

Figure 7.   A spectrum of Mn x-rays from an FE-55 source taken with an mercuric iodide detector and input FET at room temperature at two different count rates. The solid line represents 1,000 c/s, the dotted line 10,000 c/s.

research will be focused on further reducing the
electronic noise of the mercuric iodide system by means
other than cryogenic cooling. This will include a
search for less "noisy " uncooled input FETs as well
as for other alternative electrical or gas expansion
methods for cooling the input FET (i.e., by Peltier or
Joule-Thomson). Already by careful selection and
special encapsulation of the input FET we have been
able to reduce the electronic noise of totally room
temperature mercuric iodide systems to about 200 eV
(FWHM). A slight electrical cooling of the FET allowed
a further decrease of the electronic noise level to
below 175 eV (FWHM).

ACKNOWLEDGEMENTS

The authors wish to thank Christopher Bogart for
helpful discussions during preparation of the
manuscript and Jan W. Checinski for valuable technical
assistance. Support by the US Department of Energy
under contract No DE-AM03-76SF00113 and by NASA under
contract No NSG-7535 is acknowledged.

REFERENCES

1. J. S. Iwanczyk, A. J. Dabrowski, G. C. Huth, A. Del
    Duca, and W. Schnepple, IEEE Trans. on Nuclear Sci.
    NS-28, 1 (1981) 579.
2. A. J. Dabrowski, J. S. Iwanczyk, J. B. Barton, G. C.
    Huth, R. Whited, C Ortale, T. E. Economou, and A. L.
    Turkevich, IEEE Trans. on Nuclear Science, NS-28, 1
    (1981) 536.
3. J. S. Iwanczyk, J. H. Kusmiss, A. J. Dabrowski,
    J. B. Barton, G. C. Huth, T. E. Economou, and A. S.
    Turkevich, Nuclear Instruments and Methods, 193
    (1982) 73.
4. L. Ames, W. Drummond, J. S. Iwanczyk and A. J.
    Dabrowski, Advances in X-Ray Analysis, Vol. 26
    (1983) 325.
5. A. J. Dabrowski, J. S. Iwanczyk, W. M. Szymczyk,
    J. H. Kusmiss, G. C. Huth, W. Drummond and L. Ames,
    Nuclear Instruments and Methods, 213 (1983) 89.
6. A. J. Dabrowski, Advances in X-Ray Analysis, Vol. 25
    (1982) 1.
7. G. R. Ricker, J. V. Vallerga, A. J. Dabrowski, J. S.
    Iwanczyk and G. Entine, Review of Scientific
    Instruments, Vol. 53, No. 5 (1982) 700.

X-RAY TUBE IMPROVEMENTS FOR BETTER PERFORMANCE IN AN ENERGY-

DISPERSIVE X-RAY SPECTROMETER SYSTEM

Ronald Vane and Brian Skillicorn

Kevex X-ray Tube Division
P.O. Box 66860
Scotts Valley, CA 95066

ABSTRACT

Improvements in x-ray tube technology for the low-power tubes used in energy-dispersive XRF systems can provide better performance. These improvements are occurring in six areas: thin-window tubes, lower kV capability, higher kV and mA capability, pulsed tubes, smaller focal spots, and miniaturization. These improvements will lead to better excitation of the light elements, analysis of low-mass samples, higher count rates, small spot analysis, and smaller systems.

THIN-WINDOW TUBES

Low-power x-ray tubes are now available with thin 2 mil (.05 mm) thick Be windows rather than the conventional 5 mil (.127 mm) Be window. These new thin-window tubes give better transmission of low-energy x-rays, and when combined with an x-ray generator/power supply capable of operating at 3 kV, the system offers improved excitation of Na and Mg.

Na and Mg are difficult to excite for x-ray fluorescence due to their very poor fluorescent yields and the difficulty of producing primary x-rays which are close to their absorptions edges. More so than with heavier elements, the fluorescent yields of the K lines of Na and Mg are very sensitive to the energy differences between their absorption edges and the primary x-rays. To achieve the best excitation of Na and Mg, the exciting x-rays should be very close in energy to the absorption edges.

The difficulty in producing low energy x-rays for excitation lies in two areas. First, the Be window absorbs the very low energy

415

x-rays, and second, most x-ray generators and tubes cannot operate at the low potentials (less than 5 kV) which are optimum for this analysis. Beryllium foils have the relative transmission versus energy characteristics shown in figure 1.   For the standard 5 mil Be window, the 10% transmission point is at about 2 keV. A 2 mil Be window has a 10% transmission at about 1.4 keV. Thus, the thinner window allows more low-energy x-rays to pass. Figure 2 shows the scattered bremsstrahlung of 5 mil and 2 mil x-ray tubes operated at 3 kV. Note the greater intensity of low-energy x-rays from the 2 mil tube—energy available for excitation of the low Z elements Na, Mg, Al, and Si.

LOW KV CAPABILITY

Lower operating voltages for the x-ray generator also help improve the sensitivity and detection for Na and Mg. Operation at 3 kV is better than operation at 5 kV because, for real samples, there are usually other elements with lines between 2 keV and 5 keV which are excited preferentially to Na and Mg, and because there is more tube scatter from the x-ray tube at 5 kV. These other x-rays increase the count rate load on the system and also contribute background due to tailing under the Na and Mg. Operation at 3 kV typically takes place at less than 10% dead time at maximum tube current. At this voltage the real-time counts per second for Na and Mg are a greater proportion of the total spectrum counts than in a 5 kV spectrum. At 5 kV on most samples the total primary x-ray flux is great enough to produce enough counts to give 50% system dead time. Under these conditions the background is higher under Na and Mg, and it takes almost twice as long in real time to acquire the same live-time spectrum. To compare the results from 5 kV excitation and 3 kV excitation for Na and Mg, it is also necessary to compare real-time count rates and peak-to-background ratios.

Operation of the generator and the tube has not been previously allowed at 3 kV on the Kevex 0700 XRF system. Low-kV operation of a filamentary tube does not provide as much draw out potential for the electrons from the cathode filament, so the filament must run hotter to provide the same anode current. Previous designs of the x-ray generator could damage the filament of the x-ray tube if operated below 5 kV. The new design limits the filament current and allows the x-ray tube to be safely operated at 3 keV. However, this does limit the maximum current anode available from each tube at 3 kV. In the 0700, typically, the maximum anode current at 3 kV for each tube is between 1.3 and 1.6 mA, as compared with the rated maximum of the power supply at higher kVs of 3.3 mA.

Some representative samples were excited using the new thin-window tube and the conventional 5 mil window on a Kevex 0700 XRF subsystem. The 3 kV spectra were taken at maximum current (1.4 mA) in the direct mode with no collimator. The 5 kV spectra were taken with

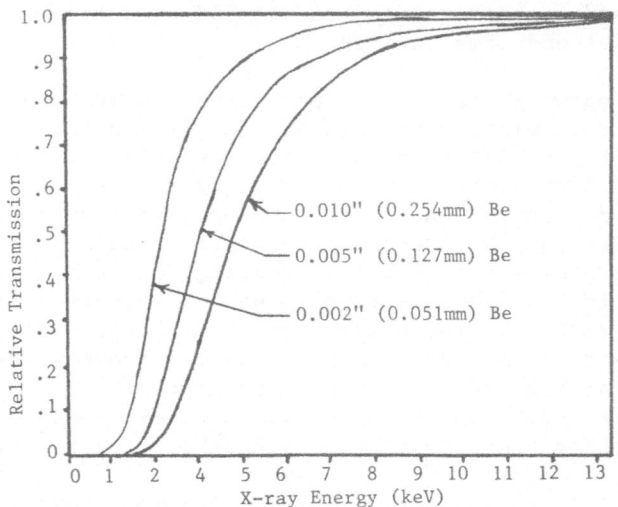

Figure 1
X-ray Transmission through Be Foil of Different Thicknesses.

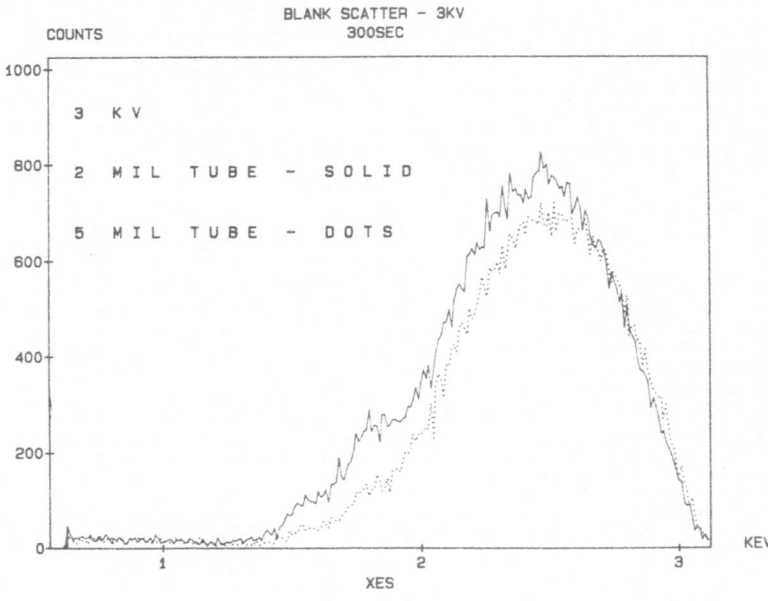

Figure 2
X-ray Production from X-ray Tubes with 2 mil and 5 mil Be Windows.

a collimator and current set to produce 50% dead time on the system
with a 8 microsecond time constant.

Figure 3 shows comparison spectra from a blank scatter sample at
3 kV from the thin-window tube and at 5 kV from a thick-window tube.
This compares the output spectra of the two tubes at the two
voltages.   The 3 kV spectrum shows many fewer counts than the 5 kV
spectrum, but due to the thinner window of the x-ray tube, it also
shows more counts on the low-energy side than the 5 kV spectrum.
These x-rays are more efficent for exciting Na and Mg than the higher-
energy x-rays in the 5 kV spectrum.   The 5 kV spectrum contains the
sharp characteristic L line peaks from the Rh tube anode at just below
3 keV and a large bremsstrahlung hump peaking above 4 keV in the
spectrum.   These higher-energy lines are less efficient at exciting Na
and Mg; however, they can strongly excite other lines in the range 2
to 4 keV, and they contribute heavily to the total spectrum background.

Figure 4 illustrates the difference between spectra of NaCl
excited at 3 and 5 kV.   At 5 keV with the thick-window tube, Cl is
very strongly excited and the Cl escape peak just below the Na peak is
nearly equal in size to the weak Na peak.   In the 3 kV thin-window
spectrum, the Na is very strong and the Cl peak is weak.   The Cl
escape peak cannot be distinguished at 3 kV.   Comparative count
rates: Na @ 3 kV = 256.9 cps, and Na @ 5 kV = 48.6 cps.

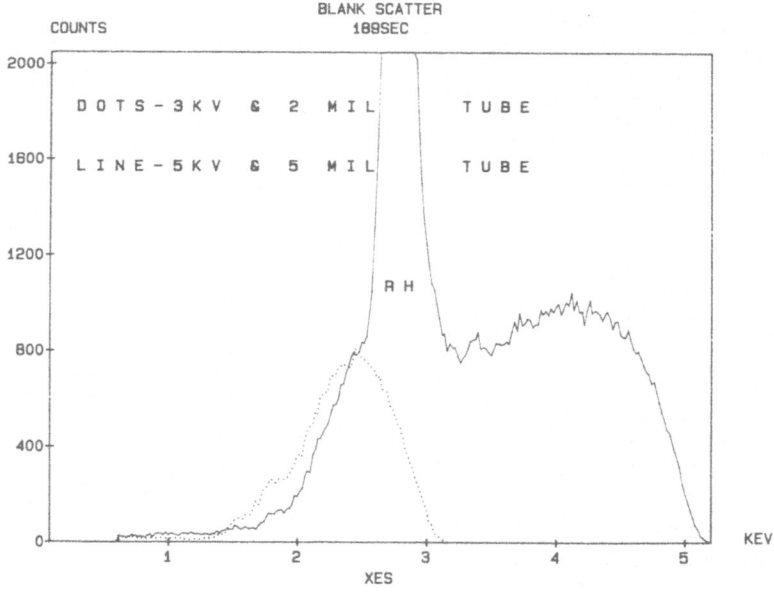

Figure 3
X-ray Production from a Thin Window Tube at 3 kV
and a Thick Window Tube at 5 kV.

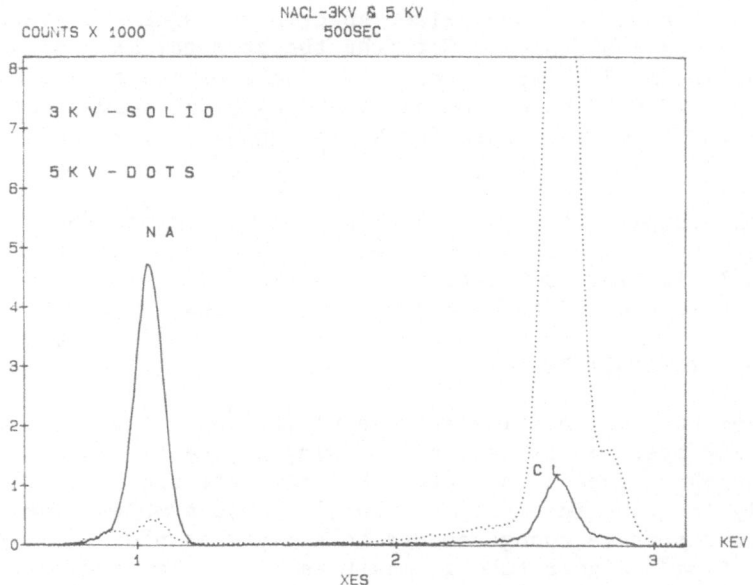

Figure 4
XRF Spectra of NaCl:
Thin Window Tube at 3 kV versus Thick Window Tube at 5 kV.

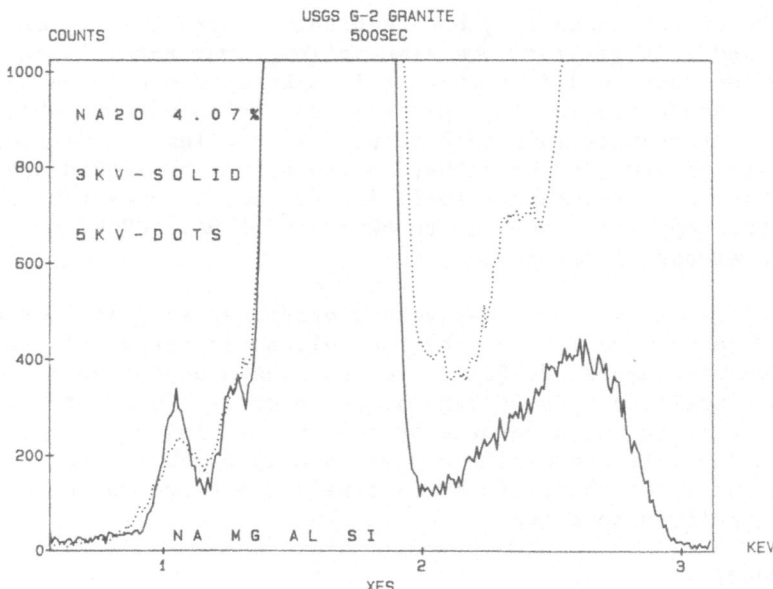

Figure 5
XRF Spectra of USGS G-2 Granite:
Thin Window Tube at 3 kV versus Thick Window Tube at 5 kV.

Figure 5 shows the comparison spectra for USGS G-2 granite. This
sample contains 4.02% Na_2O. Note the the stronger Na peak and lower
background in the 3 kV spectrum. The live-time count rates for Na are
8.97 cps @ 3 kV @ 6% dead time and 4.38 cps @ 5kV @ 50% dead time.
Thus the real-time count rate for Na is improved by a factor of nearly
four.

In the samples studied, the thin-window tube and the 3 kV power
supply gave a 2- to 6-fold improvement in Na count rates. The minimum
detection limits for 1000 second acquisition live times for Na were
found to be in the .03 to .08% range for various low-Z matrices.

HIGHER TUBE INTENSITIES

Higher tube intensities are useful for doing low-mass samples.
Most EDS/XRF systems are set up for doing high-mass, infinitely thick
samples which produce count rates that saturate the detector at
relatively low tube intensities. For low-mass samples, these systems
often do not produce enough counts to analyze samples in a reasonable
period of time. Higher tube intensities also allow for heavier
filtering of the tube output or the use of secondary targets to reduce
background and provide optimal excitation of various groups of
samples. However, higher possible tube intensities do not negate the
need for low-current operation on the same tube. A large dynamic
range of intensities is needed from the tube and generator.

Electron gun technology has been able to provide higher anode
currents and voltages for some time in low-power x-ray tubes. Some
modifications are needed to protect filaments from burning up at high
currents. Anode cooling is a problem, and tubes of more than 50 watts
maximum power require more than simple air cooling. Additional
cooling can be provided by either forced air or by liquid cooling.
Kevex is now manufacturing a small liquid-cooled x-ray tube suitable
for spectroscopy with power up to 60 kV and 5 mA. 80 kV could be
easily developed if desired.

Until recently, the x-ray generators/power supplies for these
low-power tubes have limited the possible power output of the tubes.
Several manufacturers of EDS systems have been buying on a OEM basis
the common Spellman high-voltage block which is limited to 60 kV and 2
mA. This unit is often derated by the OEMs to lower voltage and
current. Kevex is now making a power supply capable of suppling 80 kV
and 5 mA for spectroscopy in a light-weight package which is no larger
than the Spellman supplies.

PULSED TUBES

Pulsed tube systems for EDS systems have been around for over ten
years since their invention at the Lawerence Berkeley Labs.[1,2] The
first commercial tube using a grounded cathode and a control grid

electrically switched from the outside appeared in 1975.[3] Pulsed
tube systems improve the performance of EDS/XRF systems by allowing
higher stored count rates through the elimination of pulse pileup.
When an x-ray pulse is detected by the amplifier, a signal is sent to
the x-ray tube controller which turns off the x-ray production in the
tube by means of a control grid. This control grid "pulses" the tube
off for the duration of the processing time in the amplifier. Since
x-rays are not being produced in the tube, no more x-ray fluorescence
takes place in this time, and no pulse pileup occurs in the system.
Pulsed tube EDS/XRF systems can often double the maximum stored count
rates. On commercial EDS/XRF systems, real-time output count rates of
20,000 to 25,000 cps are possible. With special high-speed amplifier,
ADC, and MCA, one author (RV) achieved a output count rate of 54,000
cps with a pulsed tube system in 1978.

     Despite these advantages, pulsed tubes are not universally used
in commercial systems. Two factors contribute to this situation.
First, a grid-controlled x-ray tube with pulsing circuitry is more
complex and expensive than a filamentary controlled tube. In
addition, it is difficult to produce the popular grounded-anode tube
with a control grid, since the control grid must operate and be
switched at high potential. The grid pulse signal in this case is
optically coupled to the high-voltage section.[2] However, these
problems are not insurmountable, and more pulsed x-ray tubes can be
expected on the market for high-performance systems.

     The second factor is the complexity of dead-time corrections and
instantaneous tube current control in pulsed tube systems. A simple
dead time correction can be made where the dead time is simply the
tube off time, but this does not lead to optimum performance. A
better method is to compensate for the tube off time by increasing the
peak tube current to reduce the mean time between pulses to make up
for the tube off time. This can effectively reduce dead time to zero
at low count rate and low tube duty cycles.

MINIATURIZATION

     Smaller tubes and power supplies with the same power and voltage
capacities of older, larger tubes are now being made. Small-package
x-ray tubes are now being made by Kevex which allow the tube to be
placed closer to the sample. These small lightweight tubes also have
use in portable x-ray systems.

     A further technical innovation is the development of a portable
one-piece generator and tube system, the PXS. This system is capable
of producing 50 kV and 1 mA, is powered by 12 or 28 volts DC, and
weighs only 7 lbs. A picture of one these systems is shown in figure
6. This new x-ray source, because of its light weight and small size,
should find use in either compact or portable x-ray systems.

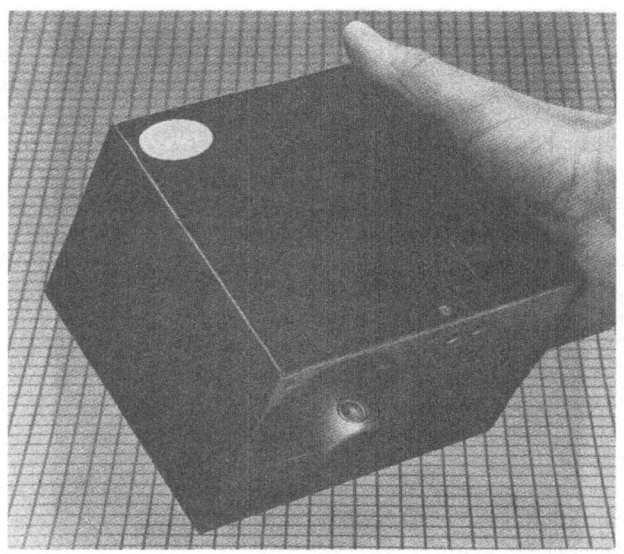

Figure 6
PXS: New Portable X-ray Source.

SMALL FOCAL SPOTS

Small focal spots on the x-ray tube anode allow the x-ray beam to
be collimated to a much smaller spot size on a sample. There is
interest in being able to collimate the x-ray beam to less than a 1/2
mm^2 spot size on a bulk sample. Currently, spot sizes of less than
3 mm^2 are difficult because of the fairly large focal spot of the
electron beam on the anode. Small focal spots require highly focused
beams with very stable spot location. Local anode cooling becomes a
problem in these designs. Technology developed for x-ray lithograghy
tubes may be used to solve this problem for spectrometry tubes.

REFERENCES

1. J.M. Jakelevic, F.S. Goulding, and D.A. Landis, "High Rate X-ray
Fluorescence Analysis by Pulsed Excitation," IEEE Trans. Nucl. Sci.
NS-19, No. 3, 392-395 (1972)

2. J.M. Jakelevic, D.A. Landis, and F.S. Goulding, "Energy Dispersive
X-ray Fluorescence Spectrometry Using Pulsed Excitation," in R.W.
Gould, ed., Adv. in X-ray Analysis, Vol. 19, 253-265, Kendall/Hunt
Publishing, Dubuque, IA (1976)

3. J.E. Stewart, H.R. Zulliger, and W.E. Drummond, "Energy Dispersive
Spectrometry at High Count Rates: Pulsed Tube Excitation and Recovery
of Resolution by Computer Processing," in: Adv. in X-ray Analysis,
Vol. 19, 153-160, Kendall/Hunt Publishing, Dubuque, IA (1976)

COMPARISON OF EXPERIMENTAL AND THEORETICAL INTENSITIES FOR A NEW

X-RAY TUBE FOR LIGHT ELEMENT ANALYSIS

John Kikkert and Graham Hendry

Philips S & I, Almelo, Holland
University of Birmingham, England

While x-ray fluorescence spectrometry is a highly sensitive and highly repoducible method of analysing samples, its one weakness is its relatively low sensitivity for light elements. This is mainly due to two problems: firstly the low fluorescent yield of the low atomic number elements, and secondly to the inherent inefficiency of exciting these elements. While it is not possible to improve the fluorescent yield, considerable improvements in light element sensitivity can be achieved by improvements in x-ray tubes.

Because of its thin (150 micron) berylium window, the Rhodium end-window-tube, allows improved light element sensitivity, but even with the thin window, only 38% of the Rhodium tube lines are transmitted through the tube window. The end-window-tube requires the use of a complex internal cooling circuit, and restricts the routine operation of the x-ray tube at voltages above 60 kV. In addition, the larger dimensions lead to a higher incident angle, a lower take-off angle, and larger sample-to-anode distances. This effect causes a loss of sensitivity for all elements, light and heavy, thus partially cancelling the theoretical gain in light element sensitivity. For this reason the side-window-tube appears to be a better solution.

To facilitate the measurement of light elements, a side-window-tube which provided high sensitivity for the light elements while still being usefull for the analysis of heavier elements, needed to be developed. Scandium, atomic number 21, appeared to be the ideal choice of anode material to provide optimal excitation for the light elements.

Some details for the relevant tube lines are shown in table 1. The mass absorption coefficient for silicon gives a first approxi-

423

mation of the excitation efficiency for silicon and other light
elements.

In order to evaluate the possible improvements in light element
sensitivity the Criss program was used[1] Intensities were calculated
for a theoretical sample consisting of carbon, with 0.01% of a large
number of other elements. These calculations were performed for a
wide range of anode angles, window thicknesses and kilovolts. As a
result of the calculations it could be concluded that the most suitable scandium tube should be made with a 26° anode angle and a 300
micron beryllium window. According to the calculations, this tube
will give an improvement factor for sensitivity of almost two in
light elements compared with the standard chromium tube sensitivity,
while being only marginally worse for the major elements. The relative sensitivities obtained with the chromium and scandium tubes were
then determined on a series of materials ranging from metals, geological materials, and some pure compounds. The result of these determinations are shown together with the calculated results in table 2.

Due to the fortunate position of the scandium tube lines with
respect to the calcium absorption edge, the scandium tube is extremely efficient for the excitation of calcium in steel. If a scandium

Table 1.   Critical data for side-window-tubes.

	Cr K	Sc K	Rh L
Wavelength (Å)	2.29	3.03	4.60
Intensity of tube lines:	89	97	23
Transmission of 300 micron Be:	78%	60%	14%
150 micron Be:	-	-	38%
Massabsorption coeff. Si:	193	426	1265

Line intensity was calculated with standard tube geometries
without window absorption at 50 kV (Cr & Sc) and 40 kV (Rh)

Table 2.   Comparison of Measured and Predicted Ratios

Elem.	Predicted Sc/Cr	Measured Sc/Cr	Elem.	Predicted Sc/Cr	Measured Sc/Cr
C	---	1.62	K	1.43	1.52
O	---	1.77	Ti	0.08	0.08
F	---	1.88	V	0.39	0.37
Na	1.86	1.78	Cr	0.97	3.41(F)
Al	1.71	1.82	Ni	0.94	0.98
P	1.73	1.80	As	0.90	0.95
Cl	1.59	1.49	Sn	0.87	0.90

tube is used, a detection limit of 0.2 ppm can be reached in 100
seconds counting time. Since the analysis of low levels of calcium
in steel is a critical problem in the steel industry, the scandium
tube can therefore be a key to the control of the steel making
process. A scan made on a sample containing 30 ppm of Ca in steel is
shown in figure 1.

Apart from the improvement in sensitivity for the determination
of all elements lower in atomic number than Ca, there is also a major
improvement for the determination of chromium and manganese since
these elements can be measured without a tube filter.

While the characteristic tubelines of all tubes interfere with
the analysis of some elements, the tubelines from the scandium tube
only impede the analysis of scandium. The titanium Ka peak is well
clear of the scandium Kb peak as shown in figure 2, therefore, titanium can be determined down to a detection limit of 7 ppm in steel.

There is also a scandium Kb third order escape peak which increases the background for aluminium. This effect doubles the background for aluminium, compared with the chromium tube, but since the
sensitivity has also improved by a factor close to two this does not
lead to a deterioration in detection limit. The measured sensitivities, background and detection limits for the analysis of steel is
shown in table 3.

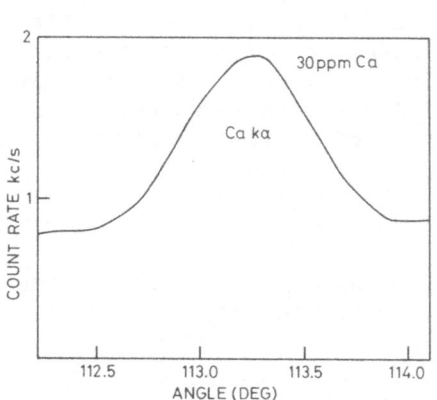

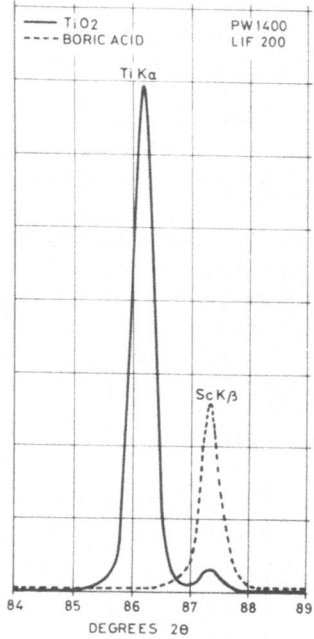

Figure 1.   Analysis of Ca   Figure 2.   Analysis of Ti with
            in Steel   the Scandium tube

Table 3.   Analysis of Low Alloy Steel with a Scandium Tube

Element	Crystal	Collimator	Sensi´ty cps/%	Background cps	LLD ppm
Al	PE	Coarse	3 740	190	11
Si	InSb	Coarse	9 330	280	5
P	Ge	Coarse	12 400	78	2
S	Ge	Coarse	26 800	30	0.6
Ca	LiF 200	Coarse	36 000	770	0.2
Ti	LiF 200	Fine	7 900	330	7
Cr	LiF 200	Fine	15 900	470	4
As	LiF 200	Fine	10 900	260	4

SPECTROMETER   -   PHILIPS PW1400

Table 4.   Analysis of Si in Ethyl Teraphtalate Film

	Rhodium tube	Chromium tube	Scandium tube
Sensitivity (cps)	4 413	5 267	10 600
Background (cps)	50	26	23
LLD (100 sec.) ppm	4.8	2.9	1.4

The scandium tube is highly efficient in exciting the light elements while the excitation of the transition elements which contribute significantly to crystal fluorescence, higher order escape peaks, scatter and other sources of background is significantly less. As a result the scandium tube gives very low backgrounds for the light element compared with other x-ray tubes, especially the rhodium tube.

These low backgrounds are reflected in the data shown in table 3. Another illustration of the low background ratio obtained with the scandium tube is seen in the analysis of silicon in thin plastic film, here the scandium tube gives significantly improved detection limits. The relevant data for the analysis of silicon in plastic film is shown in table 4.

ACKNOWLEDGEMENTS

We would like to thank Ing. C Nieuwenhuizen and Ing. J. J. de Koning for contributing some of the measurements shown in this article.

REFERENCES

1.   J. W. Criss, "Fundamental-Parameter Calculations on a Lab Micro-
      computer", Advances in X-ray Analysis, 23, 93 - 97, (1980).

USING A MICROCOMPUTER-CONTROLLED ROBOT ARM

AS A GENERAL PURPOSE SAMPLE CHANGER

John C. Russ, North Carolina State Univ., Raleigh, NC
J. Christian Russ, University of Michigan, Ann Arbor, MI
Donald E. Leyden, Colorado State Univ., Fort Collins, CO

INTRODUCTION

We have previously reported in these proceedings (1,2) on the automation of Wavelength-Dispersive X-ray Fluorescence and X-ray Diffraction Spectrometers using stepping motors, either installed in place of existing synchronous drive motors, or attached to existing gear or belt drive mechanisms that control the theta/two-theta positioning of the spectrometer. This technique is applicable to most types of instruments (successful adaptations to Philips, Siemens, GE/Diano, and Rigaku systems are working routinely at this writing).

Control of the goniometer by a microcomputer (64K Apple IIe with floppy disks and optional hard disk and graphics printer or plotter) employs suitable software to collect spectra, which are stored directly on disk (this permits very long scans, more than 10000 points being common, in up to ten segments covering portions of the spectra of particular interest). Control is via an interface card that issues and counts each stepping motor pulse (which, depending on gearing, may represent from 1/200 to 1/2500 degree two-theta), so that once initialized the exact position is always known. The card also counts pulses from the detector amplifier in a 24 or 32 bit counter, for a time period controlled by the program, before advancing the spectrometer to the next point.

Subsequent analysis of the stored data locates peaks, and for XRD spectra, reports D-spacing and normalized intensities which can then be utilized for search-match comparison to stored standards. For XRF work, the net intensities for designated elements are collected, corrected for background and dead time, and used in standard regression models (Rasberry-Heinrich, LaChance-Traill, etc.) or other quantitative models.

This makes the X-Ray Spectrometer into a highly productive tool. For some high-precision or trace phase studies, we have performed single sample scans taking more than 24 hours. In other cases, the ability to scan only preselected segments of the complete XRD spectrum, or to acquire elemental peak and background measurements for XRF quantitative analysis, means that final results are obtained in minutes. The next problem that arises, either in the case of long overnight unattended runs, or high specimen throughput, is that of providing for automatic sample changing.

Some XRF and XRD systems have existing sample changers. These can usually be controlled by the same computer, as they generally require a voltage logic level to trigger specimen advance, and provide a logic signal (or switch closure) when the next sample is in position. These devices are highly specialized and expensive, however, and not always trouble free in operation. They are not available for some existing spectrometers, and even if they exist, it is often uneconomical to add to them.

CAPABILITIES OF THE ROBOT ARM

We have adopted a general purpose robot arm to function as a sample changer. The arm (Figure 1) has six degrees of freedom: rotation of the base, elevation of the shoulder, angle of the elbow, angle of the wrist, rotation of the wrist, and closure of the gripper). Each motion has a separate stepping motor, and the interface has the additional ability to control one more motor (if a more specialized gripper motion was required) or read input from switches (for instance, attached to the fingers). This arm and others like it are also used in PC board construction, and other light duty industrial applications, as well as having an important role in teaching robotics. The particular arm described here was obtained (3) for under $700 with interface, although there are a number of roughly similar units on the market for prices ranging up to $2000, depending on speed, lifting capability, etc.

The arm can reproduce position to within 0.1 to 0.5 mm, depending upon absolute position, anywhere within a hemisphere about 40 cm. in radius. Its load capability is rated at 250 g. in the worst position (arm fully extended), but in more conservative

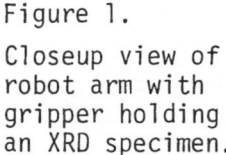

Figure 1.
Closeup view of
robot arm with
gripper holding
an XRD specimen.

positions is higher. The arm is not very fast; it takes 30 seconds or more to execute the motions required for a typical specimen change. There is no need to shut off the X-rays or close the shutter, but the goniometer must be brought to a previously determined angle setting to allow the arm to access the sample (Figure 2).

The software kernel which controls the stepping motors (counts pulses and keeps track of position) is less than 1K bytes in size, but with the complete control package (which allows points and motions to be "learned" by manually carrying them out and then storing the sequence), the program occupies about 5.6K of memory. Motion from one position in the list of points to the next is in straight lines, with all motors advancing simultaneously to their next settings. In order to clear obstructions, and provide the complex motion actually required, we introduce many points in the list (up to 40 for a typical motion from sample tray to holder); each point requires only 12 bytes of storage in a table, and previously established tables (sets of tasks) can be loaded directly from floppy disk storage when appropriate (for instance to change from one type of sample holder to another, which must be gripped differently).

The ability to learn motions (by the operator using the microcomputer keyboard to direct the arm motion from tray to holder and back) makes it practical to customize the setup for each specific spectrometer system. No hardware adaptation should be necessary, except perhaps for specialized specimen grippers as will be discussed below. The machine language arm-moving routine is called from BASIC, with the specimen number in the tray (in the

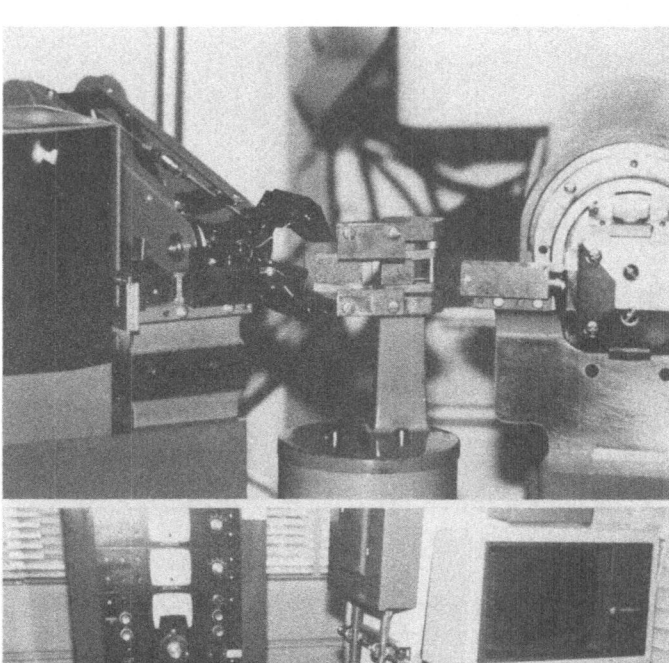

Figure 2.

Typical configu-
ration of the
robot arm used
in conjunction
with a GE XRD-5
Spectrometer.

a) Inserting a
   specimen.

b) Overall view
   (controlling
   computer at
   right).

form of its coordinates), and requires no modification for
individual spectrometers. The Basic program which collects the
X-ray intensity scans must have the set of coordinates for each
specimen in the sample tray, and of course must know how many
samples are there, but these data generally do not change
frequently.

POSSIBLE ADDITIONS

While we find the implementation shown here to be quite
satisfactory for our purposes, there are several modifications
worth considering. A funnel-shaped guide track and modified sample
holders can make the specimens easier to grasp and easier to

insert into position (a special rack to hold samples was constructed, by cutting slots in a block of wood). For some kinds of samples, a more specialized grip could be added to the arm, either in place of the existing one (available in two- and three-finger models for this arm, and in other configurations for other arms), or using an additional motorized or solenoid clamping action. Installation of microswitches to report that the grip has really found a specimen, and that a specimen is in position in the sample holder, might also be considered. More sophisticated 'touch sensitive' grippers, using a matrix of resisitive pads which report both the amount of force, and the area and shape of the contact region, are becoming available; these would in principle permit a much more intelligent program to pick up, align and insert specimens with high reliability and versatility.

We have not undertaken any of these changes for two reasons. First, because the system as it exists now fills our present needs. Second, because it is not at all clear how to program the system to take advantage of the extra information to recover from a virtually unlimited number of possible faults. For instance, at present if a sample is mishandled and drops from the gripper before analysis, we simply find it on the spectrometer table, with an empty spectrum stored on disk. The time spent on the analysis is wasted, but the sample can be re-run. Even if a program recognizes that a specimen has not been picked up, it is not likely that it will be able to do much better than deciding to go on to the next specimen. Without a sense of sight, to find the lost sample, and a considerably more versatile wrist and hand to pick it up, the only savings possible would be the time spent analyzing nothing.

OTHER USES FOR THE ARM

It is also worth noting that there are other activities in the same laboratory, some intimately associated with the X-ray analysis function and some not, which can use the same robot arm when it is not otherwise occupied. This ability to change from one task to another is the hallmark of robotics, and further improves the cost/benefit ratio over dedicated mechanisms for a single purpose. For example, aerosol filters which are analyzed by XRD and XRF must be weighed. There exist several interfaces for standard laboratory balances which report the weight to the computer, but a human is still required to handle the samples, which are typically filters mounted on square plastic holders with a pick-up tab. The use of the robot arm to automate this process requires no additional hardware, and a trivial effort to "teach" the arm the required motions. Furthermore, the weight data are then available to the computer along with the subsequent analytical results, in disk files, so that final reports can be expeditiously prepared with minimal opportunity for data transcription errors.

Other tasks worth mentioning include inserting pH electrodes into a series of specimens and recording the readings (via a suitable interface to the pH meter), sensing liquid levels by slowly raising and lowering a photodiode and LED assembly, or the collection of fractions from chromatography experiments. The fraction collectors may be tubes or cells of various configurations, each of which would ordinarily require a specialized mechanism. This, and the problem of variation in the size and shape of X-ray samples, calls for the implementation of a 'sense of sight' for the computer and robot. With current technology, a low-resolution video camera can be added to the system to allow the program to locate samples and orient the arm to pick them up. The software needed to process the image and recognize a few identifying features of a known range of sample holders, cells, etc. is not overly complex. While not essential for the present application, this is well within the capability of the computer system being used, and further illustrates the use of inexpensive technology to assist in automating high-tech instruments such as X-ray spectrometers.

CONCLUSION

The use of an inexpensive robot arm as as specimen changer, controlled by the microcomputer which also operates the basic spectrometer mecahnism, records and later processes X-ray data, has a high benefit-to-cost ratio for many laboratories. The use of standard hardware items with enough intelligence to perform specialized functions is an example of the trend toward more appropriate use of modern technology than highly specialized (ie. expensive) mechanisms designed only for one particular task.

References

1. B. B. Jablonski, D. E. Leyden, "A Microcomputer Based Wavelength Control and Data Acquisition System for Early Model WDXRF Spectrometers," Proc. Denver X-ray Conf., 1980

2. T. M. Hare, J. C. Russ, M. J. Lanzo, "X-ray Diffraction Phase Analysis Using Microcomputers," Advances in X-ray Analysis, Vol. 25, Plenum, 1982, p.237-244

3. Colne Robotics, 207 NE 33rd St., Fort Lauderdale FL 33334

LAMA III - A COMPUTER PROGRAM FOR QUANTITATIVE XRFA OF BULK SPECIMENS AND THIN FILM LAYERS

Michael Mantler*

IBM Research Laboratory
San Jose, California 95193

ABSTRACT

The previous LAMA I and II programs have been completely rewritten in this new version. Better precision and an order of magnitude increase in speed were achieved with a different numerical method and more efficient code. Fundamental parameters and/or empirical parameter calculations can be used. New routines for the analysis of oxides and other compounds of light elements, and multiple thin film layers with secondary enhancement factors are included.

INTRODUCTION

LAMA III is a package of computer programs for quantitative X-ray fluorescence analysis. In addition to the features of the previous versions of the programs, LAMA I (1) and LAMA II (2) new routines have been added for the analysis of multiple thin film layers and a convenient method for the analysis of oxides and other compounds of light elements. Fundamental parameter methods and/or empirical parameter calculations can be used. LAMA III typically performs an order of magnitude faster than the previous programs due to a new numerical approach to solve the fundamental parameter equations. This paper briefly describes the main organization of the programs, discusses the mathematical approach, and shows some results from the thin film routines.

---

*Permanent address: Technical University Vienna, Vienna, Austria.

433

ORGANIZATION OF THE PROGRAM

A main menu provides the user with several options including:

1. Permanent storage of information (such as composition, count rates, and experimental conditions) for specimens and standards under user given names. Count rates may be experimental data or can be calculated by fundamental parameter routines (Data-Library).
2. Calculation of α-coefficients from standards (Library entries) or from theoretical data. In the latter case, only the list of elements, lines, and experimental data are required, but concentration ranges can be optionally defined.
3. Analysis of unknown specimens by fundamental or empirical parameter techniques to obtain concentrations of elements or of components (if the compositions of the components have been previously defined in the system via the Data Library).

Data about specimens and/or standards are entered either manually or by reference to entries in the permanent Data Library. This allows maximum flexibility and convenience for the user, and is also an important feature if data from computer controlled instruments are processed. The dialog between programs and user is based upon full screen I/O support for the IBM 3278 terminal by the IBM 3083 computer.

NUMERICAL CONSIDERATIONS

All calculations are based upon the usual equations for primary excitation of fluorescence radiation and secondary enhancement in order to describe intensities as a function of the composition of the specimen for given experimental conditions and a given X-ray tube spectrum (3,4). McMasters' tables (5) are used for the calculation of mass absorption coefficients, and the X-ray tube spectra are based upon the experimental data from Gilfrich and Birks (6).

In order to achieve optimum performance of the program code, the equations used in LAMA I and II were analyzed to trace avoidable repetition in subroutine evaluations and to optimize the integration procedure by taking into account the magnitude of the contributions from the integration steps to the total value of the integrals.

The equations used for bulk specimens in LAMA III are given below and the symbols are defined in Table I.

$$
\begin{aligned}
r_i = \kappa(E_i)\ & \frac{\Omega}{4\pi \sin \psi_1}\ \frac{S_i - 1}{S_i}\ c_i p_i \omega_i \Biggl\{ \int_{E_{edge(i)}}^{E_{max}} I_0(E)\ \frac{\tau_i(E)}{\dfrac{\mu(E)}{\sin \psi_1} + \dfrac{\mu(E_i)}{\sin \psi_2}}\ dE \\[2ex]
& + \sum_j \frac{1}{2}\ \frac{S_j - 1}{S_j}\ c_j p_j \omega_j \int_{E_{edge(j)}}^{E_{max}} I_0(E)\ \frac{\tau_i(E_j)\tau_j(E)}{\dfrac{\mu(E)}{\sin \psi_1} + \dfrac{\mu(E_i)}{\sin \psi_2}} \\[2ex]
& \cdot \left[\frac{\sin \psi_1}{\mu(E)} \ln\left(1 + \frac{\mu(E)}{\mu(E_j)\sin \psi_1}\right) + \frac{\sin \psi_2}{\mu(E_i)} \ln\left(1 + \frac{\mu(E_i)}{\mu(E_j)\sin \psi_2}\right) \right] \Biggr\}
\end{aligned}
\tag{1}
$$

The absorption data are determined from

$$\mu = \tau + \sigma^{[c]} + \sigma^{[ic]} = \sum_i c_i \left(\tau_i + \sigma_i^{[c]} + \sigma_i^{[ic]} \right)$$ (2)

$$\tau_i = A_\tau \, \exp \left(A_0 + A_1 \, \ln E + A_2 \, \ln^2 E + A_3 \, \ln^3 E \right)$$

$$\sigma_i^{[c]} = B_\sigma \, \exp \left(B_0 + B_1 \, \ln E + B_2 \, \ln^2 E + B_3 \, \ln^3 E \right)$$

$$\sigma_i^{[ic]} = C_\sigma \, \exp \left(C_0 + C_1 \, \ln E + C_2 \, \ln^2 E + C_3 \, \ln^3 E \right)$$

The constants A, B, C are those given in McMaster's Tables (5) as a function of element and, for the A's, the absorption edges.

Table I. Symbols Used in Equations

E	Energy of radiation
E_i	Energy of fluorescence radiation from element i
E_{max}	Maximum photon energy in X-ray tube spectrum
$E_{edge(i)}$	Energy of absorption edge for line i
i	Element or line of interest*
j	Interfering line causing secondary enhancement*
$I_0(E)$	Spectral distribution of X-ray tube radiation
S_i	Absorption edge jump ratio of edge of considered line
p_i	Transition probability for observed line
ω_i	Fluorescence yield for photons of observed line
r_i	Count rate from element i (specimen of arbitrary composition)
R_i	Count rate from element i (pure element specimen)
c_i	Weight fraction of element i in specimen
c_i^*	Weight fraction of component i in spectrum
$\kappa(E_i)$	Detector efficiency for radiation E_i
$\mu(E)$	Mass absorption coefficient of compound specimen for energy E
$\mu_i(E)$	Mass absorption coefficient of element i for energy E
τ	Mass photoabsorption coefficient
$\sigma^{[c]}$	Mass scattering coefficient (coherent scattering)
$\sigma^{[ic]}$	Mass scattering coefficient (incoherent scattering)
ψ_1	Angle between specimen surface and incident tube radiation
ψ_2	Angle between specimen surface and observed radiation
Ω	Solid angle accepted by detector
p_{ij}	Weight fraction of element i in component j of a specimen

*In equation (1).

The integrals in equation (1) were solved in LAMA I and II by dividing the integration regions into a number of intervals of equal width on the wavelength scale (matching the step width in the tables of the X-ray spectral distribution) and summing the contributions from each step. Each integral for the primary intensities, secondary contributions, and pure element intensities was calculated separately. The absorption coefficients in equation (2) are the functions that are most frequently called, and the required evaluations of the large number of logarithms and exponential functions contributed significantly to the total execution time.

To avoid any unnecessary repetition of these calls, the fundamental parameter equations were slightly modified and the following integration scheme was used in LAMA III: Instead of separately evaluating the integrals for the various lines and secondary enhancement contributions (there are, for example, 12 such integrals for the count rate ratios of FeKα, NiKα, and CrKα radiations in a Fe-Cr-Ni alloy), only one integration loop is set up. At each integration step, all contributions to the intensities of primary radiations and their possible contributions TO other lines due to secondary enhancement are calculated simultaneously. This is in contrast to the contribution FROM other lines to the current line that is described in the second term of equation

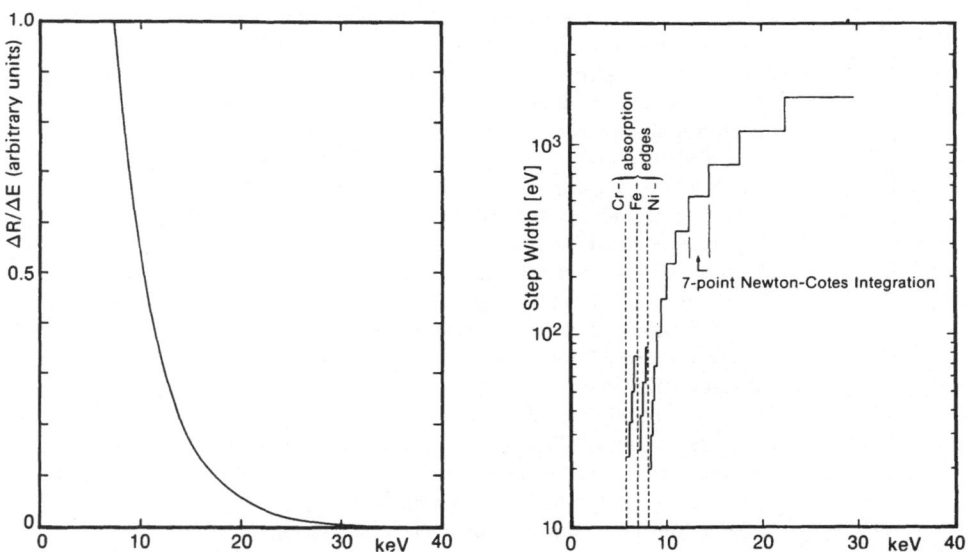

Figure 1 (left). $\Delta R/\Delta E$ as a function of energy (this represents the contribution of the individual integration steps to the total integral for R), calculated for CuKα radiation using Kramer's equation from Bremsstrahlung with 40 kV high energy limit.

Figure 2 (right). Example of efficiently selected step widths (in eV on a log scale) for the numerical integration procedure of equation (1). The step widths become narrower approaching the absorption edges.

(2).   Thus, the energy dependent factors such as absorption coefficients and interpolated values of X-ray tube spectrum (as discussed below), are calculated only once and used for all lines.   To apply this scheme the integration intervals are subdivided into regions between the absorption edges in order to include only those lines that are actually excited in each region.

The second significant improvement in the computation scheme of LAMA III is the principle of variable integration step widths.   As shown in Figure 1, the contributions of integration intervals with energies close to the absorption edge of the current line is very large compared to those at larger distances.   Since the most important contributions should be calculated with the highest numerical precision, the integration step width in LAMA III varies by more than a factor of 100 (on the energy scale) depending upon the distance of the step interval from the absorption edge.

Figure 2 shows the actual integration regions for a Cr-Fe-Ni alloy.   The main interval between the absorption edge of the lowest energy line and the high energy X-ray limit is divided into regions between the absorption edges.   Each region is subdivided into an adequate number of subregions, which are again divided into six intervals of (locally) equal width.   The boundaries of these six intervals define the seven arguments for 7-point Newton-Cotes integration.   Since there are never irregularities of the integrand in such a region (they occur only at absorption edges), this scheme is numerically far superior to the straightforward additions over the whole main integration region.   The additional time to calculate the integration boundaries is insignificant.   The integration step width in this example varies between approximately 20 eV to 2 keV, if the same total number of steps is used as in the conventional method.   The number of steps can, however, be decreased considerably and still obtain acceptable numerical precision, with further reduced computing time.

A disadvantage of this scheme is that the experimental X-ray tube spectra are usually provided in tables with constant intervals $\delta\lambda$, and interpolation is necessary to obtain the values for the given energies.   In LAMA III, interpolation is used for energies in the Bremsstrahlung region, and X-ray tube lines are treated separately.

ELEMENT AND COMPONENT ANALYSIS

Specimens often contain components (individual phases) with given compositions, such as SiO_2, MgO, or other oxides.   Results of the analysis may then be given as component concentrations c^* rather than element concentrations c.   In such cases, the number of components m is usually less than the number of elements n.   This allows the measurement of intensities of one or more (up to $n-m$) element lines to be omitted, which is convenient, if light elements like O, N, B, and others are present.

The modified α-coefficient method (α^*-method) for component analysis (7) has been included in LAMA III as well as a fundamental parameter routine based upon the same principles.   In these cases, the equations

$$c_i = \sum_{j=1}^{m} p_{ij} c_j^* \qquad i = 1...n \tag{3}$$

are additional conditions for determining the concentrations c and c^*.

In the case of the α-coefficient methods, equation (3) is used replace the c's in the conventional α-coefficient equation in order to obtain the new type of α-coefficients (α^*-coefficients) and a set of equations for the unknown component concentrations c^*:

$$-1 = \sum_{k=1}^{m} \alpha_{ik}^{*} c_{k}^{*}$$

$$\alpha_{ik}^{*} = \sum_{j \neq i}^{n} \alpha_{ij} p_{jk} - p_{ik} \frac{r_i}{R_i}$$

(4)

In the case of fundamental parameter equations, where the unknown concentrations are calculated by iterative methods rather than by solving linear systems of equations, the following procedure is applied: In the first approximation, the concentrations of the measured elements are assumed to be proportional to their count rates.   The concentrations of the unmeasured elements are obtained from equation (3) and the condition $\Sigma c_i = 1$ is used for normalization.   The new complete set of concentrations is used in the fundamental parameter routine to calculate intensity ratios.   The concentrations are then updated by comparing experimental and theoretical count rates for the measured elements, and by calculating the missing values from equation (3). This is repeated until no significant improvement is achieved.   If less than (n-m) lines have been omitted in the experiment, the set of equations (3) is overdetermined and has to be solved by least squares fit procedures.

THIN FILM ANALYSIS

Equations and computer routines have been developed to calculate intensities from multiple thin film layers.   All secondary enhancement effects from elements within the same layer as well as from elements in all other layers and the substrate are taken into consideration.   The contribution of back-scattered X-ray tube radiation from pure silica glass (SiO_2) substrates to the excitation can also be calculated.

For the calculation of concentration(s) and/or thickness(es), iterative procedures must be used depending upon the nature of the problem.   In principle, all concentrations and thicknesses can be calculated from experimental count rates, providing that no element appears two or more times in different layers, and if the concentrations in each layer truly add up to 1.   In practical applications thicknesses should be provided whenever possible.   The maximum number of unknowns is restricted by the errors in the intensities and the resulting numerical instability of the system of equations.

Figure 3 shows calculated intensities from an assumed specimen, that consists of a bulk substrate and four layers of equal thickness.   Each layer and the substrate are alloys of two elements of equal weight fractions and atomic numbers 24-32 and 47. With increasing thickness intensities first increase until the shielding effect from covering layers becomes dominant.   They practically vanish (except, of course, from the top layer) at thicknesses around 0.1 mm, in agreement with the usual assumption that bulk specimens of that thickness (from a pure element or elements of similar atomic number) are almost infinitely thick.   The higher energy $AgK\alpha$ radiation causes slightly less absorption, so that silver radiation is still observed from deeper layers until the covering layers reach several tenths of a millimeter thickness.

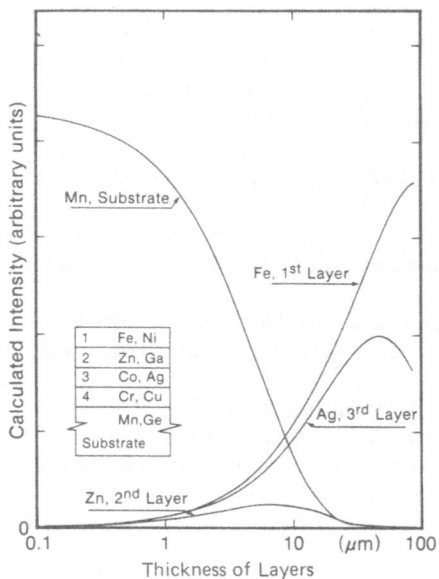 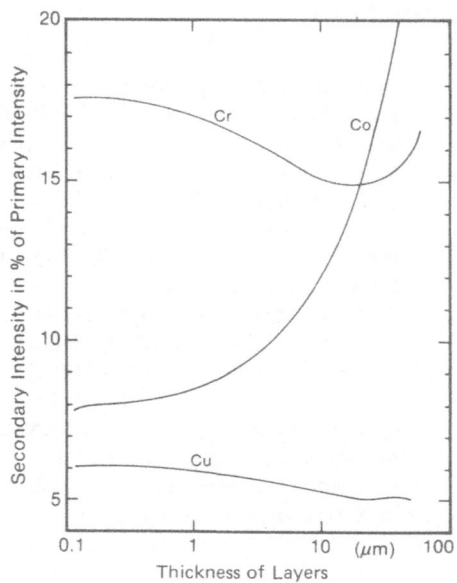

Figure 3 (left). Calculated intensities as a function of thickness for selected radiations from four thin film layers on a bulk substrate. Kramer's equation was used for the primary tube radiation spectral distribution (40 kV). All layers have the same thickness.

Figure 4 (right). Calculated intensities due to secondary enhancement, in percent of primary fluorescence intensities, as a function of layer thickness. Specimen and conditions of calculation are the same as for Figure 3.

The same assumptions about the specimen were made for the calculation of the magnitude of the interelement effect contributions as a function of thickness. Selected results are shown in Figure 4. In most cases, the contributions to secondary excitation from elements within the same layer as the observed element are much higher than those from other layers. For the case of the Cu-radiation, however, all contributions come only from other layers resulting in a comparably lower, but still considerable increase of intensity by more than 5%. The sources of the exciting radiations are probably mainly in the MnGe substrate at lower film thicknesses, and is Ag in the layer above with increasing film thicknesses.

COMPUTING TIMES

The programs have been developed and tested on an IBM 3083 computer system. Typical execution times for the program to calculate intensities in a bulk specimen are around 0.035 seconds per element. For thin films, the total execution time (for all elements) is between 0.4 seconds (two elements in one layer on a two element substrate) and 5.6 seconds (4 layers with two elements each on a substrate with two elements). The program code was written in Fortran.

REFERENCES

1. D. Laguitton and M. Mantler, LAMA I - A General Fortran Program for Quantitative X-Ray Fluorescence Analysis, Adv. in X-Ray Anal. 20:515 (1977).
2. T. C. Huang, Quantitative X-Ray Fluorescence Analysis of Thin Films Using LAMA-2, X-Ray Spectrom. 10:28 (1981).
3. J. Sherman, The Theoretical Derivation of X-Ray Intensities from Mixtures, Spectrochim. Acta 7:283 (1955); ibid. 14:466 (1959).
4. T. Shiraiwa and N. Fujino, Theoretical Calculation of Fluorescent X-Ray Intensities in Fluorescent X-Ray Spectrochemical Analysis, Japanese J. Appl. Phys. 5:10,886 (1966).
5. W. H. McMaster, N. K. DelGrande, J. H. Mallett, and J. H. Hubbell, Compilation of X-Ray Cross Section, UCRL-50174 Sec. 2-R1, Lawrence Rad. Lab., Livermore, California 94550 (1969).
6. J. V. Gilfrich and L. S. Birks, Spectral Distribution of X-Ray Tubes for Quantitative X-Ray Fluorescence Analysis, Anal. Chem. 40:7, 1077 (1968).
7. M. Mantler, A New Method for Quantitative X-ray Fluorescence Analysis of Mixtures of Oxides or Other Compounds by Empirical Parameter Methods, Adv. in X-Ray Anal. 26:351 (1983).

AN EMPIRICAL BACKGROUND CALCULATION METHOD FOR MULTI-CHANNEL X-RAY SPECTROMETERS

R. B. Kellogg

Environmental Chemistry and Emissions Research
Northrop Services, Inc. - Environmental Sciences
Research Triangle Park, NC 27709

INTRODUCTION

Several literature references[1-2] address the relationship between the reciprocal of the absorption coefficient and the Compton scatter peak intensity. Most workers use a Compton scattered tube line in the calculation of background; however, use of Bremsstrahlung scatter in the present work is also valid for this calculation. Background calculation for a fixed-angle monochromator is desirable since the background cannot be measured off peak. A method for calculating background for fixed-channel instruments by summing the calculated Rayleigh and Compton scatter intensities is presented in this paper.

THEORY AND ASSUMPTIONS

The Rayleigh and Compton scatter intensities from a flat homogeneous specimen of any thickness are given by[3]:

$$I_r = \frac{K_r I_o S_r}{\mu_2(\csc \Psi_1 + \csc \Psi_2)} [1 - EXP - \mu_2(\csc \Psi_1 + \csc \Psi_2)G] \qquad (1)$$

$$I_c = \frac{K_c I_o S_c}{\mu_1 \csc \Psi_1 + \mu_2 \csc \Psi_2} [1 - EXP - (\mu_1 \csc \Psi_1 + \mu_2 \csc \Psi_2)G] \qquad (2)$$

where

I_r, I_c = Rayleigh and Compton scatter intensities, respectively

441

K_r, K_c = overall efficiency constants,
  I_o = incident intensity on the specimen,
   G = surface density of specimen (g/cm^2),
$\mu 1$, μ_2 = absorption coefficient for incident and scattered
        radiation, respectively (cm^2/g),
S_r, S_c = Rayleigh and Compton scatter coefficients, and
ψ_1, ψ_2 = incidence and take-off angles.

The composition-dependent portion of the Rayleigh and Compton
scatter coefficients are given by[4]:

$$S_r = k_r \sum_i W_i z_i^3 / A_i \qquad (3)$$

$$S_c = k_c \sum_i W_i z_i / A_i \qquad (4)$$

where

   W_i = the weight fraction of the i-th element in the specimen,
A_i, z_i = the atomic mass and atomic number, respectively,
k_r, k_c = constants containing the energy- and
        angular-dependent portion of the scatter coefficient.

   Equation 2 is simplified for convenience by setting $\mu_1 = \mu_2$.
This approximation is valid for environmental samples because of
the unlikely nature of major element absorption edges lying between
the Rayleigh and Compton wavelengths for the elements whose
backgrounds are to be calculated (see Equipment Section).
Incorporating this approximation into Equation 2, assuming infinite
thickness and replacing S_r and S_c by Equations 3 and 4,
respectively, yields an equation in the form of Equation 5 for the
total scattered intensity or background:

$$I_{bkg} = A \ \sigma_r / \mu_2 + B \ \sigma_c / \mu_2 + C \qquad (5)$$

In Equation 5, the summation in Equations 3 and 4 was replaced by
σ_r and σ_c; $K_r k_r I_o / (csc \psi_1 + csc \psi_2)$ was replaced by A;
$K_c k_c I_o / (csc \psi_1 + csc \psi_2)$ was replaced by B; and C is a constant
inserted to allow for a non-zero intercept due to background
resulting from sources independent of the specimen matrix, such as
scattering from both specimen and crystal holder. The absorption
coefficient μ_2 is determined by the expression $\Sigma W_i \mu_{i,\lambda}$, where W_i is
the weight fraction of element i and $\mu_{i,\lambda}$ is the absorption
coefficient of element i for wavelength λ. The constants A, B, and
C are determined by a multiple linear regression analysis of data
collected on laboratory-prepared standards. These standards were
chosen to represent a broad dynamic range of the scattering
parameters σ_c / μ_2 and σ_r / μ_2.

EQUIPMENT

All measurements are made on a Siemens MRS-3 multi-channel spectrometer[5] equipped with fixed and scanning monochromators. The fixed-channel elements whose backgrounds are to be calculated are molybdenum (Mo), strontium (Sr), lead (Pb), and bromine (Br). The scanning monochromator is equipped with a sealed xenon proportional detector and has a scanning range of 11.700 keV (30.10 °2θ) to 3.52 keV (119.5 °2θ). All analyzing crystals are lithium fluoride (LiF_{100}). The K_α lines are used for all elements except for Pb where the L_β line is used. Excitation of the X-ray spectra is provided by a chromium (Cr) target tube operated at 54 kV constant potential and 44 mA for all measurements.

PREPARATION AND ANALYSIS OF STANDARD SCATTERING MATERIALS

Two sets of scattering standards are prepared to favor Rayleigh and Compton scatter, respectively. The Compton set (referred to as low z standards) consists of Fe_2O_3 in concentrations ranging from 90% to 7.5% in Carbowax 6000. The average z ranges from 11 to 3.6. The Rayleigh set (referred to as high z standards) consists of selected high-purity oxides or salts of elements of atomic numbers (z) ranging from 11 to 30. They are pressed either pure or with small (<10%) amounts of Carbowax 6000. The average z for these standards ranges from 8.4 to 18. All standards and samples are prepared greater than infinite thickness.

The intensity data for the high and low z standards are plotted in Figure 1 against their respective scatter parameters. This figure represents the experimental data graphically, and no functional relationship should be assumed between the observed intensity and the respective scatter parameters because the observed intensities have components of both Rayleigh and Compton scatter. However, the data do qualitatively illustrate the observed intensity as a function of the parameter most closely associated with the scattering process producing the intensity.

The coefficients A, B, and C in Equation 5 and their 95% confidence interval are determined by multiple linear regression analysis and are reported in Table 1. A plot of intensity measured at the analyte line position versus the calculated intensity is shown for Sr in Figure 2. All experimental data points are plotted, but data points with standardized residuals greater than two were rejected in the final determination of the regression coefficients. (See Discussion.)

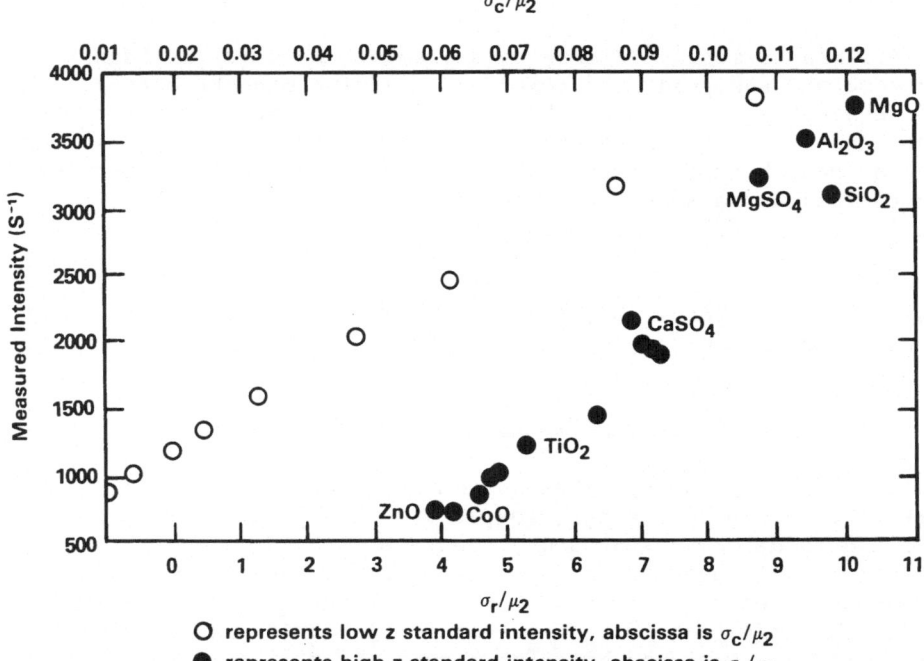

Figure 1.   Intensity of 14.14-keV X-rays (S_r).

EXPERIMENTAL RESULTS

Two related approaches were used to assess the applicability
of the method. In the first approach, the intensities are measured
on synthetic laboratory standards using combinations of the same
high-purity materials used to prepare the scattering standards but
with a different binder, boric acid. The combinations were chosen
to give widely diverse compositions and are summarized in Table 2.
In addition to synthetic standards, a standard reference material,
SRM 120b (calcium phosphate rock), is also measured. This was
chosen because a qualitative scan on a different spectrometer
capable of scanning over the analyte lines of the elements of
interest showed only Sr as an interfering element. The scatter
parameters are calculated based on the certified concentrations
from SRM 120b and the gravimetrically determined concentrations for
the synthetic standards. The backgrounds are calculated using
Equation 5 with the fitted coefficients from Table 1. A comparison
of the measured and calculated count rates is given in Table 3.

Table 1.   Regression Coefficients

Element[a]	A	B	C	r[c]	Slope[c]	Intercept[c]
Mo	70 ± 19[b]	38865 ± 1009[b]	369 ± 164[b]	0.9995	1.000 ± 0.031	0 ± 66
Sr	165 ± 32	24774 ± 1709	-77 ± 146	0.9978	1.000 ± 0.016	0 ± 70
Pb	312 ± 52	46434 ± 2377	-27 ± 186	0.9984	1.000 ± 0.025	0 ± 81
Br	368 ± 44	58606 ± 1573	-148 ± 139	0.9996	1.000 ± 0.012	0 ± 47

[a]Some data points used in the determination of A, B, and C were deleted; see Discussion for details.
[b]Represents 95% confidence interval for coefficient.
[c]Correlation coefficient, slope, and intercept from Figure 2.

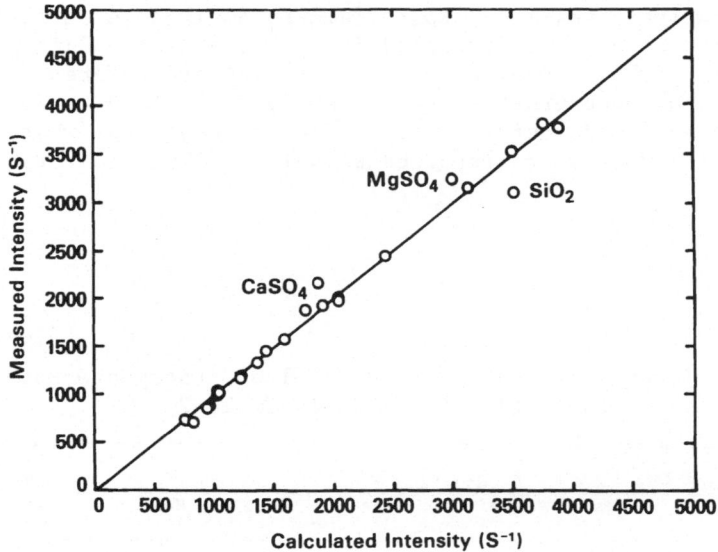

Figure 2.   Correlation of measured and calculated intensities for 14.14-keV X-rays (S_r).

Table 2.   Weight Fraction of Components of Synthetic Standards

Std	MgO	NaCl	S	Fe₂O₃	ZnO	KNO₃	H₃BO₃	Surface Density (g/cm²)	Average z
1	0	0	0.484	0	0.516	0	0	0.448	17.4
2	0.0878	0	0.435	0.477	0	0	0	0.491	14.8
3	0.0236	0.0810	0.0837	0.0347	0.0383	0.0397	0.699	0.475	5.67

The second approach to judge the method involves the X-ray fluorescence analysis of SRM 1633A (coal fly ash). In the background calculation on unknown samples, the weight fractions of all elements used to determine the absorption and scatter coefficients are determined by iteration. Major non-analyzable elements, such as oxygen in fly ash, are determined by difference, assuming that the sum of weight fractions equals unity. Table 4 shows the analysis results. Four elements analyzable on the scanning channel are also included in the data as an additional check on the method. Due to the lack of standard reference materials with concentrations sufficiently high (>100 to 200 ppm) for the elements under study, calculated and measured (where possible) backgrounds are reported as well as the concentrations.

Table 3.   Intensities (s^{-1})[a] from SRM 120b (Phosphate Rock) and Synthetic Standards in Table 2

Element	Std #1 Calc	Std #1 Meas	Std #2 Calc	Std #2 Meas	Std #3 Calc	Std #3 Meas	SRM 120b Calc	SRM 120b Meas
Mo	1621 ± 336[b]	1612	2170 ± 374	2268	5112 ± 493	5068	5824 ± 541	5748
Sr	878 ± 302	959	1159 ± 338	1258	2313 ± 443	2345	2610 ± 480	5389
Pb	1304 ± 373	1382	1687 ± 414	1815	3277 ± 531	3294	3647 ± 569	3491
Br	1249 ± 274	1327	1663 ± 303	1774	3395 ± 378	3303	3778 ± 403	3698

[a]All intensities measured at analyte line position.
[b]Represents 95% confidence interval of calculated intensity. All counting times are 100 s.

Table 4.   Trace Element Analysis of SRM 1633A[a]

Element	Certified Conc (ppm)	Peak Count Rate (s^{-1})[b]	Measured Background[c] (s^{-1})	Conc[d] (ppm)	Calculated Background (s^{-1})	Conc[e] (ppm)
Mo	29	5375	NA[f]	NA	5116 ± 504[g]	34 ± 67[h]
Sr	830	6583	NA	NA	2393 ± 459	854 ± 103
Pb	72.4	3425	NA	NA	3334 ± 545	27 ± 162
Br	–	3477	NA	NA	3511 ± 384	ND(-9)[i]
Se	10.3	694.7	677.2	ND(4)	697 ± 158	ND(-1)
Bi	–	606.9	602.3	ND(3)	600 ± 128	ND(5)
As	145	1139	535.1	179 ± 11	541 ± 222	181 ± 73
Hg	0.16	452.9	449.3	ND(-3)	435 ± 226	ND(11)

[a]Conversion of intensities to concentration is accomplished by fundamental parameters technique.
[b]Counting times for Mo, Sr, Pb, and Br are 100 s; those for Se, Bi, As, and Hg are 20 s.
[c]Measured off peak.
[d]Concentration is based on measured background.
[e]Concentration is based on calculated background.
[f]Not applicable because background cannot be measured.
[g]Error figure represents 95% confidence interval of background uncertainty.
[h]Error figure represents 95% confidence interval of all random error sources.
[i]Element not detected. The figure in parentheses is the calculated concentration and indicates the concentration resulting from the difference in the peak count rate and the background.

DISCUSSION

This technique has proven to be an acceptable approach to calculate background under the conditions described in this paper. The lower limit of applicability appears to be in the vicinity of 200 ppm and is due to the fairly broad range of the 95% confidence interval on the parameters in Table 1.

The points that deviate most from the straight line in Figure 2 are those observations made on the high atomic number standards. This deviation is expected from the data in Figure 1, which shows considerable scatter from the high z standards, possibly due to diffraction,[1] whereas diffraction effects are apparently absent in the low z standard data. There is a strong systematic bias in the SiO_2 data points, all of which had standardized residuals greater than two and were thus deleted from the data. The scatter SiO_2 was the only material that had to be ground before pressing; therefore, it could have been contaminated in the grinding process. The $MgSO_4$ data point was deleted from the Mo and Br data, and the $CaSO_4$ data point was deleted from the Sr data because their standardized residuals were also greater than two. Moisture absorption by these latter two compounds during preparation would lower the average z, thus producing higher intensity, as observed, but no efforts were made to verify this effect.

Table 4 presents information similar to that in Table 3 with the important addition of the calculated concentration and the 95% confidence interval. Almost all of the uncertainty in the concentration is due to the uncertainty in the calculated background.

This method is being extended to aerosol-loaded filters and other samples less than infinitely thick.

ACKNOWLEDGMENTS

This work was funded by the U.S. Environmental Protection Agency under Contract 68-02-2566. The author thanks the Stationary Source Emissions Research Branch, U.S. Environmental Protection Agency, Research Triangle Park, North Carolina for their continued support of this project. The author also thanks Dr. W. J. Courtney, Northrop Services, Inc. – Environmental Sciences, for discussions on data analysis, and Ms. E. G. Bryant, also with Northrop Services, Inc. – Environmental Sciences, for her skillful assistance in sample preparation.

REFERENCES

1. R. C. Reynolds, Matrix corrections in trace element analysis by X-ray fluorescence: Estimation of the mass absorption coefficient by Compton scattering, Am. Mineral. 48:1133 (1963).
2. C. E. Feather and J. P. Willis, A simple method for background and matrix correction of spectra peaks in trace element determination by X-ray fluorescence spectrometry, X-ray Spectrom. 5:41 (1976).
3. H. Meier and E. Unger, On the application of radioisotope X-ray fluorescence analysis for the solution of environmental and industrial problems, J. Radioanal. Chem. 32:413 (1976).
4. A. Markowicz, A Method of correction for absorption matrix effects in samples of 'intermediate' thickness in EDXRF analysis, X-Ray Spectrom. 8:14 (1979).
5. J. Wagman, R. L. Bennett, and K. T. Knapp, "X-Ray Fluorescence Multi-Spectrometer for Rapid Elemental Analysis of Particulate Pollutants," Report No. 600/2-76-003, U.S. Environmental Protection Agency, Washington, D.C. (1976).

COMPARISON OF X-RAY BACKSCATTER PARAMETERS FOR COMPLETE SAMPLE MATRIX DEFINITION*

K.K. Nielson and V.C. Rogers

Rogers and Associates Engineering Corporation

Salt Lake City, Utah 84107

INTRODUCTION

Equations for computing sample absorption coefficients and matrix corrections from fundamental parameters[1,2] require measurement or prior knowledge of all bulk element concentrations for samples of finite thickness. Uses of these equations in commercial software have thus required either 1) "similar" standards, 2) limited applications to metal alloys, oxides or specially-prepared matrices of known composition, or 3) user-supplied definitions of bulk light element concentrations. A fourth option, which provides superior analytical flexibility, is the use of backscattered x-ray intensities with element scatter cross-sections to define the unmeasured light-element component of the sample matrix. Backscatter intensities constitute the only spectral basis for routinely characterizing the light-element component of geological, biological and other materials which contain significant H, C, N, O, etc. They thus offer a unique basis for utilizing fundamental-parameter matrix corrections in the general case of unknown samples without standards.

Backscatter intensities have been utilized empirically since 1958;[3-9] however their general interpretation using element scatter cross-sections in fundamental parameter (FP) calculations was not reported until 1977.[10,11] The FP interpretation of scatter intensities for direct x-ray fluorescence (XRF) analysis of unknowns without standards was later extended to more diverse applications.[12-14] The method has always been limited to line excitation (isotopic or secondary) sources whose incoherent and

---

* Supported by National Science Foundation Grant CHE-8260151.

coherent scatter intensities are resolved. This paper summarizes a feasibility study for generalizing the method to also accommodate low-energy and continuum excitation sources which are commonly used in commercial XRF systems. The generalized method is designated as Complete Element Matrix Analysis from Scatter (CEMAS). The feasibility study consists of both theoretical and experimental analyses. Although the research is oriented toward energy-dispersive XRF applications, many of the basic principles also apply to wavelength-dispersive XRF analyses. Because of the dependence on scatter intensities to define sample matrices, CEMAS requires XRF systems in which backscatter from sample mountings, backings, or other system components is not significant or can be otherwise corrected. The adaptation of CEMAS to systems using direct tube excitation is considered vital to future developments of quantitative field XRF instruments as well as to advance and automate laboratory analysis methods without the present restriction to line excitation.

THEORETICAL

The theoretical basis for generalizing CEMAS for use with alternative excitation sources was examined both algebraically and numerically. The algebraic analysis of the previously-used scatter equations[10] showed that they could potentially be written for combined scatter as well as for purely incoherent or coherent scatter. The equations have the form

$$I_j = G_j \frac{1-\exp(-\Sigma Q_i \mu_{ji})}{\Sigma Q_i \mu_{ji}} (\Sigma Q_\ell \sigma_{j\ell} + \Sigma Q_h \sigma_{jh}) \qquad (1)$$

where G = calibration factor (scatter intensity per unit sample mass and per unit scatter cross-section)

Q_i = mass per unit area of element i (g/cm^2)

μ_{ji} = $\mu_{ji} csc\phi + \mu_{ei} csc\theta$ = effective mass absorption coefficient (cm^2/g) at energy j for element i; ϕ and θ are angles between sample surface and respective scattered and excitation x-ray paths.

σ_{ji} = scatter cross-sections (cm^2/g) for incoherent, coherent, or combined scatter at energy j for light (ℓ) and heavy (h) elements.

In previous applications, separate equations for incoherent and coherent scatter yielded the quantities $\Sigma Q_\ell \sigma_{inc,\ell}$ and $\Sigma Q_\ell \sigma_{coh,\ell}$, from which the light-element component of the sample mass (g/cm^2) and its average atomic number were computed,[10] and hence also the total sample mass and absorption characteristics.

For low-energy or continuum excitation, however, only a single combined intensity is usually available; hence, the light-element component of the sample must be simply described by $\Sigma Q_\ell \sigma_{comb,\ell}$. This only allows estimation of the light-element mass or its average atomic number, but not both. Whenever total sample mass is known, ΣQ_ℓ can be estimated from the difference between the total sample mass and the sum of masses of the analyzed elements, Q_h. The average atomic number of the light element component can then be calculated by comparing the light-element scatter cross-section, $\Sigma Q_\ell \sigma_{comb,\ell}/\Sigma Q_\ell$, with established values of $\sigma_{comb,\ell}$. Sample absorption characteristics can then be computed, as before, from the completely-defined sample matrix composition.

In order to preserve the greater flexibility for simultaneously computing both light element mass and atomic number, two independent scatter intensities are still required. If only continuum excitation is available, this suggests the use of dual analyses using high- and low-energy excitation. As shown in Figure 1, combined scatter at different energies may exhibit very different fractions of incoherent and coherent scatter, and thus different variations with atomic number, especially at low atomic numbers. Other applications of dual analyses may generally include the possibility for measuring separate incoherent and coherent intensities from a high-energy analysis (from characteristic tube lines, secondary or isotopic sources) and a single combined scatter intensity from a lower-energy analysis (either continuum or line source). Of the three available scatter

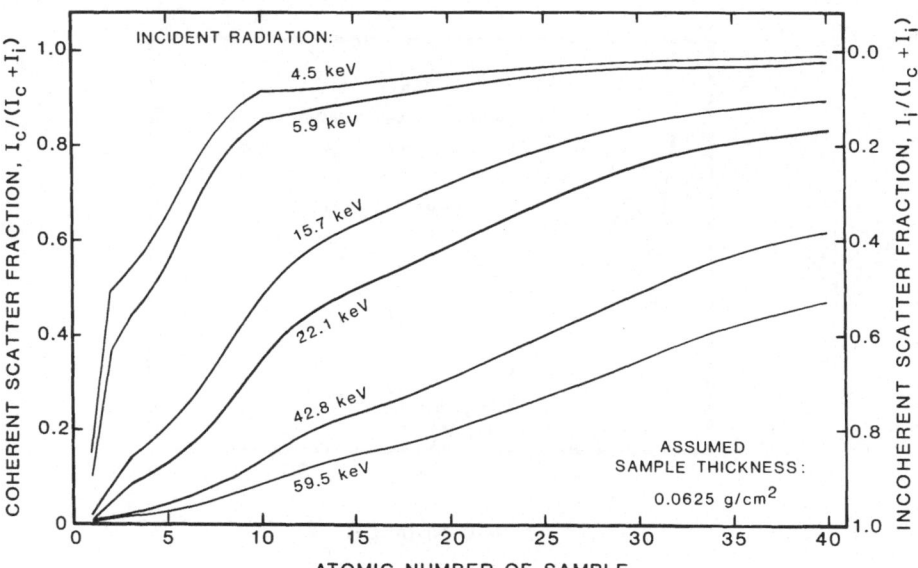

Figure 1.   Incoherent and Coherent Components of
            Combined Scatter For Varying Atomic Numbers
            and Incident Photon Energies.

intensities, only two should be required to give comparable
results to the previously-reported single spectrum case.[10]
Peaks measured from a second spectrum should extend the range of
elements analyzed, and thus improve the definition of the sample
matrix and also its absorption characteristics. Potential
utilization of all three scatter intensities suggests additional
capabilities. If the three are sufficiently independent in their
variation with atomic number and mass, it may be possible to
explicitly compute atomic numbers of two light elements, along
with their masses, instead of only their average as is presently
done.

The theoretical basis for generalizing CEMAS was also
examined by numerically evaluating the sensitivity of a wide
variety of scatter intensity parameters to the light-element
compositions of sample matrices. Primary emphasis was placed on
two-parameter methods due to their greater flexibility than
single-parameter methods for which sample mass must be known. A
combined general sensitivity parameter was derived to be

$$\frac{I_2}{I_1} \frac{d(I_1/I_2)}{dZ}$$

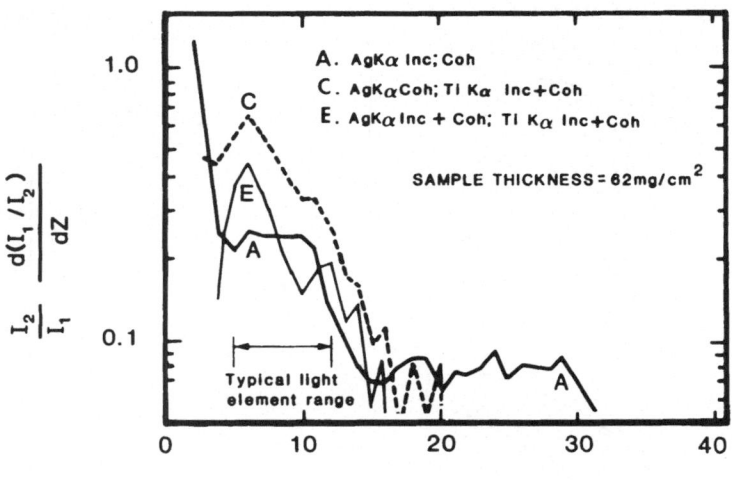

Figure 2.   Comparison of Calculated Sensitivities For
            Three Scatter Parameter Combinations For
            Light Element Selection.

where I_1 and I_2 are the two backscatter intensities defined by Equation 1, and Z is the atomic number of the component of the sample matrix to be characterized. Since scatter from heavy, measurable elements (Z>12) is computed and subtracted from measured intensities prior to estimating the light-element constituents, only the atomic number range below 12 is usually of interest. Figure 2 illustrates these sensitivities for three pairs of scatter intensity parameters.

The sensitivites shown for incoherent/coherent scatter (A) averaged 23% per atomic number unit over the range Z = 5 to 12. Variations in the assumed sample mass and in excitation energy did not appreciably alter these sensitivites. Case A corresponds to the previously used single-spectrum method[10] and thus gives a basis for evaluating the new alternative CEMAS parameters. Even greater but less uniform sensitivity was computed for combined and coherent/combined scatter intensities, with the coherent/combined case (C) being most sensitive throughout the typical light-element range. Adequate sensitivity is thus expected for light-element selection for either of these cases which are suitable for direct tube excitation.

EXPERIMENTAL APPLICATIONS

Experimental applications of the scatter parameters illustrated in Figure 2 were tested using dual analyses of a diverse suite of standard reference materials. In using the case A scatter intensities, scatter from the second, low-energy analysis was ignored. The known compositions of the materials gave a basis for measuring the accuracy of the CEMAS calculations, and the use of dual analyses allowed several two-parameter combinations of scatter intensities to be evaluated. Light element selection in these analyses was restricted to the method already developed for single-spectrum analyses.[15] The analyses utilized filtered continuum excitation (7kV, 1.7 mA, Rh tube, Al filter) for the low-energy range and silver secondary excitation (35 kV, 2mA) for higher energies. Ten-minute analyses were performed under each condition for each sample by automatically switching the analyzer between direct and secondary excitation modes (model 700, Kevex Inc., Foster City, CA). Samples were prepared by directly pressing the powdered materials without dilution, binders or other preparation under 27,000 kg to form 8 cm^2 pellets of varying (278-1200 mg/cm^2) intermediate thicknesses. The samples included five pellets of orchard leaves and single pellets of oyster tissue, animal bone, soil, coal, oil shale, and peridotite, The spectra from analysis of the oil shale sample are illustrated in Figure 3. As illustrated, the elements from Mg to Ca were determined from the low-energy spectrum and the others were determined from the high-energy spectrum.

The dual-spectrum data were successfully analyzed using the incoherent and coherent scatter parameters from the silver secondary excitation. However, attempts to utilize the combined scatter parameters shown in Figure 2 for matrix definition led to unstable iterative matrix corrections in some cases which tended to diverge. The instability appeared to result from the light element selection method, and suggested the need for a more general approach to successfully utilize these alternative parameters.

In the analyses utilizing the silver scatter peaks for matrix definition, the absorption and enhancement corrections affecting the continuum-excited peaks utilized a calculated value of 5.85 keV as their excitation energy. The silver K_α energy of 22.104 keV was used to define the excitation energy for all of the other fluorescent peaks. Mass absorption coefficient data were taken from the parameters of McMaster.[16] Since the silver scatter peaks were used for matrix definition, total scatter cross-sections for 22.104 keV were used for all elements, including the Mg to Ca range. The results of these analyses are presented in Table 1 for the 25 most frequently observed elements. Standard deviations for the five replicate orchard leaf analyses were about 8% when peak counting uncertainties were not dominant. Average relative errors ranged from 7% for orchard leaves to 18% for coal; however errors from trace elements with poor statistics dominated this estimate. The average relative bias similarly ranged from −1% for bone to +5% for oyster tissue.

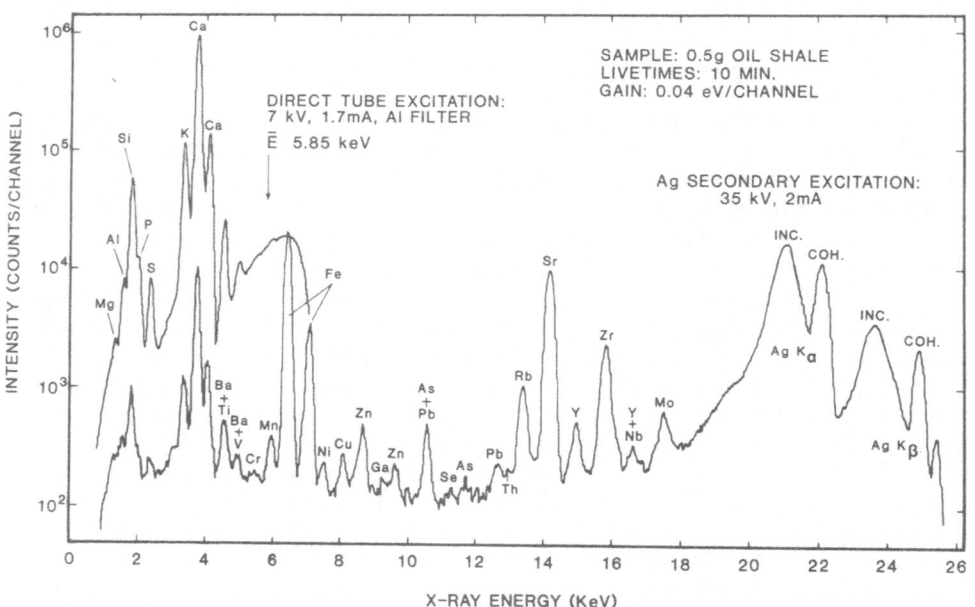

Figure 3.   X-ray Fluorescence Spectra of the Oil Shale Sample Using Low-Energy Continuum and Silver Secondary Excitation.

TABLE 1

RESULTS OF DUAL-SPECTRUM CEMAS CALCULATIONS USING INCOHERENT AND COHERENT SCATTER

	MG %	AL %	SI %	P %	S %	CL PPM	K %	CA %	TI PPM	CR PPM	MN PPM	FE %	NI PPM	CU PPM	ZN PPM	AS PPM	SE PPM	BR PPM	RB PPM	SR PPM	Y PPM	ZR PPM	NB PPM	MO PPM	PB PPM
ORCHARD LEAVES NBS-SRM-1571																									
MEAN(S)	0.82	-0.15	-0.06	0.22	0.23	740	1.43	1.94	241	-31	96	0.0313	-6	12	30	13.3	-2.1	9.9	11.5	37	-1	-1	-1	-1	45
STD.DEV.	0.21	0.00	0.00	0.02	0.01	50	0.05	0.07	48	5	15	0.0031	0	4	6	1.2	0.4	0.8	0.4	2	2	0	0	0	2
PEAK +/-	0.25	0.00	0.00	0.01	0.01	27	0.00	0.01	46	0	12	0.0008	0	3	3	1.5	0.4	0.9	1.0	1	0	0	0	0	3
REF.CON.	0.62	0.00	0.00	0.21	0.23	700	1.47	2.09	0	2	91	0.0300	0	12	25	14.0	0.1	10.0	12.0	37	0	0	0	0	45
ERROR	0.20	0.00	0.00	0.01	0.00	40	-0.04	-0.15	0	0	5	0.0013	0	0	5	-0.7	0.0	-0.1	-0.5	0	0	0	0	0	0
OYSTER TISSUE NBS-SRM-1566																									
MEAS(1)	-0.36	-0.11	0.20	0.89	0.99	9455	1.01	0.19	235	-22	-15	0.0192	-5	52	775	13.4	2.4	54.4	5.9	10	-1	-1	-1	35	-5
PEAK +/-	0.00	0.03	0.02	0.02	0.01	72	0.00	0.00	33	0	0	0.0006	0	3	5	1.1	1.0	1.2	1.1	1	0	0	0	2	0
REF.CON.	0.13	0.00	0.00	0.00	0.00	0	0.97	0.15	0	1	17	0.0195	0	63	852	13.4	2.1	0.0	4.5	10	0	0	0	0	0
ERROR	0.00	0.00	0.00	0.00	0.00	0	0.04	0.04	0	0	0	-0.0003	0	-10	-76	-0.0	0.3	0.0	1.4	0	0	0	0	0	0
ANIMAL BONE IAEA-H5																									
MEAS(1)	-0.68	-0.16	0.94	4.53	0.29	542	0.06	20.59	620	-79	-51	0.0097	-16	-12	93	-5.3	-3.3	-3.1	-2.7	102	-2	-2	-1	-1	-9
PEAK +/-	0.00	0.07	0.05	0.05	0.01	34	0.00	0.01	109	0	0	0.0018	0	0	5	0.0	0.0	0.0	0.0	2	0	0	0	0	0
REF.CON.	0.35	0.01	0.0010	10.20	0.00	550	0.07	21.20	0	0	1	0.0079	2	1	89	0.0	0.0	0.0	0.0	96	0	0	0	0	3
ERROR	0.00	0.00	0.00	-5.67	0.06	-7	-0.01	-0.61	0	0	0	0.0018	0	0	4	0.0	0.0	0.0	0.0	6	0	0	0	0	0
SOIL IAEA-SOIL5																									
MEAS(1)	1.34	8.23	29.11	1.05	0.16	332	2.03	2.76	4726	-77	967	4.1291	-40	81	368	105.8	-4.0	8.5	126.2	361	25	258	5	35	149
PEAK +/-	0.28	0.15	0.17	0.03	0.01	47	0.01	0.01	137	0	8	0.0096	0	9	8	4.2	0.0	2.1	2.8	3	2	3	1	2	7
REF.CON.	1.20	8.10	27.00	0.00	0.00	0	1.90	2.50	5200	38	900	4.8500	90	80	350	110.0	2.0	6.0	120.0	340	23	220	11	0	150
ERROR	0.14	0.13	2.11	0.00	0.00	0	0.13	0.26	-473	0	67	-0.7209	0	1	18	-4.2	0.0	2.5	6.2	21	2	38	-5	0	0
COAL NBS-SRM-1632																									
MEAS(1)	-0.34	2.10	4.04	0.24	1.41	1065	0.29	0.45	1156	-27	45	0.7120	-10	12	33	4.8	4.7	17.6	19.7	141	8	30	2	5	14
PEAK +/-	0.00	0.07	0.05	0.01	0.01	32	0.00	0.00	125	0	11	0.0024	0	3	3	1.7	1.1	1.1	1.1	3	1	1	1	1	4
REF.CON.	0.20	1.85	3.20	0.00	0.00	890	0.28	0.43	1100	20	40	0.8700	15	38	37	5.9	2.9	19.0	21.0	161	0	0	0	0	30
ERROR	0.00	0.25	0.84	0.00	0.00	175	0.01	0.02	56	0	5	-0.1580	0	0	-3	-1.1	1.8	-1.4	-1.3	-19	0	0	0	0	-15
OIL SHALE PNL-COS1																									
MEAS(1)	2.92	3.41	15.21	0.68	0.55	-82	1.49	11.87	1972	-77	407	1.9131	34	42	69	46.0	-3.8	-3.4	69.5	727	10	36	-2	28	16
PEAK +/-	0.34	0.12	0.13	0.03	0.65		0.01	0.01	125	39	34	0.0068	16	7	6	3.4	0.0	0.0	2.1	3	2	2	0	2	6
REF.CON.	3.50	3.80	14.40	0.50	0.65	26	1.62	9.90	1700	34	313	2.0100	25	38	67	44.0	2.0	0.0	72.0	679	12	38	7	21	22
ERROR	-0.58	-0.39	0.81	0.18	-0.10		-0.13	1.97	272	0	94	-0.0969	9	4	2	2.0	0.0	0.6	-2.5	48	-1	-1	0	7	-5
PERIDOTITE USGS-PCC-1																									
MEAS(1)	24.68	0.41	21.61	0.65	0.23	-72	0.00	0.44	-290	2353	897	5.0173	2074	-25	-14	-5.5	-3.3	-2.9	-2.4	-1	-1	0	-1	9	18
PEAK +/-	0.43	0.09	0.16	0.02	0.01	0	0.00	0.00	0	46	39	0.0095	29	11	36	0.0	0.0	0.0	0.0	0	0	0	0	1	5
REF.CON.	26.00	0.40	19.60	0.00	0.00	60	0.00	0.36	90	2730	959	5.8400	2340	11	36	0.1	0.0	0.0	0.0	0	0	0	0	0	13
ERROR	-1.32	0.01	2.01	0.00	0.00	0	-0.00	0.08	0	-376	-61	-0.8227	-265	0	0	0.0	0.0	0.0	0.0	0	0	0	0	0	5

[a]Results from single measurements except for orchard leaves (mean of 5 meas). Peak +/- indicates uncertainty (1-Sigma) due to peak counting statistics. Ref. Con. are the reference concentrations given by NBS, IAEA, PNL, and USGS. Error is the measured value minus the Ref. Con. Negative signs on measured or reference concentrations should be read as "less than" (<), and indicate concentrations below detection limits.

Self-absorption correction factors ranged from 1.07 for Mo in the thinnest orchard leaf sample to 470. for Mg in the thickest one.

Several significant conclusions were drawn from this feasibility study. Most importantly, quantitative analyses can be obtained without standards for unknown samples of unknown mass using CEMAS with multiple as well as single analyses. Low-energy and continuum excitation may be utilized, and can theoretically provide useful scatter intensities for CEMAS. Although the theoretical sensitivities with combined incoherent and coherent scatter are good, new light-element selection algorithms need to be developed to best utilize these parameters. The scatter intensities used for CEMAS need not be the main source of fluorescent excitation; thus scatter from other spectra, characteristic tube lines or continuua may be utilized as long as it results from the sample and not from backing materials or the sample chamber. The feasibility study thus indicates that CEMAS need not be limited to secondary source or isotopic excitation, but can potentially be applied to instruments using direct excitation by x-ray tubes.

REFERENCES

1. T. Shiraiwa and N. Fujino, "Theoretical Calculations of Fluorescent X-ray Intensities in Fluorescent X-ray Spectrochemical Analysis," Japan J. Appl. Phys., 5, 886-899, 1966.
2. C.J. Sparks, "Quantitative X-ray Fluorescence Analysis Using Fundamental Parameters," Advan. In X-ray Anal., 19, 19-52, 1976.
3. G. Andermann and J.W. Kemp, "Scattered X-rays as Internal Standards in X-ray Emission Spectroscopy," Anal. Chem., 30, 1306-1309, 1958.
4. R.C. Reynolds, "Matrix Corrections in Trace Element Analysis by X-ray Fluorescence: Estimation of the Mass Absorption Coefficient by Compton Scattering," Am. Mineral., 48, 1133-1143, 1963.
5. R.C. Reynolds, "Estimation of Mass Absorption Coefficients by Compton Scattering: Improvements and Extensions of the Method," Am. Mineral., 52, 1493-1502, 1967.
6. M. Franzini, L. Leoni and M. Saitta, "Determination of the X-ray Mass Absorption Coefficient by Measurement of the Intensity of AgK_α Compton Scattered Radiation," X-ray Spectrom., 5, 84-87, 1976.
7. C.E. Feather and J.P. Willis, "A Simple Method for Background and Matrix Correction of Spectral Peaks in Trace Element Determination by X-ray Fluorescence Spectrometry," X-ray Spectrom., 5, 41-48, 1976.

8. L. Leoni and M. Saitta, "Matrix Effect Corrections by AgK_α Compton Scattered Radiation in the Analysis of Rock Samples for Trace Elements," X-ray Spectrom., 6, 181-186, 1977.

9. R.D. Giauque, R.B. Garrett and L.Y. Goda, "Determination of Trace Elements in Light Element Matrices by X-ray Fluorescence Spectrometry with Incoherent Scattered Radiation as an Internal Standard," Anal. Chem., 51, 511-516, 1979.

10. K.K. Nielson, "Matrix Corrections for Energy Dispersive X-ray Fluorescence Analysis of Environmental Samples with Coherent/ Incoherent Scattered X-rays," Anal. Chem., 49, 641-648, 1977.

11. K.K. Nielson, "SAP3: A Computer Program for X-ray Fluorescence Data Reduction for Environmental Samples," Richland, WA: Battelle Pacific Northwest Laboratory Report BNWL-2193, 1977.

12. P. Van Espen, L. Van't dack, F. Adams and R. Van Grieken, "Effective Sample Weight from Scatter Peaks in Energy Dispersive X-ray Fluorescence," Anal. Chem., 51, 961-967, 1979.

13. K.K. Nielson and R.W. Sanders, "X-ray Fluorescence Analysis of Environmental and Geological Materials," Trans. Am. Nucl. Soc. 39, 60-61, 1981.

14. K.K. Nielson and R.W. Sanders, "Multielement Analysis of Unweighed Biological and Geological Samples Using Backscatter and Fundamental Parameters," Advan. In X-ray Anal., 26, 385-390, 1983.

15. K.K. Nielson and R.W. Sanders, The SAP3 Computer Program For Quantitative Multielement Analysis by Energy Dispersive X-ray Fluorescence," Richland, WA: Battelle Pacific Northwest Laboratory Report PNL-4173, 1982.

16. W.H. McMaster, N.K. Del Grande, J.H. Mallett and J.H. Hubbell, "Compilation of X-ray Cross Sections," Univ. of California, Report UCRL-50174, Sec. II, Rev. 1, 1969.

INVESTIGATION OF OBSIDIAN BY RADIOISOTOPE X-RAY FLUORESCENCE

Stephen B. Robie* and Ivor L. Preiss

Rensselaer Polytechnic Institute
Department of Chemistry
Troy, N.Y. 12181

INTRODUCTION

The classification of obsidian artifacts has been receiving considerable attention of late since changes in obsidian trace element composition can now be used to identify ancient trade routes[1-4]. The classification of this glassy volcanic material has been attempted using a variety of elemental analysis techniques[5-8]. The most successful and most widely employed method of non-destructive analysis has been that which employs X-ray fluorescence analysis (XRF); either wavelength dispersive (WDS), or energy dispersive (EDS).

In this study, using EDS characterization is based on the relative Zr, Rb and Sr content[9-13], with care taken to insure the maximum analytical precision and accuracy so as to arrive at substantial improvement in the certainty with which various samples can be grouped according to composition. Excitation is provided using a 2mCi ^{109}Cd source which provides nearly monochromatic exciting radiation that can be well collimated. The choice of Zr, Sr and Rb insures that the self-absorption correction within the sample by the matrix is essentially constant, as is the ionization cross-section associated with the primary radiation. There could be, in such a system a marked effect from the influence coefficient (α). However, by limiting the identification and classification to relative composition rather than absolute values, and assuming that a

---

*Present Address: Texaco Research Center
                   Beacon, N.Y. 12508

limited number of combinations of composition are unique to a given
geographical region and geological setting, the data should still be
representative of location even though not reflective of true absolute composition.

The samples under consideration could all be considered to be
of saturative thickness, but were irregular in shape. Because of
the nature of these artifacts, surfaces could not be polished as
would be required in WDS. Rather, a study was performed which determined the take off angle that produced the minimum variation in
intensity as a function of angle for the three $K\alpha$ lines of interest.
That is, irregularly shaped surfaces were placed in such a way that
the average path lengths for primary and analyte X-rays were constant
(within experimental error). The reproducibility and hence accuracy
of this system was found adequate for the study.

Of further significance is the question of assumed uniformity
of composition. If only a portion of the object is illuminated by
the primary (excitation) radiation, it must be assumed that the
volume examined is typical, or at least the average composition.
Additionally, considering that objects utilized as tools have been
worked along fracture lines, it must also be assumed that this surface is of the same composition as the bulk or even the adjacent
portion discarded by the worker as not utilitarian to his purposes.

We have examined the surfaces of objects with spacial resolution of less than $1cm^2$. We have carefully examined areas of stria-
tion and areas of uniform and non-uniform coloring. These results
indicate non-homogeneous composition with certain areas not truly
representative of the bulk.

EXPERIMENTAL

Detection System

The detection systems used are described in detail in earlier
works[14,15]. The detector-sample-source assembly, using Ag X-rays
from a 2 mCi ^{109}Cd is designed so as to allow for varying the source-
sample-detector incident and take off angles in a reproducible manner. The system also permits collimation of the primary beam to
$3mm^2$ and/or collimation of the detector itself. The sample stage
permits varying the angle of the sample surface to the detector to
be adjusted independent of the source-detector angle.

The detector, in all cases is a Si(Li) detector with 10 mm^2
area coupled to a liquid nitrogen cooled FET preamplifier with
pulsed optical feedback. Data was recorded using a Northern Model
636 multichannel analyzer with a 100 megahertz ADC with digital offset.

Resolution of the system was 170 eV at FeKα. This resolution allowed for clear separation of Kα lines of the elements of interest. Underlying Kβ's were subtracted along with system background for data analysis (see Data analysis section).

Sample Preparation

Obsidian samples used in this study were of Central American origin. In some cases, the surfaces were rinsed with de-ionized water before analysis but most were analyzed as received. No other sample pre-treatment was performed.

Collimation

Areas of the samples were chosen using collimation, to either examine specific $3mm^2$ areas or to average over larger surfaces to test for homogeneity. This collimation was achieved for small areas by "masking" the sample with a Cd shield with various aperture sizes or by actually collimating the radioisotope source itself.

Sample Positioning

The samples were mounted in the experimental apparatus at a distance of approximately 35mm from both the source and detector. The angle between the surface of the sample and the detector (incident angle) was chosen to be 38 degrees to minimize analyte intensity variations (see Results Section).

Data Reduction

The collected data was manipulated on a PDP-11 minicomputer. First, the data was smoothed using a five point quadratic fitting routine. Net integrated peak areas were calculated, first by subtracting the average of backgrounds above and below the peak. Second, the contribution of the strontium Kβ to the zirconium Kα peak had to be removed using published values for the Kα/Kβ for Sr[17]. Finally, the net peak areas for Rb, Sr and Zr were normalized to the sume of the net areas to determine the relative percentages of Rb, Sr and Zr. The values were then plotted on triangular graph paper for later graphical analysis.

RESULTS

A typical RIXRF spectrum is shown in Figure 1. Note that the Rb, Sr and Zr lines are clearly resolved except for the influence of Kβ lines from Rb and Sr which may underlie Kα lines of higher atomic number components

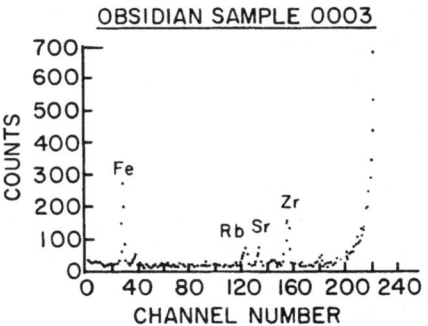

Fig. 1.   Spectrum of a typical obsidian.   The Kα lines are labelled.
The line at high energy is the Compton scatter peak from
the silver X-rays.   Count time was 2000 seconds clock time.

In order to maximize the precision and accuracy of the results,
special care had to be taken in the placement of the samples within
the experimental apparatus.   Because of the irregular surfaces of
the samples, we examined, in detail, the effect of changing take off
angle on the intensity of analyte X-rays.

We first approximated the obsidian as a flat, infinitely thick
sample with an average composition of 95% SiO_2, 5% Fe.   Because of
the low concentrations of the elements in question, we ignored all
self-fluorescence and self-absorption effects.   The changes in intensity are therefore due only to changes in the path lengths of X-
rays into and out of the sample.   If we assume the source is monochromatic, equation 1 can be derived[16] for analyte X-ray intensity,

$$I_A(E_A) = \frac{N(E_A)}{4\Pi \sin\psi_1} \frac{C(E_o) \; I_o(E_o) \; dE_o}{\mu(E_o) \; \csc\psi_1 + \mu(EA)\csc\psi_2}, \qquad (1)$$

where $I_A(E_A)$ and $I_o(E_o)$ are the analyte and source X-ray intensities,
$N(E_A)$ is the detector efficiency for analyte X-rays, $C(E_o)$ is a constant containing all fundamental parameters, $\mu(E_o)$ and $\mu(E_A)$ are the
mass absorption coefficients for source and analyte X-rays in the
matrix and ψ_1 and ψ_2 are the incident take off angles.   By normal-
izing the intensity (in this case to ψ_1 = 45 degrees), one is left
with equation 2,

$$I/I_{45} = \frac{\sin 45}{\sin \psi_1} \frac{\mu(E_o) \; \csc 45 + \mu(E_A) \; \csc 45}{\mu(E_o) \; \csc \psi_1 + \mu(E_A) \; \csc \psi_2} . \qquad (2)$$

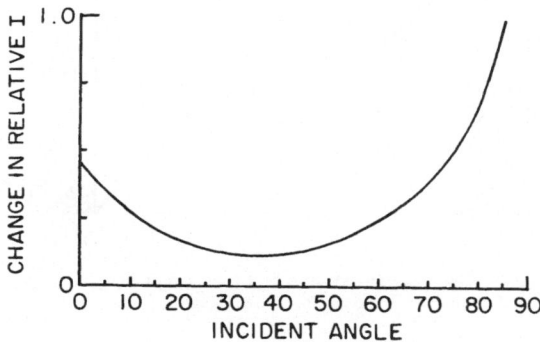

Fig. 2.   Change in strontium intensity as a function of incident X-
ray angle.

A source-detector angle of 90 degrees was used to minimize background, and fixes the sum of ψ_1 and ψ_2 at 90 degrees. Figure 2
shows the results of this equation for Sr. The observed minima can
be thought of as the point at which small changes in path length
into the sample are compensated for by small changes in the path
length out of the sample. This curve indicates that the incident
angle should be set as close to 38 degrees as possible so as to
minimize fluctuations in intensity.

In order to determine if the flat surface case was valid for
obsidians, we considered the case of a geometrically irregular surface (e.g. a sawtooth surface). Obsidian surface variations are of
low frequency so we ignored possible shadowing effects from one surface irregularity to the next. This assumption allows us to study
only one "tooth" and apply the results to the whole surface.
Figure 3 illustrates this case. For simplicity, we have used a
tooth with surfaces at 45 degrees with the incident X-rays normal
to the surface. Equations 3 and 4 apply in this case.

$$\cos \psi_2 = \frac{P_1 + \Delta P}{D + \Delta D} \qquad (3)$$

$$\cos \psi_1 = \frac{P_2 + \Delta P}{D + \Delta D} \qquad (4)$$

P_1 and P_2 are the pathlengths into and out of the sample. D is the
height of the tooth and ΔP and ΔD are the changes in path length
and height. These changes can be either positive or negative.

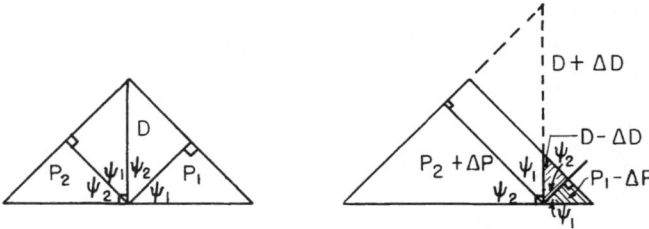

Fig. 3.   The geometry of the sawtooth case.

In this case, the change in path length into the sample is exactly compensated for by the change in path length out of the sample so that the total path length is a constant.   In the more general case, the average path length will correspond to this constant.   The most important point in this case is that the total path length in the sample is a function only of the height of the irregularity above the average surface, and not of where on the surface the X-ray enters the sample.

It can be shown that for the same total path length, the depth calculated for the flat surface case is exactly one half that calculated for the height of the "tooth".   This is the concept of the average surface which we have used in this study.   Obsidian can be modeled as a series of geometric irregularities of differing heights. The average surface is then the plane on which the sum of the heights above the plane are balanced by the sum of the depths below the plane.

Visual inspection under low magnification clearly shows non-uniform coloring or striations in the surface of the obsidian.   Such coloration indicates non-uniform composition which may originate from differential cooling of the glassy matrix[18], or remelting re-sulting in the same effect as the solid solution composition changes.

Mapping of the surface using the large collimator which aver-ages over a surface shows varying relative composition typical of the results previously reported in other studies.   A narrower range is achieved when care is taken to avoid the non-uniform areas.   In order to confirm these results, the objects were in turn analyzed using a $3mm^2$ collimator to mask the surface so only areas of colora-tion or areas free of non-uniformities are analyzed.   The results are summarized in Figure 4 for both cases.   Point A shows the rela-tive composition of an area of coloration.

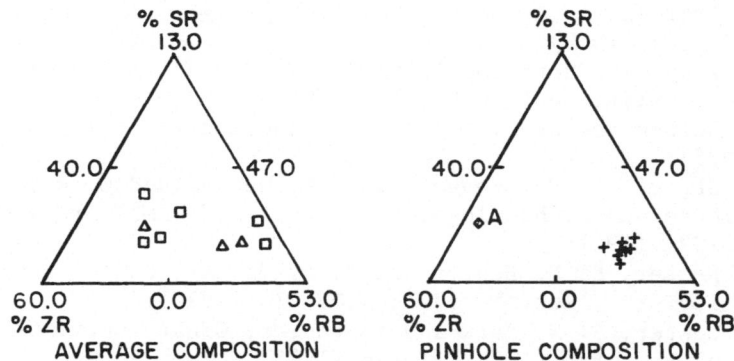

Fig. 4. Relative obsidian composition using a wide collimator to average over the sample and a 3mm^2 collimator. Point A is an area of coloration.

CONCLUSIONS

Rather substantial improvement in the ability to characterize obsidians has been achieved. Care must be taken to insure that only regions of uniform composition are illuminated. Figure 4 clearly shows that "averaging" leads to a much larger variation in the size of clusters than selecting uniform areas. The contrast in the effects of the non-uniform composition is clearly observed. Therefore, if the luxury of sample pulverization and/or surface smoothing is absent, procedures similar to those described, if employed, will considerably narrow the range of apparent composition and thus improve the ability to classify the objects.

REFERENCES

1. R. F. Heizer, H. Williams and J. A. Graham, Contr. Univ. Calif. Arch. Res. Fac., 1, 94-103 (1965).
2. D. P. Stevenson, F. H. Stross and R.F. Heizer, Archaeometry, 13(1), 17-25 (1971).
3. J. A. Graham, T. R. Hester and R. N. Jack, Contr. Univ. Calif. Arch. Res. Fac., 16, 111-115.
4. H. R. Bowman, F. Asaro and I. Perlman, Archaeometry, 15(1), 123-127 (1973).
5. J. R. Cann and C. Renfrew, Proc. Prehistoric Soc. 30, 110 (1964).

6.   J. B. Griffin, A. A. Gordus and G. A. Wright, Am. Antiquity, 34(1), 1 (1969).

7.   R. Jack and I. Charmichael, Calif. J. Mines Geol. Short Contr.Calif. Geol. S.R., 100 (1969).

8.   D. E. Nelson and J. M. D'Auria, Archaeometry, 17(1), 85 (1975).

9.   F. H. Stross, D. P. Stevenson, J. R. Weaver and G. Wyld, Science and Archaeology, R. H. Brill ed., MIT Press, 210-221 (1971).

10. T. R. Hester, R. N. Hester and R. F. Heizer, Contr. Univ. Calif. Arch. Res. Fac., 16, 105-110 (1972).

11. T.R., Hester, R. N. Jack and A. Benfer, Contr. Univ. Calif. Arch. Res. Fac., 18, 167-176 (1973).

12. J. A. Graham, T. A. Hester and R. N. Jack, Contr. Univ. Calif. Arch. Res. FAc., 16, 111-116 (1973).

13. R. H. Cobean, M. D. Coe, E. A. Perry, Jr., K. H. Turekian and D. P. Kharker, Science, 174, 666-671 (1971).

14. I. L. Preiss, S. Robie, I and E.C. 21, 676 (1982).

15. A. Frank, I. L. Preiss and A. Adyeme, Radioanal. Chem. (in press) (1983).

16. R. Jenkins, R. W. Gould and D. Aedeke, Quantitative X-ray Spectrometry, Marcel Dekker, Inc., pp37 (1981)

17. S. I. Salem, S. L. Panossian and R. A. Krause, ANDT 14, 91 (1974).

18. R. Doremus, Private Communication (1983).

A COMPARISON OF TRACE METAL DETERMINATIONS IN CONTAMINATED SOILS

BY XRF AND ICAP SPECTROSCOPIES

Douglas S. Kendall, Joe H. Lowry, Edward L. Bour
and Timothy J. Meszaros

Environmental Protection Agency
National Enforcement Investigations Center
P.O. Box 25227, Denver Federal Center
Denver, Colorado 80225

The National Enforcement Investigations Center of the EPA provides support services for the enforcement activities of the Agency. Recently, we have analyzed hazardous wastes as part of efforts to enforce the Resource Conservation and Recovery Act and the Superfund Act. Sample preparation for inorganic elemental analysis is a difficult and time-consuming step. Thus, it would be desirable to be able to use x-ray fluorescence methods which require relatively little sample preparation for the analysis of solid hazardous wastes. A major problem to be overcome is the need to calibrate for a large variety of samples. However, a compensating factor is that the error will be largely determined by the sampling error and the measurement accuracy is not quite so critical.

The study reported here was part of an effort to ascertain whether several sites heavily contaminated by mining and smelter debris should be classified as Superfund sites. A large number of samples were collected and an efficient method of analysis was necessary. The party responsible for the site wanted to use XRF for the soil analysis. The EPA needed to see if XRF was suitable for soil analysis and to learn what to expect from XRF analyses by other laboratories. Therefore, we were asked to analyze 18 of the samples for 30 elements with requested detection limits of 100 ppm.

There were a number of goals for the study. First, to ascertain whether XRF was suitable for trace metal determinations in soil.   This is basically a question of whether representative preparations could be obtained.   Second, to compare XRF with inductively coupled argon plasma optical emission spectroscopy (ICAP).   Since many elements were to be analyzed and because the samples were from an important class of samples, contaminated soils, the comparison would be valuable.   There are very few published comparisons of ICAP and XRF.   In fact, the EPA has funded comparison studies as part of efforts to compare multi-element techniques for hazardous waste analysis.   Third, to evaluate standardization using a small number of standards and using scattered radiation as an internal standard for matrix correction.   High concentrations of metals were expected and, since 30 elements were in the analytical request, the preparation of a multitude of standards was not attractive.

EXPERIMENTAL

The samples were first sieved to -10 mesh to conform to the ASTM definition of soil and then ground for two minutes in a rotary mill.   The XRF analysis was done on pressed pellets which contained ten percent boric acid as a binder.   The samples were analyzed with a Philips' model 1400 wavelength dispersive spectrometer using a tungsten tube and LiF crystals.   Both peaks and backgrounds were counted for 40 seconds.

The samples for ICAP analysis were fused with KOH.[1]   This fusion has proven to be very useful in that siliceous minerals and organic matter are rendered soluble, but the temperature is low enough that lead, cadmium, selenium and other volatile metals (except for mercury) are retained.   A simultaneous Jarrell-Ash ICAP spectrometer was used with standards matrix matched to the KOH fusion mixture.

CALIBRATION

Elements with atomic numbers greater than cobalt were standardized with USGS trace element in glass standards, essentially a blank, and three standards with nominal concentrations of 5, 50 and 500 ppm of about 40 trace elements.   Since for some samples the concentrations of some elements were well above this range, these standards were supplemented with previously analyzed soils from a smelter environment in Denver.

The method of calibration uses the scattered radiation as an internal standard to correct for matrix absorption and is based on extensive previous work.[2] [3] [4]

The elements above cobalt in the periodic were divided into three groups:

(1)   Sb, Ba, Cd, Sn
(2)   Cu, Ga, Ni, Zn
(3)   As, Br, Ge, Au, Hg, Rb, Se, Sr

Only one background was measured for each group. This was important since, with the multielement standards, it would have been difficult to find an interferencefree background near each peak.

In order to determine the relationship between the background under a peak and the group background, a series of compounds ranging in absorption coefficient from magnesium oxide to vanadium oxide was used. The compounds did not contain the element of interest and, in every case, a linear relationship existed between the group background and the background at the peak position. The group background was also used to correct for variations in absorption, under the assumption that the background is inversely proportional to the absorption coefficient. The equation for the net intensity follows:

$$I = (P - k (B) - C)/B$$

I is the net intensity, P the gross peak intensity, B the background intensity, and k and C are the coefficients relating the group background, B, to the background under the peak. Instead of dividing by B, the gross background, it would have been preferable to use the true scattered background, but the software would not permit this extension.

Table I shows the effect of the matrix correction for the six lead samples with concentrations under 500 ppm. The ICAP values are used as a point of reference. While not perfect, the data corrected with the use of scattered radiation are much improved. The last column contains the background in counts per second to show the wide variation in matrix absorption.

Table I
Effect of Matrix Correction Using
Scattered Radiation on Lead Determination

No.	ICAP (ppm)	XRF		BKGND (cps)
		Corrected (ppm)	Uncorrected (ppm)	
1	460	430	213	210
2	180	240	134	239
3	190	270	281	447
4	390	490	587	517
5	480	400	376	402
6	80	100	102	460
%RD	-	21%	40%	-

RESULTS

The high Z elements were calibrated as described above and
the low Z elements were calibrated using conventional empirical
correction methods and standards prepared from well-analyzed
soils, collected near a Denver smelter. Altogether, 29 elements
were determined by XRF and 31 elements by the simultaneous ICAP
instrument.   There were 19 elements which were determined by both
instruments.   Cobalt, nickel and selenium were almost always
below the detection limit and so offer little basis for comparison.   The four majors, aluminum, silicon, calcium and iron, compared well but will not be discussed further as such determinations are well established.   Thus, there are 12 elemental determinations in each of 18 samples which form the basis for comparison.

Figures 1, 2 and 3 are graphs of the results for three of
the elements.   Some of the higher concentrations were not plotted
in order to better show the lower concentration values.   The ICAP
and XRF values are plotted with the same scale, so that if the
results for a given sample agree, the point for that sample will
be on the 45° line.

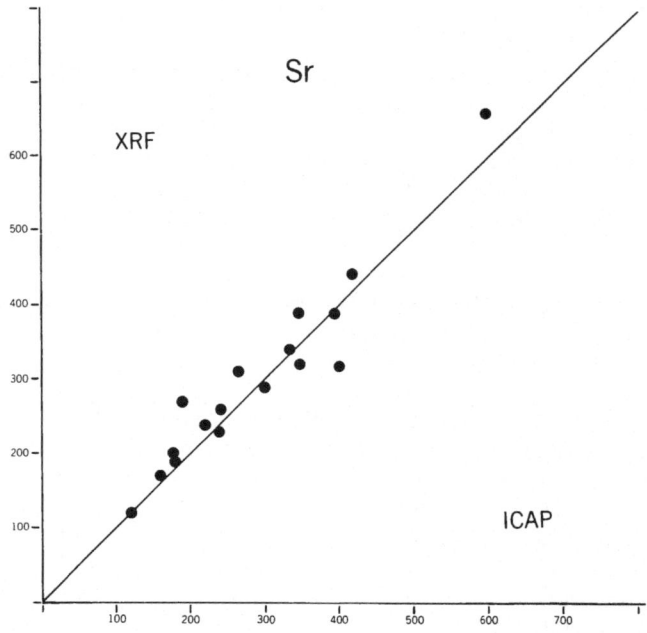

Figure 1.   Strontium comparison for concentrations to 600 ppm.

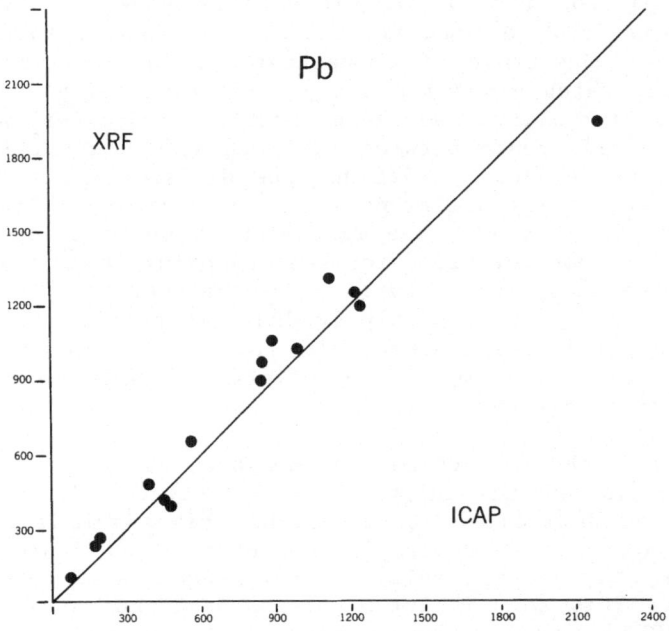

Figure 2. Lead comparison for concentrations to 2200 ppm.

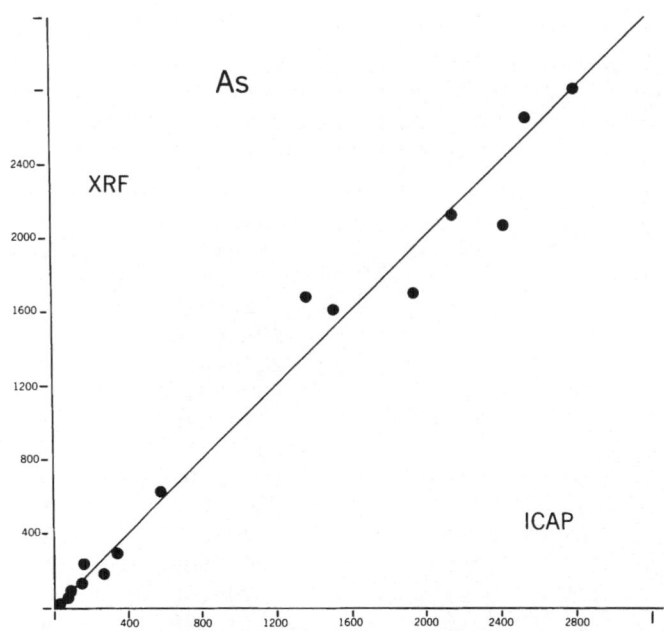

Figure 3. Arsenic comparison for concentrations to 2800 ppm.

The comparison for all 12 elements is shown in Table II. The comparison is done with relative rather than absolute differences since the range of concentrations was so large. The second column, which gives the average relative deviation, would be zero if the two methods agreed perfectly. The t test was used to see if the difference between the methods was significant at the 95% confidence limit. Although the difference was significant in some cases, the differences were relatively small. Within the goals of the study, the agreement is quite good. In our experience, the sampling error for soil sampling is at least 30%, larger than the analytical error. As can be seen from the table, many of the samples were highly loaded with metals. Since the soils were heavily contaminated, and since the requested detection limits were 100 ppm, little effort was made to push the detection limits to very low levels.

The quality checks included the measurement of standard reference materials and the analysis of duplicates. The precision data, based on duplicate determinations, gives similar results for XRF and ICAP. For one sample, measured in duplicate by both XRF and ICAP, the average relative difference was 3% for XRF and 4% for ICAP. This and similar precision data indicate the suitability of XRF for soil analysis, since it has proven possible to reproducibly prepare soil samples for analysis.

Table II

The Average Relative Deviation ($\bar{d}$), the Relative Standard Deviation (s_d), the Range of Concentrations Observed, and Whether the Difference Between the Two Methods is Significant by the t Test

Element	$\bar{d}$ (%)	Sd (%)	Range (ppm)	t Test
Sb	20	38	20-220	N
As	1	12	20-14300	N
Ba	6	21	280-1860	N
Cd	3	29	30-2010	N
Cr	21	22(11)	70-270	Y
Cu	11	21	160-14100	Y
Pb	5	19	100-12100	N
Mn	1	12	360-3200	N
Sr	6	11	120-670	Y
Ti	13	20(13)	500-3600	Y
V	11	21	20-140	Y
Zn	8	16	160-116000	N

In summary, the comparison worked out well and met the goals of the study. X-ray fluorescence spectroscopy is suitable for the rapid analysis of a large number of soil samples contaminated with metals. The analytical error would be less than the sampling error. ICAP and XRF compared well for this analysis. Calibration is easier for ICAP analysis but the KOH fusion sample preparation procedure is harder. While there was no need or effort expended to optimize detection limits, the limits of detection were similar for both methods because the ICAP fusion mix must be diluted.

REFERENCES

1. M. A. Floyd, V. A. Fassel, A. P. D'Silva, Anal. Chem. 52 (1980) 2168-2173.

2. G. Andermann, J. W. Kemp, Anal. Chem. 30 (1985) 1306-1309.

3. R. C. Reynolds, American Mineralogist 48 (1963) 1133-1143.

4. C. E. Feather, J. P. Willis, X-Ray Spectrometry, 5 (1976) 41-48.

DETERMINATION OF BARIUM AND SELECTED RARE-EARTH ELEMENTS IN
GEOLOGICAL MATERIALS EMPLOYING A HpGe DETECTOR BY RADIOISOTOPE
EXCITED X-RAY FLUORESCENCE*

J.J. LaBrecque[+]

Instituto Venezolano De Investigaciones Cientificas-IVIC
Apartado 1827, Caracas, Venezuela

I.L. Preiss

Rensselaer Polytechnic Institute-RPI
Troy, New York 12181 (USA)

INTRODUCTION

The laterite material (geological) from Cerro Impacto was
first studied by air radiometric techniques in the 1970's and was
found to have an abnormally high radioactive background. Further
studies showed this deposit to be rich in thorium, columbium,
barium and rare-earth elements (mostly La, Ce, Pr and Nd). A
similar work (1) has been reported for the analysis of Brazil's
lateritic material from Morro do Ferro to determine elemental compositions (including barium and rare-earth elements) and its relationship to the mobilization of thorium from the deposit using a
Co-57 radioisotope source.

The objective of this work was to develop an analytical method
to determine barium and rare-earth element present in Venezuelan
lateritic material from Cerro Impacto. We have employed a method
(2) before,employing a Si(Li) detector,but due to the low detection
efficiencies in the rare-earth K-Lines region (about 30 KeV - 40
KeV), we have decided to study the improvement in sensitivities
and detection limits using an hyperpure germanium detector.

*This work forms part of UNESCO's IGCP-129 Project and was funded
 in part by an international grant between CONICIT (Venezuela) and
 NSF (USA).

+This author was on sabatical leave at RPI where this work was
 performed.

TABLE I

The comparison of the certified values for the standard reference
materials and measured values by this method n = 1 trials.

Element	Value*	SY-3	SY-2	GSP-1	BRC-1	G-2
Ba	C	440	460	1300	680	1850
	M	523	418	1487	587	1780
La	C	1400	85	200	25	100
	M	1398	80	220	<dl	86
Ce	C	2000	210	390	54	150
	M	1998	209	415	83	99
Nd	C	500	70	190	29	60
	M	499	53	195	22	73

*C = Certified; M = Measured
<dl = less than detection limit

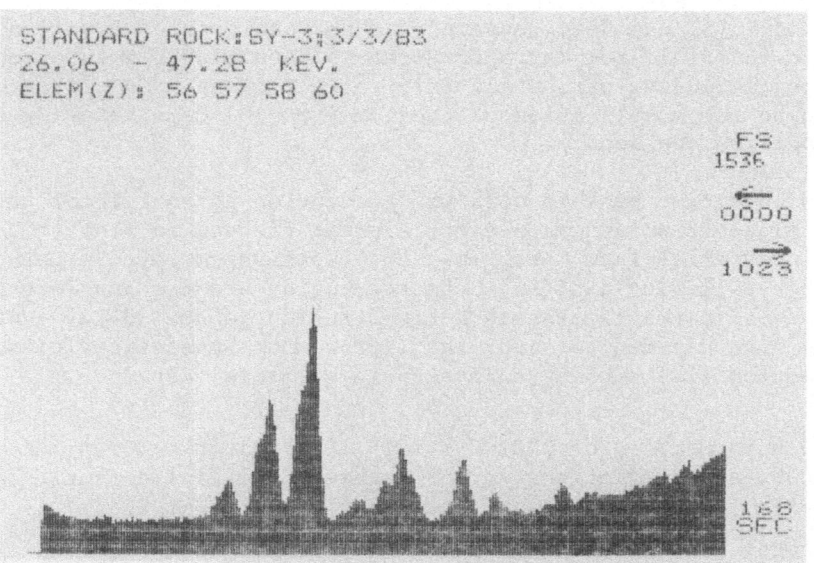

Figure 1.   The K-lines from Ba, La, Ce and Nd from
            SY-3 standard rock material.

EXPERIMENTAL

Sample Preparation

   Calibration curves were constructed for barium, lanthanium,
cerium and neodynium by measuring five standard rocks. These
reference materials were pressed into 1 cm diameter pellets using
a conventional RI die applying 6 tons of pressure for 10 minutes.
About 350 mg of each standard rock was weighed after it was dried
at 102º + 5ºC for one day. Another set of pellets of the same
five standard rocks (SY-3, SY-2, GSP-1, G-2 and BRC-1) was pressed
in the same manner to be used as unknowns. Finally in the case of
the laterite materials, they were first ground to less than 200
mesh particle size before drying. Each pellet was mounted separately with two pieces of Scotch magic tape in a 2 x 6 cm piece of
IBM card with a 1.5 cm aperature for presentation to the X-ray
fluorescence spectrometer.

Instrumentation

   The samples and standards were excited by a 30 mCi Am-241
point source (disc) and the secondary X-rays were collected with a
hyperpure germanium detector (HpGe) with a resolution of about
180 eV FWHM at 5.9 KeV (MnKα). A spectrum of the rock standard,
SY-3 is given in Figure 1 and a typical spectrum of a Venezuelan
laterite from Cerro Impacto is shown in Figure 2; both were
fluorescent for about one hour.

   It can be seen in Figure 2, that the low energy region (<25
KeV) was discrimated and offset. This was achieved by employing
a bias amplifier in addition to a spectroscopy amplifier since this
feature is not available on the ADC (the Nucleus) and to utilize an
Apple II + micro processor as a multichannel analyzer. The hardware of this system is described elsewhere in full (3,4). The
software used was supplied by Dapple Systems (Sunnyvale, CA).

   Since the K-shell X-rays were employed for these determinations, for which the absorption effects due to the difference in
matrix density are minimum and the elements of interest are in
small amounts, we have applied a linear least square fit calibration program only, thus not taking into account absorption or
enhancement corrections. The peak intensities were calculated
directly from the spectra by intergrating the 1.2 x FWHM region
for the Kα-peaks without background subtraction.

TABLE II

The precision of the Ba, La, Ce and Nd determination of the SY-3
(standard reference material) for five independently prepared
samples for one hour fluorescent time.

Element	Ba	La	Ce	Nd
X (intensity)	4763	8716	12924	6769
± standard deviation	350	277	801	304
X (ppm)	472	1395	1998	498
± standard deviation	35	44	124	25
Relative standard deviation (%)	7.35	3.17	6.20	5.02
Certified value	440	1400	2000	500

Figure 2.   The K-lines from Ba, La, Ce and Nd
from a laterite from Cerro Impacto.

TABLE III

Comparison of the results for the determinations of Ba, La, Ce and Nd in Venezuela laterite materials from Cerro Impacto by this X-ray fluorescence method and an internal standard X-ray fluorescence method.

Sample	BaO (ppm)		La_2O_3 (ppm)		Ce_2O_3 (ppm)		Nd_2O_3 (ppm)	
	This Work	Internal Standard	This Work	Internal Standard	This Work	Internal Standard	This Work	Internal Standard
A	7301	7391	213	211	1215	1138	107	119
B	6842	7802	120	141	1072	1118	78	89
C	9108	8988	181	183	1380	1292	87	100
D	7492	7462	178	164	1362	1146	82	97
E	7315	7641	176	198	1224	1217	102	106
F	8271	8155	203	221	1262	1280	136	125
G	4235	3371	598	652	949	1621	426	327
H	5346	4618	894	1026	1571	1442	454	388
I	6118	6089	671	648	809	691	109	81
J	5401	5588	544	571	745	598	123	153
K	8014	8204	323	348	673	604	623	402
L	2176	2304	498	519	873	1020	233	251
M	4806	4992	187	167	922	1203	198	238
N	3309	3417	192	208	1642	1514	279	309
O	4262	4002	566	598	1832	2106	261	241
P	3919	3798	325	309	1961	2364	138	108
Q	1706	1971	431	462	1983	1999	180	204

RESULTS AND DISCUSSION

A comparison of the concentrations of barium, lanthanium, cerium and neodynium by this method for the following standard reference rocks: SY-2, SY-3, GSP-1, G-2 and BRC-1 with their cer- tified values are presented in Table I. The accuracy of these determinations are very acceptable for the concentrations repre- sented. The accuracy should improve if an average of 3-5 measure- ments on different sub-samples were performed rather than one measurement. Again it should be noted that the X (intensity) also includes the background counts in the 1.2 x FWHM region.

The total precision of the barium, lanthanium, cerium and neodynium determinations for SY-3 (reference rock) from five independently prepared samples is given in Table II. Again the precision is given in Table II. Again the precision is adequate for our geological analysis. It can be seen that in the case of barium that the average value of 472 ppm is more accuracy than the 523 ppm value found when only one determination was performed in Table I.

Finally, it should be noted that this technique was employed to analyze Venezuelan laterite samples from Cerro Impacto and the results were in good agreement (as can be seen in Table III) with those obtained from an internal standard XRF technique (2) even though the matrix in these samples is much different then those of the standard reference rocks used for the calibration curve.

REFERENCES

1. G.R. Lauer, J. Furfaro, M. Carlos, W. Lei, R. Ballao, and T.J. Kneip, Advances in X-Ray Analysis, 25 (1981) 201.

2. J.J. LaBrecque, W.C. Parker, and D. Adams, J. of Radioanal. Chem., 59 (1980) 193.

3. D. Hale, J.C. Russ and D.E. Leyden, Microbeam Analysis, (1982) 473.

4. K. Borowski, I.L. Preiss, J.J. LaBrecque, and C. Pauley, Utilization of an Apple II + Microcomputer as a Multichannel Analyzer for X-ray Fluorescence Spectrometry, J. of Computer Enhanced Spectroscopy (in press).

A VERSATILE XRF ANALYTICAL SYSTEM

FOR GEOCHEMICAL EXPLORATION AND OTHER APPLICATIONS

T. K. Smith

Institute of Geological Sciences
London, United Kingdom

For more than a decade the Institute of Geological Sciences has carried out large-scale geochemical analysis in pursuit of mineral exploration and regional geochemical reconnaissance programmes for the Department of Industry. Over this period an XRF analytical system has been developed to meet part of this requirement for multielement data. The elements of interest range from the fourth period to uranium, and have become more numerous as exploration and analytical techniques have expanded. In particular, additions have been made of those elements such as arsenic, molybdenum and tungsten which are more difficult of determination by conventional methods.

The original concept was that XRF analysis would be employed for samples panned from stream sediments and that the concentration range for many elements would be 0-1%. This process increased geochemical contrast and brought some concentrations above the detection limit but made others so high and sample size so small that analysis was made more difficult. In any case at least one more degree of freedom was introduced into the system. The very highest accuracy was unnecessary where the main requirement was location of mineralisation since exploration was to a large extent based on statistical treatment of large numbers of samples. Improvements in hardware and software, however, led to higher sensitivity and accuracy and, together with a shift of emphasis in the programmes, led to the inclusion of other sample types of lower analyte concentration. Moreover, consideration could then better be made of elemental inter-relationships.

Sample preparation methods evolved simultaneously from inefficient mechanical mortars through vibratory disc to planetary ball mills using agate medium, as molybdenum,chromium and cobalt were added to the scheme. The pressed powder pellet remained the most satisfactory form of sample preparation for speed and sensitivity, but the adhesive employed was developed from an organic solution of a methacrylate polymer or an aqueous solution of polyvinyl alcohol to a high purity copolymer requiring neither solvent nor curing.

Early instrumentation comprised a Philips PW1220C computer-controlled sequential Bragg spectrometer with 24 mechanically-set channels. Its capacity was later tripled by purchase of a Torrens sample loader with 108 positions. The necessity for alternation between tubes of tungsten and molybdenum anode was removed by purchase of an additional instrument, a PW1450/10. Its 72 channels were controlled by thumbwheels and with its 60-position sample loader it provided a further increase in throughput by a factor of two. The measurement sequence was however somewhat restricted and was extended by an electronic modification (Smith,1980). A microcomputer (Diehl Alphatronic 332) with a specially designed interface allowed on- and off-line operation. Sample size was increased at this time from 8 to 12 grams to maintain the thickness of larger diameter pellets. Improvements in detection limits from the wider optics and higher stability encouraged replacement of the PW1220C system by a computer-controlled PW1450/20 with 60-position sample loader.

Interelement correction, initially by mass absorption coefficients applied to approximate determination of major elements, was later changed to use of scattered radiation measured high on the continuum in order to improve speed of analysis, to allow some extension of the method to materials other than silicates, and to provide for the changing requirements of mineral reconnaissance, where specific areal targets were reflected in restricted but diverse requests for elemental analysis. Software was provided for these requirements, with on-line calculation replacing remote processing of paper tape intensities into punched card concentrations.

THE CURRENT SYSTEM

The aim has been to provide a versatile scheme for handling large numbers of samples for a wide range of elements. Currently a capacity exists for 150000 elemental determinations per annum for up to 28 elements.

Almost any geological solid material can be handled but where analyte concentrations exceed maximum limits of calibration the samples may be diluted with silica. 3.0±0.1g of Elvacite 2013 (Dupont & Co's methyl n-butyl methacrylate copolymer) and 12.0±0.1g of sample (reduced to a suitable size by jaw-crushing if required) are ground together at maximum speed in a Fritsch P5 planetary ball mill equipped with eight agate pots. The resultant -200 mesh powder is momentarily pressed at 25 tons load in a 40mm die. It does not

adhere to the mill or the die and curing or drying is not required.
The discs, which are relatively robust and have good resistance to
X-ray damage, are permanently stored in individual aluminium cups as
a reference collection.

The instrumentation remains as previously described except that
a Hewlett Packard HP87XM system now replaces the Diehl Alphatronic.
The configuration, which continues the provision for on- and off-line
calculation, includes 256 kbyte user memory, additional ROMs, dual
$3\frac{1}{2}$ inch flexible discs, matrix printer and A4 two-pen plotter. A
further enhancement is terminal emulation on the Natural Environment
Research Council/Science and Engineering Research Council network of
large mainframes. Direct distribution of data to remote customers
is therefore provided for.

Three principal types of standards are used in calibration.
Because of the difficulty of obtaining well-analysed natural materials of suitable concentration range synthetic standards are prepared
(in the same way as for unknowns) from high purity compounds using
matrices composed of calcium carbonate, ferric oxide, silica and
similar materials. Sympathetic variation is avoided in multi-analyte
standards in order to prevent further degradation of intensity data.
Aluminium alloy cylinders, used to contain 36mm diameter pellets in
order to improve their durability, are not seen by the detectors
because of a mask in the X-ray path at the entrance to the primary
collimator A similar arrangement is employed for international
standards, which are used to check calibrations, and for the third
type, which fills the requirement for external ratio and high analyte
concentration discs where the integrity must be such that no material
is shed in the spectrometer. A cylinder is placed on a sheet of
polyethylene on a flat surface and filled with a mixture of the required composition and an epoxy resin which cures rapidly at low
temperature. The polyethylene sheet, which does not adhere to the
resin, may be removed after hardening has taken place.

Many papers exist on the use of scattered radiation for mass
absorption correction e.g. the early work of Andermann and Kemp(1958).
This correction may use Rayleigh or Compton peaks (gross or net),
background at, or some distance from the analyte peak, or some
function of these. For this application an energy was selected high
on the continuum at 24.14 keV and inverse internal ratio used in
addition to external ratio for all elements with analytical energies
greater than the principal iron absorption edge (Hower,1959). Only
approximate concentrations of other elements are required. Analytical parameters are shown in Table 1. Additional line overlap corrections may be applicable in some cases. Advantages exist in that no
background subtraction is necessary in order to obtain a net scatter
peak as used by some authors and that the intensity is sufficiently
high to obtain a precise ratio in a short time. The algorithm for
conversion of intensities to concentrations is given in Smith (1983).

Table 1. Analytical Parameters.

Atomic No.	Line	Energy (keV)	Anode	kV	mA	Crystal (LiF)	Detector	Stop count (k)	Stop time (sec)	Internal ratio	Line overlap	LLD (Qtz; 3σ; ppm)
20	CaKα	3.690	W	20	10	(200)	F		10	*		
22	TiKα	4.508	W	60	45	(200)	F		40	*		
57	LaLα	4.648	W	60	45	(200)	F		40	*		3
23	VKα	4.949	W	60	45	(200)	F		40		Ti	1
24	CrKα	5.411	W	60	45	(200)	F		40		V	1
25	MnKα	5.894	W	60	45	(200)	F		40		Cr	1
26	FeKα	6.398	W	60	20	(200)	S	10				
27	CoKα	6.924	W	60	45	(200)	F+S		40		Fe	1
26	FeKab	7.110	W	60	45	(200)	F+S	10		*		1
28	NiKα	7.471	W	60	45	(200)	F+S	10		*		1
29	CuKα	8.040	W	60	45	(220)	F+S	20		*		2
74	WLα	8.396	Mo	30	60	(200)	F+S	200	100	*		3
30	ZnKα	8.630	Mo	60	45	(200)	F+S	10		*		1
83	BiLα	10.84	Mo	80	30	(200)	S	100	40	*		2
33	AsKβ	11.72	Mo	40	60	(200)	F+S	400	400	*		2
82	PbLβ	12.62	W	30	60	(200)	F+S		200	*		3
90	ThLα	12.97	W	80	30	(200)	S		100	*		2
37	RbKα	13.37	W	80	30	(200)	S		100	*		1
92	ULα	13.61	W	80	30	(200)	S		100	*		2
38	SrKα	14.14	W	80	30	(200)	S		100	*		1
39	YKα	14.93	W	80	30	(200)	S		100	*	Rb	1
40	ZrKα	15.74	W	80	30	(200)	S		40	*	Sr	2
41	NbKα	16.58	W	80	30	(200)	SS		40	*	Y,U	2
42	MoKα	17.44	W	40	60	(220)	S		100	*	Zr,U	2
47	AgKα	22.10	W	80	30	(200)	S		200	*		2
28	NiKab (I.R.)	24.14	W	80	30	(200)	S		40	*		1
50	SnKα	25.19	W	80	30	(200)	S		400	*		2
51	SbKα	26.27	W	80	30	(200)	S		400	*		3
56	BaKα	32.06	W	80	30	(220)	S		100	*		8
58	CeKα	34.57	W	80	30	(220)	S		100	*		10

Table 2. PW1450/20ASP Measuring Program Structure and Enhancements.

PW1466 Sample Loader (5 discs each with 12 positions)

Disc	Measuring Program Set (max. 31)
1	Measuring program subset (max. of any 6)
2	" "
3	" "
4	" "
5	" "

Measuring Programs (*)

Channel Set (max. 63)	High Speed/Normal/High Precision		
Channel subset A (max. of any 31)	*	***	***
" B "	*	***	***
" C "	*	***	***

It is applicable to other matrices such as vegetation.

In addition to enhancement of the standard software to permit internal as well as external ratio, and two line overlaps as well as two background subtractions per analyte, the major modification was to permit selection of any element subset from any measuring program. The current arrangement is shown in Table 2. For each of the 5 discs of the loader any 6 (from 31) measuring programs may be selected, each with any channel subset. Triplication of some measuring programs permits differing channel subset selection from the same element group within the loader. In addition programs are provided for measurement of high concentrations of elements such as titanium, barium and cerium. In this way a flexibility has been provided to meet changing customer requirements over the years with respect to analyte and concentration range.

ACKNOWLEDGEMENT

This paper is published with the approval of the Director, Institute of Geological Sciences (NERC).

REFERENCES

Andermann, G. and Kemp, J. W., 1958, Scattered X-rays as internal standards in X-ray emission spectroscopy, Anal. Chem., 30:1307.

Hower, J., 1959, Matrix corrections in the X-ray spectrographic trace analysis of rocks and minerals, Amer. Mineral., 44:19.

Smith, T. K., 1980, An electronic modification of an automatic X-ray fluorescence spectrometer with sample loader to increase the flexibility of the measurement sequence, X-Ray Spectrometry, 9:2.

Smith, T. K. and Ball, T. K., 1983, XRF analysis of vegetation samples and its application to mineral exploration, Adv. X-Ray Anal., 26:409.

XRF ANALYSIS OF LOW LEVEL CATION CONCENTRATION OF SODIUM

SILICATE SOLUTIONS

E. M. Sabino, M. R. Derolf and J. L. Bass

The PQ Corporation
Research and Development Center
P.O. Box 258
Lafayette Hill, PA   19444

INTRODUCTION

Sodium silicate solutions have been manufactured commercially for over one hundred years. Such solutions have been used in the past as major constituents in waxes, polishes and adhesives, and in detergents. A current additional use is their application in enhanced oil recovery. In many uses of sodium silicate solutions trace metal levels adversely effect performance. Depending on the source of the raw materials used in the silicate production and on processing conditions iron and vanadium in particular may be found at the 50-200 ppm level. X-ray fluorescence is used as a rapid method for analysis of iron in sand, one of the raw materials used in silicate manufacture. In order to use an instrument already at the plant site, we decided to develop an XRF method for metals analysis in the silicate solutions as well.

These solutions are characterized by their solids content and SiO_2/Na_2O ratios, and high pH in the 12-13 range. Until recent work with ^{29}Si NMR not much was known about the molecular arrangement of the (SiO_2) x radicals. Now there is evidence of cyclic dimers, trimers and four silicon cyclic arrangements existing simultaneously in the same solution. There is no evidence for long polymeric chains. The other species of interest are assumed to be adsorbed onto the silicate polymer.

Because of the high pH and instability of these solutions standard addition methods could not be used. Sols can be formed from these solutions by careful acidification, and there are

several published methods for commercial preparation of stable
silica sols. However, these all involve the ten to twenty fold
dilution of the silicate, too dilute for XRF analysis when the
initial concentrations are in the range of 50–200 ppm. We have
developed a method of forming sols in which the silicate dilution
is not more than 3.5 fold, and the sols have stable liquid
phases, depending on the total electolyte concentration, of two
to thirty hours. These have pH values of less than 2; thus
standard addition methods using readily available AA reference
standards are easily employed. In addition, since liquid silica
sols gel without a change in refractive index, density, and local
variations in silicate to water ratios we also investigated the
gelled sol as a specimen presentation mode for analysis.

SOL FORMATION

Literature references[1] state that a water clear sol with
sol particles less than 7 nm will be formed if the silica
(SiO_2) content is approximately ten percent. Available
apparatus limited the total weight of sol formed to 72 g. Using
the product specification sheets we calculated the amount of
silicate solution needed to provide 9.4 g of silica. Next we
calculated the weight of ten percent HCl needed to neutralize the
Na_2O listed in the specifications. A ten percent excess of
equivalents was actually used. If this was to be a spiked
specimen the weight of the spike (appropriate commercially
available AA reference standard) was added to the acid. In order
to reach the final 72 g sol weight we added water to dilute the
silicate, thus reducing viscosity which aids in sol formation.
All weights were to the nearest milligram.

The weight of acid or acid plus spike was weighed into a
four ounce specimen jar which had a threaded top for adding a
tight lid. This jar was then placed on a lab jack and raised
into position under a plastic three blade stirrer on a plastic
coated shaft. The stirrer was fitted on a variable speed motor.
Next we placed the cylinder of a plastic 50 cc syringe fitted
with a special needle keeper made from a #1 rubber stopper
upright in a four ounce plastic bottle and tared this assembly
along with a blunted #18 needle and the plunger from the 50 cc
syringe. Into the syringe cylinder we weighed the required
amount of silicate solution and diluting water. The syringe was
assembled, shaken to thoroughly mix the water and silicate, the
needle keeper was removed and the blunted #18 needle attached.
The acid was rapidly mixed to form a vortex in the acid. The
syringe was then used to direct the stream of the added diluted
silicate solution against the bottom of the vortex avoiding the
blade and shaft. Five to ten seconds are needed to empty the
syringe at the proper rate. This is the crucial step

in the formation of the sol. The silica in this system will gel
in a fraction of a second between pH 5-9, therefore the mixing
and rate of addition of the diluted silicate must be adjusted so
that the silicate is not in the critical pH range long enough for
gel to form. If this is not done correctly bits of gel and light
flock will form resulting in an inhomogeneous liquid, and this
preparation must be discarded.

EXPERIMENTAL

Polypropylene cups fitted with polypropylene windows were
filled with the liquid sols and run in Philips 1410 wavelength
dispersive spectrometer. The following instrument conditions
were used: kv 50, mA 50, tube Cr, atmosphere He, crystal LiF,
detector proportional, counting time 20 seconds, Fe K α
$\pm 1°$ 2θ, V K α $\pm 1°$ 2θ.

RESULTS AND DISCUSSION

The analysis was done on two types of production samples:
the first, manufactured by a glass solution method had no
detectable vanadium, detectable iron and less than one percent
total impurities (samples 1-6); the second manufactured by an
acid extraction method had detectable vanadium, low levels of
iron and total impurities in the three to five percent range
(samples L, 11-14).

Table 1

Sample #	ppm Fe		AA
	XRF		
	Liquid	Gel	
1	67		92
2	75		85
3	77		91
4	70		66
5	87		86
6	120		92
L	35	30	33
11	16	14	20
12	20	22	18
13	17	22	18
14	11	15	17

2, 3 analyzed using line 1
5, 6 analyzed using line 3
  12 analyzed using line 11

Table 2

	ppm V		
Sample #	Liquid	Gel	AA
11		223	210
12		252	210
13	235	249	210
14	203	253	210
L	212	235	210

12 analyzed using line 11

The detection limit for iron at the three sigma level was 9 ppm in the original solution. Peak to background ratios were used to plot the typical standard addition lines and a HCl blank was used to correct for the iron contribution from the tube spectrum. Only the high impurity group was run as the sol in gel phase as well as the liquid since these gelled in one to eight hours whereas the low impurity group needed twenty-four to thirty-six hours to gel. The comparison to the AA results is quite acceptable. Table 1 gives the results obtained by our XRF analysis and by atomic absorption for iron.

The limits of detection for vanadium at the three sigma level was 6 ppm in the original solution. Because of the impurities in samples 11 and 12 they gelled before they could be run as liquid phase sols. The comparison to the AA values is much better if only the liquid phase results are examined. From this data we conclude that the specimens presented to the x-ray beam should only be liquid phase specimens. Table 2 gives the results obtained by our XRF analyses as well as atomic absorption values for vanadium.

CONCLUSIONS

We feel these data prove that reliable results can be obtained using this method of analysis. It is universally applicable provided the analyte intensity can be measured. It does also avoid the sodium interference experienced with AA. However careful technique is needed for good sol formation. In the case of silicate solutions with relatively high impurities it is especially important to analyse the sols immediately after preparation.

REFERENCE

1.   Iler, Ralph K. "The Chemistry of Silica", John Wiley & Sons, New York, N.Y. 1979.

A RAPID, LOW COST, MANUAL FUSION SAMPLE PREPARATION TECHNIQUE

FOR QUANTITATIVE X-RAY FLUORESCENCE ANALYSIS

Gerald D. Bowling, Iris B. Ailin-Pyzik,
and David R. Jones IV

Owens-Corning Fiberglas
Technical Center
Granville, OH   43023

INTRODUCTION

This study compares the quality of the fused samples obtained by three separate methods. The first set of samples was prepared by the method used at USGS in Denver and reported by Taggart and Whalberg (1). The second set was fused by our manual method and cast in graphite molds. The third set was fused in the Herzog HAG-12 automated fusion device.

The manual fusion technique requires the use of a muffle furnace capable of 1100°C (2100°F) and graphite molds. No release agents such as KBr and LiBr are required since the disks release easily from the graphite. The 25mm diameter center of the "fire-polished" upper surface of the disk is used for analysis without further surface preparation. This method has been shown to be suitable for preparation of a wide variety of glasses and raw materials including burned dolomite, silicates, high zircon materials such as BCS-388, calcined alumina, and alumina refractories.

EXPERIMENTAL

The x-ray intensity data for all samples were obtained on a Philips PW1600 simultaneous x-ray spectrometer with a rhodium target tube operated at 50 kV and 50 mA. A 25mm diameter mask is used in the x-ray analysis.

Five samples of a -200 mesh nepheline syenite were fused by each fusion technique resulting in a total of 15 individual fused disks.

The first set was fused by the USGS procedure (1). The second set was prepared by weighing 1.5000–1.6000 grams of powdered sample into a 4 oz. glass bottle. Exactly 10 times the sample weight of a –60 mesh Li2B4O7 was weighed into the jar and thoroughly mixed. This homogeneous powder was then pressed to form a pellet sturdy enough to be handled, and placed in a 10mm deep graphite crucible. Using long tongs and gloves, it was placed in a 1100°C muffle furnace for 15 minutes, and removed to a specially designed vacuum device (Figure 1) to draw out the bubbles through the bottom of the graphite crucible. It was allowed to cool, and run on the x-ray without further surface preparation.

The third set was prepared in the same way as the second set except that the pressed pellets were loaded in the Herzog automated fusion device and allowed to complete the fusion cycle unattended.

In addition, six commercially available SRM glasses were fused by the manual fusion technique described above to establish calibration curves. These were EC 1.1, EC 1.2, NBS 620, NBS 621, NBS 1830 and NBS 1831.

STATISTICAL EVALUATION

The results of the reproducibility study can be seen in Table 1. Each prepared disk was recycled in the x-ray spectrometer three times to provide three 50 second counting times. This provided fifteen values for statistical evaluation. In addition, one disk from each set was recycled 15 times to provide a measure of the short term instrumental drift.

The tables list:

1. The oxide concentration in the sample prior to flux dilution and fusion.

2. The coefficient of variation for five disks and for one disk.

   COV = (STANDARD DEVIATION/AVERAGE) X 100%

3. The intensity (counts per second).

4. The counts per second per percent of oxide.

5. The theoretical instrument variation which might be expected based on the total number of counts obtained (COV = $1/\sqrt{N}$).

Table 2 lists the correlation coefficients for the calibration curves using the six SRM glasses. (At the time of this study, there was some question about the reliability of the silica fixed channel. The correlation coefficient for SiO2 may therefore be higher, had there been no such difficulty.)

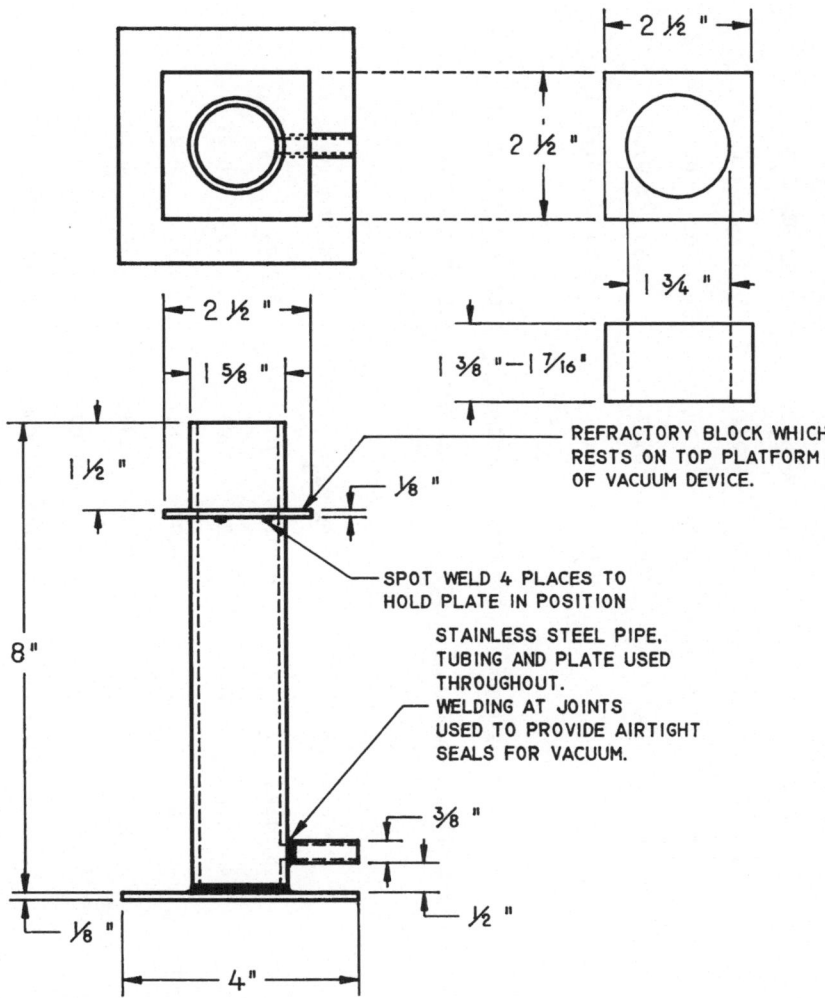

Fig. 1. Vacuum device for manual fused disk preparation.

Table 1

	COV 60.4% SiO_2		COV 0.35% Fe_2O_3		COV 23.0% Al_2O		COV 0.59% CaO	
	5 Disks	1 Disk	5 Disks	1 Disk	5 Disks	1 Disk	5 Disks	1 Disk
USGS	0.15	0.25	0.74	0.21	0.20	0.12	0.70	0.45
MANUAL FURNACE GRAPHITE	0.21	0.18	0.39	0.17	0.33	0.10	0.61	0.51
HERZOG GRAPHITE	0.26	0.22	0.54	0.21	0.23	0.13	0.73	0.44
THEORETICAL COV		0.15		0.22		0.14		0.45
	9000 COUNTS/SECOND		4000 COUNTS/SECOND		10,000 COUNTS/SECOND		1000 COUNTS/SECOND	
	150 COUNTS/SECOND/%		11,150 COUNTS/SECOND/%		450 COUNTS/SECOND/%		1500 COUNTS/SECOND/%	

	COV 0.09% MgO		COV 10.0% Na_2O		COV 5.1% K_2O	
	5 Disks	1 Disk	5 Disks	1 Disk	5 Disks	1 Disk
USGS	1.56	1.67	1.20	1.05	0.22	0.21
MANUAL FURNACE GRAPHITE	1.68	1.51	0.97	0.82	0.21	0.26
HERZOG GRAPHITE	2.15	1.23	0.91	0.98	0.24	0.20
THEORETICAL COV		1.69		1.00		0.22
	70 COUNTS/SECOND		200 COUNTS/SECOND		4000 COUNTS/SECOND	
	700 COUNTS/SECOND/%		20 COUNTS/SECOND/%		800 COUNTS/SECOND/%	

Table 2

COMMERCIALLY AVAILABLE STANDARD REFERENCE
GLASSES (EC 1.1, 1.2, NBS 620, 621, 1830, 1831)

	OXIDE RANGE WT. %	CORRELATION COEFFICIENT
SiO_2	71.94 – 73.08	0.9796
Fe_2O_3	0.04 – 0.12	0.9952
TiO_2	0.011 – 0.04	0.9667
Al_2O_3	0.12 – 1.81	0.9994
CaO	7.12 – 10.71	0.9993
MgO	0.27 – 4.01	0.9982
Na_2O	12.71 – 14.40	0.9848
K_2O	0.03 – 2.01	0.9996

CONCLUSIONS

The choice of the use of an automated fusion device versus a manual method depends on a number of factors, three of which are (1) required sample throughput, (2) capital investment, and (3) manpower availability.

The advantages and disadvantages of the USGS fusion method (which uses platinum alloy crucibles and molds) versus fusion in graphite were covered in a previous paper (2). The advantages of the graphite crucible method were that there was no need to invest in platinum alloy with the attendant cleaning and maintenance, and no need for a release agent. This discussion will therefore cover only the manual fusion versus the automated fusion in graphite.

The disadvantages of manual furnace graphite fusion versus automatic Herzog graphite fusion are:

1. More labor intensive than with Herzog. Requires someone be available at beginning and end of fusion time to load, unload, and place sample on vacuum device.

2. Must handle crucible with molten mass of material.

3. If spilled in the furnace, you must have furnace refractory available to replace damaged tiles.

The advantages are:

1. Low capital investment.

2. Minimum number of specialty items required.

3. Multiple samples can be fused simultaneously rather than sequentially, depending on furnace capacity and number of vacuum devices.

REFERENCES

1. J. E. Taggart, Jr., and J. S. Whalberg, New Mold Design for Casting Fused Samples, Advances in X-Ray Analysis, Vol. 23, 257-261, (1980).

2. I. B. Ailin-Pyzik and G. D. Bowling, Evaluation of Two Fusion Sample Preparation Techniques for Quantitative X-ray Fluorescence Analysis, Paper presented at the 1983 Pittsburgh Conference on Analytical Chemistry and Applied Spectroscopy.

QUANTITATIVE X-RAY FLUORESCENCE ANALYSIS FOR FLY ASH SAMPLES

Scott Schlorholtz and Mustafa Boybay

Iowa State University Engineering Research Institute
Materials Analysis and Research Laboratory
Ames, Iowa

INTRODUCTION

The disposal of fly ash from coal burning power plants is rapidly becoming an environmentally complex problem. Recently though, the attitude towards fly ash use has been changing from a disposal oriented point of view to a more rational position which considers fly ash as a resource to be recycled. One major hinderance of fly ash use has been the extreme variability of composition that exists between fly ashes produced at different power plants. This variability makes the analysis of fly ash very important.

The most common methods currently used for fly ash analysis are atomic absorption or wet chemistry methods defined in ASTM C311[1]. Both methods tend to be expensive, time consuming, and sample preparation is both tedious and critical for some elements. In this study X-ray fluorescence (QXRF) is used for the quantitative analysis of the major and minor elements found in "typical" fly ashes. The method, which is computer controlled, is quick, reliable, and requires minimal sample preparation.

CALIBRATION CURVE DEVELOPMENT

The relationships between the intensity of the fluorescent X-radiation and the concentration of an element in a sample can be expressed by several equations[2]. In this study the computer was used to build a mathematical model based on a multielement equation that performs matrix corrections (including absorption, enhancement, and line interferences)[3]. The general equation is as follows:

$$X_i = A_i + B_i I_{fi} + C_i I_{fi}^2 + \sum_{\substack{j=1 \\ j \neq i}}^{n} m_{ij} I_{fj} + \sum_{\substack{j=1 \\ j \neq i}}^{m} K_{ij} I_{fi} I_{fj} \tag{1}$$

where

X_i = concentration of the ith element
A_i = background
B_i = slope
C_i = rate of change of the slope (second derivative)
I_{fi} = analyte line intensity of the ith element
I_{fj} = analyte line intensity of the jth element
m_{ij} = line interference coefficient of element j on element i
K_{ij} = absorption or enhancement coefficient of element j on element i
$n-1$ = number of line interferences
$m-1$ = number of absorptions and/or enhancements
where $n + m \leq 9$.

The Siemens program uses multiple regression to determine the calibration coefficients for any one of the elements measured in the standard samples[3]. The accuracy of the calculated coefficients can be judged based on the relative error of the measurements and the mean square deviation which are calculated by the program. The program has the capability to make drift corrections (based on a reference sample) for the short or long term changes in the instrument.

SAMPLE PREPARATION

The most critical aspect of this research hinged on the development of a representative set of standards to be used for generating reliable calibration curves for each of the elements present in fly ash. It is very important in calibration curve type analysis methods (particularly for the matrix correction) to choose a set of standard samples that have elemental compositions covering wide enough ranges to include the "expected" composition of the unknown samples. This task is by no means a simple one for fly ash due to its high variability. To overcome this difficulty two fly ash samples of extremely different elemental compositions were chosen as standards and they were mixed with small amounts of various pure, stable compounds to extend the calibration curve over a wide range.

The first standard fly ash chosen for mixing was obtained from the National Bureau of Standards (NBS standard #1633a). The NBS fly ash is designated Class F (ASTM C 618 classification) and was subjected to quite rigorous analysis at the NBS which yielded a reliable

standard. The second standard fly ash chosen for mixing was obtained
from Lansing Power Plant located in Lansing, Iowa. The Lansing fly
ash is designated Class C (ASTM C 618 classification) and it was
chosen because of its high elemental calcium content. Note, that
current power plant design coupled with the burning of low sulfur
western caol typically produces fly ashes that are similar to the
Lansing fly ash (Class C). Thus, even though the NBS does not currently have a standard Class C fly ash it was mandatory to include
one in the calibration curve. The Lansing fly ash was subjected to
general chemical analysis by two independent laboratories which used
various techniques to determine its elemental composition. The compounds used for blending with the NBS and Lansing fly ashes were
Fe_2O_3, Al_2O_3, SiO_2, S, $CaCO_3$, TiO_2, MgO, $NaHCO_3$, sodium feldspar
(NBS standard 99a), and portland cement (NBS standard 635). All
samples were ground for four minutes in a shatterbox and then dried
to a constant weight in a 100°C oven; different portions of the
mixtures were weighed with an analytical balance. The samples were
then thoroughly mixed by hand to insure homogeneity.

Four other standard samples were included in the calibration
curves for certain elements. Three of the samples were fly ashes
that were analyzed at Ames Laboratory at Iowa State University
(I.S.U.) while the remaining sample was a rock sample obtained and
analyzed by the ceramic engineering department at I.S.U. Ultimately
24 standards were available for use in the calibration curves for the
10 major and minor elements present in fly ash. Table 1 indicates
the ranges of the 10 measured elements present in the 24 fly ash
standards.

Samples for fluorescent analysis were prepared by taking four
grams of fly ash, pressing it into a pellet and then pressing the
fly ash pellet into a wax supporting material at 25 tons load.

INSTRUMENT DESCRIPTION AND SETTINGS

Specimen analyses were performed on a Siemens SRS-200 sequential
X-ray spectrometer controlled by a PDP 11/03 microcomputer. The
spectrometer is equipped with a 10 sample controlled atmosphere
specimen chamber. A chromium tube was used as an X-ray source and
it was operated at 50 KV and 50mA. The Siemens spectrometer makes
use of a proportional counter (flushed with argon/methane) in series
with a scintillation counter so both counters were operated throughout the analyses. All analyses were performed with the sample
chamber and spectrometer under vacuum. The measuring parameters
(i.e., crystal type, collimator, filter, etc.) were selected to minimize line overlaps and to give optimal spectral resolution and counting rates.

Table 1.   The Ranges of Fly Ash Standards Available for Use in the Calibration Curves.

| Element | Range in Percent | |
	Minimum	Maximum
Na	.07	3.71
Mg	.10	4.96
Al	5.28	14.76
Si	9.12	28.93
P	.00	.49
S	.04	3.25
K	.24	3.83
Ca	1.00	24.46
Ti	.16	2.29
Fe	1.32	17.23

CALIBRATION

The intensities and concentrations were correlated for each element by using the Siemens program. The calibration parameters were defined by introducing possible interferences and minimizing the relative error of the calibration curve generated for each element while maintaining a reasonable number of degrees of freedom. No significant line interferences have been found for any of the elements being investigated. The elements in a given sample that can emit secondary fluorescent X-radiation with wavelengths approximately equal to the absorption edge wavelength of the element being analyzed were investigated for enhancement interferences.

Absorption interferences appear to be of major importance in fly ash analysis. Figure 1 illustrates the importance of absorption effects on the predictability of the calibration curve for iron. In Figure 1a we assumed that there were no absorption interferences while in Figure 1b we corrected for a calcium absorption interference. Note, in Figure 1, how the 95% confidence intervals for a single response narrow significantly after correcting for the calcium absorption interference. Also, the regression yields a line that is very well behaved (i.e., meaningful slope and intercept, high R^2 value) when compared to the line generated when we ignored the absorption interference.

The problem concerning the evaluation of possible absorption interferences from elements present in the calibration standards was difficult to deal with. Initially, graphs of mass absorption versus atomic weight were made for each element in fly ash to help indicate

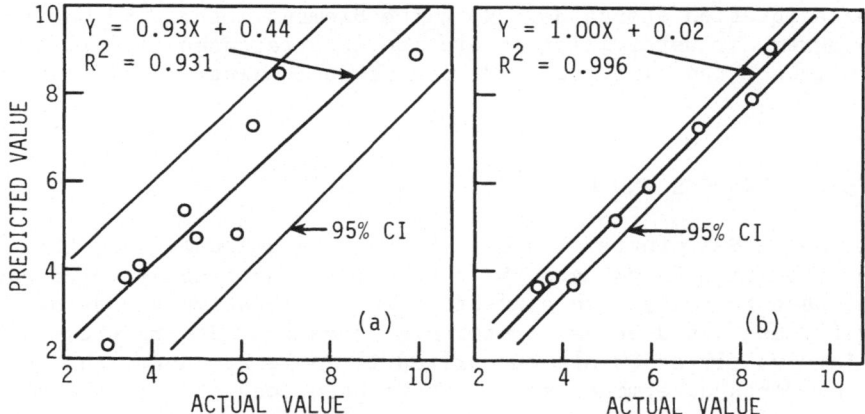

Fig. 1. Influence of absorption on the calibration curve
         for iron: (a) % Fe uncorrected; (b) % Fe corrected
         for calcium absorption.

possible absorption interferences. Although this method helped to
define possible absorbers it failed to indicate which element or
elements would dominate absorption in a sample used as a calibration
standard. A computer program was developed to eliminate this
problem. Since the calibration standards had known compositions the
program was designed to calculate the mass absorption of analyte
fluorescent radiations by all elements present in each sample. The
total mass absorption of each calibration standard can then be obtained by summing up the absorptions of the individual elements. The
procedure is quick and can easily be repeated for every sample used
for a calibration standard. Direct graphical output cannot be obtained from the computer program at this time but should be available in refined versions. One of the strongest assets of the computer program is that it indicates the relative magnitude of absorption by the various elements in a calibration standard and thus helps
to shorten the trial and error procedure used to develop reliable
calibration curves. It should be kept in mind that when the computer
program was developed it was assumed that the standard samples consisted only of the ten elements (in their oxide forms) listed in
Table 1. The mass absorption coefficients for the various elements
were obtained from published tables[4]. Coefficients of the final calibration equations for each of the ten major and minor elements
present in fly ash were determined. Linear fits were used whenever
possible to obtain the simplest model with meaningful coefficients.
Siemens has also developed software to help evaluate absorption interferences for each element in the calibration standards[5]. It also
calculates the maximum variation in absorption for each element in

all the calibration standards. Both the Siemens method and our computer program for determining absorption interferences are simply
tools to reduce the time required to produce meaningful calibration
curves.

ANALYSIS OF UNKNOWNS AND DISCUSSION

     Siemens MESRX program was used to analyze all unknowns; this
program automatically changes the spectrometer settings to correspond
to those used to analyze an element in the calibration standards.
Lansing fly ash was used as a reference standard. The results obtained for two fly ashes are summarized in Table 2. Note, that the
various elements determined by QXRF have been converted to their
stable oxide forms to allow comparison with the results reported by
the Analytical Services Lab (ASL) at I.S.U. The ASL performed wet
chemical analysis as described in ASTM specifications C 311 and C 114
and they also used more common methods of analysis which consisted of
atomic absorption, flame emission, and spectrophotometry.

     With a few exceptions the results obtained by QXRF compare very
favorably with those obtained by the other methods. The major discrepancies appear to be in the analyses of silicon in fly ash 1,
aluminum in fly ash 1, and titanium in both fly ashes.

     The reason for the wide variation in the measured silicon among
the three types of analyses is unknown at this time. One must keep
in mind that the wet chemistry (ASTM) procedure is complex and very
prone to operator errors while both atomic absorption and QXRF are
less complex and thus should be less difficult for routine analysis.

     The discrepancy on the aluminum determination is not as severe.
The ASTM method used for determining aluminum is indirect; it uses
results from the analysis of other elements in the unknown sample.
On the contrary the determination of aluminum by QXRF or atomic absorption is a direct measurement, not prone to cumulative errors.
The results for aluminum determined by QXRF and atomic absorption
agree within about 2% of each other which may indicate that the ASTM
method has overpredicted the actual weight percent of aluminum in
fly ash 1. Note, that the ASTM method also finds more aluminum in
fly ash 2 than the other two methods.

     Titanium determination had a different problem; the ASTM pro-
cedure (taken from ASTM C 114) does not work for some fly ashes
because a large amount of the titanium present is not soluble in
hydrochloric acid. We did not expect the ASTM method to work for
all fly ashes because the specification (ASTM C114) was written for
the analysis of portland cement. Blindly using the test procedure
defined in ASTM C 114 is wrong but one must remember that there is
no standard method for determining the amount of titanium in fly ash,

Table 2. Results of Fly Ash Analysis

Fly Ash 1: North Omaha Power Plant
Method of Analysis

	ASTM	ASL	QXRF			
			Mean	Day 1	Day 7	Day 30
SiO_2	47.78	53.09	58.46	59.80	57.82	57.76
Al_2O_3	23.52	16.66	18.98	19.18	18.93	18.83
Fe_2O_3	19.00	18.90	18.37	18.36	18.47	18.34
CaO	---	3.75	4.48	4.46	4.50	4.48
MgO	---	1.00	1.44	1.38	1.38	1.53
SO_3	1.15	---	1.24	1.34	1.25	1.13
K_2O	---	2.13*	2.13	2.14	2.13	2.13
Na_2O	?	?	0.59	0.58	0.61	0.58
TiO_2	<0.01	0.64	0.89	0.90	0.89	0.89
P_2O_5	0.09	---	0.10	0.10	0.10	0.11
		Total =	106.68			

Fly Ash 2: Sherburn Power Plant
Method of Analysis

	ASTM	ASL	QXRF			
			Mean	Day 1	Day 7	Day 30
SiO_2	43.87	41.52	44.16	44.36	44.30	43.83
Al_2O_3	19.68	17.34	17.59	17.70	17.50	17.54
Fe_2O_3	6.72	6.62	6.07	5.65	6.27	6.28
CaO	---	18.06	18.99	19.04	18.98	18.95
MgO	---	3.17	3.64	3.58	3.62	3.71
SO_3	1.98	---	2.34	2.40	2.32	2.31
K_2O	---	1.01*	0.97	0.98	0.96	0.98
Na_2O	?	?	3.17	3.38	3.09	3.05
TiO_2	0.55	0.85	0.69	0.70	0.69	0.69
P_2O_5	0.41	---	0.37	0.40	0.36	0.36
		Total =	97.98			

*
 extrapolated value

as a matter of fact several of the minor constituents of fly ash
have been totally ignored by ASTM specifications. The amounts of
titanium determined by the QXRF and ASL methods do not agree well
for either fly ash 1 or 2. A systematic error may be present
because both analyses yield repeatable results. The lack of data
for K_2O and Na_2O in Table 2 should be noted. The measurements made

by using QXRF are our best estimates of the total amounts of potassium and sodium in the two fly ashes. The ASL experienced severe problems obtaining reasonable results from three different analytical techniques. The values for the percent K_2O listed in Table 2 under the ASL column were obtained by applying a correction to the raw data based on the results of the NBS standard. The numbers are only approximate but they are in good agreement with the values determined by QXRF.

Table 2 shows the precision of the QXRF method when a reference standard is used to correct for spectrometer drift. The table shows that the QXRF method is quite stable as a function of time. Table 2 also shows how prone the sample preparation technique is to different technicians. The samples in Table 2 denoted as "day 1" were prepared and analyzed by a lab technician. The samples denoted as "day 7" were prepared and analyzed by a different lab technician one week later. After analysis the day 7 samples were allowed to stay at room temperature and humidity for about three weeks and then they were reanalyzed and are denoted as "day 30" in Table 2. Lansing fly ash was used as a reference standard for all of the analyses. The samples appear to be more prone to preparation error than to spectrometer drift or ambient conditions.

SUMMARY AND CONCLUSIONS

With the help of the computer program calibration curves for silicon, aluminum, iron, calcium, magnesium, phosphorous, sodium, potassium, titanium, and sulfur were developed. The accuracy of the QXRF analysis method is reasonable when compared to other methods of analysis that are commonly used. The QXRF method shows good precision when a reference standard is used to correct for machine drift.

REFERENCES

1.   American Society for Testing and Materials, 1981 Annual Book of ASTM Standards. Part 14. Philadelphia: ASTM, 1981.

2.   Bertin, Eugene P., Principles and Practices of X-ray Spectrometric Analysis, 2nd edition, New York: Plenum Press, 1975.

3.   Plesch, R. and G. Thiele, "Fundamentals of the Siemens Computer Programs for X-ray Spectrometry," Siemens X-ray Analytical Application Note, No. 33, Sept. 1977.

4.   Leoux and Think, "Revised Tables of X-ray Mass Attenuation Coefficient," Corporation Scientifique Claisse, Inc. Quebec, Canada.

5.   Beard, D. W. and B. G. Reed, "Matrix," Siemens X-ray Application for SRS 200. February 1982.

POLARIZATION OF X-RAYS BY SCATTERING FROM THE INTERIOR OF A CYLINDER

I. Single Scatter

John D. Zahrt[1] and Richard Ryon[2]

[1]Chemistry Department, Northern Arizona University
 Flagstaff, AZ
[2]Lawrence Livermore National Laboratory
 Livermore, CA

INTRODUCTION

It is of interest today to use polarized X-rays in X-ray secondary fluorescence as a means of improving signal to noise ratios in the analysis of trace elements. Current experimental design makes use of two mutually perpendicular scatterings from plane parallel materials. Radiation with the electric field vector in the scattering plane (scattering angle = 90^0) will be annihilated. Hence, after the mutually orthogonal, 90^0 scatterings no source X-rays should reach the detector. In practice source X-rays will only be greatly reduced at the detector[1] due to such things as multiple scatter[2] and collimator divergence[3]. An experimental problem associated with this design however is the reduced intensity of the signal because of the scatterings with concomitant increase in analysis time.

An experimental design using the interior of a B_4C cylinder as the scatterer of MoK_α X-rays should enhance the signal many fold. The justification of this statement resides in the elementary geometry theorem that any triangle constructed from three points on a circle, two of which define a diameter, is a right angle. The cylinder design thus places the source at one end of the cylinder diameter and the analyte at the other end. A baffle is inserted so that there is no direct excitation of the sample by the source. The detector is directly above the sample (i.e. 90^0 to the source-analyte plane). This design, shown in Figure 1, allows all X-rays in a "thick" plane to be utilized rather than just a pencil of radiation. This will give rise to greater excitation intensities while maintaining the polarization. Such an instrument has in fact been built[4] by is currently plagued by stray scattered radiation. The theory has been pursued in an effort to improve the experimental design.

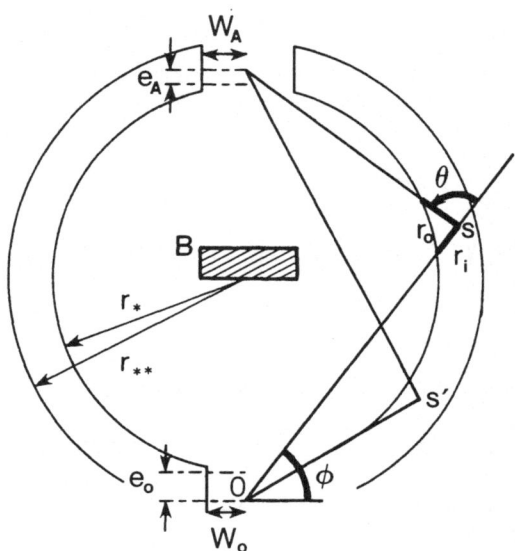

Figure 1. A schematic cross section of the cylindrical scatterer.
The existing instrument parameters are given in the text.

THEORY

An X-ray will be emitted from the source, O, at a take off angle
ϕ measured from the source plane. It will penetrate the cylinder wall
a distance r_i before scattering at some point s. If the scattered
photon is to be seen by the sample it must also travel a distance r_0
before exiting the cylinder wall. Notice that such penetration <u>into</u>
the cylinder wall before scattering destroys the ideal geometry
necessary for complete polarization. Because there is no unique
penetration depth at which <u>all</u> scattering takes place the ideal
geometry can never be recovered. However, it can be compensated for
by placing the source and analyte into the cylinder wall by amounts e_0
and e_A respectively. These distances can be varied independently to
optimize the polarization.

We will consider the source of X-rays and the analyte as points O
and A respectively. We make the following concession to the real
instrument design, however. While keeping the source and analyte as
points we acknowledge that in reality they occupy space and therefore
we cut gaps in the cylinder of an amount $2 w_0 = 1.4$ cm at the source
and $2 w_A = 1.6$ cm at the analyte. The maximum take off angle is not
$\pi/2$ but 1.3112 due to the baffle B. The inside and outside radii of
the B_4C cylinder are 3.175 cm and 3.81 cm respectively. The cylinder
height is 0.953 cm.

The combination of these two modifications, recessing the points and cutting the cylinder, allows for the possiblility that a photon may exit the cylinder shell and reenter it on its way to the scattering point and/or on its way to the analyte from the scattering point. (This situation is demonstrated by path OS'A shown in Figure 1.)

Three Dimensional Differential Scattering Probability

A photon emitted from O has a probability of $d\phi/1.3112$ of striking the cylinder surface between ϕ and $\phi + d\phi$ $0 \leq \phi \leq 1.3112$. It then has a probability of $\sigma_T e^{-\sigma_T r_i} dr_i$ of entering the cylinder wall to a dr_i neighborhood of r_i and a dz neighborhood of z. σ_T is the total absorption coefficient. In this neighborhood the probability of scattering by a given mechanism (Compton, Rayleigh or both)[1] is f. The intensity (the normalization of the probability distribution) changes according to the scattering angle by $3(1 + \cos^2\theta)/16\pi$. This is due to the polarization phenomena. The probability of reaching the analyte is then $e^{-\sigma_T r_0}$. Combining all of these independent events we arrive at the differential scattering probability given by

$$dP = \frac{2d\phi}{1.3112} \sigma_T e^{-\sigma_T r_i} dr \frac{dz}{2Z_*} \cdot f \cdot \frac{3}{16\pi} (1 + \cos^2\theta) e^{-\sigma_T r_0} \qquad (1)$$

where $2Z_*$ is the height of the cylinder, σ_T is the total linear absorption coefficient and f is σ_s/σ_T where σ_s is the linear scattering coefficient. ($\sigma = 0.5$ cm^{-1} and $\sigma_T = 0.932$ cm^{-1} for MoK_α).

It should be remembered that r_i, r_0 and θ are all functions of ϕ. The three dimensional scattering angle θ is easily computed from

$$\theta = \cos^{-1} \frac{\vec{r}_i \cdot \vec{r}_0}{|\vec{r}_i||\vec{r}_0|} \qquad (2)$$

Degree of Polarization

For one path which changes direction by an angle θ due to a single scatter event in otherwise free space, the degree of polarization (DOP) is defined as

$$DOP = \frac{1 - \cos^2\theta}{1 + \cos^2\theta} \qquad (3)$$

In the present context we have calculated the DOP by three different methods. Only one of these is physical, while the other two calculations allow some comparison and contrasting of the overall method of computation. As will be seen below we sample only a small number of determined paths. It might be thought that this finite analytic sampling might bias the calculation of DOP. To check this we have calculated

$$P_s = \frac{1}{n} \sum_{i=1}^{n} \frac{1 - \cos^2 \theta_i}{1 + \cos^2 \theta_i} \qquad (4)$$

This in effect measures the DOP of the sampling points themselves. Secondly, we have computed the average scattering angle $\langle \theta \rangle$ by

$$\langle \theta \rangle = \frac{1}{n} \sum_{i=1}^{n} |\theta_i| \qquad (5)$$

and then computed

$$\bar{P} = \frac{1 - \cos^2 \theta \, \langle \theta \rangle}{1 + \cos^2 \theta \, \langle \theta \rangle} \qquad (6)$$

Thirdly, we have computed

$$P_W = \frac{\sum_{i=1}^{n} e^{-\sigma_T (r_i + r_0)} \dfrac{1 - \cos^2 \theta_i}{1 + \cos^2 \theta_i}}{\sum_{i=1}^{n} e^{-\sigma_T (r_i + r_0)}} \qquad (7)$$

where we have weighted each individual DOP by the probability of getting to A from 0 by scattering through an angle θ_i. This is the DOP which would be measured at A.

METHOD OF COMPUTATION

     To obtain the integrated probability or total probability we must evaluate a triple integral (see eqn. (1)). Only the integrals over ϕ and z are independent as the upper limit of the r_i integration is a function of ϕ and z. Also, looking ahead to part II of this study[5] where we will discuss double scattered X-rays and which will involve 6-fold integrations of non-independent variables, we choose the method of Gaussian quadratures[6]. In this method the integrand is sampled at a small number of points each analytically determined and weighted to minimize the error of the approximation. We have found that 4 or 5 points per variable gives results to much better then 0.1%.

     A one dimensional integral is evaluated by

$$\int_a^b f(x)dx = \frac{b-a}{2} \sum_{i}^{N} f(x_i) \, W_i \qquad (8)$$

with

$$x_i = \frac{b-a}{2} Z_i + \frac{b+a}{2}$$

where Z_i is the ith zero of the Nth degree Legendre polynomial and W_i is the appropriate weighting factor.

Applying this method to equation (1) we see that N^3 scattering points are sampled (if N = 4 only 64 points need to sampled to obtain better than 0.1% accuracy). In particular

$$P = 2 \, \frac{3}{32\pi(1.3112)Z_*} \, \sigma_s \, \sum_{i=1}^{N} \sum_{j=1}^{N} \sum_{k=1}^{N} W_i \, W_j \, W_k$$

$$(1 + \cos^2\theta(\phi_i, z_j, r_k))\exp\{-\sigma_T[r_k(\phi_i, z_j)$$

$$+r_0(\phi_i, z_j, r_k)]\} \cdot \frac{1.3112}{2} \cdot Z_* \cdot \frac{r_{max}}{2} \qquad (9)$$

For a given ϕ, r_{max} is the maximum distance a photon could travel in the wall before exiting (no scattering).

RESULTS

We present the results of this study in the following two figures. In Figure 2 the degree of polarization (▲) and the probability P (☉) are plotted against the wall thickness. The curves are monotonic. That is, one can continue to obtain better polarization by making the cylinder wall thinner and thinner, with a corresponding loss of intensity and increasing time of analysis.

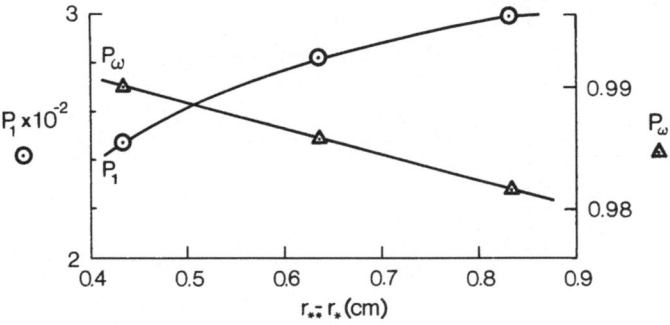

Figure 2.   The probability of single scatter P_1, and degree of polarization P_w as a function of wall thickness. $r_* = 3.175$ cm, $Z_* = 0.476$ cm, $e_0 = e_A = 0.2$ cm.

Figure 3 shows the optimized value of the degree of polarization
($\blacktriangle$) and P (O) against the cylinder height, $2Z_*$. Again the curves
are monotonic so no optimization with respect to Z_* is possible.

Table 1 presents values of P_s, $\overline{P}$, and P_w (see eqns. (4), (6) and
(7)) for various sets of instrumental parameters. $\overline{P}$ was optimized in
all cases by varying e_0 and e_A. Although the differences are small
$e_o = e_A = 0.2$ cm gave the best polarizations.

Table 1. Values of the average, unweighted and weighted degrees of
polarization for various instrument parameters. r_* was set to 3.175 cm
and e_o and e_A were both set to 0.2 cm.

$r_{**}-r_*$	Z_*	P_s	$\overline{P}$	P_W
0.435 cm	0.476 cm	.989	.998	.990
0.635	0.476	.976	.992	.985
0.835	0.476	.960	.983	.982
0.635	0.376	.977	.993	.986
0.635	0.576	.975	.992	.986

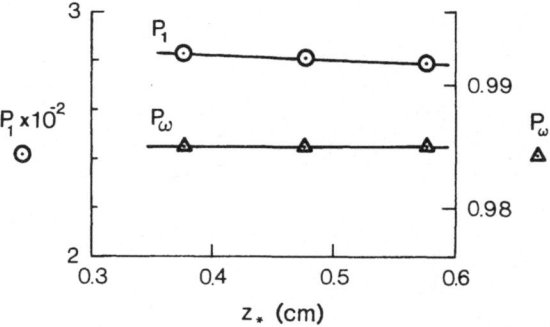

Figure 3. The probability of single scatter P_1 and degree of
        polarization P_w as a function of cylinder height. $r_* =$
        3.175 cm, $r_{**} = 3.81$ cm, $e_o = e_A = 0.2$ cm.

CONCLUSIONS

For a point source and point analyte the degree of polarization can be made essentially unity by various instrumental designs. There is no unique way to accomplish this and no design that simultaneously optimizes intensity (probability) and degree of polarization. A trade-off between intensity and polarization must be made.

It is interesting to note that while both wall thickness and wall height function in this design as collimators, the degree of polarization is more sensitive to the wall thickness.

Since we are sampling the cylinder wall at a small number of points there was concern that this would limit the possible values of the degree of polarization that were computed. Table 1 shows, however, that the physically significant degree of polarization is always larger than that obtained from the sampled points themselves. That is, scattering in some parts of the cylinder is more important than scattering in other regions of the cylinder. We observe no limitation as a result of the method of evaluation.

In comparison to the rectilinear system with say 0.952 cm diameter collimators the surface area of the cylinder is about 24x larger. The probability of single scatter is about 4x larger for the cylinder design. A point divergent source in a cylindrical system should have about 100x the intensity at the analyte as a perfectly parallel source (of the same source intensity) rectilinear system. As mentioned in the introduction this has yet to be realized in practice. There also is the danger of comparing a point divergent source to a perfectly parallel source. A finite source-finite analyte is currently being modelled and preliminary computations indicate that the probabilities are smaller by about 2.5x than those of the point source-point analyte and the degree of polarization is depressed about 4% compared to the numbers reported here.

REFERENCES

1. Richard W. Ryon, Adv. in X-Ray Anal. 20, 575 - 590 (1977).
2. John D. Zahrt and Richard W. Ryon, Adv. in X-Ray Anal. 24, 345 - 350 (1981).
3. John D. Zahrt, unpublished manuscript.
4. Richard W. Ryon and John D. Zahrt, Adv. in X-Ray Anal. 22, 453 - 460 (1979).
5. John D. Zahrt, Adv. in X-Ray Anal. 27 (following paper) (1983).
6. S. Chandrasekhar, "Radiative Transfer" (see particularly Chapter II), Dover, New York, 1960.

POLARIZATION OF X-RAYS BY SCATTERING FROM THE INTERIOR OF A

CYLINDER   II. Double Scatter

John D. Zahrt

Chemistry Department
Northern Arizona University
Flagstaff, Arizona

INTRODUCTION

In the previous paper[1], ZRI, we have discussed the properties of single scatter radiation emanating from a point source, displaced a distance e_o from the inside radius r_* of a cylinder wall, scattering inside the wall, and reaching a point analyte on a diameter line with the source but removed a distance e_A from the inside radius. Figure 1 shows the instrumental design. It is to be recalled that a triangle inscribed by a circle with two of its verticles defining a diameter is always a right triangle. Thus all such scattering events described above are approximately 90°. This in turn almost annihilates the electric field vector in the scattering plane and creates essentially plane polarized X-rays at the analyte. If the analyte-detector line lies perpendicular to the plane of the cylinder, scattering of the polarized source X-rays by the analyte into the detector is also a 90° event, and being perpendicular to the first scattering plane, should essentially annihilate the source X-rays at the detector.

The words almost and essentially are used because of two fundamental processes which lessen the degree of polarizations. First, the X-rays will not scatter from the inside radius of the cylindrical shell but from somewhere within the shell. This means that the scattering angle will no longer be 90° but close to it. The reason then for recessing both the source and the analyte is to regain, as much as possible, highly polarized radiation. This has been described in ZRI. Second, polarization will be lowered because of the multiple scattering of X-rays within the shell. This too can not be eliminated since it is a fundamental process, but it can be reduced by instrumental design. This report concerns itself with the leading term in multiple scatter, that is double scatter.

513

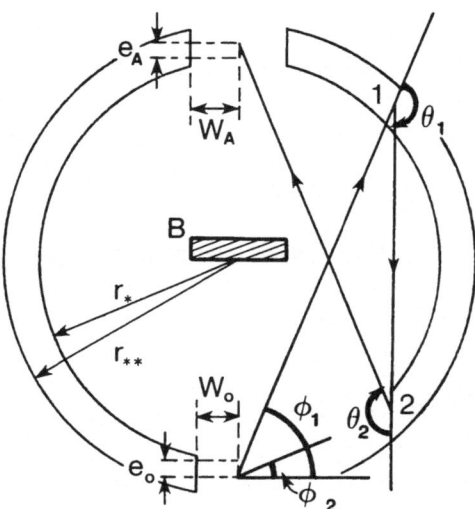

Figure 1.   A schematic cross section of a double scatter event in the
cylinder wall. The existing instrumental parameters are
$r_* = 3.175$ cm, $r_{**} = 3.81$ cm, $Z_* = 0.476$ cm, $W_o = 1.4$ cm,
$W_A = 1.6$ cm, $e_o = e_A = 0$. B is a baffle to exclude direct
excitation of the analyte.

THEORY

The experimental parameters and results described in ZRI all
pertain and are used to obtain the first scattering point. Because of
the method of evaluating integrals, namely Gaussian quadrature, each
scatter point is uniquely defined, so we only need to add to our
considerations the path from the first scatter point to any of the
other first scatter points, and the path from there to the analyte.
The integrations increase from three fold to six fold which poses no
real problem. The two difficulties which do arise are these: a) Is
the path from the first to the second point contained in the
cylindrical shell or does it exit the shell and reeenter it?)   b) The
state of polarization of the radiation reaching the analyte must now
be worked out in detail. There is no simple formula like eqn. (3) of
ZR1 to describe the polarization.

The angles from 0 on the source plane to the first and second
scattering points are set as are the z coordinates. The distances
into the cylindrical shell are then determined. The rays from 0 to 1,
1 to 2 and 2 to A are examined and the path length in the cylindrical
shell are determined. The scattering angles are found from

$$\cos \theta_1 = \frac{\vec{r}_1 \cdot \vec{r}_{12}}{|\vec{r}_1||\vec{r}_{12}|} \tag{1}$$

$$\cos \theta_2 = \vec{r}_{12} \cdot \vec{r}_2 / |\vec{r}_{12}||\vec{r}_2| \tag{2}$$

Three Dimensional Double Scatter Probability.

The differential probability for observing double scatter at A
(Figure 1) is given by

$$dP_2 = 2 \frac{d\phi_1}{1.3112} \frac{dz_1}{2Z_*} \sigma_T e^{-\sigma_T r_1} dr_1 f_1 \frac{3}{16\pi}(1+\cos^2\theta_1)$$

$$\sigma_T e^{-\sigma_T r_{12}} \frac{d\phi_2}{1.3112} \frac{dz_2}{2Z_*} dr_2 f_2 \frac{3}{16\pi}(1+\cos^2\theta_2) e^{-\sigma_T r_2} \tag{3}§$$

The terms and symbols have the same meaning as in ZRI.
A one dimensional integral can be evaluated approximately by
Gaussian quadrature. The working formulae are

$$\int_a^b f(x) dx = \frac{b-a}{2} \sum_i^n f(x_i) w_i \tag{4}$$

with

$$x_i = \frac{b-a}{2} z_i + \frac{b+a}{2} \tag{5}$$

and where z_i is the ith zero of the nth degree Legendre polynomial and
w_i is the appropriate weight factor. Applying eqn. (4) to eqn. (3)
for each integration leads to

$$P_2 + 2(\frac{3\sigma_s}{32(1.3112)\pi z_*})^2 \sum_i^N \sum_j^N \sum_k^N \sum_l^N \sum_m^N \sum_n^N (\frac{1.3112}{2})^2 z_*^2$$

$$\frac{R_1 R_2}{4} w_{i1} w_{j1} w_{k1} w_{12} w_{j2} w_{k2} [(1+\cos^2\theta_1)(1+\cos^2\theta_2)-1]$$

$$e^{-\sigma_T(r_1 + r_{12} + r_2)} \tag{6}$$

N = 6 was used in these calculations.

---

§While writing eqn. (3) in this manner for convenience and
physical interpretation, it must be remembered that the Thomson
scattering cross sections are polarization dependent and hence
$(1+\cos^2\theta_1)(1+\cos^2\theta_2)$ must be replaced by $2[(1+\cos^2\theta_1)(1+\cos^2\theta_2)-1]$.
See J. DuMond, Phys. Rev. 36, 1685–1701 (1930).

RESULTS

The results of several computer calculations are shown in
Table 1. The double scatter amounts to about 3% of the single scatter
reaching the analyte (see also ZRI). This is in general much higher
than for plane slabs. This is explicable on the basis of the absence
of collimators acting to channel the radiation to the analyte.

The amount of unfavorable radiation (electric field vector
parallel to the cylinder plane) is unusually high and is no doubt due
to the "wrap-around" aspect of the cylinder along with the absence of
collimators to reduce this effect.

Figure 2 shows both the probability of double scatter and the
percent of unwanted (parallel) radiation as a function of wall
thickness and cylinder height respectively.

Table 1. Variations of double scatter probabilities and the parallel
component to P_2 (undesirable radiation) with wall thickness
and height. $r_* = 3.175$ cm and e_o and e_A are 0.2 cm.

$r_{**}-r_*$	z_*	$P_2 \times 10^{-4}$	$P \times 10^{-4}$	P /P_2
0.435 cm	0.476 cm	5.97	1.80	0.302
0.635	0.476	7.97	2.40	0.301
0.835	0.476	9.24	2.78	0.301
0.635	0.376	8.23	2.37	0.288
0.635	0.576	7.73	2.42	0.313

CONCLUSIONS

All computations have been made utilizing a vertical symmetry
plane defined by the analyte and source. Hence we have not considered
double scattering from one half-cylinder to the other. Such
scattering, we believe, would only increase the amount of unwanted
radiation. We therefore suggest the placement of an additional baffle
along the analyte-source plane.

While the total radiation reaching the analyte is much increased
by a factor

$$\frac{\text{area of cylinder } (P_1 + P_2)}{\text{area of slab } (P_1 + P_2)}$$

for the cylinder over the slab, the unwanted radiation is also much
higher.

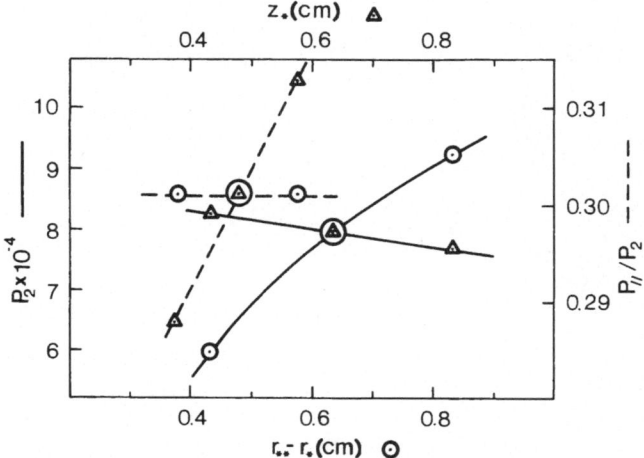

Figure 2.   The variations of the probability of double scatter (solid
line) and the fraction of undesirable double scatter
radiation (dashed line) as a function of cylinder wall
thickness (⊙) and cylinder height (▲). For the $r_{**}-r_*$
abscissa $Z_* = 0.476$ cm while for the Z_* absciss $r_{**}-r_* =$
0.635 cm. In both cases $r_* = 3.175$ cm.

A further study including the surface area of the source and
analyte may be necessary to sort out other effects concerning the
advantages of the cylindrical scatterer. If intensity is the main
concern, the cylinder has the advantage by at least a factor of about
ten, while if polarization is the main concern, we presently suggest
the rectilinear system. Although this is still undecided, we suggest
that part of the stray radiation problem of the existing instrument
may be due to multiple scattering.

REFERENCES

1.   John D. Zahrt and Richard Ryon, Polarization of X-Rays by
     Scattering from the Interior of a Cylinder, I. Single Scatter
2.   S. Chandrasekhar, "Radiative Transfer" (Chapter II), Dover, New
     York, 1960.

APPLICATION OF A POLARIZED X-RAY SPECTROMETER FOR ANALYSIS OF ASH FROM A REFUSE-FIRED STEAM GENERATING FACILITY

William E. Maddox

Department of Physics and Astronomy
Murray State University
Murray, Kentucky

ABSTRACT

The Refuse-Fired Steam Generating Facility (RFSGF) funded jointly by NASA, the U.S. Air Force, and the City of Hampton is presently in operation at the NASA/Langley Research Center in Hampton, Virginia. The facility burns approximately 200 tons/day of refuse and supplies approximately 170×10^3 tons/year of steam at 350 psig to the Langley Center. Concentrations of trace elements in the bottom ash and in the ash from the electrostatic precipitators were determined using the Murray State Polarized X-Ray Fluorescence Spectrometer (PXFS). The PXFS uses x-rays from a Phillips PW1140/96 x-ray generator in a double scattering process to make quantitative measurements on elements in pressed briquette samples. The double scattering process is used to produce polarized x-rays for excitation of the samples. Minimum detectable limits (MDL) of 1-3 ppm are achieved for elements with Z = 26 to 42. Lower Z elements have significantly higher MDL's; the lowest Z element detected, sulfur, has an MDL of 100 ppm. Elements with Z's higher than 42 have MDL's in the range of 4-10 ppm. Elements detected in the RFSGF ash were S, Cl, K, Ca, Ti, V, Mn, Fe, Cu, Zn, Br, Rb, Sr, Zr, Sn, Sb, and Pb. The concentrations ranged from a few ppm to several mg/g.

INTRODUCTION

Energy dispersive x-ray fluorescence has been used for many years to determine trace concentrations of many elements in a variety of materials. Murray State University first started doing this type of analysis using a 2-MV Van de Graaff accelerator to do proton induced x-ray fluorescence. Although this method achieved excellent sensi-

519

tivities, the target preparations and total run times allowed only
a few samples a day to be analyzed. Considering the type and number
of samples being analyzed, it was decided to construct an x-ray induced fluorescence system at MSU. It was desired to have a system
that could be easily calibrated, that covered a broad range of elements in a single run and one that could achieve sensitivities in
the ppm range.

After considering a number of possibilities, it was decided to
build a spectrometer using polarized x-rays as an excitation source.
A number of papers[1,2,3] have demonstrated that the low backgrounds
and the broad polychromatic excitation radiation typical in a polarized x-ray system could give us the capabilities we were inter-
ested in.

A polarized x-ray spectrometer, as depicted in Fig. 1, utilizes
a 90° scattering process to produce x-rays nearly 100% polarized.
These polarized x-rays are used to excite elements present in the
sample. If the fluorescent x-rays from the sample are detected
along a direction perpendicular to the first scattering plane, then
the probability of an x-ray originating in the x-ray tube and ending
up in the detector is minimized, thus reducing the background.

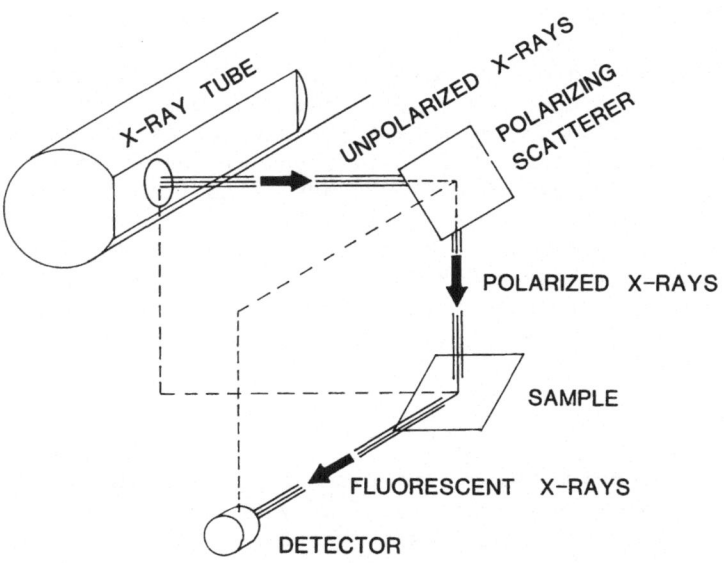

Fig. 1.   Geometry for a polarized x-ray spectrometer.

EQUIPMENT AND TARGET PREPARATION

The polarized x-ray fluorescence spectrometer utilizes a 3000 watt tungsten anode tube for the source of x-rays and an amorphous graphite plate as the polarizing scatterer. The samples are thin, pressed briquettes and the detector is a 160-eV Si(Li) detector. The analyzer is a Nuclear Data ND66 8000 channel multichannel analyzer interfaced to a DEC 1103 computer for data storage and analysis.

In an effort to keep the background as low as possible, the samples were thin, pressed briquettes. The sample material was ground to a fine powder and mixed with 10% boric acid to help bind the material. Approximately 0.4 g of this material was pressed into a pellet of 1.8 cm diameter and approximately 1.5 mm thick. This was as thin as the samples could be made without having a high percentage of breakage occurring during the preparation. The samples were mounted in a supportive assembly as shown in Fig. 2. This produced a sample that could be easily handled and stored yet was still thin as needed for the low backgrounds.

ANALYSIS OF ASH

To demonstrate the use of the spectrometer, we chose to analyze ash samples taken from a refuse burning facility. The ash samples were supplied by the NASA Langley Research Center in Hampton, Vir-

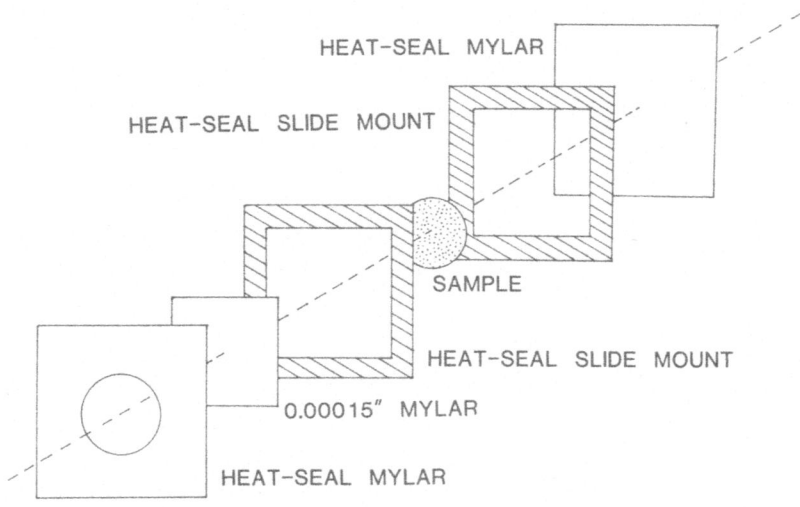

Fig. 2. Supportive assembly for mounting samples.

ginia, which operates a refuse-fired steam generating facility at
this center. The facility burns approximately 200 tons per day of
refuse from the City of Hampton, the NASA Center and the Langley Air
Force Base. We were supplied with ash samples that had been collected at weekly intervals over a 3 month period from the two electrostatic precipitators and from the ash removed from the bottom of
the furnaces. A typical spectrum of the electrostatic precipitator
ash is shown in Fig. 3. Twenty-two elements may be identified in
this spectrum. The energy range in this spectrum goes up to 32 keV
allowing the K x-rays of elements up to Ba to be analyzed. Except
for Pb, all elements detected could be covered by one calibration
curve obtained from seven different elements as shown in Fig. 4. For
normalization purposes, each sample was doped with 1000 ppm Se.

As a check on the linearity of the system, several targets from
the same sample were prepared with different concentrations of Cr
added to each. As can be seen in Fig. 5, the normalized counts in
the Cr peak are directly proportional to the concentration of the Cr
in the sample. The total number of counts in the Ti peak in each of
these targets was noted to be independent of the Cr concentration.
This indicated no absorption or enhancement corrections are needed
even at these low energies where the target is probably close to, if
not infinitely thick.

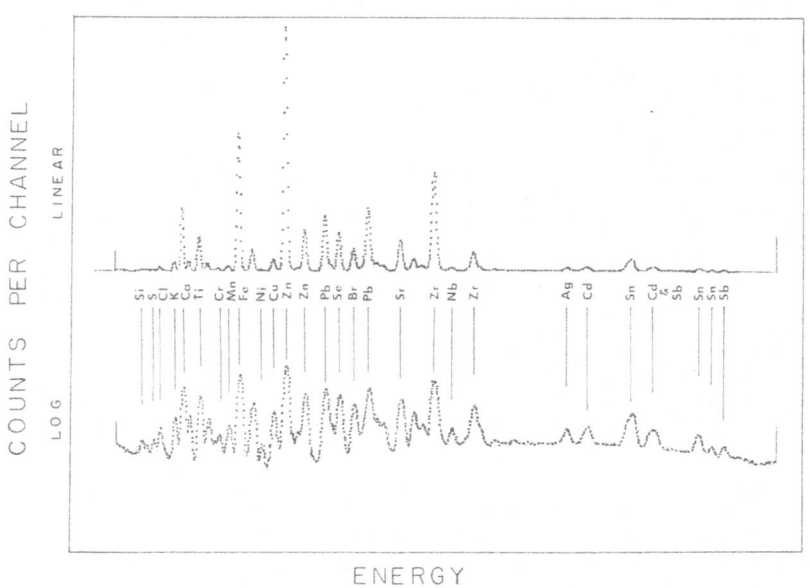

Fig. 3. Typical spectrum of the electrostatic precipitator ash.

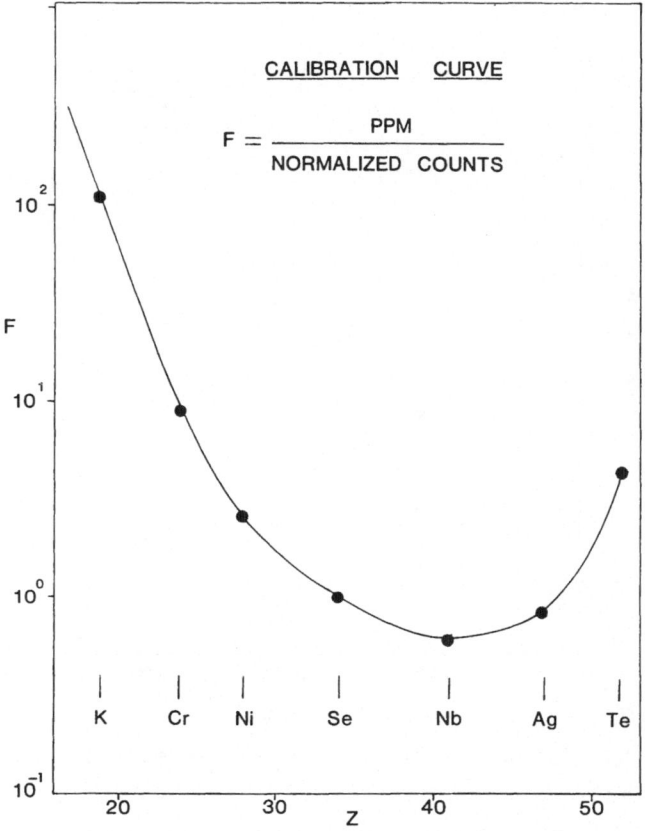

Fig. 4. Calibration curve for K x-rays.

The minimum detectable limits (MDL) obtained in these runs are shown in Fig. 6. For all elements higher in Z than V, the MDL's are below 10 ppm with elements Fe - Ag being less than 3 ppm. The MDL's are for approximately 55 minute run times and are calculated using two standard deviations of the background.

The results of the measurements are given in Table 1. Each of these measurements should be in error by less than 20% of the measured value.

DISCUSSION

The results of the analysis of the complex ash samples indicate that the polarized x-ray spectrometer can be used to make simultan-

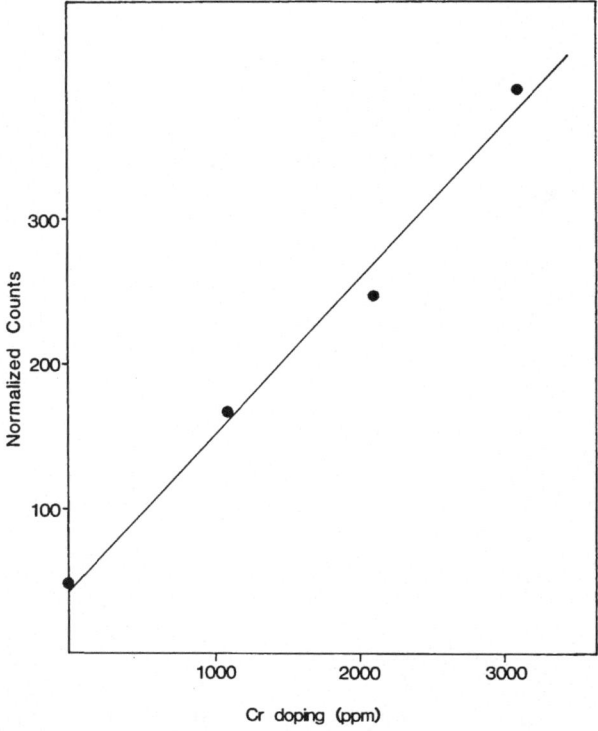

Fig. 5.   Linearity of the system response for Cr.

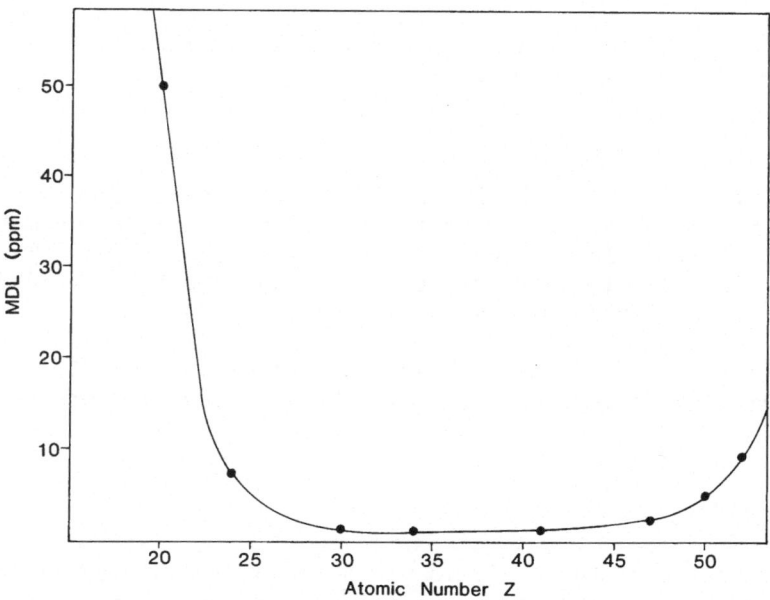

Fig. 6.   Minimum detectable limits for the K x-ray elements.

Table 1. Measured intensities for trace elements in the two types
of ash. The first two numbers give the range of values
for all samples and the number in parentheses is the
average.

Z	ELE	UNITS	BOTTOM ASH	ESP ASH
16	S	mg/g	11- 39(23)	9- 26(12)
17	Cl	mg/g	11- 39(23)	18- 58(29)
19	K	mg/g	21- 40(32)	17- 35(21)
20	Ca	mg/g	24- 448(258)	67- 102(78)
22	Ti	mg/g	20- 36(26)	12- 17(14)
24	Cr	ppm	675-1220(901)	345- 701(463)
25	Mn	mg/g	1.3- 2.7(2.0)	0.7- 1.3(0.9)
26	Fe	mg/g	40- 130(72)	15- 17(16)
28	Ni	ppm	65- 435(200)	42- 278(125)
29	Cu	mg/g	0.8- 2.5(1.5)	0.5- 0.9(0.6)
30	Zn	mg/g	5- 19(9)	9- 27(12)
35	Br	ppm	19- 293(121)	325-1460(675)
37	Rb	ppm	20- 74(50)	
38	Sr	ppm	430- 578(490)	320- 720(490)
40	Zr	mg/g	0.5- 1.8(0.7)	1.1- 2.3(1.6)
41	Nb	ppm	13- 33(21)	20- 58(31)
47	Ag	ppm	5- 26(16)	20- 110(45)
48	Cd	ppm	6- 69(23)	110- 560(230)
50	Sn	ppm	17- 337(222)	172-1750(820)
51	Sb	ppm		23- 185(85)
82	Pb	mg/g	1.5- 2.1(1.7)	2.3- 6.2(3.2)

eous measurements on all elements above Si. Being able to use the
K x-rays for the analysis of Ag - Ba allows accurate and highly sensitivity measurements to be made on these elements. Trying to analyze these elements using their L x-rays is almost impossible when
there are high concentrations of the elements S, Cl, K and Ca in the
samples.

The run times in this work were quite high (55 minutes) because
the number of samples to be analyzed was low and time was no problem.
Run times of 10 to 15 minutes could have been used to analyze this
ash without a large change in the accuracy of the measurements.

ACKNOWLEDGEMENTS

The author wishes to thank Mr. Robert Young at Virginia Tech and
Mr. Warren Kelliher at NASA Langley Research Center for the assistance in supplying the ash samples used in this work.

REFERENCES

1. T. G. Duzubey, B. V. Jarrett, and J. M. Jaklevic, "Background
    Reduction in X-Ray Fluorescence Spectra Using Polarization,"
    Nuclear Instruments and Methods 115:297-299 (1974).

2. R. W. Ryon, "Polarization Radiation Produced by Scatter for
    Energy Dispersive X-Ray Fluorescence Trace Analysis," Ad-
    vances in X-Ray Analysis 20:575-590 (1977).

3. R. W. Ryon and J. D. Zahrt, "Improved X-Ray Fluorescence Capa-
    bilities by Excitation with High Intensity Polarized X-Rays,"
    Advances in X-Ray Analysis 22:453-460 (1979).

APPLICATIONS OF ROOM TEMPERATURE ENERGY DISPERSIVE
X-RAY SPECTROMETRY USING A MERCURIC IODIDE DETECTOR

D.E. Leyden , A.R. Harding and K. Goldbach

Department of Chemistry
Colorado State University
Fort Collins, Colorado   80523

INTRODUCTION

Energy dispersive X-ray spectrometry has been used extensively for the rapid, simultaneous determination of elements in a variety of sample types.[1] Excitation of the analytical sample can be by either X-ray tube, secondary targets, or radioactive isotopic sources. Tube sources have the advantage of convenient control of the excitation conditions, whereas an isotopic source or secondary target must be physically replaced by another to affect an excitation change. The use of primary filters between the sample and X-ray tube can greatly enhance the flexibility of the excitation conditions.[2]

After the sample has been irradiated and the characteristic X-rays of the elements present are emitted, the detection system becomes the most important component. The typical detection system of the modern X-ray spectrometer consists of a lithium drifted silicon crystal protected by a thin beryllium window. The crystal must be reverse-biased and cryogenically cooled. The charge generated in the detector by electron hole pairs formed by photoionization is converted to a voltage pulse by the Field Effect Transistor (FET) preamplifier. These voltage pulses are input into an amplifier where they are shaped, amplified and passed onto a multichannel analyzer where the voltage pulses are first sorted and then stored over time.[3]

Because of the requirement of continuous cryogenic cooling for the Si(Li) detector, development of a detector capable of room temperature operation is desirable.   The use of mercuric iodide (HgI_2) crystals for a detector of this type is ideal because of the high atomic numbers of the elemental components, and the large bandgap (2.1 eV) associated with electronic transitions.[4]   This paper will discuss some of the applications of a HgI_2 detector-based energy dispersive X-ray spectrometer to analytical samples that are routinely analyzed using X-ray spectrometric methods.

The form of mercuric iodide crystal used is the low temperature tetragonal crystalline structure.   The crystal is red in color and undergoes a destructive phase change to beta-HgI_2 (yellow) at 128°C.   As mentioned earlier, the energy bandgap of the crystal is 2.1 eV which classifies it as an insulator or semi-insulator.   The ionization efficiency for the mercuric iodide detector is 4.2 eV per electron hole pair formed.   In comparison, the value given for the Si(Li) crystal is 3.8 eV per electron hole pair.   This difference results in a larger charge collection when a Si(Li) detector is used.   A better maximum resolution is achievable using Si(Li) as compared to mercuric iodide.   To fabricate the HgI_2 detector, a large crystal is grown (up to 100 grams) and cleaved along a lamellar plane to a thickness of 100-500 μm.   Use of such a thin section is made possible because of the highly efficient X-ray absorbance of the HgI_2 crystal. The sides of the thin crystal are painted with a carbon coating, wire contacts are embedded into this coating and finally the entire crystal is sealed with an encapsulant used for electronic components. In the system to be described, not only the detector but also the FET is operated at room temperature.[4]

The potential advantages for a HgI_2 detector based X-ray spectrometer are two-fold.   First, the freedom from liquid nitrogen cooling allows for greater flexibility in use of the system.   Second, the potential for miniaturization of the detector hardware further aids the development of a portable energy dispersive X-ray spectrometer. The main disadvantage of the room temperature detector and FET is the poor resolution as compared to a Si(Li) system with cooled detector and FET.   The particular system described here exhibits a resolution of 500 eV with no detector or FET cooling.   A resolution of 199 eV has been achieved with cooled mercuric iodide crystals and associated FET.

Typical Si(Li) systems have resolution on the order of
145 eV. Resolution is measured as the FWHM of the Mn
K$_\alpha$ peak. With the advent of smaller, low power X-ray
tubes, the outlook for a truly portable energy
dispersive X-ray spectrometer is good.

APPARATUS

The detector used for this system was fabricated
by the University of Southern California, Medical
Imaging Science Group, under the direction of Dr. A.J.
Dabrowski. Physical dimensions of the HgI$_2$ crystal/FET
enclosure are as follows: 9 cm width, 9 cm height and
14 cm length. A battery bias supply is also included
in the enclosure to maintain a bias of 100 to 500
volts/cm across the crystal. The crystal is masked to
expose only 3 mm^2 to incoming X-rays and is otherwise
protected by an aluminum window. A 5 mm Pb collimator
is mounted on the front of the detector to reduce X-ray
scatter in the interior of the detector. In the
spectrometer used, the detector is mounted, as shown in
Figure 1, at an adjustable take-off angle of near 45°
to the sample holder, which can also be adjusted. The
detector is 4.5 cm from the sample surface. The sample
holder is a machined aluminum block that allows the
reproducible positioning of samples. Excitation of the
sample is accomplished by using a Kevex X-ray tube
(Kevex Corp., Tube Division, Scotts Valley, CA) 50 kV,
1.0 mA, silver anode target, with a beam collimator of
7 mm. The tube is mounted directly above the
adjustable sample holder. High voltage was provided to
the tube by a Watkins-Johnson (Watkins-Johnson Co.,
3333 Hillview Ave., Palo Alto, CA) 50 kV, 1.0 mA DC

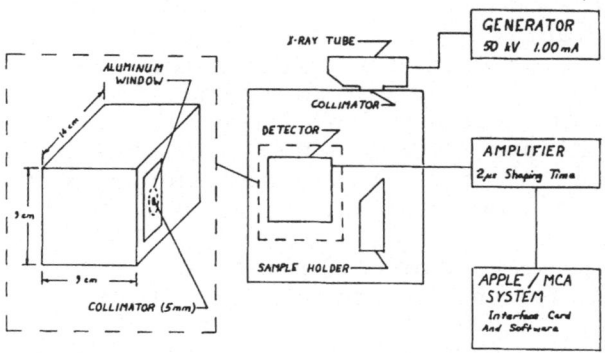

Figure 1: Schematic of Mercuric Iodide detector based
X-ray spectrometer.

high voltage source. For the paint pigment analyses, a
smaller 30 kV, 0.3 mA X-ray tube power supply replaced
the Watkins-Johnson generator. The X-ray source used
is a Rh target 30 kV, 9 watt X-ray tube. The physical
dimensions of both the power supply and tube are much
smaller than the 50 kV system. An Ortec (EG & G Ortec,
100 Midland Rd., Oakridge, TN) 416A amplifier was used
at a pulse shaping time of 2 us. Longer shaping times
adversely affected the peak resolution due to charge
buildup in the detector. The amplifier output is input
to The Nucleus Model 811 multichannel analyzer (MCA)
card (The Nucleus Inc., 461 Laboratory Road, Oakridge,
TN) which is mounted in a buss slot of an Apple II+
microcomputer (Apple Computer Inc., 10260 Bandley
Drive, Cupertino, CA). Software for the computer
system is MC-APPLE-YZER II available from Scientific
MicroPrograms (213 Merwin Road, Raleigh, NC). The
software/MCA card system allows acquisition of X-ray
spectra in 1024 channels and subsequent storage to
floppy disk. Calibration of the analyzing system is
accomplished by acquiring the spectrum of a reference
Cu-Ti sample with iterative adjustment of the amplifier
gain and MCA zero. The resultant calibration (eV per
channel) and the apparent zero is stored in a permanent
disk file. This particular analyzer system is useful
for the low cost accumulation, storage and manipulation
of X-ray spectra.

RESULTS

Nickel Alloys

Analysis of alloys by X-ray spectrometry is
routinely performed in the metals industry for quality
control purposes. Alloys are excellent X-ray
spectrometry samples because of their solid phase and
homogeneous fabrication.[5] The alloys analyzed were for
the most part provided by Precision Castparts Corp.
(4600 S.E. Harney Drive., Portland, OR). Two of the
six alloys were standard Ni base alloys obtained in-
house. All alloys were greater than 50% by weight Ni.
Of the six alloys only three had known, certified Ni
values. The alloys contained, in addition to nickel,
various amounts of Cr, Co, Ti, Fe, Mo and Nb.
Qualitatively, all these elements could be determined.
Figure 2 shows the spectrum of alloy 17218 which
contains Cr, Fe, Ni, Mo, and Nb. By changing the
excitation conditions to those shown in Figure 3, the
overlapped Mo and Nb peaks are seen at the high end of
the energy (keV) axis. Quantitation was carried out

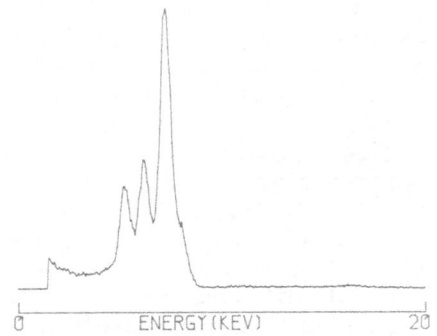

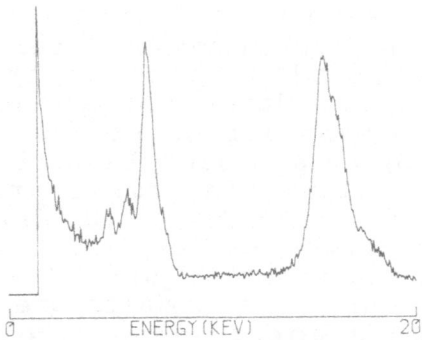

Figure 2: Nickel alloy
(17218) spectrum: Ag X-ray
tube, 20 kV, 0.04 mA, no
filter, 250 s. counting time.
VFS = 2700 counts. 19% Cr,
15% Fe, 53% Ni, 3.1% Mo and
5% Nb.

Figure 3: Nickel Alloy
(17218) spectrum: Ag
X-ray tube, 38 kV, 0.12
mA, 0.127 mm Ag filter,
500 s. counting time.
VFS=12,200

for the nickel peak by counting the alloy sample at a
specific excitation condition and time period (500
sec).   A reference Ni peak was generated similarly
using a sample of pure nickel.   A fit was done by
utilizing a software command that would multiply the
reference Ni counts in each channel by an analyst
entered factor (<1) and then overlay the adjusted
reference spectrum with the alloy spectrum.   When, by
the analyst's judgment, a match was realized (i.e., the
alloy and reference spectra overlapped exactly), a
subtraction was done and the residual scrutinized.
Requisites for a good fit were, taking into account
background, a minimum number of counts in each channel
about the Ni peak remaining after the adjusted
reference peak was subtracted from the unknown. Since
the minimization of the residual was done by visual
inspection, an error was introduced in that actual
differences in residual could not be visually detected
unless a change of +0.05 in the factor was introduced.
This translates into an error of 5% in the mass
fraction determination of nickel.   Work on a
mathematical fitting technique is underway in this
laboratory.

Because only three of the alloys had known Ni
values, an independent technique was chosen to monitor
the results obtained by the X-ray spectrometric
analysis described above.   Atomic absorption (AA)
determination of Ni was performed on a Varian AA 5
using an air-acetylene flame.   The atomic absorption
conditions and sample preparation for nickel alloy
analysis are described elsewhere.[6]   Results of the X-
ray determination of Ni versus those obtained by AA
are plotted in Figure 4.   Analysis of the results shows
that all X-ray results are within 10% of the AA values.
Interelement effect correction methods such as
Rasberry-Heinrich (R-H) were attempted but did not
improve the data.   Because only three standard Ni
alloys were analyzed, R-H absorption/enhancement
coefficients, of necessity, came from published
literature.[8]   Undoubtedly, upon the development of
better fitting techniques, the results will become more
reliable.   Nonetheless, the analysis shows the
feasibility of qualitative and quantitative
determination of an element in an alloy.

Preconcentrated Trace Ions

Preconcentration of trace elements from aqueous
solutions prior to X-ray spectrometric determination
has been studied previously.[9,10]   One successful method
for trace metal determination is the addition of a
precipitation agent, such as sodium dibenzyldithio-

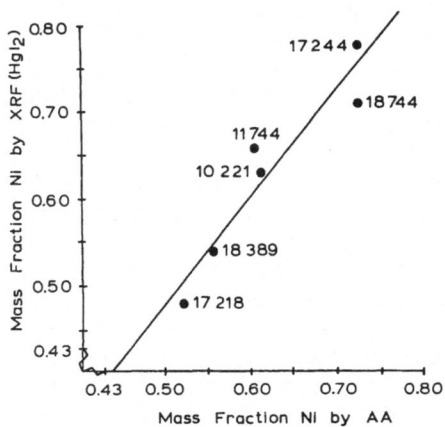

Figure 4:   Plot of mass fraction nickel determined by
mercuric iodide-based energy dispersive X-ray
spectrometry versus those found by AA: slope = 1.22,
intercept = - 0.13

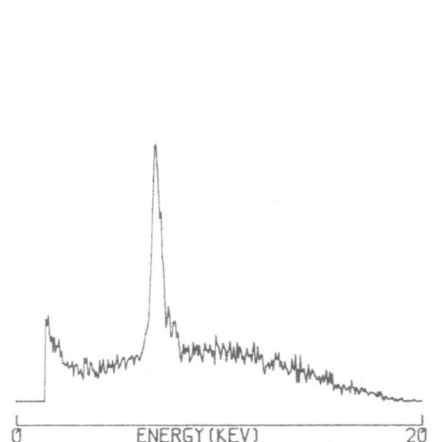

Figure 5: X-ray spectrum of 150 micrograms of Ni precipitated by DBDTC and collected on a membrane filter. Ag X-ray tube, 20 kV, 0.04 mA, no filter, 500 s. counting time. VFS = 275 counts.

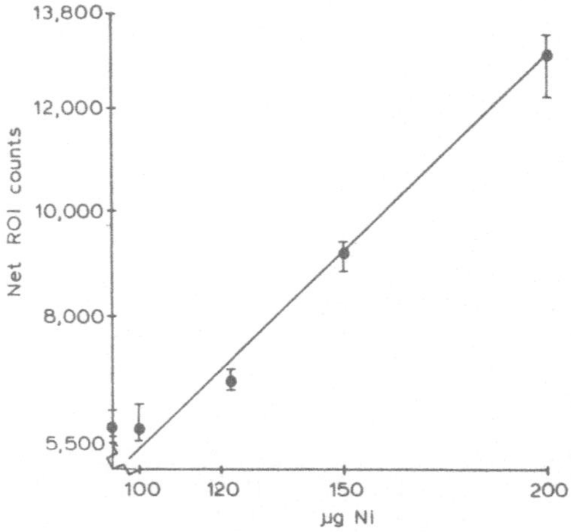

Figure 6: Plot of counts for Ni precipitate collected on a membrane filter, versus μg of Ni. Slope= 0.149 cps/μg, LLD=100 μg.

carbamate (DBDTC), to the solution. The precipitate is then collected on a 0.45 μm pore diameter membrane filter. The filter carrying the metal-DBDTC precipitate is then analyzed using X-ray spectrometry. In this particular study, aqueous solutions of Ni(II) were prepared at concentrations of 100, 120, 140 and 200 μg/100 ml. These solutions were precipitated and filtered as described above. The filter media used is a Nuclepore (Nuclepore Corp., 7035 Commerce Circle, Pleasanton, CA) polycarbonate membrane. Nuclepore filters result in much lower background scatter compared to other support media. Conditions for excitation of the Ni-DBDTC samples are 20 kV, 0.04 mA, no radiation filter. Counting times are 500 seconds. Figure 5 shows a typical Ni-DBDTC spectrum of 150 μg of Ni. Figure 6 shows a calibration curve determined by plotting net counts in a region-of-interest (ROI) window vs. μg Ni on the filter. Net counts are determined from the gross counts by taking the ROI (6.38-9.22 keV) minus background counts (6.38-9.22 keV) as determined from a blank. By defining a lower limit of detectability (LLD) as the amount of Ni which yields a net count greater than three times the standard

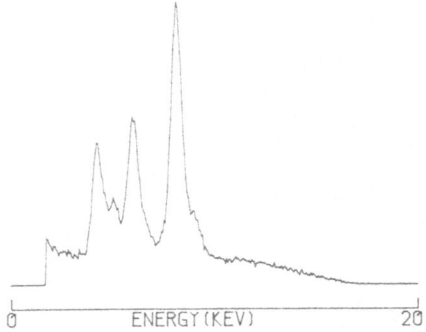

Figure 7:   Spectrum of paint pigment: Rh tube, no
filter, 17 kV, 0.08 mA, 500 s. counting time.   VFS =
2300 counts.   23.1% Ti, 13.8% Cr, 9.2% Fe, 2.4% Zn.

deviation of the background counts, the minimum
detectable amount is 100 µg Ni. The importance of
water analysis for trace metal ions is well documented.
The above example shows that a potential application of
a refined portable HgI_2-based X-ray spectrometer would
be for the quantitation of trace metal in water with a
minimum amount of sample handling and transport.[11]

Paint Pigment Analysis

        Another application of the room temperature HgI_2
X-ray spectrometer system is the determination of Fe
and Zn in a Fe_2O_3 based paint sample. Paint samples
were provided by Sigma Coatings, Amsterdam, the
Netherlands. The standard paints were prepared for XRF
as follows: five compounds, titanium dioxide, barium
sulfate, iron oxide, zinc chromate and zinc phosphate
were mixed in a shaker. A hardener was added and the
paints were spread on a support material. Circles of
2.45 cm were cut and mounted into XRF sample cups
between mylar sheets.[12] Nine different mixtures of the
five compounds were prepared and duplicate circles were
cut. A qualitative spectrum is shown in Figure 7.
Conditions of excitation are as listed.

        Thin film standards were prepared in two
thicknesses, 112.5 µm and 150 µm using a Bird
applicator. Quantitative analysis of the samples began
by fitting the paint spectrum to reference peaks
obtained from pure element standards using a linear
least squares routine. The resulting intensity ratios
(k-ratios) were then input into an empirical correction

program executed in the Apple II+ computer in which
absorption/enhancement coefficients were iteratively
determined.   After the fit of the data, the
coefficients are stored for use with data from unknown
samples.   Iron and zinc were determined in both the
112.5 µm and 150 µm samples. The lower limit of
determination is 0.77% for Zn, and 3.65% for Fe in the
112 um thick samples. The value for iron is much
higher than that for zinc due in part to the overlap
with the Cr K_β peak that occurs because of the
relatively low resolution of the HgI_2 detector. Figure
8 shows a plot of the percentage zinc found versus that
reported for a set of standards. This line has a least
squares fitted slope of 0.973, an intercept of 0.05,
and a correlation coefficient of 0.986. Figure 9 shows
the results from the same set of standards analyzed
with a commercial Si(Li) based system. The line shown
in Figure 9 has a least squares fitted slope of 0.987
an intercept of 0.02, and a correlation coefficient of
0.993. In the case of the zinc determinations, one
value was found to deviate more than 10% from the least

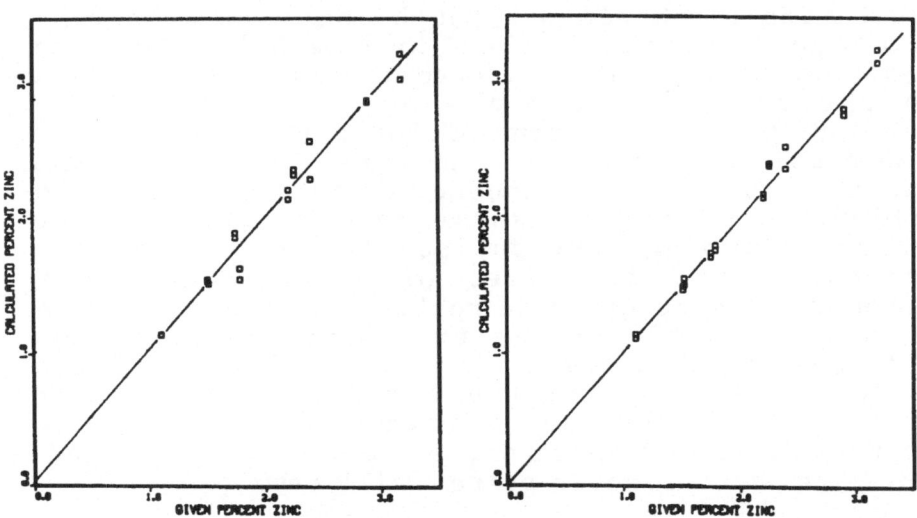

Figure 8: Plot of percent Zn
in paint pigment standard
versus that found by mercuric
iodide based X-ray
spectrometry.

Figure 9: Plot of
percent Zn in paint
pigment versus that
found by Si(Li) based
X-ray spectrometry.

squares line with HgI_2 detection, and none more than 7% from the line with Si(Li) detection. Results of the iron determination show no deviation greater than 15% for HgI_2 detection, and 5% for Si(Li) detection.

The precision of the determination of zinc and iron in the paint pigment samples is not a good as that obtained with a full commercial Si(Li)-based spectrometer. However, the hardware configuration and sample presentation in the current system are not well refined. As these improve, substantial improvements in sensitivity, as well as the accuracy and precision, are expected.

CONCLUSION

This report has shown that an energy dispersive X-ray spectrometer system based on a room temperature detector is capable of providing reliable analytical results on routinely analyzed XRF samples. There were three such samples: Ni based alloys, preconcentrated aqueous nickel, and Fe_2O_3 based paint pigment. Qualitative and quantitative analysis were performed; spectra are presented for the former, calibration plots for the latter. In all three instances the quantitation is obtained through somewhat different data manipulation but all show the feasibility for such analysis by a HgI_2 based spectrometer/analyzer system. With the advent of low power, small size X-ray excitation tubes, the potential for miniaturization of the spectrometer system from that in Figure 1 is good. Further, due to the independence of the detector from liquid nitrogen and the subsequent omission of the necessary plumbing, the outlook for portability is enhanced. The advantages of a portable X-ray spectrometer system could be realized in those analyses that require little sample transport and where the results must be obtained rapidly and at low cost for a variety of elements in the sample. Upon further development of analyzer system software and the configuration of the spectrometer, a low cost, compact, useful X-ray spectrometer system will result.

ACKNOWLEDGEMENT

Financial support from the AMAX Foundation and the Phillipson Chair, University of Denver, is gratefully acknowledged. The cooperation of Professor A.J. Dabrowski and Tracor XRay, Inc. is appreciated.

REFERENCES

1. Drummond, W.E.; Stewart, W.D., "Automated Energy-Dispersive Fluorescence Analysis," American Laboratory,November 1980, p 71.
2. Bertin, E.P., Introduction to X-Ray Spectrometric Analysis, Plenum, NY, 1978, Chapter 1.
3. Bertin, E.P., Introduction to X-Ray Spectrometric Analysis, Plenum, NY, 1978, Chapter 6.
4. Huth, G.C.; Dabrowski, A.J.; Singh, M.; Economou, T.E. Turkevich, A.L., Advances in X-Ray Analysis,22, 461, (1979).
5. Bertin, E.P., Introduction to X-Ray Spectrometric Analysis, Plenum, NY, 1978, Chapter 10.
6. Kinson K.; Belcher, C.B., Anal. Chim. Acta, 30, 64, (1964).
7. Rasberry, S.D., Heinrich, K.E.J., Anal Chem., 46, 81, (1974).
9. A.T. Ellis, D.E. Leyden, W. Wegscheider, B.B. Jablonski, and W.B. Bodnar, "Preconcentration Methods for the Determination of Trace Elements in Water Using X-Ray Fluorescence Spectrometry. Part I -Response Characteristics," Anal. Chim. Acta, 142, 73-87,(1982).
10. A.T. Ellis, D.E. Leyden, W. Wegscheider, B.B. Jablonski, and W.B. Bodnar, "Preconcentration Methods for the Determination of Trace Elements in Water Using X-Ray Fluorescence Spectrometry. Part II - Interference Studies," Anal. Chim. Acta, 142, 89-100, (1982).
11. Singh, M., Nuclear Inst. and Methods, 193, 135, (1982).
12. Sigma Coatings, Amsterdam, The Netherlands, Personal Communication,1982.

WALL EFFECT AND DETECTION LIMIT OF

THE PROPORTIONAL COUNTER SPECTROMETER

Marja-Leena Järvinen and Heikki Sipilä

Institute of Physics
Outokumpu Oy
Espoo, Finland

ABSTRACT

The detection limit of the proportional counter spectrometer is determined by the low-energy background caused by the wall effect. The choice of the filling gas of the proportional counter has been studied to minimize the background. Detection limits (3σ) for different elements have been calculated. Theoretical results have been compared with the experimental value obtained with Cd-109 and Am-241 excitation. The results are applied to the analysis of metal concentrations in oil samples and mapping the contents of zinc in the bark of living trees with a portable XRF-analyzer.

INTRODUCTION

The wall effect in the proportional counter is caused by incomplete charge collection. The traces of energetic photoelectrons terminate on the walls of the counter before the ionizing electron has lost its energy into detector gas. There are also other phenomena causing background such as escape peaks, escape of photons after undergoing Compton scattering within the detector, and impurity lines from the probe construction. However, except for the wall effect, these phenomena can be easily minimized or even avoided.

The low energy background determines the detection limit of the proportional counter spectrometer if the fluorescence spectrum of the sample is dominated by the back scattering of the exciting radiation. This is the case in organic samples with low impurity concentrations.

There are different methods to eliminate the background. The signal to background ratio can be improved by a factor of twenty by

electric means using rise time analysis /1/. Another method is to
use pulsed excitation and gating the pulse collection time. In both
of these methods a dead zone is formed near the detector wall where
the ionization events are ignored. The wall effect can also be
reduced using for instance a wall-less anticoincidence multiwire
counter /2/. However, these methods are too complicated to be used
in simple systems. Therefore the choice of the filling gas of the
proportional counter has been studied to minimize the background.

WALL EFFECT

     Sealed cylindrical counters were used in the experiments. The
diameter of the counter was 2.3 cm. The absorption of photons was
assumed homogenous and the ejection of photoelectrons isotropic.
In the cylindrical detector the photoelectrons emitted in the volume
adjacent to the wall with depth equal to maximum range are able to
strike the detector wall. Results of accurate treatment of the
problem with the assumptions above are tabulated in reference 3.
With track length less than half of the diameter of the counter the
fraction of traces terminating on the detector walls to all traces
can be estimated as

$$f = \frac{r}{D} \tag{1}$$

where   r is the range of photoelectrons
        D is the diameter of the detector.

     The range of electrons is obtained for 1.5 - 25 keV from
expression

$$r = 0.572 \; E^{1.67} \quad (g/cm^2) \tag{2}$$

where   E is the energy of photoelectrons in MeV /3/.
For other energies it is obtained from tables in reference 4.

     In Fig. 1. there is shown the spectrum of Fe-55 source with a
chromium filter measured with a proportional counter filled with
neon-argon up to a pressure of 7 bar. There can be seen the escape
line of argon and the K-line of aluminium evoked from the inner shell
of the detector. The high voltage is 520 V. The absorption length
of 1.49 keV in this filling is 0.6 mm. This shows that the charge
collection efficiency is good and the wall effect is not caused by
back diffusion of electrons into the walls of the detector.

     The theoretical values were compared with the experimental
values measured with a counter filled with argon, pressure ranging
from 1 bar to 5 bars. The result given by expression (1) decreases
faster with increasing pressure as the experimental value. For 1 bar
counter the deviation is about 6 % but for counter filled to 5 bar

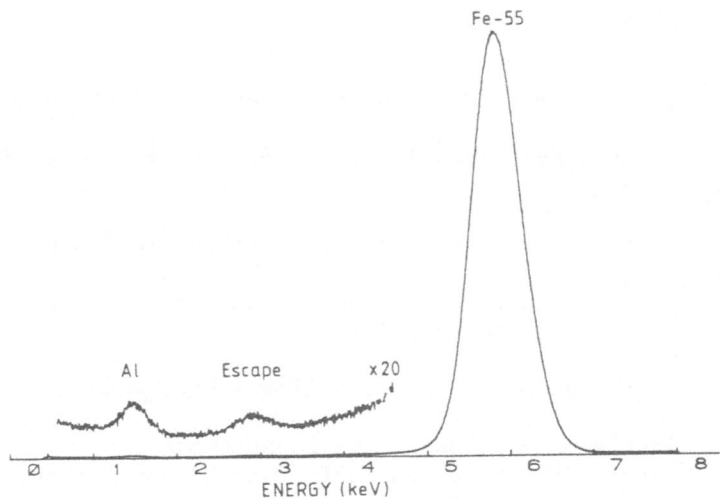

Fig. 1. A spectrum of iron-55 with a chromium filter. K-line of
          aluminium is evoked from the inner shell of the counter.
          The filling gas is neon-argon and the filling pressure is
          7 bar.

it is 30 % for energies less than 25 keV and 50 % for 60 keV. Obviously, this difference is due to inhomogenous absorption and the
inhomogeneity of ejection photoelectrons. The emission of electrons
is peaked forwards as the energy of photons increases. Also the
γ-background increases with increasing pressure.

EXPRESSION FOR DETECTION LIMIT

     The detection limits C_{DL} (3σ) have been calculated for water
solutions. The sample has been assumed to be thick. The following
expression has been derived for computing C_{DL} value

$$C_{DL} = \frac{3 \cdot \sqrt{A \cdot f}\ \varepsilon_1}{\varepsilon_2 (1-\omega_d)}\ \frac{\mu_1 + \mu_2}{\mu_1 + \mu_3} \cdot \frac{\sigma}{\omega q s t}\ \frac{1}{\sqrt{N}} \qquad (3)$$

where ε_1, ε_2 are the efficiencies of the detector to the scattered
             radiation and fluorescence radiation of the element
             under study,

     ω_d is the fluorescence yield of the detector gas,

μ_1, μ_2, μ_3 are the mass attenuation coefficients for exciting radiation, scattering radiation and fluorescence radiation,

σ is the Compton attenuation coefficient of the sample,

τ is the photoelectric attenuation coefficient of the sample,

ω is the fluorescence yield of the element under study,

g   is the proportion of the line under study to the total emission of the shell,

s   is the proportion of the line in the measurement channel,

A   is the proportion of the measurement channel of the low energy background channels, the channel is set to FWHM of the element line,

N   is the number of counts during the measurement.

It may be seen that C_{DL} is determined by two things: first, the ratio of the efficiencies of the detector to scattering radiation to the fluorescence peak intensity and second, the fraction of photoelectrons absorbed incompletely into the detector volume. The count rate of the system is limited to 5 kHz. It is assumed that this value can be achieved in all cases by choosing suitable source intensity.

RESULTS

   For Cm-244 and Cd-109 excitation neon, argon and xenon would be usable. Krypton is omitted due to its strong absorption of scattered radiation in this energy region. The ratio of efficiency to fluorescence radiation to scattering radiation increases from neon to argon and still to xenon. As the filling pressure of the detector is increased the ratio of efficiencies increases and the wall effect decreases. The optimum range as a function of filling pressure is broad. The optimum value is achieved as the decrease of the wall effect cannot beat the change in the ratio of efficiencies. Each element would have its optimal filling pressure. One filling pressure has been chosen which corresponds the optimal value of the most common applications.

   The resolution of the counter is not assumed to change as the pressure is increased. This can be achieved using Penning mixtures of neon and argon. Also the voltage is low. For instance the high voltage of the detector filled with neon-argon up to 7 bar is 500 - 600 V and argon up to 5 bar is 900 V. When ordinary gas mixtures are used in the proportional counter the high voltage will increase

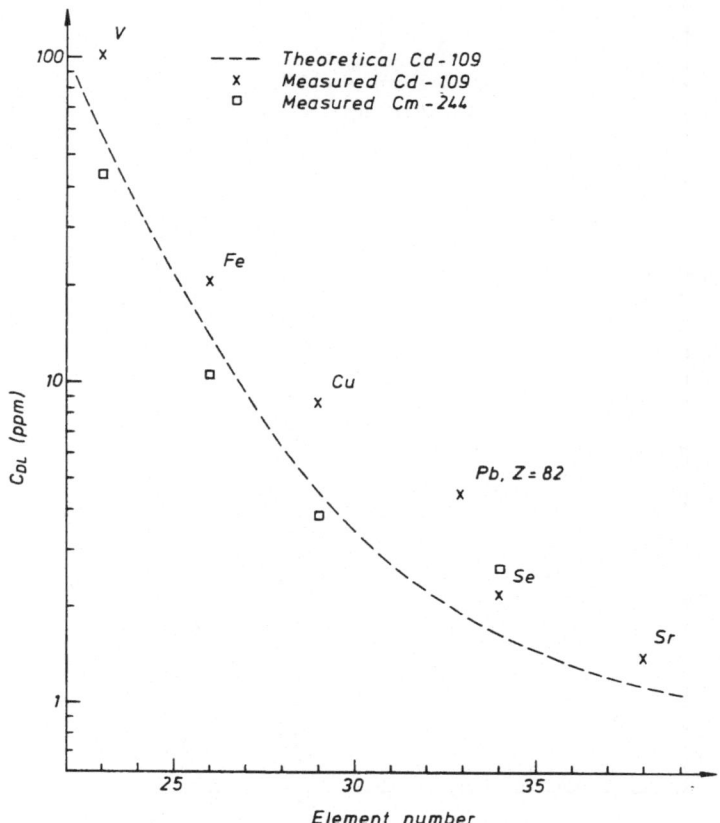

Fig. 2. Calculated and measured detection limits C_{DL} (3σ) for water
solutions measured with 5 bar argon detector. The sample
thickness is 9 mm. The measurement time is 4 minutes and
the count rate 5 kHz. A sample is excited with Cd-109 and
Cm-244 in central source geometry.

strongly and the resolution of the spectrometer will get worse as the
pressure is increased.

Neon filling would give the best result due to its low efficiency
to scattered radiation. In practice counters filled with neon are
useful with Cm-244 excitation and for lighter elements than copper.
An efficiency of 50 % for iron can be easily achieved. The use of
strong sources increases the background caused by γ-radiation of the
source.

Counters filled with argon will provide a result worse by a
factor of ~ 2 than those filled with neon. In Fig. 2. there are
shown the results calculated for Cd-109 excitation and measured
values for some elements for Cd-109 and Cm-244 excitation. The

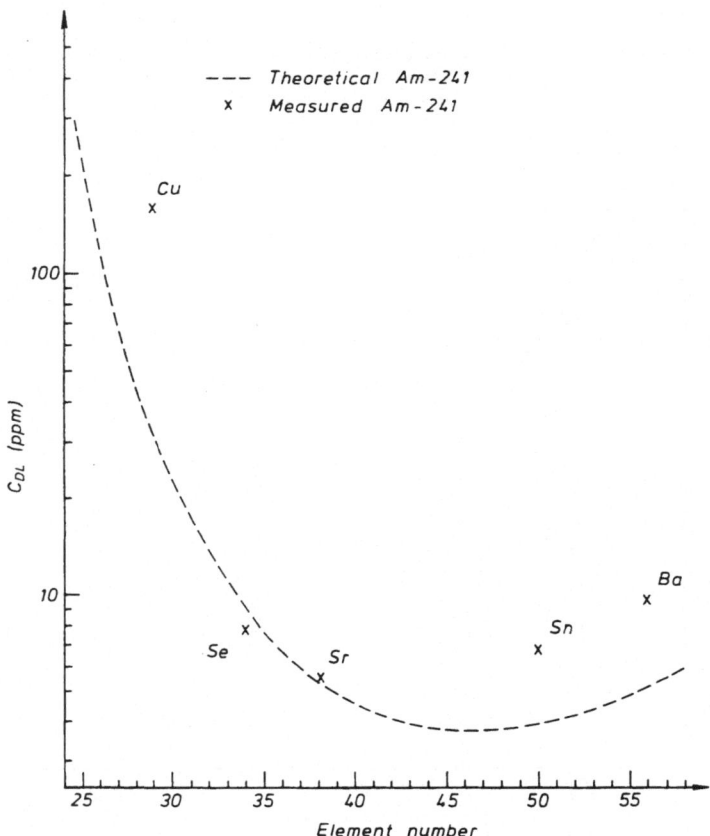

Fig. 3.   Calculated and measured detection limits C_{DL} (3σ) for water
         solutions measured with 5 bar argon detector.   The sample
         thickness is 9 mm.   The measurement time is 4 minutes and
         the count rate 5 kHz.   A sample is excited with Am-241 in
         central source geometry.

filling pressure of the detector is 5 bar.   This is the optimal
value for fluorescence lines around 10 keV.   A central source geometry has been used in the measurements.   The correspondence of the
theoretical value to the experimental value is reasonably good.

    For Am-241 excitation argon and krypton fillings are useful.
Out of these two fillings, argon would give better results.   The
optimal pressure for heavier elements than strontium is above 10 bar.
No big sacrifications are made if the filling pressure of argon
is chosen to 5 bar.   Another nice feature of argon detectors is the
low escape peak intensity.   For 5 bar filling it is only 5 % of the
principal peak intensity.   In Fig. 3. are the results calculated
and measured using Am-241 excitation and a counter filled with
argon.

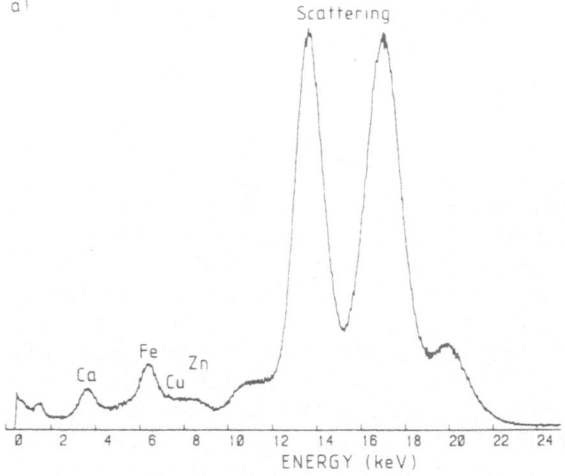

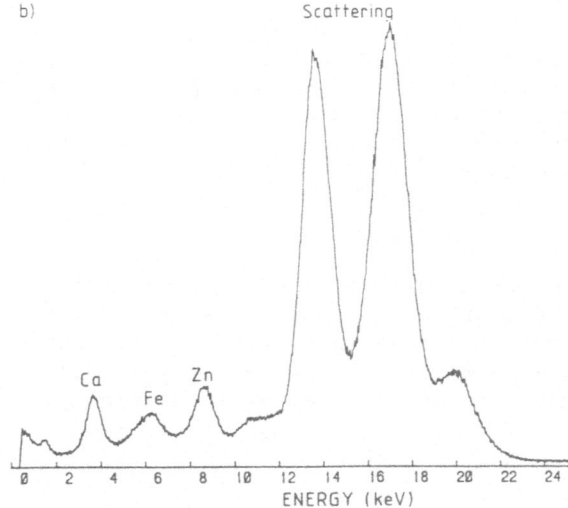

Fig. 4. Spectra obtained from bark of aspen with Cm-244 excitation
and argon counter. The measurement time is 300 seconds.
a) The concentration of zinc is 60 ppm, which corresponds
the background level in aspen.
b) The spectrum obtained from a bark of aspen growing on
the ore deposit. The concentration of zinc is 285 ppm.

The bad correspondence of the result for copper is due to the
background caused by Compton electrons. This phenomenon has been
ignored in the calculation.

The resolution can be improved optimizing the thickness of the
water sample and using side source geometry. A factor of 2 - 4 can

be easily obtained in this way, but also higher source intensities
are needed.

APPLICATIONS

Zinc is one element that accumulates in plants. Its concentration may be anomalous in plants growing on ore deposits /5/. Outokumpu portable XRF-analyzer was used in mapping the contents of zinc
in the bark of living trees on a known ore deposit. The calibration
was made with water solution. Using Cm-244 excitation and a counter
filled with argon the detection limit in the measurement was 16 ppm
for a measurement time of 15 seconds. There were aspens growing on
the deposit. The highest concentrations measured were 300 ppm. The
background level in the bark of aspen was found to be between 60 - 80
ppm. In Fig. 4 a and b spectra from a tree corresponding to the
background level and an anomalous tree are shown. The method was
found sensitive enough to separate anomalous contents of zinc in the
bark of living trees. About a hundred trees could be measured in an
hour.

Another application is to measure metallic impurities from oil
samples. Detection limits less than 2 ppm have been obtained for
iron and copper using Cm-244 excitation and a counter filled with
neon up to a pressure of 7 bar. The measurement time in the experiment has been 5 minutes /6/.

CONCLUSION

It has been shown that extremely low detection limits can be
obtained with a proportional counter spectrometer and isotope source
excitation without special sample preparation. Optimized proportional
counters with low background are especially useful in portable X-ray
fluorescence analyzers and in special industrial gages.

REFERENCES

1. H. Sipilä and E. Kiuru, Advances in X-ray analysis, Vol. 20
    (1977) 555.
2. R. W. Fink, Chap. 5 in B. Crasemann, Editor, Atomic Inner-
    Shell Processes, Vol. II, Academic Press (1975).
3. G. E. Kocharov and V. V. Petrov, Kosm. Luchi, No. 10 (1969)
    167 (in Russian).
4. D. B. Brown, "Electron Range and Electron Stopping Power",
    in J. W. Robinson, Editor, Handbook of Spectroscopy,
    Vol. I., p. 249, CRC Press (1974).
5. A. L. Kowalevskii, "Biogeokhimicheskie Poiski Rudnykh
    Mestorozhdenii", Moskow 1974, Trans. "Biochemical Explo-
    ration for Mineral Deposits" NTIS TT 76-52029 (1979).
6. L. Packer, private communications.

DETERMINATION OF OXYGEN AND NITROGEN IN VARIOUS MATERIALS

BY X-RAY FLUORESCENCE SPECTROMETRY

T. Arai and S. Ohara

Rigaku Industrial Corporation

Osaka, Japan

INTRODUCTION

Spectrographic analysis for carbon and boron using fluorescent x-rays has been studied over the past few years; principles and applications for using those ultra-soft x-rays were described, based on the combination of total reflection and filtering rather than on the wavelength dispersive method of Bragg reflection (1, 2). However, oxygen and nitrogen, with x-ray wavelengths of 23.71Å and 31.60Å, respectively, cannot be detected as easily because of their high absorption by the detector window materials such as polypropylene, polyester or formvar films.

Reports describing the measurement of oxygen and nitrogen x-rays employing both electron and x-ray excitation, have been published. Relationships between x-ray spectra and chemical bonds in various materials have been described by Henke (3), Mattson and Ehlert (4), Fisher (5), Perera and Henke (6) and Uchikawa and Numata (7). The feasibility of the determination of oxygen in various materials was shown by Uchikawa (8) and the analysis of oxygen in silicon nitride was described by Sugizaki (9). It is the purpose of this paper to discuss the determination of oxygen and nitrogen in various materials by means of improved instrumentation.

EXPERIMENTAL

A Rigaku S/Max sequential spectrometer, equipped with a RH-target end-window x-ray tube (OEG-75) with a thin beryllium window, was employed for this study. The fluorescent x-rays of oxygen and nitrogen were monochromatized, under vacuum, by using the combination of a primary coarse collimator, secondary soller slits and flat analyzing crystals. A TlAP crystal was employed for O-Kα x-rays (11, 12). For N-Kα x-rays either a multilayer soap crystal,

547

LSD (13), or a total reflection mirror was employed. The monochromatized fluorescent x-rays were detected by a gas flow proportional
counter equipped with a specially prepared thin polyester film window (Mylar®), covered with a thin layer of aluminum metal.

Figure 1 shows the absorption characteristics of two thin film
window materials. The high absorption for carbon and the advantage
of the Mylar® window, which contains 11.3% oxygen, over one made of
polypropylene are clearly shown, thus allowing effective detection
of oxygen and nitrogen x-rays.

X-RAY EMISSION AND ABSORPTION

When N-Kα x-rays are to be detected, the overlapping interferences from weaker L, M and N series x-rays of heavy elements had
to be investigated before measurements could be made. Also, higher
order x-rays from short wavelength radiation interfere with the
x-rays to be measured. These interferences are between 2nd order
Na-Kα as well as 7th order Ca-Kα x-rays and the O-Kα line and between 2nd order Co-Lα x-rays and the N-Kα line.

Because of the high absorption in the soft and ultra-soft x-ray
regions the analyzing depth becomes shallow and matrix absorption
effects become remarkable. Oxygen compounds of C, Ca, Ti, V, Zr or
Mo exhibit decreased intensities of O-Kα x-rays because of strong
absorption effects whereas those of B, Mg, Al, Si, Cr, Mn, Fe or
Ba exhibit relatively high intensities because of relatively low
absorption of O-Kα radiation. Since N-Kα x-rays are located within
high absorption regions of C, Ca, K and Pd, lower intensities can
be expected when N-Kα is measured in compounds containing these elements.

When the method of monochromatization involves the combination
of total reflection and filtering, interferences caused by overlapping of longer wavelength x-rays with the x-rays to be measured can
be found. Examples are overlapping of C-Kα on O-Kα and N-Kα and
overlapping of Cr-Lα and Ti-Lα on N-Kα.

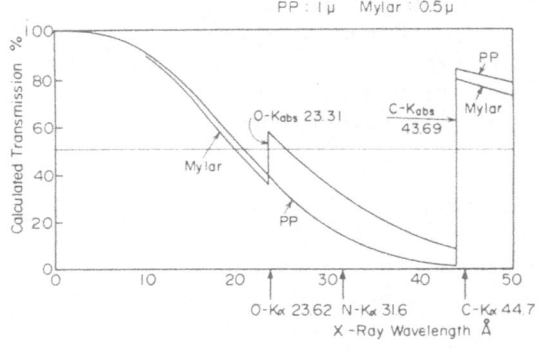

Figure 1 — X-Ray Transmission of
Polypropylene and Mylar Film

EXPERIMENTAL RESULTS AND DISCUSSION

1. Monochromator

In comparing monochromators, the following parameters were investigated: Reflecting intensity, resolution of interfering x-rays, thermal expansion of lattice constant, emission of fluorescent x-rays from the monochromator surface, life under use, etc. In the accumulation of experimental data for elemental analysis with soft and ultra-soft x-rays it is most important to obtain the highest possible intensities and, for practical applications, to have long-term reproducibility of the measured results. Table 1 shows a comparison of reflected peak intensities for various monochromators. When using total reflection by a glass mirror, the width of the monochromatized x-ray beam was narrower than the horizontal width of the primary soller slit because of the limitations imposed by the length of the glass mirror and the small scattering angles. If the geometric conditions of the optical system were the same for all characteristic x-rays, the measured intensity reflected from the glass mirror would become 2 or 3 times larger than the experimental results of total reflected intensity shown in Table 1.

A TlAP crystal was used for the shorter wavelength O-Kα x-rays because of its high resolution and relatively high intensity. A total reflection mirror, combined with an appropriate filter, had to be used for N-Kα x-rays in order to obtain higher intensity.

2. Oxygen Analysis

For purposes of investigating the general characteristics of oxygen analysis in chemical compounds and coals, the experimental relationship between maximum peak intensity of the O-Kα line and oxygen content is shown in Figure 2. Where the analyzed samples contain elements which heavily absorb O-Kα x-rays (TiO_2 and coal)

Table 1 — Comparison of Reflected Peak Intensity of Various Monochromators

X-Rays	C-Kα 44.7	N-Kα 31.6	O-Kα 23.62	F-Kα 18.32	Na-Kα 11.91	Note
Emitter	Graphite	Si₃N₄	SiO₂	NaF	NaCl	
Analyzer						
Glass mirror	7731 c/s at 8.0°	330 c/s at 7.0°	610 c/s at 6.0°	4855 c/s at 5.0°	2403 c/s at 4.0°	Width of X-ray beam / Mirror material
TAP 2d=25.763Å	—	—	1552 at 132.7°	9830 at 90.6°	79259 at 54.9°	Reflecting surface
LSD 2d=100.7Å	947 at 52.6°	24.1 at 34.0°	294 at 31.0°	3074 at 20.9°	12026 at 13.5°	Number of layers / Life

S/MAX   X-Ray conditions : OEG-75 Rh-Target 40 KV 70 mA
Soller slit : 0.52°

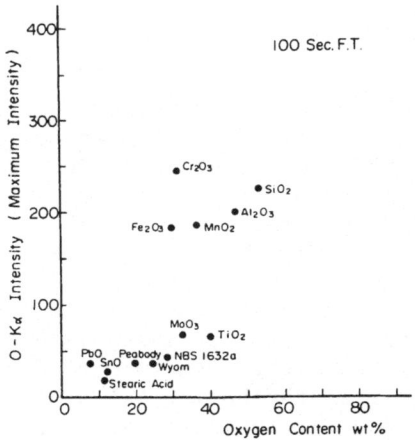

Figure 2 — Measured Intensity of
O-K$_\alpha$ X-Rays from Chemical Oxygen
Compounds against Oxygen Content

the oxygen x-ray intensity is low; on the other hand, samples containing low absorbers for O-Kα x-rays (Al$_2$O$_3$, MnO$_2$, etc.) show relatively higher oxygen x-ray intensity. For the compounds investigated, the 2θ value of the O-Kα peak maximum varied within a range
of almost one degree; the intensities of the peaks varied by as
much as 10% from that of fused SiO$_2$ (Table 2). In order to examine
the scatter of data as found in Figure 2, a plot was made of peak
intensities versus the ratio of oxygen content to mass absorption
coefficient for O-Kα x-rays, as shown in Figure 3. Even though
the positions and intensities of the measured peaks are influenced
by chemical bonding conditions, the measured intensity of the O-Kα
x-rays shows, as a first approximation, a linear relationship to
the ratio of oxygen content to mass absorption coefficient of the
sample for O-Kα x-rays.

Table 2 — X-ray shift ($\Delta 2\,\theta$) of peak maximum and intensity reductions
arising from chemical bonding conditions.

Oxide Materials	2θ of Peak Maximum	Intensity at 2θ of peak maximum as % of SiO$_2$ (fused) intensity
SnO$_2$	133.30°	92%
Cr$_2$O$_3$	133.10°	93%
Fe$_2$O$_3$	133.05°	95%
Al$_2$O$_3$	133.00°	97%
MnO$_2$	132.95°	98%
TiO$_2$	132.85°	99%
SiO$_2$ (fused)	132.75°	100%
MoO$_3$	132.75°	100%
BaO	132.75°	100%
Coal (NBS1632a)	132.70°	100%
PbO	132.65°	100%

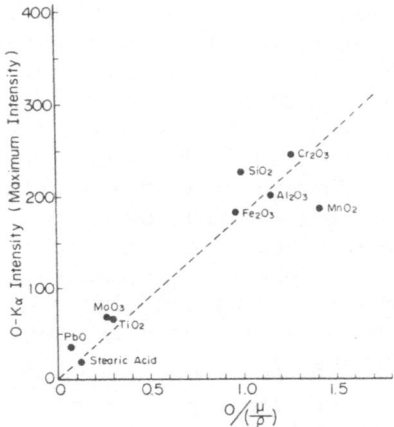

Figure 3 — Relation between O-K$_\alpha$
Intensity and O$/(\frac{\mu}{\rho})$ of Various
Oxygen Compounds

When measuring the O-Kα intensity of iron ores, no decrease was found for iron ores containing hematite, magnetite and small amounts of manganese oxide, silicon oxide, limonite, etc.; however, the O-Kα intensity becomes very low when ores contain small amounts of calcium oxide or titanium oxide. Figure 4 shows typical measured O-Kα intensities from iron ores; also included are the intensities for MnO_2, Fe_2O_3, TiO_2 and $Ca(OH)_2$. Using a simple mathematical correction formula with CaO and TiO_2 factors, $W = KI (1+ \alpha W_{TiO2} + \beta W_{CaO})$, x-ray oxygen contents were determined and plotted in Figure 5. The accuracy was found to be about 1 wt% in the 28-36 wt% oxygen range.

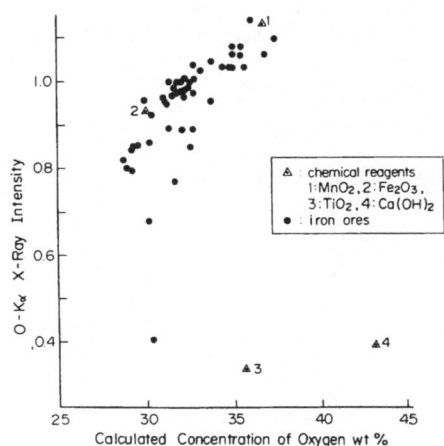

Figure 4 — Measured x-ray intensities
from powder sample against oxygen
concentration

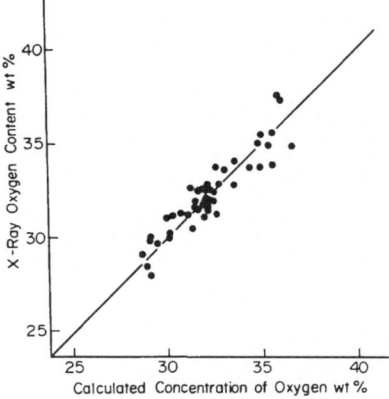

Figure 5 — Corrected x-ray intensities
from powder samples against oxygen
concentration

In order to show examples of the most complex samples for x-ray
analysis, oxygen determinations in coals were carried out. Coal is
a mixture of fine particles of both organic and mineral matters in
both of which oxygen is a component. Since carbon has a large absorption coefficient for O-Kα, x-ray emission from organic matter is
low while that from the particles of mineral matter is high because
of the relatively lower absorption of the mineral matter for O-Kα
x-rays. Since, considering the shallow analyzing depth and the
particle size of powdered coal, only one or two layers of particles can be analyzed, the measured intensity is the sum of low O-Kα
intensity from the organic matter and high O-Kα intensity from the
mineral matter, as determined by the characteristics of the coal
being analyzed (see Figure 6).

Table 3 shows the relationship between O-Kα x-ray intensity
and oxygen content in coal. The high temperature ash (HTA) value
is determined at about 800°C by ASTM or JIS methods. The value
1.33 is the average ratio (14) of low temperature (150-200°C) to
high temperature ash, and the product of HTA and 1.33 is the expected total mineral matter in the coal. The oxygen content of
mineral matter, 0.47, was found by neutron activation analysis (15)
of LTA samples. The total expression, i.e., the product of HTA,
1.33 and 0.47, represents the total amount of oxygen in mineral
matter. Therefore, ASTM or JIS oxygen content of coal is the total
of the oxygen volatilized in high temperature ashing of coal which
consists of the organic oxygen and that part of the oxygen from the
mineral matter which is released when it reacts during the ashing
process. The expected organic oxygen was derived from ASTM or JIS
values of oxygen content.

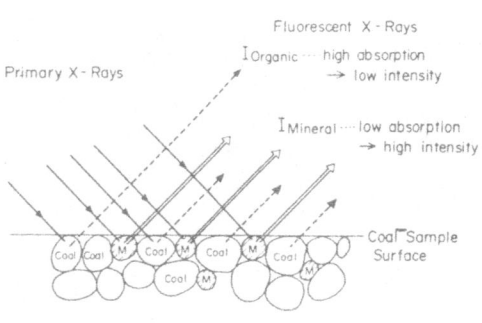

Figure 6 — Schematic Model of
Analyzing Surface of Coal Powder
Sample

Table 3 — Relation between O-K$_\alpha$
Intensity and Oxygen Content in Coal

Measured Intensity

$$I_{Measured} = I_{Mineral} + I_{Organic}$$

X-Rays from Mineral Matter   X-Rays from Organic Matter

Oxygen in Coal

$$O_{Total} = O_{Mineral} + O_{Organic}$$

$$O_{Mineral} = (HTA) \times 1.33 \times 0.47$$

by ASTM or JIS   LTA/HTA   Oxygen in Mineral Matter

Mineral Matter in Coal

$$O_{Organic} = O - (HTA) \times \frac{0.33 \times 0.47}{0.16}$$

by ASTM or JIS   Oxygen Escaping from Mineral Matter by High Temperature Ashing

Figure 7 shows the measured O–Kα intensity versus expected oxygen content for coking and fuel coals. The O–Kα x-ray intensities are low for NBS1632A (bituminous) and NBS1635 (sub-bituminous), whose oxygen contents were measured by fast neutron activation analysis. Using regression techniques with a correction term consisting of the ratio of oxygen to carbon, analytical results were obtained for oxygen in coal, as shown in Figure 8. The analytical accuracy was found to be 1.7wt percent, a figure 8 times larger than the standard deviation found when making repeat measurements on the same coal sample. If no correction is made for the ratio of oxygen to carbon, the analytical accuracy is 2.2wt percent.

3. Thickness Measurement of Thin Chromate Films on Aluminum Metal

The experimental results shown in Figures 9 and 10 were obtained from Al metal plate covered wtih a very thin coating of a

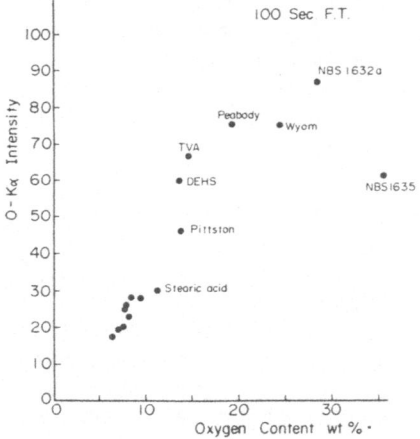

Figure 7 — Relation between O-Kα Intensity and Oxygen Content in Coal

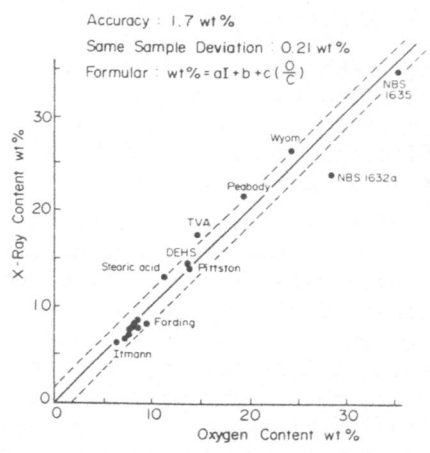

Figure 8 — Oxygen Determination in Coal by X-Ray Fluorescence Analysis

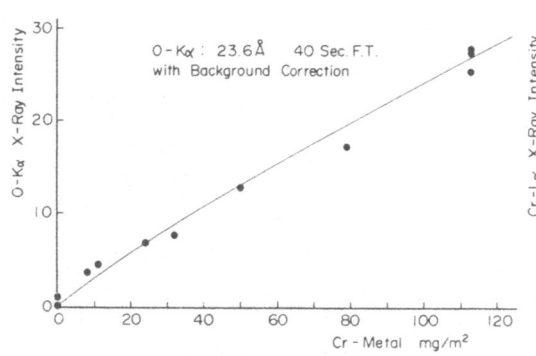

Figure 9 — O-Kα Intensity Measurement of Chromate Treatment Film on Aluminum Metal Plate

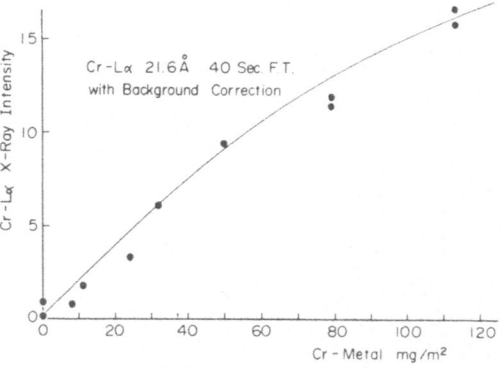

Figure 10 — Cr-Lα Intensity Measurement of Chromate Treatment Film on Aluminum Metal Plate

chromium-oxygen complex, which is used to bind paints to the aluminum. The coating thickness varies between 50 and 300Å and was
measured using O-Kα x-rays at 23.62Å and Cr-Lα x-rays at 21.6Å.
Using simple calculations, the estimated thickness for saturation
intensity of O-Kα and Cr-Lα x-rays in chromium oxide are approximately 2.2μ and 7,400Å, respectively. The O-Kα intensity measured
was 20 times higher than the Cr-Lα intensity. An increase in Cr-Lα
intensity could be found when the aluminum substrate contains small
amounts of chromium and the chromate layer was thin. Figure 11
shows the measured relationship between O-Kα and Cr-Lα x-rays; it
clearly shows that the oxygen/chromium ratio remains constant with
changes in coating thickness.

4. Measurement of N-Kα X-Rays

Since measured intensities for nitrogen are very low, nitrogen
analysis should be carried out using the glass mirror total reflection technique and a 0.5μ Mylar® window on the detector. Figure 12
shows calculated intensity versus glancing angle curves for C-Kα,
N-Kα, O-Kα, Fe-Lα and Na-Kα, based upon Henke's formula (17). It
is difficult to separate N-Kα x-rays from those of C-Kα and O-Kα;
since the glass plate contains oxygen as a main component, the total
reflected intensity of O-Kα becomes low because of anomalous scattering by the oxygen atoms.

Figure 13 shows the relationship between N-Kα intensity and
nitrogen content for various compounds. It was found that the
observed intensity from titanium and chromium nitrides is very high
because of the overlap of their L-series x-rays with N-Kα x-rays.
In order to eliminate the interferences from carbon and oxygen in
the fertilizer samples, a correction, using regression methods, was
made to the N intensity measured at 6 degrees. This correction used
intensity measured at 8 degrees, which contains the totally reflected
intensity of the interfering x-rays.

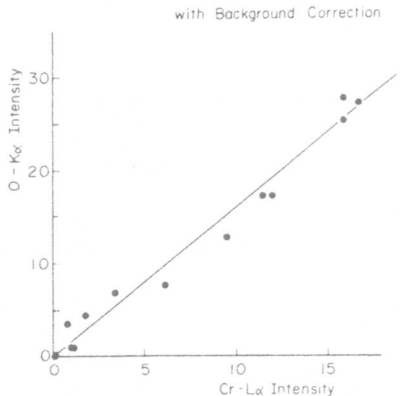

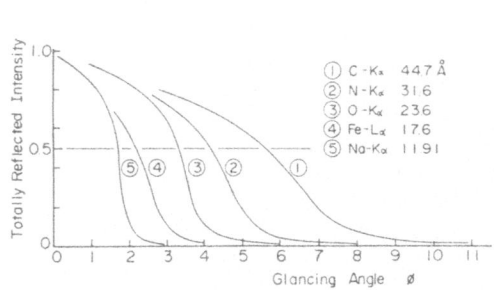

Figure 11 — Relation between O-K$_\alpha$
and Cr-L$_\alpha$, X-Rays of Chromate
Treatment Film on Aluminum Metal
Plate

Figure 12 — Calculated Intensity —
Glancing Angle Curve of Plate Glass
for Various Characteristic X-Rays

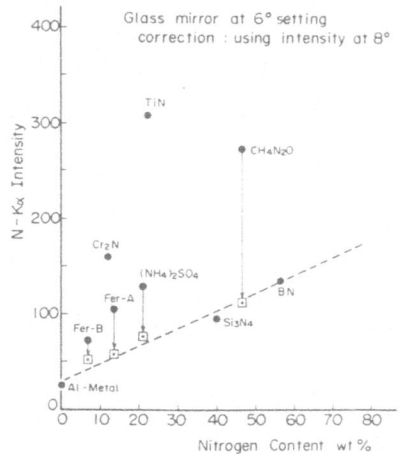

Figure 13 — Relation between N-K$_\alpha$
Intensity and Nitrogen Content in
Various Compounds

CONCLUSION

The x-ray analytical problems of determining oxygen content
in various materials, such as chemical compounds, coal, iron ore,
etc., have been studied; detection was carried out using a TlAP
crystal and a flow proportional counter with a thin Mylar® window.
The measured x-ray intensity of O-Kα is affected by several causes,
such as matrix absorption of O-Kα x-rays, fine structure of the O-Kα
line arising from chemical bond effects, overlapping influences
from various other x-rays, heterogeneity of the analyzed samples
and shallow analyzing depth stemming from the high absorption of
the samples for O-Kα x-rays. In the analysis of coal, a most complex sample for x-ray analysis, investigations were carried out on
the relationship between the emission of O-Kα x-rays and the state
of the oxygen contained in the samples, which is a function of the
heterogeneous nature of coal.

Investigations on the detection of nitrogen in various substances found that the intensity of the N-Kα line was not high
enough for determination of the nitrogen content of the materials;
additional instrumental improvements will be required to increase
this intensity. For example, when thinner materials will be able
to be used for proportional counters, progress will be made in the
analysis of nitrogen by x-ray fluorescence and the development of
new materials, such as ceramics, will be aided.

ACKNOWLEDGEMENT

The authors are indebted to Dr. Kazuo Taniguchi of Osaka
Electrocommunication University and express their gratitude for
kind consideration given by Mr. Hikaru P. Shimura, President of
Rigaku Industrial Corporation.

REFERENCES

1. Arai, Japanese J. Appl. Phys., Vol. 21 (1982), p. 1347

2. Arai, Sohmura, and Tamenori, Advances in X-Ray Analysis, Vol. 26 (1983), p. 423

3. Henke, Advances in X-Ray Analysis, Vol. 7 (1963), p. 460

4. Mattson and Ehlert, Advances in X-Ray Analysis, Vol. 9 (1965), p. 471

5. Fischer, J. Chem. Phys., Vol. 25, No. 11 (1965), p. 3814; J. Appl. Phys., Vol. 41 (1970), p. 3922

6. Perera and Henke, X-Ray Spectrometry, Vol. 9, No. 2 (1980), p. 81

7. Uchikawa and Numata, Yogyo-Kyokai-Shi, Vol. 81, No. 5 (1973), p. 189 (in Japanese)

8. Uchikawa and Numata, Japan Analyst, Vol. 20 (1971), p. 1207 (in Japanese)

9. Sugizaki, Makino, and Fuse, Bunseki Kagaku, Vol. 31 (1982), p. 885 (in Japanese)

10. Kirkendall and V'Aradi, Analytical Chemistry, Vol. 39, No. 11 (1967), p. 1342

11. Viele Sage and Grubis, X-Ray Spectrometry, Vol. 2 (1973), p. 189

12. Barrus and Blake, X-Ray Spectrometry, Vol. 10, No. 1 (1981), p. 48

13. Henke and Lent, Advances in X-Ray Analysis, Vol. 12 (1969), p. 480

14. O'Gorman et al., Fuel, Vol. 50 (1971), p. 135

15. Hamrin et al., Fuel, Vol. 58 (1979), p. 481

16. Nadkarni, Anal. Chem., Vol. 52 (1980), p. 299

17. Henke, Phys. Rev., Vol. 6 (1972), p. 94

TRACE ELEMENT ANALYSIS BY SYNCHROTRON RADIATION EXCITED XRF

S.T.Davies[1],D.K.Bowen[1,2],M.Prins[3] and A.J.J.Bos[4]

[1]University of Warwick, [2]Daresbury Laboratory,U.K.,
[3]Technische Hogeschool Eindhoven, [4]Vrije Universiteit,
Amsterdam, The Netherlands

INTRODUCTION

The ability rapidly and quantitatively to detect trace elements in a host matrix is of great importance in many areas of science and engineering. This fact is underlined by the considerable amount of effort that has been put into developing such techniques as electron probe microanalysis, proton induced x-ray emission (Pixe), the laser microprobe mass analyser (Lamma) as well as conventional x-ray fluorescence methods. Recently the availability of electron storage rings partially or wholly dedicated to producing intense beams of synchrotron radiation has provided a new tool to complement the above mentioned techniques. This paper reports on work at Daresbury Laboratory on synchrotron x-ray fluorescence (SXRF) for quantitative trace element analysis.

A comprehensive account of the advantages of synchrotron radiation for elemental analysis has been given by Sparks (1). The low energy deposition in the sample, non-destructive nature, rapidity, simultaneous multi-element detection capability and low detection limits can be listed as some of the attractive features. Clearly the use of synchrotron radiation, which for most users is only available at a remote facility, can only be contemplated if some significant bonus is to be gained. Lower minimum detectable limits for a given analysis time, high spatial resolution, element mapping and element discrimination and quantitation that is difficult by other means have all been put forward as reasons for developing SXRF as an analytical tool.

Considering here the minimum detectable limit we adopt the criterion given by Currie (2) and express the minimum detectable limit (MDL) as:

$$MDL = 3.29C\sqrt{B}/N$$

Where C is the mass fraction in parts per million
       B is the background count
   and N is the analyte line count after background subtraction.

The intensity, spectral range and linear polarisation of synchrotron radiation make it a suitable agent for exciting x-ray fluorescence and reducing the MDL.

SYNCHROTRON RADIATION

     The flux available at the station used at the SRS is of order 10^{12} photons/sec/mrad horizontal/100MA/0.1% bandwidth integrated over all vertical opening angles for a wavelength interval from 0.7 A to ~ 5A. This is some two orders of magnitude more than obtainable from the characteristic x-rays from a conventional source in a comparable bandwidth and geometry. Synchrotron radiation is plane polarised with the electric vector in the plane of the electron orbit around the storage ring.   The differential cross-section for Compton scattering on a free electron may be expressed

$$d\sigma = \frac{1}{2}\left(\frac{e^2}{mc^2}\right)\left(\frac{\lambda_0}{\lambda_1}\right)^2\left(\frac{\lambda_1}{\lambda_0} + \frac{\lambda_0}{\lambda_1} - 2\cos^2\theta\right)d\Omega$$

where λ_0 is the wavelength of the incident radiation
        λ_1 is the wavelength of the scattered radiation
and θ is the angle between the polarisation vector and the direction of the scattered beam.

At low energies $\lambda_1 \rightarrow \lambda_0$ and hence$\left(\frac{\lambda_1}{\lambda_0} + \frac{\lambda_0}{\lambda_1} - 2\cos^2\theta\right) \rightarrow 0$

for scattering parallel to the polarisation vector.   Hence collecting the fluorescent radiation at 90° to the incident beam means that primary Compton scattered radiation is greatly reduced in comparison to that from an unpolarised source.   In practice Compton scattering cannot be eliminated as secondary scattering will take place and also the beam is only completely polarised in the median plane. The source also has a finite size and integrating over all emission angles results in a figure of about 95% polarisation in the median plane for the SRS. In addition the finite solid angle subtended by the detector means that elliptically polarised radiation is admitted.   It is possible however to obtain significant improvements and it has been shown (3) that reductions by as much as a factor of 10 in the Compton scattered radiation is possible with corresponding improvements in the signal to background ratio.

     Synchrotron radiation is also continuously tunable, over a wide spectral region.   For machines operating at several GeV the range is appropriate for exciting the K or L lines of all elements in the periodic table.   The sensitivity for a particular element may be maximised by tuning the incident radiation just above its K or L edge absorption edge.   A compromise must be made between a broad band-pass monochromator giving high flux to excite the line of interest and a narrow band-pass device giving much reduced flux but superior sensitivity.

EXPERIMENTAL TECHNIQUE

A photograph of the apparatus used at Daresbury Laboratory for the trace element analysis experiments is shown in Fig. 1.

Fig. 1 Apparatus used for trace element analysis

The experiments were conducted at the double crystal topography station on beam line 7 of the SRS at a distance of 60 m from the tangent point.   The white beam is defined by a set of horizontally and vertically adjustable lead slits before being incident on a plane graphite monochromator crystal.   The hot-pressed pyrolitic graphite crystal has mosaic spread of ˜ 0.5° and diffracts approximately kinematically with $\Delta E/E$ of 1KeV at about 20 KeV.   The beam is diffracted in the vertical plane and the broad-band monochromatic radiation is admitted into the vacuum chamber through a thin (25 μm) kapton window.   The rough vacuum of ˜ 10^{-1} torr serves to eliminate the air-scattered radiation and hence further improve the signal/noise ratio.   The diffracted beam is further defined by a set of lead slits inside the chamber and finally by a perspex collimator, which serves to eliminate any contamination of the observed fluorescence from the sample by the surroundings. Samples are normally mounted on a thin kapton backing or sandwiched between thin kapton sheets attached to perspex frames and oriented at

45° to the incident beam.   The fluorescent radiation is collected at 90° to the incoming beam.

The detector used was a lithium–drifted silicon solid state detector with a beryllium window of thickness 8 μm.   The active area was 12.6 mm^2 and the measured resolution was 180 EV.   The output from the preamplifier was processed by standard NIM and CAMAC modules controlled by a PDP11/04 computer.   It was possible to display, plot and store spectra for subsequent analysis.   In addition the software provides for easy energy calibration, hard copy of channel contents, peak area evaluation, background subtraction and smoothing.   Energy calibration was checked at the start and end of each shift and was found to be stable to within 0.1%.

APPLICATIONS

Much interest has been shown in using SXRF for trace element analysis of biological samples and detailed comparisons have been made between SXRF and alternative methods of fluorescence excitation for such materials (3.4).   Experimental determinations of minimum detectable limits using SXRF have also been made (5,6) covering a wide range of Z values for elements deposited on a low Z support matrix.

In contrast this work focusses on the applicability of SXRF to assessing semiconductor materials in respect of quantitation of implanted species in semiconductor substrates and determining the trace element content in nominally high–purity materials.

Fig. 2 shows the spectrum of an arsenic implanted silicon specimen.   The silicon has been implanted with (6±1) x 10^{15} AS+ ions cm^{-2} at 150 KEV and subsequently annealed for 100 hrs at 720°C producing a roughly Gaussian distribution of implant ions with mean penetration 0.2 μm.   The sample size was ˜ 10 x 20 mm and ˜ 250 μm thick.   The beam cross–sectional area was 1 mm^2.   An excitation energy of 17 KeV was selected, corresponding to a Bragg angle of 6° for the graphite monochromator.   With the SRS running at 2 GeV and 190 mA circulating current the total counting rate was 2000 counts s^{-1}. Assuming a Gaussian profile for the implanted species the implant dose was calculated from the SXRF spectrum (17) and found to be 3 x 10^{15} ions cm-2 which is in reasonable agreement with that calculated from the implanation parameters.

The spectrum of a cadmium telluride specimen which has been implanted with (3±0.5) x 10^{15} copper ions cm^{-2} is shown in Fig. 3. The sample was in the form of a disc of area 100 mm^2.   The incident beam cross–section was 1 mm^2 and of energy 17 KeV with the SRS running at 2 GeV and 115 mA.   The total counting rate was 5000 counts s^{-1}.   The cadmium Kα line has been excited by the second harmonic.

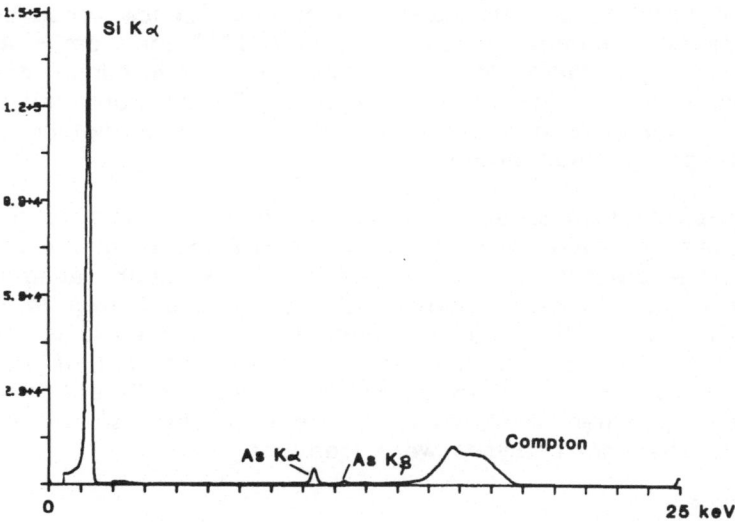

Fig. 2   Spectrum of arsenic implanted silicon

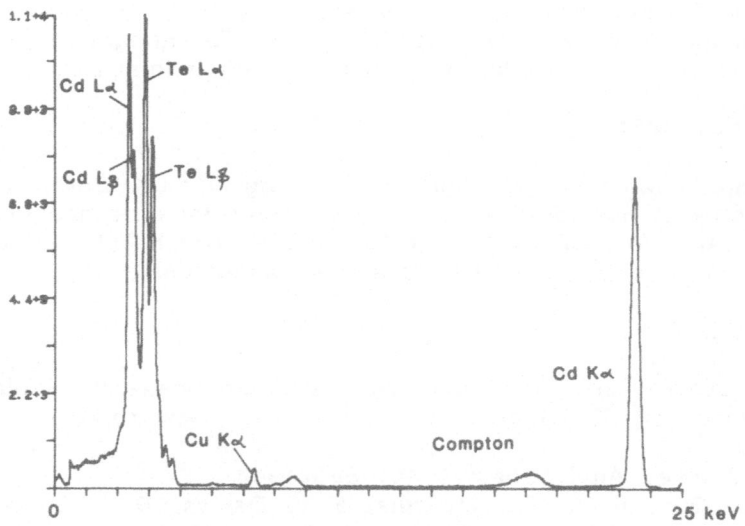

Fig. 3   Spectrum of Cu-implanted cadmium telluride

Using the minimum detectable limit criterion defined above the smallest amount of arsenic and copper that could be detected in 100s counting times was 6×10^{13} cm^{-2} and 2×10^{14} cm^{-2} respectively. The experimental conditions were not optimised for detection of either of these elements and an improvement of a factor of 10 on these values would seem possible on the basis of the fluorescence cross-sections. Ion implantation commonly involves doses > 10^{13} ions cm^{-2} and thus the method is suitable for assessing the uniformity, dose and reproducibility of ion implanted species.   Furthermore no radiation damage is produced and this level of sensitivity could not be achieved without damage by other methods.

High-purity semiconductor grade gallium and arsenic have also been analysed for trace element content.   The object of the study was to characterise precursor materials for molecular beam epitaxy growth of III-IV devices.   Different sources of arsenic are known to result in differences of up to 10^3 in carrier concentration and there is interest in relating this to the type and amount of impurity content of the source material.   It has already been shown that clearly different 'fingerprint' spectra were obtained from these different sources even though the macroscopic chemical analyses were identical.

CONCLUSIONS

It is clearly feasible to detect elements at ppm levels or below in a heavy matrix using energy-dispersive SXRF.   If simultaneous multi-element detection is not a priority then the problem of detector saturation may be overcome by using a wavelength dispersive system. In specific cases it may be possible to use an order sorting monochromator and avoid exciting the matrix lines altogether although a penalty is paid for this in terms of flux and hence total analysis time.

ACKNOWLEDGEMENTS

We are indebted to Dr. Christopher Hogg and Dr. Bernard Lunn of the University of Hull for providing the cadmium telluride specimen, and to Dr. T. Ambridge of the British Telecom Research Laboratory for the As-doped Si specimen and the Ga and As specimens.

REFERENCES

1.   C.J. Sparks, Jr., in Synchrotron Radiation Research, edited by H. Winick and S. Doniach (Plenum Press, New York) p.459, (1980).
2.   L.A. Currie, Anal. Chem., 40, 586(1968).
3.   A.J.J. Bos, R.D. Vis, M.Prins, S.T. Davies, D.K. Bowen, J.Makjanic and V. Valkovic, Nuc.Inst. and Methods, (in press).
4.   M. Prins, J.A. Van der Heide, A.J.J. Bos, D.K. Bowen, and S.T. Davies, I.E.E.E., Trans.Nuc.Sci., NS-30(2), 1243(1983).
6.   J.V. Gilfrich, E.F. Skelton, S.B. Quadri, J.P. Kirkland and D.J. Nagel, Anal.Chem.,55,187(1983).
7.   A. Knochel, W. Petersen and G. Tolkiehn, Nucl.Instr. and Methods 208, 659 (1983).
8.   D.K. Bowen and S.T. Davies (to be published).

X-RAY FLUORESCENCE SPECTROMETRIC ANALYSIS OF UNCONTAMINATED AND CONTAMINATED TROPICAL PLANT MATERIALS FOR TRACES OF HEAVY METALS

Hasso Schorin and Laura Piccioni

I.V.I.C., Centro de Ingeniería

Apartado 1827, Caracas 1010-A, Venezuela

INTRODUCTION

52 elements of the periodic system are considered as potential-ly usable by living organisms[1]. Most of them, as some heavy metals, are required only in trace amounts. If a certain concentration of a given trace element is exceeded in an organism it becomes toxic. Plant analyses are considered in agronomy as a diagnostic tool and as a guide to fertilizer practice as the plant uptake is the best criterion of nutrient and micronutrient availability[2]. But now-adays also toxic metals may be assimilated by plants due to the increasing contamination by inorganic compounds containing these toxic elements[3]. In geosciences plant analyses may be of impor-tance because indicator plants are employed for the geochemical prospection of certain heavy metals[4].

X-ray fluorescence was applied by several authors for the quantitative determination of several light but also some heavy elements [5-12]. In this paper the precise and accurate analysis for Si, K, Ca, V, Cr, Mn, Fe, Co, Ni, Cu, Zn, As, Mo, Cd, Sb and Pb in tropical plant materials is presented. The precision, the accura-cy, the detection limit and the distinction limit are given for each element. Finally, the contents of these elements in some tropical plant materials are presented.

EXPERIMENTAL

To set up the calibration standards for the elements to be determined microcristalline cellulose for thin layer cromatography was employed. After washing the powder with 0.1 molar EDTA solu-tion[10] the corresponding concentrations of the different elements

563

were added as solutions to 10g portions of cellulose. The concentration ranges were 0-1000ppm for Si, 0-500ppm for Fe, 0-250ppm
for Mn, 0-200ppm for Zn, 0-100ppm for Cu, 0-50ppm for Mo, Cr, Sb,
As, Pb, Ni, Cd and 0-20ppm for Co and V. After drying at 80°C for
about 48h the cellulose was homogenized and thereafter briqueted in
a press at 10t. As Ca and K are considered to influence the intensities of the K-lines of Cr, Mn, Fe, Ni, Cu, Zn and Pb by adsorption
a second group of calibration standards was set up with the above
mentioned trace metal contents adding 0-6% K and Ca as carbonates
respectively. To establish from which concentration Ca or K affect
the intensities of the elements Mn, Fe, Cu and Zn briquets with 3%,
1.5%, 0.75%, 0.3% and 0% of Ca or K and 500ppm Fe, 200ppm Mn, 20ppm
Cu and 50ppm Zn were prepared.

The standard reference materials from the National Bureau of
Standards Nº 1567, 1568, 1569, 1571, 1573 and 1575 as well as the
analyzed natural materials (dried at 80°C to weight constancy) were
briqueted employing 8g of sample and 2g of wax (Hoechst Wachs C) for
better consistency. To verify the precision 5 cellulose discs were
prepared with the following element concentrations: Fe (300ppm), Mn
and Si (250ppm), Zn (50ppm), Cu (20ppm), Ni, Sb, Co, As, Pb, Cd, V
and Cr (10ppm) and Mo (5ppm).

All samples were analyzed with the Siemens SRS100 X-ray spectrometer equipped with a logic controller. The target excitation of
the Cr tube was 55KV and 40mA. For all elements the intensities
of the $K\alpha$ lines were measured, for Pb the $L\alpha_1$ line was used. The
flow counter was employed for the intensity-measurement of the
lines of Si, Ca and K while the scintillation counter was applied
for all other elements. The counting times were 1sec. for K and
Ca- 40sec. for Cu, 80sec. for Fe and Zn, 100sec. for Mn and Sb and
200sec. for the remaining elements.

RESULTS

The parameters for the calibration standardizations were calculated by linear regression analysis using the program P88[13] and
a PDP11/45 computer. Additionally, to the slope and the y-axis
(intensity) intercept the standard deviation for the calibration
is computed. It comprises all possible errors and is therefore
an indicator for the quality of the calibration[14]. Based on this
standard deviation the distinction limit and the detection limit
for a level of confidence of 95% were calculated[14]. The data are
shown in Table 1.

In Table 2, the data on the precision of the method are summarized.

Table 1. Standard Deviation for the Calibrations, Detection
          Limit (95% Confidence Interval) and Distinction
          Limit (95% Confidence Interval)

Element	Si (ppm)	K (wt%)	Ca (wt%)	V (ppm)	Cr (ppm)	Mn (ppm)	Fe (ppm)	Co (ppm)	Ni (ppm)
Standard deviation	±16	±0,13	±0,05	±0,3	±1,4	±3,4	±3,4	±0,12	±0,7
Detection limit	38	0,30	0,12	0,7	3,5	7,7	7,7	0,3	1,6
Distinction limit	±50	±0,39	±0,15	±1,0	±4,6	±11	±11	±0,5	±2,1

Element	Cu (ppm)	Zn (ppm)	As (ppm)	Mo (ppm)	Cd (ppm)	Sb (ppm)	Pb (ppm)
Standard deviation	±1,7	±1,5	±0,4	±0,4	±0,9	±0,8	±0,7
Detection limit	3,8	3,6	0,9	0,9	2,7	1,9	1,6
Distinction limit	±5,4	±4,8	±1,2	±1,2	±4,1	±2,5	±2,2

Table 2. Precision of the Method

Element	Si (ppm)	V (ppm)	Cr (ppm)	Mn (ppm)	Fe (ppm)	Co (ppm)	Ni (ppm)
	283	8,7	9,1	243	323	8,6	9,0
	274	8,3	8,8	245	320	8,2	8,4
	270	9,2	9,3	254	327	8,5	9,8
	284	8,9	9,1	251	317	8,0	9,7
	281	8,9	8,8	241	313	8,7	8,0
Mean $\bar{X}$	278	8,8	9,0	247	320	8,4	9,0
Stand. Dev. S	± 6	±0,3	±0,2	± 5,5	± 5,4	±0,32	±0,8

Element	Cu (ppm)	Zn (ppm)	As (ppm)	Mo (ppm)	Cd (ppm)	Sb (ppm)	Pb (ppm)
	19,5	49,4	9,7	6,9	10,5	10,1	7,8
	19,4	50,2	9,3	4,3	10,8	11,9	7,3
	19,8	49,5	9,9	7,8	8,1	13,3	7,2
	20,5	49,3	9,8	6,3	9,9	12,7	
	18,7	49,5	10,2	4,4	9,3	14,0	
Mean $\bar{X}$	19,6	49,6	9,8	5,9	9,7	12,4	7,4
Stand. Dev. S	± 0,7	± 0,4	± 0,3	±1,6	± 1,1	± 1,5	±0,3

Table 3.   Data on the Accuracy of the Method

National Bureau of Standards
Standard Reference Materials

Element		1567	1568	1569	1571	1573	1575
Si	given	/	/	/	/	/	/
	found	/	115	0	1193	0	758
K	given	0,136±0,004	0,112±0,0002	1,55*	1,47±0,03	4,46±0,03	0,37±0,02
	found	0	0	1,94	1,70±0,13	4,46±0,13	0,38±0,13
Ca	given	0,019±0,001	0,014±0,002		2,09±0,03	3,00±0,03	0,41±0,04
	found	0	0	0,36	2,92±0,05	4,16±0,05	0,73±0,05
V	given	/	/	/	0,6±0,2*	1,3*	0,4*
	found	/	/	/	1,5±0,3	1,5±0,3	0,4±0,3
Cr	given	/	/	2,12±0,05	2,6±0,3	4,5±0,5	2,6±0,2
	found	/	/	3,1±1,4	3,4±1,4	7,7±1,4	0,1±1,4
Mn	given	8,5±0,5	20,1±0,4	7,0*	91±4	238±7	675±15
	found	9,9±3,4	20,4±3,4	10,6±3,4	89±3,4	215±3,4	696±3,4
Fe	given	18,3±1	8,7±0,6	707*	300±20	690±25	200±10
	found	11,2±3,4	4,1±3,4	751±3,4	317±3,4	610±3,4	211±3,4
Co	given	/	0,02±0,01		0,2	0,6	0,1
	found	/	0		0	0,3	0,25
Ni	given	0,18	0,16	/	1,3±0,2	1,2	3,5
	found	0	0	5,4±0,7	1,7±0,7	2,2±0,7	2,6±0,7
Cu	given	2,0±0,3	2,2±0,3	11*	12±1	11±1	3±0,3
	found	3,0±1,7	2,5±1,7	21±1,7	16,7±1,7	11,6±1,7	5,9±1,7
Zn	given	10,6±1	19,4±1	70*	25±3	62±6	61*
	found	9,9±1,5	18,7±1,5	67±1,5	17±1,5	57±1,5	62±1,5
As	given	0,006	0,41±0,05		10±2	0,27	0,21±0,04
	found	0	0	0,20	5,9±0,4		0,72±0,4
Mo	given	0,4	1,6		0,3±0,1	/	
	found	0,31±0,4	1,7±0,4	3,31±0,4	0	/	0,4
Cd	given	0,032±0,007	0,029±0,004		0,11	3	<0,5
	found	0	0		0	0,6	0
Sb	given	0,04*	0,005*	0,005*	2,9±0,3		0,2
	found	0	1,1±0,8	0,65±0,8	1,6±0,8		0
Pb	given	/	/	/	45±3	6,3±0,3	10,8±0,5
	found	/	/	/	57±0,7	4,2±0,7	9,5±0,7

Ca and K in wt%; all other elements in ppm

*not certified values

The analysis of the cellulose discs with the different additions of K or Ca have established that concentrations of K or Ca
below 0,3wt% do not affect noticeably the intensity of the Kα-line
of Mn (200ppm) while the intensities of the lines of Zn (50ppm) Cu
(20ppm) and Fe (500ppm) are not influenced by K or Ca concentrations
below 0,75wt%. Higher concentrations make necessary an additional
correction for absorption by these two elements.

The analytical results of the NBS-standards 1567 (wheat flour),
1568 (rice flour), 1569 (brewer's yeast), 1571 (orchard leaves),
1573 (tomato leaves) and 1575 (pine needles) are summarized in table
3. The presented method for the analysis of plant materials may be

considered as accurate if no difference between the certified concentration[15] and the obtained analytical results can be ascertained.
This is the case for a confidence interval of 99,7% if the difference
between the certified and analytical data for a given element in a
SRM is smaller than the triple of Sd which is defined by the equation[12]

$$S_d^2 = S_e^2 + S_x^2$$

S_e = Standard deviation of the calibration standardization

S_x = Standard deviation of the certified concentration value.

Additionally, Mn, Fe, Ni, Cu, Zn and Pb were determined quantitatively by the following methods in NBS 1573: standard addition
method and dilution method[16]. The obtained results together with
the concentrations determined with the calibration standardization
method as well as the certified values are given in Table 4.

In the case of Mn, Cu, Ni and Zn good agreement among the results can be ascertained. For Fe the content obtained by the
standard addition method is somewhat high, while the consistent
values of the calibration standardization and dilution method are
too low as compared to the certified value. For Pb no conformity
among the concentration data could be established.

Finally in Table 5 the content of heavy metals in some tropical plant materials from Venezuela are presented.

Table 4.   Concentration data obtained by Different Methods
           Applied to NBS 1573

Method	Element	Mn (ppm)	Fe (ppm)	Ni (ppm)	Cu (ppm)	Zn (ppm)	Pb (ppm)
Standard Addition		230	767	4,2	13	71	3,7
Dilution		228	640	4,0	10,9	57	10,5
Calibration Standardization		215	610	2,2	11,6	58	4,2
Certified Value		238	690	1,2	11	62	6,3

Table 5.   Contents of trace metals in tropical plant materials from Venezuela in (ppm); Ca and K in wt%, based on dry material

Element	Si	K	Ca	V	Cr	Mn	Fe	Ni	Cu	Zn	As	Mo	Cd	Sb	Pb
Plant Material															
Papaya	193	5,3	0,8	/	8,4	10,1	56	1,9	7,2	17,5	/	0,4	0,7	3,9	1,8
Medlar	165	1,2	0,1	1,4	2,4	3,5	48	0,7	2,7	2,0	/	/	/	0,9	6,4
Black bean	209	2,4	0,1	0,1	3,7	15,6	114	1,1	11,2	33	/	6,1	/	/	0,1
Lettuce	1295	7,5	1,9	1,9	18,6	28,5	675	1,9	19	54	/	/	/	/	3,4
Rice	59	/	/	0,2	0,3	13,7	105	/	3,12	16,3	0,3	1,3	1,3	/	/
Corn	61	0,3	/	0,4	2,3	15,2	216	/	1,7	25	/	0,3	/	/	/
Pineapple	1417	0,8	0,3	/	2,6	135	84	2,6	8,7	9,7	/	/	0,7	2,5	0,8
Cassava	434	1,4	/	1,2	3,2	34	71	0,6	4,2	8,6	/	0,3	/	3,5	/
Indian Mango	465	1,0	0,3	0,2	1,8	25	67	/	9,0	4,8	0,7	/	/	/	/
Melon	1362	3,9	0,4	/	5,9	21	102	0,8	3,7	20	nd	nd	nd	nd	nd

nd = Not Determined

CONCLUSIONS

Light and heavy metals can be determined accurately and precisely in plant materials employing cellulose discs as calibration
standards and applying the regression analysis (Program P88) and
a computer to set up the calibration parameters.

If the concentrations of K or Ca in the sample materials lie
above 0,3wt% these elements have to be added to the calibration
briquets. The intensities of the lines of Mn, Fe, Cu, Zn and Pb
have then to be corrected for adsorption by K and/or Ca.

For analyses of the elements in the ppb range in dried materials
special enrichment procedures should be considered.

ACKNOWLEDGEMENTS

The author wants to thank Gustavo Mejías for the purification
of the cellulose and the sample preparation.

REFERENCES

1.   E. Frieden, Los Elementos Químicos de la Vida, in "Química y
     Ecósfera", Herman Blume Ediciones, Madrid, (1976).

2.   F.G. Viets, "Soil Testing and Plant Analysis, Part II, Plant
     Analysis", Soil Science Society of America, Inc., Publisher,
     Madison, Wisconsin, U.S.A. (1967).

3.   R.A. Horne, "The Chemistry of Our Environment", John Wiley &
     Sons, New York (1978).

4.   A. Levinson, "Introduction to Exploration Geochemistry" Applied
     Publishing Ltd., Calgary (1974).

5.   J. Kubota and V.A. Lazar, Routine X-Ray Emission Spectrographic
     Analysis of Common Forage Plants. In "Soil Testing and Plant
     Analysis, Part II, Plant Analysis". Soil Science Society of
     America, Inc., Publisher, Madison, Wisconsin, U.S.A., 93 (1967).

6.   A. Murdoch and O. Murdoch, Analysis of Plant Material by X-ray
     Fluorescence, X-Ray Spectrometry 6 : 215 (1977).

7.   K. Norrish and J.T. Hutton, Plant Analysis by X-Ray Spectrometry
     I, X-Ray Spectrometry 6 : 6 (1977).

8.   J.T. Hutton and K. Norrish, Plant Analysis by X-Ray Spectrometry
     II, X-Ray Spectrometry 6 : 12 (1977).

9.   L. Beitz, Die Anwendung der Matrixkorrektur bei der Analyse von
     pflanzlichen Produkten, Mitt. Gebiete Lebensm. Hyg. 68 : 451
     (1977).

10. G. Crössmann und H. Rethfeld, Anwendung der RFA bei der Unter-
     suchung von Pflanzen und Futtermitteln auf Schwermetalle, 1.
     Colloquium "Anwendungsmöglichkeiten der Röntgenfluoreszenz-
     analyse", 104 (1977).

11. L. Beitz, L. Müller und R. Plesch, Die Analyse von Futtermitteln
     mit dem Sequenz – Röntgenspektrometer SRS 200, Analysentechni-
     sche Mitteilung Nr 195, Siemens AG.

12. L. Beitz, R. Plesch und H. Rethfeld, Röntgenanalyse auf Schwer-
     metalle in pflanzlichen Produkten, Landwirtschaftliche
     Forschung, 33 : 30 (1980).

13. R. Plesch, Analytische Grundlagen der Siemens Rechenprogramme
     für die Röntgenspektrometrie, Siemens Zeitschrift 48 : 355
     (1974).

14. R. Plesch, Praktische Fehlertheorie der Röntgenspektrometrie,
     X-Ray Spectrometry 7 : 56 (1978).

15. E.S. Gladney, "Compilation of Elemental Concentration Data for
     NBS Biological and Environmental Standard Reference Materials",
     National Bureau of Standards, Washington (1981).

16. E.P. Bertin, "Principles and Practice of X-Ray Spectrometric
     Analysis", Plenum Press, New York, (1975).

AUTHOR INDEX